The Handbook of Peer Production

Handbooks in Communication and Media

This series provides theoretically ambitious but accessible volumes devoted to the major fields and subfields within communication and media studies. Each volume provides experienced scholars and teachers with a convenient and comprehensive overview of the latest trends and critical directions, while grounding and orientating students with a broad range of specially commissioned chapters.

The Handbook of Children, Media, and Development, *edited by Sandra L. Calvert and Barbara J. Wilson*
The Handbook of Crisis Communication, *edited by W. Timothy Coombs and Sherry J. Holladay*
The Handbook of Internet Studies, *edited by Mia Consalvo and Charles Ess*
The Handbook of Rhetoric and Public Address, *edited by Shawn J. Parry-Giles and J. Michael Hogan*
The Handbook of Critical Intercultural Communication, *edited by Thomas K. Nakayama and Rona Tamiko Halualani*
The Handbook of Global Communication and Media Ethics, *edited by Robert S. Fortner and P. Mark Fackler*
The Handbook of Communication and Corporate Social Responsibility, *edited by Øyvind Ihlen, Jennifer Bartlett and Steve May*
The Handbook of Gender, Sex, and Media, *edited by Karen Ross*
The Handbook of Global Health Communication, *edited by Rafael Obregon and Silvio Waisbord*
The Handbook of Global Media Research, *edited by Ingrid Volkmer*
The Handbook of Global Online Journalism, *edited by Eugenia Siapera and Andreas Veglis*
The Handbook of Communication and Corporate Reputation, *edited by Craig E. Carroll*
The Handbook of Media and Mass Communication Theory, *edited by Robert S. Fortner and P. Mark Fackler*
The Handbook of International Advertising Research, *edited by Hong Cheng*
The Handbook of Psychology of Communication Technology, *edited by S. Shyam Sundar*
The Handbook of International Crisis Communication Research, *edited by Andreas Schwarz, Matthew W. Seeger, and Claudia Auer*
The Handbook of Peer Production, *edited by Mathieu O'Neil, Christian Pentzold, and Sophie Toupin*

Forthcoming
The Handbook of Strategic Communication, *edited by Carl Botan*
The Handbook of Public Relations Theory and Methods, *edited by W. Timothy Coombs, Sherry J. Holladay and Melissa Dodd*

The Handbook of Peer Production

Edited by

Mathieu O'Neil
Christian Pentzold
Sophie Toupin

WILEY Blackwell

This edition first published 2021

Registered Office
John Wiley & Sons, Inc., 111 River Street, Hoboken, NJ 07030, USA

Editorial Office
111 River Street, Hoboken, NJ 07030, USA

For details of our global editorial offices, customer services, and more information about Wiley products visit us at www.wiley.com.

Wiley also publishes its books in a variety of electronic formats and by print-on-demand. Some content that appears in standard print versions of this book may not be available in other formats.

Library of Congress Cataloging-in-Publication Data

Names: O'Neil, Mathieu, editor. | Pentzold, Christian, editor. | Toupin, Sophie, editor.
Title: The handbook of peer production / edited by Mathieu O'Neil, University of Canberra, Christian Pentzold, Leipzig University, Sophie Toupin, McGill University.
Description: Hoboken, NJ : John Wiley & Sons, Inc., 2021. | Series: Handbooks in communication and media | Includes bibliographical references and index.
Identifiers: LCCN 2020025498 (print) | LCCN 2020025499 (ebook) | ISBN 9781119537106 (cloth) | ISBN 9781119537144 (paperback) | ISBN 9781119537113 (adobe pdf) | ISBN 9781119537090 (epub)
Subjects: LCSH: Open source software. | Computer file sharing. | Shareware (Computer software) | Information commons.
Classification: LCC QA76.76.O62 H36 2021 (print) | LCC QA76.76.O62 (ebook) | DDC 005.3–dc23
LC record available at https://lccn.loc.gov/2020025498
LC ebook record available at https://lccn.loc.gov/2020025499

Cover Design: Wiley
Cover Image: © LaylaBird/Getty Images

Set in 9.5/11.5pt Galliard by SPi Global, Pondicherry, India

10 9 8 7 6 5 4 3 2 1

MIX
Paper from responsible sources
FSC www.fsc.org FSC® C013604

Contents

List of Figures

List of Tables

Notes on Contributors

Nicholas Anastasopoulos is an architect, researcher, and assistant professor at the School of Architecture, National Technical University of Athens. He holds a PhD from the same university in alternative communities and sustainability and a MArch from Yale. He has taught at Patras University, Parsons School of Design (NYC), and elsewhere. As postdoctoral Prometeo Researcher (IAEN, Ecuador, 2014) he contributed to the FLOK Society project and conducted research on aspects of Buen Vivir and sustainability. Nicholas has conducted extensive research and collaborated with architects, artists, and researchers in Ecuador, Europe, and South America. He initiated the MET workshop and the Ports in Transition Workshops (Europe and South America) for spatial policies. His research interests concern the commons, communities, systems theory, ecology, and complexity. He is currently academic representative and senior researcher in charge on behalf of NTUA for SoPHIA, a H2020 consortium research program aiming to create a Social Platform on Holistic Impact Assessment of European Cultural Heritage.

Panayotis Antoniadis is the co-founder of NetHood, a Zurich-based non-profit organization that combines research and action in the development of tools for self-organization and conviviality, bringing together different forms of commoning in the city such as community networks, complementary currencies, and cooperative housing. http://nethood.org/panayotis/

Michel Bauwens is a Peer-to-Peer (P2P) theorist and a writer, researcher, and conference speaker on the subject of technology, culture, and business innovation. He is a theorist in the emerging field of P2P theory and the director and founder of the P2P Foundation, a global organization of researchers and activists working collaboratively to explore peer production, governance, and property. He has authored a number of books and essays, including his seminal essay "The Political Economy of Peer Production." In 2014, Michel was the research director of the transition project towards the social knowledge economy, an official project in Ecuador (see floksociety.org). This project produced a first integrated Commons Transition Plan for the government of Ecuador, in order to create a "social knowledge economy." In 2016, Michel was Honorary Fellow/Visiting Scholar with the Havens Center at UW-Madison, as an "activist in resident" funded by the Link Foundation.

Yochai Benkler is the Berkman Professor of Entrepreneurial Legal Studies at Harvard Law School and the Berkman Klein Center for Internet and Society at Harvard University, USA. Since the 1990s he has played a role in characterizing the role of information commons and decentralized collaboration to innovation, information production, and freedom in the networked economy and society. His books include *Network Propaganda: Manipulation, Disinformation, and Radicalization in American Politics* (Oxford University Press, 2018) and *The Wealth of Networks: How Social Production Transforms Markets and Freedom* (Yale University Press, 2006), which won academic awards from the American Political Science Association, the American Sociological Association, and the McGannon award for social and ethical relevance in communications. His work is socially engaged, winning him the Ford Foundation Visionaries Award in 2011, the Electronic Frontier Foundation's Pioneer Award for 2007, and the Public Knowledge IP3 Award in 2006. Benkler has advised governments and international organizations on innovation policy and telecommunications, and serves on the boards or advisory boards of several nonprofits engaged in working towards an open society. His work can be freely accessed at benkler.org.

Benjamin J. Birkinbine is an Assistant Professor of Media Studies in the Reynolds School of Journalism and Center for Advanced Media Studies at the University of Nevada, Reno, USA. His research is grounded in the critical political economy of communication tradition with a specific focus on the digital commons and free and open source software. He is the author of *Incorporating the Commons* (University of Westminster Press, 2020) and the co-editor (along with Rodrigo Gómez and Janet Wasko) of *Global Media Giants* (Routledge, 2017). He is currently a Vice Chair of the Political Economy section of the International Association for Media and Communication Research.

Peter Bloom holds a BA in Urban Studies from the University of Pennsylvania, USA, and a Master's degree in Rural Development from the Autonomous Metropolitan University in Xochimilco, Mexico. He was the founder in 2002 and ex-director of Juntos, the first organization in Philadelphia dedicated to organizing and defending the human rights of Latino immigrants. In 2009 Peter began working in Nigeria as a development consultant and media maker and lived in the Niger Delta region of Nigeria for two years, co-founding the Media for Justice Project based outside of Port Harcourt. Since 2011 Peter has been coordinating Rhizomatica, an organization he started to promote new communication technologies that helps run the first community-owned and managed cell phone network in the Americas.

Yana Boeva is a postdoctoral researcher at the Department of Sociology of Technology and Environment, Institute for Social Sciences, as well as at the Cluster of Excellence on Integrative Computational Design and Construction for Architecture based at the University of Stuttgart, Germany. She has studied makerspaces and fab labs in Western Europe and Canada focusing on the socio-political and historical dimensions of digital fabrication in design towards de-professionalization of design practice, concepts of expertise, and notions of re-industrialization. Her current research explores the transformation of design, architectural practice, and different user perceptions with the inclusion of active matter and automation in contemporary fabrication models. She holds a PhD (2018) in Science and Technology Studies from York University, Toronto, and an MA (2011) in Media Studies from Humboldt University Berlin.

Margie Borschke is the author of *This is Not a Remix: Piracy, Authenticity and Popular Music* (Bloomsbury Academic, 2017). She is a Senior Lecturer in Journalism and Media in the Department of Media, Music, Communication and Cultural Studies at Macquarie University in Sydney, Australia.

Kat Braybrooke is a designer and digital anthropologist whose work explores the critical implications of the spaces and practices of creative digital communities in places like Europe and China as sites of social and environmental transformation. She is Research Fellow in the School of Engineering and Informatics at the University of Sussex, UK, and Visiting Researcher in the Department of Digital Humanities at King's College London. Web: http://codekat.net.

Sébastien Broca is a sociologist. He is currently Associate Professor in the Media and Communication Department at Université Paris 8 Vincennes Saint-Denis, France. He works on digital capitalism and on the digital commons, at the crossroads of political economy and critical theory. He has published *Utopie du logiciel libre* (Le passager clandestin, 2013). He is currently involved in the research projects EnCommuns and TAPAS and is co-editor of the scholarly journal *Anthropology&Materialism*. sebastien.broca@univ-paris8.fr

Stéphane Couture is an assistant professor at University of Montreal, Canada. Trained in computer science, sociology, and communication studies, he has completed several research projects on the collaborative practices and cultures of free and open source software. He has also studied and been engaged in initiatives related to alternative media, digital activism, and the relationship between democracy and technology more broadly.

George Dafermos is a postdoctoral researcher at the Heteropolitics research project at the Aristotle University of Thessaloniki, Greece, where he explores the transformative potential of commons-based peer production. He has been involved in several of the projects mentioned in the chapter "Prophets and Advocates of Peer Production": he was a participant in the Oekonux Project (2002–2013) and a core member of the FLOK Society Project in Ecuador (2013–2014). He is also a founding member of the *Journal of Peer Production* and a research associate of the P2P Foundation.

Maitrayee Deka is a Lecturer in Media and Social Theory at the Department of Sociology, University of Essex, UK. Her research focuses on economic sociology, new media, consumption cultures, and social theory. She is currently working on her monograph, *Traders and Tinkers: The Popular Economy of the Bazaar* that is based on an ethnographic account of Delhi's electronic marketplaces. Her most recent publication is "Embodied Commons: Knowledge and Sharing in Delhi's Electronic Bazaars," *The Sociological Review*, 66(2), 365–380, 2018.

Wolfgang Drechsler is Professor of Governance at the Ragnar Nurkse Department of Innovation and Governance, Tallinn University of Technology, Estonia, Honorary Professor at University College London in the Institute of Innovation and Public Purpose, UK, and Associate and member of the advisory board at Harvard University's Davis Center, USA. In civil service, he has been Advisor to the President of Estonia, Executive Secretary with the German Wissenschaftsrat, and, as an APSA Congressional Fellow, Senior Legislative Analyst in the United States Congress. Wolfgang's main interests are Public Management, Technology, and Innovation; Non-Western Public Administration, Governance, and Economics (especially Buddhist, Confucian, and Islamic); and Public Management Reform generally.

Mélanie Dulong de Rosnay, PhD in law, has been an associate research professor at the French National Centre for Scientific Research (CNRS) since 2010 and is the director of the Center for Internet and Society of CNRS (UPR 2000) which she co-founded with Francesca Musiani in 2019. Her research focuses on digital commons, regulation by technology, information technology law and policy. She has co-edited four open access collective books on the digital public domain and digital commons. She is a founding member and was the legal lead for Creative Commons France, a fellow at Science Commons, a staff member of Creative Commons Netherlands, a staff member of the Communia European Thematic Network, and a founding member of the Communia association for the digital public domain, which she represented at the World Intellectual Property Organization (WIPO) as the first chair of its administrative council. She was also member of the board and vice-president of the scientific board of OpenEdition scientific publishing platform (2010–2019). melanie.dulong@cnrs.fr; @melanieddr

Adam Fish is an associate professor and Scientia fellow in the Faculty of Arts and Social Sciences, School of Arts and Media, at the University of New South Wales. He is a cultural anthropologist, documentary video producer, and interdisciplinary scholar who works across social science, computer engineering, environmental science, and the visual arts. Dr. Fish employs ethnographic, participatory, and creative methods to examine the social, political, and ecological influences of new technologies. He has authored three books including: *Hacker States* (2020 MIT Press with Luca Follis), about how state hacking impacts democracy; *Technoliberalism* (2017 Palgrave Macmillan), an ethnography of the politics of Internet and television convergence in Hollywood and Silicon Valley; and *After the Internet* (2017 Polity Press with Ramesh Srinivasan), which reimagines the Internet from the perspective of grassroots activists, citizens, and hackers on the margins of political and economic power. He is currently completing a book, *Drone Justice,* with MIT Press.

Jutta Haider is Professor of Information Studies at the Swedish School of Library and Information Science, University of Borås, and Reader in Information Studies at Lund University, both in Sweden. She holds a doctorate from City, University of London, UK (2008). Her research concerns information practices and infrastructures in relation to digital cultures' conditions for production, use, and distribution of knowledge and information. This includes research on information inequalities and on knowledge institutions, such as encyclopedias, search engines, and the scholarly communication system. She is author of *Invisible Search and Online Search Engines: The Ubiquity of Search in Everyday Life* (Routledge, 2019).

Rebecca Karp is a PhD candidate at Boston University Questrom School of Business, Boston, USA. She is a field researcher who studies how firms and collectives learn, adapt, and upend their strategies for growth and sustained competitive advantage when confronted with how different market actors interact with their innovations. Her research specifically considers how innovators deploy resources to gain market acceptance, economic traction, and grow market relevancy for their innovations.

Vasilis Kostakis is the Professor of P2P Governance within the Ragnar Nurkse Department of Innovation and Governance at TalTech, Estonia. He is also a Faculty Associate within the Berkman Klein Center for Internet and Society at Harvard University, and a Visiting Professor within the Institute of Environmental Science and Technology at Autonomous University of Barcelona. He is the founder of the P2P Lab. In 2018, Vasilis was awarded a four-year grant from the European Research Council, to study the convergence of the digital commons with local manufacturing technologies. Along with an interdisciplinary team of scholars, activists, and social entrepreneurs, Vasilis focuses on how to create an economy based on locally sustainable communities that are digitally interconnected. His work has appeared in 15 languages.

Mariam Mecky is an Egyptian feminist researcher. She received an MA in Gender Studies and Law at SOAS, University of London in 2018 as a Chevening Scholar and a BA in Political Science from the British University in Egypt in 2013. Mecky has worked as a researcher, journalist, and NGO worker focusing broadly on gender issues and politics in Egypt and the MENA region at large. Alongside her current research work, she is the Communication Unit head at HarassMap, an anti-sexual harassment NGO in Egypt. Her interests include but are not limited to: feminist mobilization and activism, gender-based violence, body politics, resistance, and legal reform.

Morgan Meyer is a research director at the French National Centre for Scientific Research (CNRS). He holds a PhD in Sociology (University of Sheffield) and has been a visiting professor at the University of Vienna (Department of Science and Technology Studies) and a visiting researcher at the University of Edinburgh (Genomics Forum). His research concentrates on three main topics: (1) participation and co-production of knowledge (natural history, do-it-yourself biology, open source agriculture), (2) new configurations and communities in biology (synthetic biology, gene editing), (3) intermediation, translation, and representation of knowledge.

Stefania Milan is a digital sociologist interested in new forms of political participation facilitated by technological innovation. Stefania is Associate Professor in New Media and Digital Culture at the University of Amsterdam, the Netherlands, where she leads a research project exploring the design and governance of technology standards from a human rights perspective, funded by the Dutch Research Council. Previously she worked at the European University Institute, Central European University, Tilburg University, the Citizen Lab at the University of Toronto, and the University of Oslo. In 2015–2021 she was the principal investigator of two projects financed by the European Research Council exploring data- and algorithmic-mediated forms of civic engagement (data-activism.net and algorithms.exposed). She is the author of, among others, *Social Movements and their Technologies:*

Wiring Social Change (Palgrave Macmillan, 2013/2016), and co-author of *Media/Society* (Sage, 2011). Stefania enjoys creating bridges between research, activism, and policymaking, and experimenting with methodological innovation. Website: stefaniamilan.net

Amisha Miller is a PhD candidate at Boston University Questrom School of Business, Boston, USA. She studies how entrepreneurs and their early-stage firms interact with actors from outside the firm to create value. More specifically, her field research explores how diverse entrepreneurs and early-stage firms can interact with institutions, collectives, and communities that support entrepreneurs, to create valuable firms and products, and more inclusive institutions.

Francesca Musiani (PhD, socio-economics of innovation, MINES ParisTech, 2012), has been an associate research professor at the French National Center for Scientific Research (CNRS) since 2014. She is Deputy Director of the Center for Internet and Society at CNRS, which she co-founded with Mélanie Dulong de Rosnay in 2019. She is also an associate researcher at the Center for the sociology of innovation (i3/MINES ParisTech) and a Global Fellow at the Internet Governance Lab, American University in Washington, DC. She is the author and editor of several books including *Nains sans géants. Architecture décentralisée et services Internet* (Dwarfs Without Giants: Decentralized Architecture and Internet Services, Presses des Mines, 2013 [2nd ed. 2015], recipient of the French Privacy and Data Protection Commission's Prix Informatique et Libertés 2013).

Sven Niederhöfer is a research assistant at the Chair of Management and Digital Markets at the University of Hamburg, Germany. He studied Information Systems at Technische Universität Darmstadt and Tampere University of Technology in Finland. His research interests cover innovation management, digital platforms, and business ecosystems. Specifically, he is researching aspects related to orchestrating digital platform ecosystems.

Helen Nissenbaum is a professor at Cornell Tech and in the Information Science Department at Cornell University, USA. Her research takes an ethical perspectives on policy, law, science, and engineering relating to information technology, computing, digital media, and data science. Topics have included privacy, trust, accountability, security, and values in technology design. Her books include *Obfuscation: A User's Guide for Privacy and Protest*, with Finn Brunton (MIT Press, 2015) and *Privacy in Context: Technology, Policy, and the Integrity of Social Life* (Stanford, 2010).

Ory Okolloh currently serves as Managing Director, Luminate Group. Luminate is a global philanthropic organization that funds non-profit and for-profit organizations that help people participate in and shape the issues affecting their lives, and make those in power more transparent, responsive, and accountable. Based in Nairobi Ory is also tasked helping the drive the growth of Omidyar Network's overall investment portfolio in Africa. Prior to this, Ory was Google's policy manager for Africa. Previously, Ory was at the forefront of developing technology innovation as a founding member of Ushahidi. She served as the organization's executive director from inception until December 2010. Ory is also the co-founder of Mzalendo, a web site that tracks the performance of Kenyan MPs. In 2011 she was named a Young Global Leader by the World Economic Forum, and one of Africa's most Powerful Women by *Forbes Magazine*. In 2014 she was named Time 100's most influential people in the world. She serves on the Board of Directors of the Thomson Reuters Founders Share Company, is an Aspen Global Leadership Network (AGLN) Fellow, and an advisory board member to Twiga Foods and Endeavor Kenya. Ory earned a JD from Harvard Law School and a BA in political science from the University of Pittsburgh.

Siobhán O'Mahony is the Feld Family Professor, Innovation and Entrepreneurship in Strategy and Innovation at the Boston University Questrom School of Business and the Academic Director of Innovate@BU, Boston, USA. She received her PhD in Management Science and Engineering from Stanford University. Professor O'Mahony's research examines organizing processes in community and project forms and how creative and technical collectives organize for innovation, creativity, and growth. Her work has been published in *Administrative Science Quarterly*, *Organization Science*,

Academy of Management Journal, *Research Policy*, *Research in Organizational Behavior*, *Research in the Sociology of Organizations*, *Industry and Innovation*, the *Journal of Management and Governance* among other edited volumes. somahony@bu.edu

Mathieu O'Neil is Associate Professor of Communication at the University of Canberra's News & Media Research Centre, where he leads the Critical Conversations Lab, and Honorary Associate Professor of Sociology at the Australian National University, where he co-founded the Virtual Observatory for the Study of Online Networks. Mathieu's research examines the policy and organizational aspects of commons-based peer production, as well as risk issue diffusion and the adoption of causes and innovations in the online environment. His book *Cyberchiefs* (2009) was the first systematic examination of governance in peer production projects. He is the founder and maintainer of the *Journal of Peer Production*. His work has been published in *Social Networks*, the *Journal of Peer Production*, *Réseaux*, *Information, Communication & Society*, *Organization Studies*, and *New Media and Society*, amongst others. mathieu.oneil@canberra.edu.au

Alekos (Alexandros) Pantazis is a core member of the P2P Lab, an interdisciplinary research collective focused on the commons, and a junior research fellow at the Ragnar Nurkse Department of Innovation and Governance, Tallinn University of Technology, Estonia. Moreover, Alekos has 20 years of involvement in international civil movements, focusing on agrarian indigenous populations and the commons. He is pursuing a PhD on the convergence of convivial technologies, commons, and non-formal education.

Alex Pazaitis is a core member of the interdisciplinary research collective P2P Lab, a spin-off of the Ragnar Nurkse Department of Innovation and Governance, Tallinn University of Technology, Estonia, and of the P2P Foundation, and Junior Research Fellow and PhD candidate at the Ragnar Nurkse Department. Alex is involved in numerous research activities and research and innovation projects. He has professional experience in project management and has worked as a consultant for private and public organizations. His research interests include technology governance; innovation policy; digital commons; open cooperativism and distributed ledger technologies.

Christian Pentzold is Professor of Media and Communication in the Department for Communication and Media Studies at Leipzig University. Before that, he worked in the Centre for Media, Communication and Information Research at the University of Bremen and at Chemnitz University of Technology. He is broadly interested in the construction and appropriation of digital media and the roles, information and communication technologies play in modern society. His work in communication research and media analysis links to insights coming from cultural sociology, linguistics, as well as science and technology studies. Currently he is looking at the public understanding of big data, the organization, and governance of peer production, as well as the interplay of time, data, and media. Beyond that, he is interested in applying theories of practice to the study of media and communication and in linking qualitative with quantitative methods.

Gwen Shaffer is an associate professor in the Department of Journalism and Public Relations at California State University, Long Beach, USA, where she teaches courses on Internet regulation and communication law. Her research on topics such as broadband connectivity and data privacy explore ways in which digital exclusion and algorithmic bias compound existing challenges. Her research has been published in the *Journal of Information Policy*, *Media, Culture & Society*, *First Monday*, and the Association for Computing Machinery's *Transactions on Internet Technology*, among other journals and book chapters. She has also co-authored policy papers on topics such as mobile phone privacy and digital inclusion. Gwen Shaffer chairs the City of Long Beach's Technology and Innovation Commission, which advises the mayor and City Council on relevant policy issues.

Adrian Smith is Professor of Technology and Society at the Science Policy Research Unit at the University of Sussex, UK. He is involved in interdisciplinary research projects investigating the politics of technology and innovation for sustainable development. He has a particular interest in grassroots involvement in this politics, and questions of democracy. This has included studies of makerspaces in Europe, Latin America, and Asia.

Sebastian Spaeth holds the Chair of Management and Digital Markets at the University of Hamburg, Germany. He studied Business and Engineering at the Karlsruhe Institute of Technology and graduated at the Institute of Technology at Linköping University, Sweden. He received his doctorate in Business Administration from the University of St. Gallen, Switzerland where, cooperating with Georg von Krogh, he examined open source software development projects. He conducted research on collaborative open innovation and collaborative business models as a postdoctoral fellow at ETH Zurich, Switzerland. In 2013, he founded the Chair of Management and Digital Markets at the University of Hamburg to focus on the challenges and opportunities arising from the digitalization of society for citizens and organization. He has published in the *Strategic Management Journal*, *Management Science*, *Research Policy*, *Information Systems Research*, and *MIS Quarterly*.

Michael Stevenson is Associate Professor of New Media and Digital Culture at the Media Studies Department, University of Amsterdam, the Netherlands. His research interests include a range of topics in Internet history, software studies, and digital culture. He is working on a book called *Making Media New*, about the historical development of social media and the broader new media field in the 1990s and 2000s. In 2019 he made *Geeks in Cyberspace*, a web documentary about a group of open source software enthusiasts who created the influential websites Slashdot, Everything2, and PerlMonks.

Olof Sundin is Professor in Information Studies at Lund University, Sweden. He holds a doctorate from the University of Gothenburg, Sweden (2003). His work primarily concerns the configuration of information in contemporary society, the construction of public knowledge in relation to trust, and information searching and use. He has investigated Wikipedia, as well as commercial encyclopedias, from a perspective of use and findability as well as in terms of how encyclopedic information is constructed. In 2019, he published the book *Invisible Search and Online Search Engines: The Ubiquity of Search in Everyday Life* (Routledge).

Abraham Taherivand studied Business Informatics, Information Management and Engineering and Design Thinking. He is a multiple award-winner at national and international business plan competitions, holds various patents, and has been a serial entrepreneur since 2008 in the tech, Internet, and consumer sector. He has worked as a strategy consultant in various projects for companies and organizations. Since 2012 he has worked for Wikimedia Deutschland. He is the co-founder of the first worldwide free structured knowledge database Wikidata and was awarded the Open Data Award by Tim Berners-Lee and his teams. Since December 2016, Abraham Taherivand has been the Managing Director of Wikimedia Deutschland e.V.

Nathaniel Tkacz is a reader in digital media and culture and Deputy Director of the Centre for Interdisciplinary Methodologies at the University of Warwick, UK. He has written widely and critically on the topics of peer production, currency, money, and software culture. In 2012, he co-founded the MoneyLab network in Europe with Geert Lovink and is co-editor of the first MoneyLab Reader (2015). He has edited and authored a number of other monographs, including Wikipedia and the *Politics of Openness* (2015).

Sophie Toupin is a Fonds de recherche du Québec - Société et culture (FRQSC) postdoctoral fellow at the University of Amsterdam where she explores the linkages between feminism, data and infrastructure. She completed her PhD in the Department of Art History and Communication Studies at McGill University, Montreal, Quebec, Canada. Her doctoral research examined the relationship between technology and anti-colonialism during the South African anti-apartheid struggle. Her work has been published in New Media and Society, Feminist Media Studies, Canadian Journal of Communication, Ada: A Journal of Gender, New Media, and Technology, and Journal of Peer Production, among others.

Peter Troxler is a research professor at Rotterdam University of Applied Sciences, the Netherlands. He studies how emerging ways of designing and manufacturing by "fabbers" and "makers," enabled by readily available direct digital manufacturing technologies, challenge and influence incumbent practices in the creative and manufacturing industries. His research interests are the intersections and interdependencies of people, technology, business, society, and the environment. He also investigates how new paradigms of cooperation and business models are developed based on network patterns, self-organization, lateral governance, and open source principles. He holds a PhD (1999) in Technology, Management and Economics and an MSc (1993) in Industrial Engineering, both from ETH Zurich. He has worked internationally in the energy industry, as a consultant to private and public sector enterprises, as a researcher and research manager in organizational psychology and artificial intelligence, and as a director and producer in various artistic projects in music, theater, literature, and social arts.

Pablo Velasco González is an assistant professor in the Department of Digital Design and Information Studies at the University of Aarhus, Denmark, where he co-directs the Center for the Study of Technological, Emerging, and Ethical Methods. His work has focused on digital culture and politics of technical devices, in particular blockchain technology and cryptocurrencies. He has collaborated on these and other topics in two Moneylab Readers (2015, 2018), the *Metaphilosophy* and *APRA* journals, and several book chapters. More information about his work can be found at pablov.me

Stefano Zacchiroli is Associate Professor of Computer Science at the University of Paris Diderot, France, and is currently on leave at Inria. His research interests span formal methods, software preservation, and free/open source software engineering. He is co-founder and current CTO of the Software Heritage project. He has been an official member of the Debian Project since 2001, and was elected to serve as Debian Project Leader for three consecutive terms from 2010 to 2013. He is a former Board Director of the Open Source Initiative (OSI) and recipient of the 2015 O'Reilly Open Source Award.

Preface

The *Handbook of Peer Production* has emerged from a community of researchers, practitioners, and activists who share a belief in the virtue of open collaboration. We are grateful for this community's existence and hope we have been equal to the task of documenting its work. Academic books are always collective endeavors; we are immensely thankful to those who have helped us in bringing this project forward.

The *Handbook* chapters were peer reviewed by the three editors. In addition, Chapter 1 ("The Duality of Peer Production"), Chapter 29 ("What's Next?"), and Chapter 30 ("Be Your Own Peer!") which we co-wrote, were reviewed by Sébastien Broca. His insightful comments played a substantial part in making our arguments more coherent, so we express our heartfelt thanks for his contribution.

Our congratulations to all the authors who worked hard to follow our timeline and recommendations for a job well done – especially those who came to the project late as others had to drop out. We are also grateful that Yochai Benkler and Helen Nissenbaum allowed us to reprint their article "Commons-Based Peer Production and Virtue." Thanks to Nils Borchers, George Dafermos, Chris Giotitsas, Vasilis Kostakis, Arwid Lund, and Denise Thwaites for their help during the course of the project.

At Wiley, our thanks go to Haze Humbert who encouraged us to embark on this three-year endeavor, to Kartiga Ramalingam and Janani Govindankutty who helped us navigate legal issues, as well as to Kelley Baylis, Todd Green, Skyler Van Valkenburgh, and Andrew Minton who supported the volume through various stages of production. Finally, our thanks to Rosemary Morlin for copy-editing and to Aijth Kumar for putting together the volume's index.

Mathieu would like to thank Stefan Merten and the Rosa Luxemburg Foundation for flying him to the United Kingdom from Australia to attend the 4th Oekonux conference in Manchester back in the day. In Manchester I met for the first time Michel Bauwens, George Dafermos, Johan Söderberg, Athina Karatzogianni, Stefan Merten, and many other researchers and activists. The conference was so much fun and so inspiring that during the closing plenary session I proposed setting up a journal, which eventually became the *Journal of Peer Production*. On the way to Manchester from London I had made a stop at the Oxford Internet Institute to attend a workshop on online governance organized by two (at the time) emerging researchers: Malte Ziewitz and Christian Pentzold. Years later, Christian and I sometimes bumped into each other at academic conferences and after I suggested we collaborate, he came up with the idea for this *Handbook*. Sophie Toupin, who by then had joined the *Journal of Peer Production* editorial team, had complementary research interests and she proved the perfect choice to join us. Working with Christian and Sophie on this labor of love has been a wonderful and enriching experience.

Thanks to my colleagues at the University of Canberra's News and Media Research Centre for their unwavering support and sound advice. Thanks also to my research associate Xiaolan Cai as well as to my research partner Rob Ackland at the Australian National University's Virtual Observatory for the

Study of Online Networks for helping us map the online network of peer production actors. Thanks to the Australian National University's School of Sociology for its continuing support. I am grateful to the University of Canberra's Faculty of Arts and Design for supporting my Outside Studies Program in 2019, which enabled me to focus on this project and to work with colleagues overseas: George Dafermos (Integral Cooperative, Heraklion), Laure Muselli (Télécom Paris), Giacomo Poderi (IT University of Copenhagen), and last but not least Mélanie Dulong de Rosnay and Francesca Musiani (Centre Internet et Société, CNRS). My parents Mike and Charmian kindly hosted during my time in Paris. At the time of writing (April 2020) my father is fighting Covid-19. My parents' politics and humor inspired me. I dedicate this *Handbook* to them.

Christian would like to thank his colleagues at Bremen and later at Chemnitz University. Both places provided the intellectual environment and academic workplace necessary to pursue such long-term project. I owe special thanks to my student assistants Lea Neubauer who helped us with the transcriptions of the interviews and Jessica Giesa who compiled the lists of tables and illustrations. Editing this volume and collaborating with Mathieu and Sophie has been an inspiring learning process and I thank them for their dedication, inquisitiveness, and their eye for all the many details that were necessary to make this book happen.

Sophie would like to thank the professors and graduate students in the Communication Studies Department at McGill University who have continued to give her invaluable advice, intellectual support, and community. Joining Mathieu and Christian in editing this *Handbook* at the beginning of my PhD was at first daunting, but it has proven to be very rewarding. Now that these two long-term projects are coming to an end – the PhD and the volume – I remain inspired by the many collectives and groups who use the practice of peer production to reimagine the world they want to live in. I dedicate this *Handbook* to them. Finally, heartfelt thanks to Christian and Mathieu; it has been a delight working with you!

Chapter Summaries

Part I Introduction

1 The Duality of Peer Production: Infrastructure for the Digital Commons, Free Labor for Free-Riding Firms

Mathieu O'Neil, Sophie Toupin, & Christian Pentzold

This introductory chapter examines a series of productive tensions located in and around peer production: we interrogate the meaning of peer-to-peer infrastructure models and find that some forms of peer infrastructure have thrived, whilst others were effectively banned. We review Yochai Benkler's influential theorization of "commons-based peer production," and ask to what extent it embodies Western, first-world assumptions. We evaluate the relationship of peer production to the dominant economy, considering the rich scholarship on peer production's transformational potential, which was inspired by Benkler's model and is often imbued with utopian overtones. At the same time peer infrastructure plays a central role in the digital economy, and we critically examine political economy understandings which hold that peer production has been recuperated by capitalism and enabled new forms of labor exploitation. We also analyze the organizational form which facilitated the emergence of hybridization between commercial firms and communal projects, outline the aims of this *Handbook*, and summarize its structure and content.

Part II Concepts: Explaining Peer Production

2 Grammar of Peer Production

Vasilis Kostakis & Michael Bauwens

In 2005, Michel Bauwens published *The Political Economy of Peer Production*, which discussed the principles, characteristics, and the future of the then nascent ecosystem of peer production. This chapter revisits Bauwens' 2005 article, adopting, expanding, and refining the operational concepts – or "grammar" – he used to define peer-production projects and the institutional ecosystems that sustain them. Our aim is to provide a framework that would give a theoretical underpinning to the transformative practices of peer production. In a time of deep environmental, social and political crisis, it is important to understand how a new kind of society, based on the centrality of the commons and within a reformed market and state, is possible. This chapter, thus, discusses and introduces new paradigmatic ways of value creation that have the potential to be more radically inclusive and sustainable.

3 Political Economy of Peer Production

Benjamin J. Birkinbine

This chapter provides a framework for understanding the political economy of peer production. As such, I interrogate the intersection of peer production and capitalism along two axes. First, I contextualize the rise of peer production within broader structural changes occurring within capitalism and evaluate the extent to which peer production contradicts, or reinforces, these global economic trends. Second, I draw from more recent theories of commons value circuits to position peer production as dialectically situated between capital and the commons, which highlights the ways in which communities of peer producers intersect with circuits of capital accumulation. The question I explore in the latter part of the chapter is whether the emergent cultural practices within peer production have the capability to subvert the prevailing tendencies of capitalism and offer a path toward a post-capitalist future.

4 Social Norms and Rules in Peer Production

Christian Pentzold

The regulation of peer-production projects is achieved by the users themselves. These forms of self-organization and self-management depend on shared social norms and have generated, in turn, sets of rules. Some of them characterize the larger population of peer-production projects; others seem to be an attribute of particular projects. The chapter provides an overview and comparison of peer production's signature norms and rules, it traces their origins and describes their implications for collaboration and editorial work.

5 Cultures of Peer Production

Michael Stevenson

How can we make sense of cultures of peer production, which exist in diverse national, cultural, and language contexts, span several industries and domains, and comprise a range of different organizational structures? To set the groundwork for such an understanding, this chapter argues that it is necessary to see that peer production is, by and large, a form of cultural production, and thus bears structural similarities to existing cultural fields like art, literature, and journalism. The chapter shows how (1) peer-production projects are clearly embedded in existing cultural fields, and often represent an autonomous form of production that seeks to resist certain economic and political pressures in favor of core values such as meritocracy and openness and (2) such autonomy is achieved through the enactment of those core values, which are in turn related to the social hierarchies, forms of exclusion, and other limitations that characterize these projects and the groups of people who populate them.

6 Commons-Based Peer Production and Virtue (reprint)

Helen Nissenbaum & Yochai Benkler

Commons-based peer production is a socio-economic system of production that is emerging in the digitally networked environment. Facilitated by the technical infrastructure of the Internet, the hallmark of this socio-technical system is collaboration among large groups of individuals, sometimes in the order of tens or even hundreds of thousands, who cooperate effectively to provide information, knowledge, or cultural goods without relying on either market pricing or managerial hierarchies to coordinate their common enterprise. While there are many practical reasons to try to understand a novel system of production that has produced some of the finest software, the fastest supercomputer and some of the best web-based directories and news sites, here we focus on the ethical, rather than the functional dimension. What does it mean in ethical terms that many individuals can find themselves cooperating productively with strangers and acquaintances on a scope never before seen? How might it affect, or at least enable, human action and affection, and how would these effects or possibilities affect our capacities to be virtuous human beings? We suggest that the emergence of peer production offers an opportunity for more people to engage in practices that permit them to exhibit and experience virtuous behavior. We posit: (a) that a society that provides opportunities for virtuous behavior is one that is more conducive to virtuous individuals; and (b) that the practice of effective virtuous behavior may lead to more people adopting virtues as their own, or as attributes of what they see as their self-definition. The central thesis of this chapter is that socio-technical systems of commons-based peer production offer not only a remarkable medium of production for various kinds of information goods but serve as a context for positive character formation. Exploring and substantiating these claims will be our quest, but we begin with a brief tour through this strange and exciting new landscape of commons-based peer production and conclude with recommendations for public policy.

Part III Conditions: Enabling Peer Production

7 Prophets and Advocates of Peer Production

George Dafermos

From the beginning, boosters of peer production portrayed it as heralding a better way of life. Since then activists and researchers have detected in peer production the seeds of a post-capitalist society (Oekonux Project, P2P Foundation) or worked to help policy makers and governments transition towards commons-based models (FLOK Society Project). Others have attempted to engage critical intellectuals inside and outside academia (*Journal of Peer Production*) and to establish peer production as a promising research field in the social sciences (P2P Lab). This chapter retraces the history of these attempts, teases out their differences and convergences, and evaluates their impact.

8 Virtue, Efficiency, and the Sharing Economy

Margie Borschke

Peer production is assumed to be virtuous and public-spirited, a networked socio-economic system of production, that is efficient, promotes individual agency, harnesses collective knowledge, creates robust technologies and information and contributes to sustaining the public domain in the Internet era. This organizational innovation is also often associated with the rise of social networking technologies, practices, and platforms in the 2000s and an ethos of participation, sharing, and remix. Yet at the heart of key conceptualizations of peer production is a tension between virtue and pragmatism, between a belief that particular kinds of networked spaces and practices can enable the development of personal and social virtues and also be more efficient than other forms of production. This tension becomes more visible in the metaphor, practices, and platforms of the sharing economy where ethical debates about agency, property, privacy, and collective rights abound and where utopian rhetoric acts as a cover for the corporate drive for efficiency over ethical concerns. This chapter considers these tensions and how the ideals of peer production were shaped by their social and material histories.

9 Open Licensing Peer Production

Mélanie Dulong de Rosnay

This chapter traces the evolution of legal conditions meant to support the production and flourishing of digital, knowledge, intellectual or information commons by facilitating access and reuse while preserving them from enclosure. Licenses have been drafted and fine-tuned in order to subvert and adapt copyright rules designed to reserve rather than to grant rights. Different legal options and conditions set up by peer-production platforms, or single creators, to users and potential audience can ensure different levels of openness, leading to the construction of informational, cultural, knowledge, or digital commons. From free and open source software to creative works, including scientific articles, cultural heritage, public sector information, and open data, the nature of works which can be peer produced and subjected to an open license extended to functional works such as databases and tangible output, such as open hardware and Internet infrastructure. Reflecting political debates and ideologies in the digital commons sphere, licensing options oscillate between public domain, copyleft, and the reservation or the control of commercial use and derivative rights.

10 User Motivations in Peer Production

Sebastian Spaeth & Sven Niederhöfer

Peer-production systems often attract larger communities of paid and unpaid volunteers, who contribute to their respective projects. This chapter examines different underlying motivations that fuel these contributions. It thereby takes a tripartite form and summarizes current literature on (1) individual motivations to participate, (2) selection of tasks, and (3) participation in peer production as a social practice. In the first two parts, we draw on self-determination theory, which discusses various intrinsic (the joy performing the task itself), extrinsic (rewards such as pay), and internalized extrinsic motives (internalized mores and values). The discussed literature shows that contributors are motivated not by a single motive, but by a whole range of interacting intrinsic, internalized extrinsic, and extrinsic motives with different magnitudes. It further shows that peers' motivation partly determines the type of task they will self-allocate, whereby (internalized) extrinsic motives seem to play a crucial role in impelling individuals to perform mundane tasks. In the third part, we view peer-production systems as social practices, conceptualizing these systems as collectives of contributors with shared general principles, whose lives increasingly become intertwined with these communities. Reviewed literature suggests that motivation may be influenced by factors such as social exposure and institutional frameworks.

11 Governing for Growth in Scope: Cultivating a Comparative Understanding of How Peer-Production Collectives Evolve

Rebecca Karp, Amisha Miller, & Siobhán O'Mahony

Scholars have been fascinated by the rise of peer-production collectives and how governance mechanisms can foster or inhibit the growth of new contributors. Few scholars have attended to what explains changes in the scope of innovation activities peer-production collectives assume. We identify five roles that peer collectives can play in the innovation lifecycle, from idea generation to post-production review, and compare 12 mature peer collectives to understand how their scope evolved over time. Our provisionary comparative analysis suggests that peer collectives that expanded their scope were more likely to distribute governance rights to contributing participants through a more collaborative mode of production. We offer a framework for analyzing why peer collectives grow differently and articulate a research agenda that embraces a dynamic approach to examining how scope and governance evolves.

Part IV Cases: Realizing Peer Production

12 Free and Open Source Software

Stéphane Couture

Free and open source software (FOSS) was formally launched in the 1980s by Richard Stallman in opposition to proprietary or closed software. Its speed of development was boosted by the rise of the Internet in the 1990s, and Linux became the original example of crowd-sourced "bazaar" production whereby the number of eyes "makes all bugs shallow." This chapter presents the history of FOSS, its modes of production, and its impact on both the infrastructure of the network society and the culture and practices of peer production projects. It ends by addressing some of the challenges FOSS is facing, in particular in terms of developing sustainable models, enhancing diversity in participation, and negotiating its growing integration into market processes.

13 Wikipedia and Wikis

Jutta Haider & Olof Sundin

Wikis are often considered to be the core platform of peer production. This chapter brings together a broad range of research on wikis and Wikipedia from different disciplines. It delineates the central design principles and affordances of wikis and also pays attention to their historical development and embeddedness in society. Wikis are described as content management systems that allow for flexible collaboration without a defined content owner or leader. Users can modify the content and structure of documents directly in their web browser. Edits are usually archived and open to revision. This chapter pays particular attention to the most successful wiki-based system, the non-profit, online encyclopedia Wikipedia. As the chapter explains, Wikipedia actually contains a broad range of more or less individual wiki projects and has inspired a plethora of other endeavors, both open source and proprietary. The particular peer production model employed in Wikipedia is elucidated, and in the course also complicated, as is the role of Wikipedia in the contemporary commercial Internet. The chapter concludes by highlighting several tensions emerging from a wiki-based peer-production model, between amateurs and experts, human editors and bots, lay knowledge and academic knowledge, and the shaping of trust through external actors.

14 Participatory Cartography: Drones, Countermapping, and Technological Power

Adam Fish

An investigation into participatory cartography – amateur and collaborative mapmaking – exhibits how theories of peer production often neglect how collective collaboration is dependent upon the state and fuels corporate technological development. This chapter investigates four case studies of participatory cartography, using theories of new materialism and technological power to expose the links between volunteerism and capitalism. For example, the participatory mapping project OpenStreetMap and its links to technology companies such as Bytemark and CloudMade, the early days of the US drone hacker network transforming into the for-profit companies 3DR and DJI, and the problems of inequality related to the Indonesian Dayak community countermapping projects offer vivid examples of how the autonomy of peer production can be limited by elite technological power. Also, the Global Positioning System, the most important technology for participatory mapping, is a highly costly platform financed by the state. The new materialistic approach taken in this chapter emphasizes the entanglements of technology, culture, and politics and the movement of power across these different domains. That peer production is dependent on extant forms of technological power does not extinguish its political potential. On the contrary, the materialism approach invites proponents of peer production to better respond to the inequities associated with elite domination of technological power.

15 P2P Learning

Panayotis Antoniadis & Alekos Pantazis

In this chapter we identify a wide variety of learning projects, platforms, tools, and methodologies which could be characterized as "peer-to-peer" and present their main characteristics along three core dimensions (curriculum selection, learning process, and knowledge abstraction). We then discuss how p2p learning processes can be encouraged, facilitated, and supported by digital and physical infrastructures, keeping a critical outlook on the often-hidden power asymmetries that are always present at the infrastructure level. We pay particular attention to the case of small intentional groups of adult learners, and to four exemplary case studies – two digital platforms and two physical spaces – with similarities and differences, which help us to deconstruct and critically analyze the different dimensions of identified p2p learning.

16 Biohacking

Morgan Meyer

Over the past few years, a plethora of terms have emerged to describe scientific activities in the life sciences that happen outside traditional institutions: biohacking, garage biology, DIY biology, DIY genetics, DIY medicine, DIY science, and so on. These movements have attracted a growing number of enthusiasts, from young students to professional scientists, from artists to aspiring entrepreneurs, from people with no technological background to computer hackers. This chapter reviews and discusses the key literature on the topic. It first provides an overview of the history, and the various practices of DIY biology. Thereafter it offers a few examples of DIY medicine. The chapter then discusses and analyzes some of the key issues of DIY biology: the openness of the movement; what the "yourself" in do-it-yourself stands for; and concerns with ethics and governance. The penultimate section of the chapter looks at the economic aspects and the valuations of DIY biology. The conclusion will argue that further academic work could look into the geographies and fragilities/instabilities of DIY biology, as well as its relationship with the public.

17 Makers

Yana Boeva & Peter Troxler

Makers have engaged with different practices of peer production for the last two decades. Given access to electronic prototyping tools, digital fabrication tools, biohacking equipment, as well as different craft techniques and materials, makers exchange their knowledge on- and offline signaling bottom-up processes of design, production, and education. The so-called "maker movement" would bring about the democratization of technology and user emancipation by its prospect of personal fabrication and peer collaboration. Given the multiplicity of maker identities and what making has achieved, in this chapter, we tentatively define makers as individuals creating idiosyncratic artifacts in the form of most often hard- and software fixes to actual problems and challenges. The chapter provides a critical overview of the different cultures of making, their motivations as well as the socio-technical infrastructures that encourage these forms of peer production. We outline a succinct and incomplete history of making in peer production. The chapter then explores the ways maker rhetoric and its diverse proponents have shaped making's development and perception across the world. It discusses how the prospects of transforming users into makers and producers as well as the idea of makers representing a global social movement have unfolded since the year zero of making.

18 Blockchain, or, Peer Production Without Guarantees

Pablo Velasco González & Nathaniel Tkacz

This chapter provides an overview on blockchain technology in its relation with peer production. Blockchain refers to an open yet secure distributed ledger that secures a robust workflow, authenticated by mass collaboration. As such, the principles and technologies of blockchain have inspired a range of aspirations and implementation in close relation to the principles of peer production. The chapter locates the emergence of the peer-production model in relation to the organizational models of hierarchies and markets, and discusses the role and potential of blockchain technology to generate "common good" within a fair distribution of production, while also participating as an extractive model. An overview of different projects, interrogated under the light of modes of production, is offered in the second part of the chapter. These projects distinguish four aspects of blockchain initiatives that offer different readings of what may be a peer within blockchain configurations: peer production, peer-development, peer-governance, and peer-exchange. The hybridity of blockchain-based projects fluctuates the degree and roles of participation, and complicates a narrow definition of a common. While these and other blockchain-based projects can then be understood as instances of peer production, they are primarily an opportunity to rethink the emergence of this mode of production and reflect on what it can become.

19 Community Wireless Networks

Gwen Shaffer

This chapter explores the history of community wireless networks (CWNs) through diverse case studies in Europe, Africa and the United States. Mesh networks are a prime example of a peer-production practice. Anyone willing to place a router on his or her rooftop or window can play a role in helping a mesh network thrive. CWNs emerge for a variety of reasons. For some, a grassroots network is the sole option for Internet connectivity. For individuals concerned about the erosion of online privacy, security and anonymity, peer networks enable confidential online communications. Compared to vulnerabilities inherent in a centralized Internet design, mesh networks are disaster proof – whether the disaster is caused by a totalitarian government cutting off Internet access, or by a hurricane. Ideology also drives the current resurgence in mesh networks. Many CWN participants value the possibilities associated with sharing resources and deciding for themselves how – and where – to deploy Internet infrastructure. Rather than rely on Internet service providers, peer networks represent a bottom-up scheme of governance open to anyone with a desire to contribute. This chapter also examines how these projects serve as a testbed for related technologies, such as FreedomBox's secure software. The case studies highlighted in this chapter suggest that developing open source applications, infrastructure, and platforms – on which competitive providers may offer content and services – is as important to community wireless projects as connectivity itself.

20 Commoning the Urban

Nicholas Anastasopoulos

This chapter seeks to illuminate facets of the urbanized world that most of humanity now lives in, as an introduction to the alternatives that may have existed for the longest time or are being created as we speak. The urban commons are examined as a distinct domain, a category of their own, taking into account cultural traditions and commoning activities in an urban context. Their production, political nature and cultural identity, the dilemmas involving their governance, as well as current trends regarding the impact of creative practices, architecture and of digital worlds are also examined, and a variety of case studies serves as a reference to these topics.

Part V Conflicts: Peer Production and the World

21 Peer Production and Social Change

Mathieu O'Neil & Sébastien Broca

This chapter examines the claim that peer producers can meaningfully oppose the social harms of neoliberal capitalism and the democratic failures of the state in liberal democracies. We distinguish "political" actions from "economic" ones, as separating terms enables a clearer delineation of recurrent tensions between social change today and tomorrow, between grassroots activism and electoral politics, between the commons for capital and the commons as the "germ" form for a post-capitalist society. We critically engage with this radical vision, as peer production projects are often absorbed within neoliberal digital capitalism: what if the commons now enable the metastasizing of a ***capitalism beyond the commodity form*** into the heart of peer production projects? Inside some free and open source projects, financial rewards are rejected on the grounds that they might distort the self-directed means of determining the relative value of project goals. Not only does this rejection reproduce class and gender inequalities, it also hinders the evolution of the commons into a sustainable mode of production. We conclude that the capacity of peer production principles to inform a believable alternative depends on peer producers' capacity to make the commons economy more connected to wider society and more robust, challenging the structural imbalance between what digital capitalism obtains from the commons and what it gives back.

22 Peer Production and Collective Action

Stefania Milan

Over the last three decades, social movements around the world have embraced peer-production principles such as collaboration, co-production, and self-organization. From Anonymous to the Spanish 15M movement, from the global #Occupy mobilization to the ultra-conservative Alt-Right, peer-production mechanisms have empowered movements to generate models of organizing for ensuing protests to appropriate. They have been used to create shared normative references and collective action frames, outreach to novel audiences, and mobilize other like-minded individuals. This chapter investigates the consequences of peer production for social protest, looking at how peer-production reshuffles social change activism today and exploring the convergences and tensions between peer networks and social movements. First, the chapter traces the historical trajectory of peer production, linking distinct approaches to organizing to technological innovation. Second, it reflects on the social affordances of digital infrastructure and their role in fostering specific modes of creativity. Third, it explores three consequences of peer production for social movements: cultural production and norm change, collective identity, and the commons. It finally examines three tensions that might emerge while embedding peer-production mechanisms and values in collective action: individual vs. collective engagements, peer networks vs. social movement organizations, and self-organized vs. commercial infrastructure.

23 Feminist Peer Production

Sophie Toupin

Feminist(s) have not only criticized certain aspects of peer-production practices, but have come to "do" feminist peer production. This chapter examines the feminist criticisms of peer production, which they argue has rendered invisible the analytical categories of gender and race. It then focuses on new practices that have emerged out of feminist peer-production sensibilities, practices grounded in feminist objectivity and which reflect specific socio-technical realities.

24 Postcolonial Peer Production

Maitrayee Deka

What are postcolonial peer-production practices? This chapter aims to answer this question by examining the ways in which postcolonial peer production has emerged as a way to provincialize the dominant Euro-American versions. The proposition that imagines the West as the unique origin of peer-production practices is disrupted in this chapter, which traces the histories of peer production in postcolonial contexts and highlights their main characteristics. The main focus of the chapter is to centralize embodied knowledge systems and ad hoc techno-social assemblages that form a large part of creative and productive practices in the Global South. By highlighting popular bases of knowledge creation, one diverts attention from the dominant Euro-American peer communities in their reliance on creative figures onto a more chaotic universe of creative practices.

25 Gaps in Peer Design

Francesca Musiani

From a technical standpoint, systems based on peer design are often deemed to be superior to proprietary and more centralized systems, because they value long-term robustness over cost-effective commercial expedience. Yet, in several cases, peer projects are unable to compete with proprietary systems. A number of factors may be the cause of this phenomenon, including the difficulty in providing proper quality of service in the early phases of a system that relies exclusively on users' contributions; the volunteer development model, which oftentimes lacks incentives for performing routine albeit necessary tasks; the difficulty in equaling user-friendly, sometimes non-technical aspects of proprietary systems, such as attractive and comfortable design; or finding a straightforward business model that can successfully be associated with decentralization. Drawing on two case studies, this chapter addresses the issue of the "gaps" in peer-based design of the technical architecture of Internet-based services; although net architecture will be our primary focus, we will see how dynamics of motivation/incentives to participate in peer-based systems and their attractiveness/usability are fundamentally linked to the constraints and opportunities of different architectural designs. Indeed, the chapter shows how these gaps are grounded in a mix of technical, social, and economic factors, and contributes to explain why, while user-controlled, peer-based decentralized alternatives to Internet-based services have been regularly put forward in preference to Internet giants, their developers have found it challenging to compete with proprietary market leaders.

26 Makerspaces and Peer Production: Spaces of Possibility, Tension, Post-Automation, or Liberation?

Kat Braybrooke & Adrian Smith

Makerspaces are open community workshops for peer production which offer the technical tools and training to experiment with making, learning, and hands-on participation around material cultures. Makerspace networks provide people with skills and access to versatile design and fabrication technologies, as well as traditional hand-tools; and they provide social spaces that foster communities who share an open and collaborative ethos interested in the possibilities that democratized design and fabrication technologies might offer personally, socially, economically and culturally. However, as well as being spaces of creative and transformational possibility, makerspaces also experience many of the tensions of our current conjuncture. Some makerspaces have become synonymous with neoliberal business-as-usual, where a kind of entrepreneurial citizenship is prototyped through the exploitation of precarious labor by businesses and institutions. Peer production in makerspaces sits in tension with pressures to enclose, commodify, and compete to provide profitable inputs into global manufacturing circuits. This chapter explores the dynamics of makerspaces as spaces of possibility, tension, post-automation, and liberation, examining in particular how institutional encounters prompt makerspaces to interpret, reinforce, and challenge prevailing socio-technical regimes in society.

27 Peer Production and State Theory: Envisioning a Cooperative Partner State

Alex Pazaitis & Wolfgang Drechsler

This chapter theoretically examines the concept of the partner state, as a new form of symbiosis between state and civil society, based on the principles and practices of peer production. The general stance of peer production advocates is almost intrinsically anti-state. However, state theory arguably reveals that the examination of the state and its institutions actually helps us understand the position and potentials of peer production for broader social and economic transformation. A tentative union between Hegelian and Gramscian thought delineates why and how the state can, and arguably should, embrace and support peer production. Finally, a preliminary framework for the prefigurative institutions of the partner state is offered through the analysis of the concept of open cooperativism.

Part VI Conversions: Advancing Peer Production

28 Making a Case for Peer Production: Interviews with Peter Bloom, Mariam Mecky, Ory Okolloh, Abraham Taherivand, & Stefano Zacchiroli

Peer production is first and foremost a practical affair. In these interviews, practitioners involved in setting up, developing, and maintaining diverse peer-production projects share their experiences and insights: Peter Bloom talks about the Rhizomatica project, Mariam Mecky speaks about the endeavor to create HarassMap; Ory Okolloh reflects on Ushahidi; Abraham Taherivand discusses the role of Wikimedia, and Stefano Zacchiroli provides insights into the Debian FOSS community.

29 What's Next? Peer Production Studies?

Mathieu O'Neil, Sophie Toupin, & Christian Pentzold

This chapter re-examines the dual contribution of peer production to productive efficiency and social justice. We first interrogate each of these concepts' potential for future research. Next, the chapter reflexively evaluates peer production as an object of study by mapping a network of peer-production researchers and by considering whether a field of "peer production studies" has emerged, as well as how such a field is structured. The questions that animate this chapter are: How should we think about peer production? How are other people thinking about it? What is it good for, in analytical terms, exactly? Should peer production become a field of study?

30 Be Your Own Peer! Principles and Proposals for the Commons

Mathieu O'Neil, Sophie Toupin, & Christian Pentzold

At the heart of peer production lies the impulse to take control of one's life and produce something independently of the authority of social and professional hierarchies. This is also a deeply moral impulse, as the fruits of one's labor are meant to be shared with others: the commons are both the resource being produced and the means to produce it. In this way peer production challenges the dominant societal model based on the solitary consumption of perishable items which are always produced *somewhere else*. Peer production is therefore organically connected to issues such as sustainable development, the re-localization of the economy, and "degrowth." This final chapter builds on the conceptual breakthroughs discussed in the *Handbook of Peer Production* to reflect on the existential import of peer production as a set of alternative ethical life-choices. It sets out to define legal, economic, and policy initiatives required to grow the commons. Readers are presented with practical suggestions to shape the future by collaborating with others to create common goods.

Part I

Introduction

1

The Duality of Peer Production

Infrastructure for the Digital Commons, Free Labor for Free-Riding Firms

Mathieu O'Neil, Sophie Toupin, and Christian Pentzold

1 Introduction

There never was a "tragedy of the commons": Garrett Hardin's overgrazing farmers were victims of a tragedy of *self-management*, as they failed to collectively regulate, as equals, their common pasture. When Elinor Ostrom was awarded the Nobel Prize in Economics in 2009, the immemorial notion that there are only two types of goods in the world – private and public, coordinated by markets or the state – was finally put to rest. In the most general terms, peer producers are people who create and manage common-pool resources together. It sometimes seems as if "peer production" and "digital commons" can be used interchangeably. Digital commons are non-rivalrous (they can be reproduced at little or no cost) and non-excludable (no-one can prevent others from using them, through property rights for example). So, practically speaking, proprietary objects could be produced by equal "peers," however we argue that peer production has a normative dimension, so that what chiefly characterizes this mode of production is that "the output is orientated towards the further expansion of the commons; while the commons, recursively, is the chief resource in this mode of production" (Söderberg & O'Neil, 2014, p. 2). Though there are many historical antecedents, the term "peer production," as an object of public and scientific interest, is historically situated in the early 2000s.[1] The meanings associated with a term that is deeply connected to the Internet as it was 20 years ago are bound to change. Today, "peer production" describes a vast array of self-organized collaborative ventures and distributed work arrangements, from the collective practice of peers who advocate for an issue through a hashtag on social media or evaluate restaurants and holiday accommodation on dedicated websites, to participation in hacklabs and makerspaces. This introductory chapter to the *Handbook of Peer Production* focuses on peer production's original incarnations, such as free and open source software and Wikipedia, which depended on the open Internet's affordances for distributed communication, production, and organization. Non-Internet mediated forms such as work in shared machine shops or the development of mesh networks are covered extensively in other chapters in this *Handbook*. We will refer to them if necessary, but they are not our prime concern here. In part, this is because the original forms and understandings of peer production are most relevant to a media and communication audience. But we also choose to focus on Internet-based peer production in order to explore the term's genesis: what kind of "production"? And why is it called "peer"?

To answer these questions, this chapter examines a series of productive tensions located in and around peer production. We begin in our second section by interrogating the meaning of infrastructure for peer-to-peer models, and find that some forms of peer infrastructure have thrived, whilst others were effectively banned. We next consider Yochai Benkler's influential theorization of "commons-based peer production," and ask to what extent it embodies Western, first world assumptions. Our fourth section explores at length the relationship of peer production to the dominant economy. It begins by outlining claims about peer production's transformational potential, which are inspired by

The Handbook of Peer Production, First Edition. Edited by Mathieu O'Neil, Christian Pentzold, and Sophie Toupin.

Benkler's model and often imbued with techno-utopian overtones. It then focuses on the organizational and cultural characteristics which enabled co-optation by, and hybridization with firms. We conclude this central section on peer production's political economy by reviewing the literature which suggests that peer production, despite its alleged utopian potential, has been recuperated by capitalism and enabled new forms of labor-exploitation. Our fifth and final section explains the aims of this *Handbook* and summarizes its structure and content.

2 Peer-to-Peer Infrastructure

In the early years of the second millennium, the word "peer" became widely known because of the conjunction of two distinct understandings, one scientific, the other popular. On the scientific side, legal scholar Yochai Benkler (2002) proposed in his journal article "Coase's Penguin, or Linux and the Nature of the Firm" a seminal understanding of free and open source software (FOSS) as a form of "commons-based peer production" whose productive efficiency, based on the ease and speed of incorporating multiple contributions to an object, surpassed that of firms and markets. Meanwhile in the Global North more generally, the notion of "peer-to-peer" generated wide public interest. This derived from the popularity of practices enabled by the non-centrally controlled, or distributed, structure of the early mass Internet, prior to its subsequent enclosure by proprietary social media platforms (see Birkinbine, this volume; Kostakis & Bauwens, this volume). Such practices, whose archetype was the Napster file-sharing service, included torrenting or exchanging files online for free. What was truly original about Napster is that files available for download were not located in a central computer: these files were stored on the user's machines, who made them available to others through Napster's (centrally hosted) software. Each node, wherever it was located in the world, was accessible and contributed to the peer-to-peer system.

This collaborative production and exchange of content, knowledge, and systems involved participants with varying degrees of ownership and control of the (software/hardware) means of production. A system like Napster relied on participants to function, and they in turn could use the service for free, but Napster soon became a for-profit company (Alderman, 2001). Now, in the second millennium's third decade, we face a somewhat different situation. Peer-to-peer practices such as torrenting have been almost completely criminalized out of existence, but the Napster model of using and contributing to an online service for free became widespread in the mid-2000s, with Facebook an emblematic example. In terms of architecture, for many people the Internet is now a content delivery model on closed platforms such as social media or entertainment streaming networks, not a system allowing users to perform effective peer-to-peer networking. To be sure, peer production emerged in the 1990s and 2000s despite the network's physical infrastructure – the fiber-optic submarine links, terrestrial cables, data centers, cloud storage and Internet of things – being privately owned. A similar paradox concerns the principle of net neutrality, the idea that Internet Service Providers (ISPs) and governments should treat all data equally – instead of charging users differentially or limiting access to certain platforms or applications. It is less of a surprise that net neutrality has lasted this long if we understand it as an example of the neoliberal principle of free and undistorted competition (Cohen, 2019).

It now becomes necessary to distinguish an expansive definition of infrastructure as pervasive digital arrangements, from a narrow one that focuses on physical and material settings only. In restrictive terms, when it comes to peer-to-peer physical infrastructure, or "built networks that facilitate the flow of goods, people, or ideas and allow for their exchange over space" (Larkin, 2013), the potential for non-corporate users to autonomously own and control a global network has been neutralized. In contrast, when it comes to a more expansive definition, the situation is reversed. It should be noted that "infrastructure" is not solely limited to material components: "beyond bricks, mortar, pipes or wires, infrastructure also encompasses more abstract entities, such as protocols (human and computer), standards, and memory" (Bowker et al., 2010, p. 97).

Peer produced digital infrastructure, that is to say free and open source software, is ubiquitous online. Let us consider the foundational LAMP open source web application acronym (Linux, Apache, MySQL, Perl/PHP/Python): Google owes its dominance to Linux (used in Android and Chrome

OS); Apache powers 40% of the Internet's web servers; without the MySQL database, there would be no online commerce (Paypal, Amazon), social media (Facebook, Twitter, LinkedIn), or "sharing economy" (Uber, Yelp); Perl/PHP/Python are also highly popular programming languages. As for Wikipedia, it is no longer an unreliable joke, but a legitimate source of correctness in the age of networked disinformation – in fact, it is among the most popular websites in the world (van Dijck, 2013). The narrow definition of peer produced infrastructure – such as torrenting – was effectively banned; the expansive one – such as updating the Linux kernel – permitted, and put to work for the global communication network.

What might still be possible in the future, despite the platformization of the Internet and arbitrary regulatory mechanisms? Furthermore, what kind of resilient infrastructures will foster people's ability to participate in peer production without over-consuming natural resources and contributing to the destruction of the biosphere?

3 The Exclusive Attraction of Commons-Based Peer Production

We consider first what peer producers are building right now. Many continue to focus on autonomous infrastructure in order to oppose the technological giants and offer an alternative to civil society. Examples include distributed physical infrastructure at the local level, in the form of mesh or wireless community networks (such as Guifi[2] in Catalonia, Freifunk[3] in Berlin, and many others; see Shaffer, this volume); peer-to-peer encrypted messaging and forums (such as Briar;[4] see also Velasco Gonzáles & Tkacz, this volume); hackerspaces, hacklabs, and biohacklabs (see Boeva & Troxler, this volume; Meyer, this volume) community telecommunications infrastructure such as Rhizomatica in Mexico, Columbia and Brazil (see interview with Bloom, this volume; Shaffer, this volume); free digital libraries where copyrighted material can be found and uploaded, such as Memory of the World,[5] Library Genesis,[6] and Monoskop;[7] tech collectives that peer produce services geared towards activists such as VPNs, file sharing, server space, and many others. At the hardware level, peer production projects have developed open source machines, tools, and infrastructure which fight against pollution, such as precious plastic,[8] who make available blueprints showing how to build plastic-recycling machines (see also Braybrooke & Smith, this volume).

Many others have embraced the expansive definition of infrastructure by contributing to Wikipedia (see Haider & Sundin, this volume), project Gutenberg (a volunteer-run text digitization project), or by uploading code commits to GitHub (a code sharing, publishing service and social networking site for programmers whose "social coding" has proved wildly popular with the FOSS community). Why has distributed collaboration between volunteers proved so successful? The promise that peer production is always radically decentralized, collaborative, and nonproprietary (Benkler, 2006, p. 60) has not eventuated: the most technologically advanced forms of peer production have hybridized with the market, as detailed in our next section. But Benkler did not just define peer production's infrastructural characteristics: he also addressed the moral benefits of sharing resources and of self-determination. Peer production, as a way of working collaboratively with peers, can only thrive if people treat each other with respect and dignity. The cumulative impact of non-exploitative micro-actions is profound; a normative model, based on autonomy and the sharing of common-pool resources for the good of humanity, emerges (see Borschke, this volume; Nissenbaum & Benkler, this volume). Further, deliberations are meant to be based on "the authority of the better argument." Stephen Levy ([1984] 2010) showed how this was translated into hacker language in his book *Hackers: Heroes of the Computer Revolution* when he defined hacker ethical principles, such as the commitment to the free access of computers and information, the mistrust of centralized authority, and the insistence that hackers be evaluated meritocratically, and not by "bogus" criteria (age, degrees, etc.) but rather by how well they can hack (Levy, [1984] 2010). In a world dominated by dispossession and exploitation, these "do-ocratic" attributes proved attractive. This explains why the expansive definition of peer produced infrastructure has blossomed into 1,000 flowers, including peer learning (see Antoniadis & Pantazis, this volume), cartography (Fish, this volume), and collective action (Milan, this volume). The *Handbook of Peer Production* showcases this great diversity of peer projects. What unites them is a common practice: collective control over production and creation processes, recursively intertwined with the means and ends of this self-governed practice, the commons.

That being said, we do not intend to suggest that peer production is truly inclusive. Despite Levy's influential principle, "bogus" criteria have historically shut the door to women and people of color, who have in turn advocated for the importance of recognizing that barriers to entry do exist when it comes to learning how to code and to being accepted in white and male-dominated techno-cultures. Further, racist assumptions of deviant behaviors (such as scamming or spamming) have led various institutions to block access to both corporate and non-corporate platforms, including Wikipedia, in several African countries (Burrell, 2012), reducing the possibility for locals to take part in such projects (Burrell argues in her book *Invisible Users* that the racist interaction which Africans experienced online in the mid-2000s led, in part, to practices such as scamming). Being aware of this history helps to understand how peer production has developed, and who has the opportunity to take part without significant barriers. Verrips and Meyer (2001) describe the collective maintenance of technologies such as automobiles by all available means in a country like Ghana: while peer production, as a desire for autonomy, may occur in contexts of commodity affluence and disposable income, the reappropriation of work and technology can also stem from a need for survival, in the South or in disadvantaged sectors of the North. As influential as Benkler's definition of commons-based peer production was, it seems to reflect the assumptions of settler colonial worlds (see Deka, this volume; Toupin, this volume).

When dealing with social participation, it is always sound practice to ask: who can take part? In this case we should reflect on who the peer producers are, or to put it differently: where can they thrive; what are the material requisites? Specifically, we must consider the context and conditions in which peer production occurs in the North and how they might differ from the Global South, particularly outside of elite circles. Examples of these assumptions include a constant flow of electricity, minimal infrastructural breakdown, and easy access to computers (rather than only cell or mobile phones through which the majority of people access the Internet in the South). In the North, obstacles preventing entry into peer production projects do exist – for women and people of color in particular – but in terms of digital infrastructure, Internet access, and access to computers, barriers are usually low, except for Indigenous people on reserves and to some extent in remote rural areas. In the Global South access may be restricted by class to the elite and middle class; in India, by caste. Conflict might also come into play, when a government decides to cut off access to the Internet as occurred in Kashmir, Baluchistan, Ambazonia (the English-speaking part of Cameroon) or in many other regions experiencing contestation around elections. The point here is not to delight in critical self-flagellation (the *Handbook*'s editors and many of its contributors are located in the Global North), but to be aware of the situated quality of the produced knowledge. Technological development is one of the Global North's enduring ideologies, serving to naturalize domination. The industrial revolution occurred not simply after the slave trade, but thanks to it (James, [1938] 1989; Robinson, [1980] 2000).

Technological power is therefore historically intermingled with processes of dispossession. In more recent cases – such as the subject of this *Handbook* – technologically advanced projects could be framed as constituting, in the Global North, micro-enclaves of privilege. Beyond the reproduction of social domination through restricted access to the free time, cultural capital, and social networks necessary to take part in peer production, what role do the digital commons play in the capitalist development process? The relationship of peer production to market forces forms the subject of the next section.

4 The Digital Commons and Capitalist Production

4.1 Post-Capitalist Imaginaries

Peer production's position in the wider political economy is contradictory: De Angelis and Harvie (2014) note the "ambiguity" between commons-within-and-for-capital and commoning-beyond-capital. This is particularly the case when it comes to peer production's historically preeminent exemplar, the production of free and open source software. FOSS now plays a central role in the digital economy, because innovation through open collaboration is the new standard, and because enrolling the free labor of scores of volunteers reduces production costs. Yet since the slyly subversive General Public License (GPL) or "copyleft" was introduced in 1989, FOSS has been described by analysts and advocates as portending or prefiguring a postcapitalist future (see Birkinbine, this volume; Couture, this volume; Dafermos, this volume; Dulong de Rosnay, this volume; O'Neil & Broca, this volume).

In the 1990s peer production politics extended beyond communication and deliberation in that they were portrayed as distinct from a capitalist mode of production based on exclusive private property rights. FOSS licenses set up a legal environment in which contributors could entrust their intellectual properties to individuals with whom they had no prior personal contact (Lee & Cole, 2003; see also Spaeth & Niederhöfer, this volume). In other words, workers in peer production abrogate their exclusive property rights over the product of their labor. Many authors have connected this relinquishment of control by peer producers to a future-facing socio-technical imaginary, as well as to earlier models of human cooperation which were historically just as prevalent as competitive market models. Notable examples are activists from the Oekonux network and the Foundation for Peer to Peer Alternatives. In the case of Oekonux, peer production was envisaged as the "germ form of a new mode of production beyond capitalism" (Meretz, 2012), signifying that there is a fluid interplay between the emerging new mode of production and organization and the old model of capitalism and hierarchy: since peer production is understood as dialectically co-constituted by its "other" (e.g., developed as a reaction to and as part of capitalism), in many of its iterations, such as free and open source software, it also advances capitalist interests (for an extended discussion see Euler, 2016; see also O'Neil & Broca, this volume). In recent years a wealth of books (Bauwens et al., 2019; Bollier & Helfrich, 2019; Mason, 2015; Srnicek & Williams, 2016) have also argued that peer production and peer-to-peer infrastructure create digital and non-digital commons and thereby the foundation of a postcapitalist economy, beyond the current economic system. For Bollier and Helfrich (2019), cooperation is a natural human impulse which is stymied by society. They present a range of principles meant to help develop sustainable ventures. In their book *Peer-to-Peer: The Commons Manifesto*, Bauwens, Kostakis, and Pazaitis (2019) show how peer-to-peer is essential for building a commons-centric future. Whilst they do not refer to a postcapitalist imaginary, their commons-based future centered around people and nature gestures toward it. What they set to demonstrate is that peer production all at once encompasses social relations, infrastructure, and new modes of production and property ownership, and that these elements create the conditions for a transition to an economy geared towards people and nature (see also Kostakis & Bauwens, this volume; Pazaitis & Drechsler, this volume).

Journalist Paul Mason (2015) is another author who detects (re)generative power in peer production. In *PostCapitalism: A Guide to Our Future*, he defines peer production as the production of "free stuff that drives out commercially produced commodities" (2015, p. 138). Mason cites Wikipedia as an example of a peer produced space where commercial interests cannot operate. In his view, society has to *design* the transition to postcapitalism (2015, p. 140). To engender this transition, he suggests paying everyone a basic income while automating as many tasks as possible and freeing people to contribute to a peer production economy. Along the same lines, we can cite Nick Srnicek and Alex Williams' (2016) book *Imagining Life After Capitalism*. For them, a postcapitalist economy will liberate us from work; it is through the development of technologies that our freedoms are expanded. Another recent book mobilizing similar tropes and arguments is Aaron Bastani's (2018) *Fully Automated Luxury Communism*. Bastani anticipates the end of societies based on waged work: thanks to activist leftist government's use of technology, society will succeed in mastering our planetary crises.

These authors all herald the advent of a postcapitalist society through changes in wage labor. Most also espouse a deep belief in the transformational capacity of peer production as a practice and in the notion that technology, particularly automation, will save humanity. In short, they all adopt utopian socio-technical imaginaries. According to Sheila Jasanoff and Sang-Hyun Kim (2016), technological development, like science fiction, operates in constant interaction with the social context that inspires and supports its production. And indeed, although the abovementioned depictions of peer production are postcapitalist, it could be queried whether economic growth and constant technological innovation are truly the best way to tackle the environmental and social crises, and whether peer production should not be put to work in a more localized and simple manner, oriented towards "degrowth," for example (for a discussion of related ideas readers are invited to consult the final chapter of this *Handbook*, "Be Your Own Peer! Principles and Policies for the Commons").

Some Black scholars, who do not refer to a peer production framework, also question the techno-utopian assumptions that many of these accounts of postcapitalist futures espouse (Benjamin, 2019; Noble, 2018). For example, in *Race after Technology*, Ruha Benjamin (2019) criticizes naïve assumption of access to computers and the Internet as a solution to inequality. Further, it is doubtful whether elaborations of postcapitalist futures sufficiently take into account commons-based peer production's role in the present-day capitalist economy. This role is itself a subject of debate: does the fact that firms are benefiting from the free labor of volunteer code developers situate FOSS and peer production more generally within the same exploitative historical trend exemplified by the rise of the so-called "sharing economy" (where, under the guise of increased freedom and flexibility, the social rights of individuals are in effect stripped away, since Uber drivers and others are contractors whose working conditions are precarious, rather than employees benefiting from social protections and rights)? This at any rate was the crux of Kreiss et al.'s (2011) virulent critique of peer production, which in their view represents a step backward for workers' rights. We explain why this critique is only partly justified in the remainder of this section, starting with the organizational structure of peer projects. We then examine the process whereby a labor relation predicated on voluntary participation and the self-selection of tasks has become subsumed into the capitalist economy.

4.2 The Organizational Structure of Peer Projects

The intricacies of self-governance have been a prime focus of research into peer production (Arazy et al., 2019; Auray, 2005; Dafermos, 2012; O'Mahony & Ferraro, 2007; O'Neil, 2009, 2014; Pentzold, 2018). For example, Wikipedia has variously been described as "anarchic" (Reagle, 2005), "democratic" (Descy, 2006), "polycentric" (Mindel et al., 2018), and "meritocratic" (Bruns, 2008). It has also been called a "hybrid of different governance systems" (Holloway et al., 2007), a "self-governing institution" (Spek et al., 2006), a form of "collective governance" (Aaltonen & Lanzara, 2015), an "adhocracy" (Konieczny, 2010), and an "ethical-modular organization" (O'Neil, 2015). From a historical perspective, peer production projects can be likened to self-run organizations such as cooperatives and kibbutzim. However, their most clear antecedents are what were previously known as "voluntary associations" and "collectivist organizations." During the 1960s, the rejection of traditional societal institutions as well as of colonialism and imperialism led to the rise of countercultural groups in the Global North, which explicitly rejected what Max Weber called "traditional" and "legal-rational" forms of authority (Weber, 1947). Social-scientific interest in communes and cooperatives accordingly increased. The workings of collectivist organizations were analyzed by Rothschild-Whitt (1979), who defined them as alternative institutions which "self-consciously reject the norms of rational-bureaucracy" (p. 509). Collectivist organizations are groups in which decisions become authoritative to the extent that all members have the right to full and equal participation. There are no established rules of order, formal motions and amendments, or votes, but instead a "consensus process, in which all members participate in the collective formulation of problems and negotiation of decisions" (pp. 511–512).

Following on from the 1960s concern for more inclusive and participatory forms of activism and politics (Kaufman, 1969), the necessity of rules was recognized in collectivist groups in order to avoid what feminist activist Jo Freeman (1972) called the "tyranny of structurelessness": the absence of explicit rules facilitates power being monopolized by informal cliques who manipulate communications (by approving the declarations of their fellows and ignoring or disparaging those of others) and decisions (by deliberating secretly). In line with this concern – how to democratically organize cooperation amongst volunteers – researchers have examined rule-making in self-organized peer projects (Auray, 2005; O'Mahony & Ferraro, 2007; see also Karp et al., this volume; Pentzold, this volume). Authors who examined the emergence and evolution of FOSS communities such as Debian (Coleman, 2012; O'Mahony & Ferraro, 2007; O'Neil, 2014) and FreeBSD (Dafermos, 2012; Jørgensen, 2007) focused on the "succession problem," that is, on the evolution from an informal mode of legitimacy, organized around the figure of the charismatic founder, to a more formalized and democratic mode.

Online collectivist organizations comprise typically bureaucratic mechanisms such as the maintenance of archives of all decisions, and precise rules (O'Neil, 2009), which are nowadays increasingly formalized in Codes of Conduct. In order to introduce a bureaucratic basis of authority into a community form, members must design democratic mechanisms to limit that basis of authority (O'Mahony & Ferraro, 2007). In Debian the power of the project leader is limited in four ways: leaders must defer to the project; they have limited authority over technical matters; members can recall leaders; and the authority of leaders is counterbalanced by that of a Technical Committee (see also interview with Zacchiroli, this volume). Wikipedia editors who demonstrate their willingness to work for the common good can become "administrators," "stewards," or "bureaucrats" and hence exercise varying degrees of control over other participants. They can be replaced by other Wikipedians: meritocracy and the separation of roles and persons are other key characteristics of bureaucratic systems, in contrast to charismatic and traditional leadership (Weber, 1947). The difference with traditional bureaus is that rules are supposed to be generated transparently and democratically (O'Neil, 2009; Haider & Sundin, this volume). Wikipedia is thus both a formal bureaucracy with a structured system of rules and roles and a "deliberative bureaucracy" as decisions are made through "consensus with universal and equal participation, and care is taken over time to create mutual understanding and agreement" (Joyce, Pike, & Butler, 2013).

In sum, protection from abuse in peer production is provided by the community, rather than by state-backed contracts, so recourse procedures may be unstable; nonetheless, Kreiss et al.'s (2011) assessment that peer organizations offer no protection from unjust domination is incorrect. It remains the case that the legal protection of (for example) the GPL applies to the result of labor, not to workers themselves. Peer organizations are not legally responsible for the welfare of participants, so do not offer the same level of support as formal bureaucracies do (O'Neil, 2015).

How do contemporary collectivist organizations fit into the wider political economy? Table 1.1 maps contemporary configurations of work according to their extrinsic or intrinsic logic (vertical axis) and to the degree of control workers are afforded over their labor (horizontal axis).

Organizations where workers sell their labor for wages and where decision-making rests with a restricted leadership are situated in quadrant A. In contrast to employees of capitalist-centralized firms, independent workers such as informal networks of tradespeople in quadrant C have much greater latitude as to whether to work on a particular job. Quadrants B and D comprise communal forms of labor. The community validation and self-fulfillment of domestic labor are lesser than those accrued by participants in voluntary/collectivist associations, but they exist nonetheless. Indeed, Marxist feminist scholar Kylie Jarrett (2016) draws a comparison between the affective or reproductive work of traditional housewives and the unpaid labor performed by contributors to digital platforms. But where workers in quadrant D have control over their production, workers in quadrant B who freely engage in consumption work (or its more interactive variants "prosumption" and "co-creation") must accept that their contributions to product development or networked communication are subject to an external authority's approval or disapproval, with no possibility of redress: this distinction is perhaps not sufficiently drawn out in Jarrett's (2016) otherwise excellent book, which conflates FOSS labor with that of Facebook users for example.

Table 1.1 Organizational governance and logic: a typology.

	Capitalist logic ("alienated" labor)	*Ethical logic* ("communal" labor)
Centralized governance	A private firms, public administration, non-governmental organizations	B consumption work, co-creation, prosumption
Modular governance	C independent workers, freelancers, contractors	D domestic labor, voluntary/collectivist organizations

Source: O'Neil, M. (2015). Labour out of control. The political economy of capitalist and ethical organizations. Organization Studies, 36(12): 1627–1647. © 2015 SAGE Publications.

In contrast, maintainers in ethical free and open source software projects must take pains not to antagonize contributors, who may not only vote with their feet and exit the project, but have the capacity to use all produced resources and therefore reproduce and "fork" the project into a different direction. Actors may move from one quadrant to another: hobbyists can strike it big and grow into large centralized firms (D=>C=>A), as in the case of computer operating systems or search engines developers. Domestic labor can be contracted out (D=>A/C); firms may attempt to glom onto the energy and creativity of consumers (B=>A) or of collectivist organizations (D=>A) to foster innovation and reduce costs.

4.3 How Capitalism Co-opts Peer Production: The Case of Free and Open Source Software

Early critical understandings of Californian Internet culture (Barbrook & Cameron, 1996), of online communities (Terranova, 2000), and of computer hacking (Wark, 2004) took it for granted that capitalist interests would seek to capture autonomous online labor, though these accounts were in the main written when this co-optation was an interesting novelty, rather than a central component of IT firms' business model, as it is today. But how did the integration of communities of peer producers into the "ecosystems" of firms come about?

Boltanski and Chiapello (2005) had presciently argued in their book on the "new spirit of capitalism," originally published in 1999, that capitalism uses critique to rejuvenate itself, by integrating the 1960s countercultural critique of tradition, boredom, and hierarchy. This helped to justify the freeing of capital, the deployment of anti-welfarist ideology, the weakening of the state and the erosion of organized labor by emphasizing personal liberation rather than social emancipation, which alleviates exploitation. To these insights Fisher (2010) added the legitimizing function of a technological discourse in which hackers are a central productive force. Similarly for Barron (2013), FOSS exemplifies a particularly pure form of the "new spirit of capitalism": a post-Fordist regime of accumulation, arranged around lean firms working as networks with a multitude of participants, organizing work in the form of teams or projects, intent on customer satisfaction, and a general mobilization of workers thanks to their leaders' vision.

In an article detailing how hacker practices and innovations are adopted, adapted, and repurposed by corporate and political actors, Delfanti and Söderberg (2018) reprise the notion that assimilated critiques serve to legitimize capitalism and suggest that hacking itself is being hacked, as "the very idea that tinkering offers a way to subvert the agendas of the powers-that-be has become a foundational myth of contemporary capitalism" (p. 461). Beyond repurposing, technical innovations such as modularized production and distributed mesh networks and retrieval systems are now "integrated in the material infrastructure of capitalism" (p. 476), and coupled with distrust for incumbent actors, aka "disruption."

Christopher Kelty's (2008) influential definition of FOSS projects as "recursive" is key to understanding how what was once perceived as a force resisting privatization has been integrated into dominant circuits of capital. Hackers have extremely divergent politics, but they all agree that proprietary software and intellectual property rights, as well as surveillance and censorship, should be rejected. This stems from the fact that such an opposition constitutes the techno-legal preconditions for the hacker public to exist as such: "recursive politics" aim to consolidate and grow the material conditions for the survival of this public. In contrast issues such as feminism and workers' rights are not "recursive" in the sense that hackers "perceive them to be unrelated to what really matters to them the most, computers and Internet freedom" (Delfanti & Söderberg, 2018, p. 463). This was the key for the disruptive potential of FOSS to be tamed: all firms needed to do was to adopt hackers' core demand (providing access to code through "open" licenses) to ensure that participants could continue to help their environments thrive. The ethical logic of self-fulfillment and the focus on technical excellence did the rest, imbuing projects with a propensity to accept any valid contribution (irrespective of whether it originates from a commercial or communal setting) and an aversion to discussing questions of subsistence ("who can afford to take part?"), as such discussions complicate the notion that contributions are solely evaluated on merit.

The integrated firm-project economic model raises the issue of the sustainability of projects. As Nadia Eghbal observed in her *Roads and Bridges* report: "fundamentally, digital infrastructure has a free rider problem" (2016, p. 106). This tension had been noted since the beginning of FOSS development, when firms were described as harvesting the altruism of volunteer developers (Haruvy et al., 2003), resulting in a relationship between altruistic individuals and selfish firms (Bonaccorsi & Rossi, 2004). That firms are benefiting from the free labor of volunteers situate FOSS and peer production more generally within the same exploitative historical trend exemplified by domestic work. This is a move made by Kylie Jarrett (2016, 2019) who argues, in line with past feminist critiques (Dalla Costa & James, 1972), that unpaid domestic labor is crucial for capitalist production and reproduction. The binary between production and reproduction work renders essential forms of labor invisible.

Lund and Zukerfeld (2020) suggest that profit deriving from what they call the "enclosures model" seeks to increase the price of outputs, whilst profit deriving from the "openness model" seeks to decrease as much as possible the price of inputs. Copyright-based production processes exploit productive activities during labor time, whilst "profit from openness is to a greater extent based on the exploitation of productive activities during leisure time" (Lund & Zukerfeld, 2020, p. 23). For these authors, firm adoption of FOSS is part of an emerging "Openness Ideology" representing a shift from a "profit from enclosures" model (based on the rhetoric of individuals, property, and exclusion) to a "profit from openness" model which extols the virtues of communities, inclusion, and freedom: the peer production of software opened the way for the wholesale capture by commercial interests of free labor. This was facilitated by the introduction of algorithms, which enable social media platforms to extract valuable behavioral data from participants' "digital labor," that is to say labor which does not think of itself as labor (Frayssé & O'Neil, 2015). A concrete example of the process whereby a commons is turned into a commodity can be seen through what Schöpf (2015) called the "commodification of the couch": the commons-oriented Couchsurfing hospitality exchange platform, which enabled the conversion of private households into shareable commons, attracted funds from venture capitalists in 2011, and subsequently started gathering and selling data on the activities of its users.

This brings us back to the critique articulated by Kreiss et al. (2011): in the end peer production amounts to little more than ultra-exploitation. Yet the notion that unpaid participants are necessarily always being exploited should be reassessed. Red Hat, whose Fedora project combines waged and volunteer labor, has a business model which is only possible because products are created at a much lower cost than a fully-waged workforce would entail. Does this constitute exploitation? Firms such as Red Hat are not appropriating FOSS code, which is accessible to all. Benjamin J. Birkinbine (2020) advances the notion of "incorporation" over that of "enclosure" which typically refers to the imposition of higher excludability on the common resource. Instead, "corporations have developed unique ways of transforming the products and processes of commons-based peer production into commercial offerings without placing restrictions" (Birkinbine, 2020, p. 24) on the community's access to their collective resources. As pointed out by Sébastien Broca (2013), this is quite different from the situation of proletarians who are dispossessed from the fruits of their labor. Here the exchange seems to be mutually profitable, even if the goods being swapped – economic profits for the firms, self-realization for the developers – are different. It could even be argued that Red Hat creates an environment where developers can play with passion (Lessig, 2008).

Peer production projects originate from within what could be called a hobbyist sphere, characterized by two inter-related traits: in contrast to salaried work, active users – not just CEOs and managers – can assume concrete strategic and operational control over projects; but these projects do not enable participants to make a living from their volunteer work. The exception to this rule is of course FOSS, where developers were either held to be able to convert the cultural capital acquired in communities into economic capital in the form of employment (Lerner & Tirole, 2002), or – with the adoption of FOSS by firms – simply be recruited to produce firm-oriented free and open source software alongside volunteers. The fact that firms are paying the salaries of developers in FOSS projects was originally found to be notable (González-Barahona & Robles, 2013; Mansell & Berdou, 2010; Riehle et al., 2014) whereas the wholesale integration of FOSS into dominant industrial circuits, partly through waged work, is now deemed complete (Eghbal, 2016; O'Neil et al., 2020a, 2020b).

The hybridity of this collective development (some people are paid, some are volunteers) has generated new institutional formations. Studies of organizational dynamics suggest that firms and FOSS projects are organized around different institutional principles (or "logics"): commercial logics for firms, communal logics for projects (O'Neil et al., 2020a). In order for these organizations to cooperate, discursive legitimation and concrete practical arrangements are required. An online survey of Debian project participants and interviews with Debian Developers found that a first phase of legitimation centered around licenses, in effect since the early 2000s, aimed to erase the distinction between work performed in firms and projects. In contrast a second phase of legitimation, centered around waged labor, put forward the notion that developers should be paid for work performed in Debian. Another finding was that firms were increasingly seeking to configure software to collect analytics in the form of statistics about which buttons users are pressing and which product features are being used, thereby potentially violating FOSS privacy principles, and laying the foundation for future conflicts and forks (O'Neil et al., 2020a).

The rise of the integrated firm-project code development ecosystem can be partly attributed to the rise of the GitHub repository, though some projects have always been reluctant to use GitHub because of its proprietary status. This integration is evident in the business practices of IT giants and in the increasing propensity of end-user firms – not just IT firms – to create Open Source Offices (OSPOs) which will act as open source advocates within the firm, and as firm liaisons with the code-producing open source "communities" (O'Neil et al., 2020c).

5 The *Handbook of Peer Production* Aims to be Inclusive and Political

Like any academic *Handbook* worth its salt, this volume seeks to establish the state of the art of research in a given field of activity; to map origins, manifestations, achievements, and contradictions; to gather a group of contributors who are both knowledgeable and passionate. In addition, our approach is purposely inclusive and political.

When we say we wish this *Handbook* to be "inclusive" we mean that the commons-based and oriented peer production approach towards generating and circulating all kinds of information goods, which fundamentally differs from individualistic models, from the competition of all against all, is occurring in an era where myriad forms of organizing and exploiting collective digital labor are in operation. It is hence necessary to pin down core elements of this alternative model of cooperation and governance such as cooperation and trust, transparency in production, collective democratic decision-making, and the like. However, we also believe it makes sense to chart areas where some, though perhaps not all, these tenets have been adopted. So making peer production "inclusive" means casting the net wider and including a comprehensive range of cognate endeavors. In consequence, the *Handbook* traces peer production's resonance in a broad number of fields, from a wide variety of perspectives.

When we say the *Handbook* is "political," we mean that the chapters investigate and discover new possibilities for political action by re-thinking concepts such as crowdsourcing, making, urban commons, and the partner state, or by conceptualizing the contradictions of autonomous production. Albeit in different guises, all chapters share a concern with how peer producing is intertwined with political issues such as hierarchical power, capitalism, gender, and race. By framing peer production as "political," this *Handbook* offers the possibility of critically exploring the assumptions underpinning, and the contradictions animating, commons-based and oriented peer production; and it also attempts to move beyond critique, towards praxis.

The *Handbook of Peer Production* is divided in six parts. Part I is the shortest, being made of this chapter, in which we (editors Mathieu O'Neil, Sophie Toupin, and Christian Pentzold) set the scene. Part II (chapters 2 to 6) outlines the key *Concepts* which help to make sense of peer production: Vasilis Kostakis revisits a famous Michel Bauwens article to define the core elements, or *Grammar of Peer Production* projects and ecosystems. Benjamin J. Birkinbine defines the *Political Economy of Peer Production*, Christian Pentzold outlines the *Social Norms and Rules of Peer Production*, and Michael Stevenson discusses the *Cultures of Peer Production*. We conclude this part with this volume's sole reprint, Yochai Benkler and Helen Nissenbaum's 2006 article on *Commons-Based Peer Production and Virtue*.

Part III (Chapters 7 to 11) defines the necessary *Conditions* for peer production to exist: George Dafermos defines the *Prophets and Advocates of Peer Production* and Margie Borschke explores *Virtue, Efficiency, and the Sharing Economy*. Next, Mélanie Dulong de Rosnay analyzes *Openness and Licensing*, and Sebastian Spaeth and Sven Niederhöfer focus on *User Motivations in Peer Production*. We conclude this part with an analysis of the relationship between project governance and scope, *Governing for Growth in Scope: Cultivating a Dynamic Understanding of How Peer Production Collectives Evolve* by Rebecca Karp, Amisha Miller, and Siobhán O'Mahony.

Part IV (Chapters 12 to 20) presents *Cases* of peer production practice, including *Free and Open Source Software* by Stéphane Couture, *Wikipedia and Wikis* by Jutta Haider and Olof Sundin, *Collective Cartography: Drones, Countermapping, and Technological Power* by Adam Fish, and *Peer Learning* by Panayotis Antoniadis and Alekos Pantazis. Morgan Meyer analyses *Biohacking*, Yana Boeva and Peter Troxler present *Makers*, and Pablo Velasco Gonzáles and Nate Tkacz critically assess *Blockchain, or Peer Production Without Guarantees*. Finally, Gwen Shaffer retraces the history of *Community Wireless Networks* and Nicholas Anastasopoulos of *Commoning the Urban*.

Part V (Chapters 21 to 27) maps out areas where peer production practices and projects enter into *Conflict* with internal and external structures of power. Mathieu O'Neil and Sébastien Broca address *Peer Production and Social Change*, and Stefania Milan *Peer Production and Collective Action*. Sophie Toupin discusses *Feminist Peer Production* and Maitrayee Deka *Postcolonial Peer Production*. Francesca Musiani assesses *Gaps in Peer Design* whilst Kat Braybrooke and Adrian Smith interrogate *Makerspaces and Peer Production: Spaces of Possibility, Tension, Post-Automation, or Liberation?* We conclude this section with Alex Pazaitis and Wolfgang Drechsler's *Peer Production and State Theory: Envisioning a Cooperative Partner State*.

Having perused this list of authors and chapters, we trust our readers will agree with us that the *Handbook of Peer Production* has succeeded in accounting for the diversity of peer production histories, endeavors, theories, and contradictions. But when dealing with this topic, which is centered around do-it-yourself values such as individual empowerment, cooperation amongst equals, and engagement and participation, this is not enough. It would be impossible to write a comprehensive volume about peer production, and not contribute in some way. We must go further, by connecting peer production to a progressive social agenda.

Part VI (Chapters 28 to 30) therefore defines *Conversions*, elements that will advance peer production. In *Making a Case for Peer Production* we present interviews with practitioners Peter Bloom (*Rhizomatica*), Mariam Mecky (*HarassMap*), Ory Okolloh (*Ushahidi*), Abraham Taherivand (*Wikimedia*), and Stefano Zacchiroli (*Debian*). In *What's Next? Peer Production Studies?*, we (editors Mathieu O'Neil, Sophie Toupin, and Christian Pentzold) outline promising avenues for new research into the contribution of peer production to productive efficiency and to social change. We also reflect on whether the *Handbook* fits into a field of "peer production studies," which necessarily involves a brief discussion of the political economy of academic publishing. Our final chapter, *Be Your Own Peer! Principles and Policies for the Commons* situates the participatory and democratic qualities of peer production in the wider context of current political, health, and ecological crises, and outlines strategic principles and policy proposals aiming to increase sustainability and fairness.

Creating the first *Handbook of Peer Production* has been a privilege. We hope it is of use.

Notes

1 The advent of commons-based peer production in the late 1990s is the outcome of overlapping historical, social, economic, and technological factors, including, but not limited to: (a) the *traditional self-management of common-pool resources* identified by Elinor Ostrom. The bulk of Ostrom's work was applicable to finite natural resources (such as fish stocks in a river) administered by a local community. Such rival resources raise issues linked to their long-time preservation; in contrast digital commons raise issues concerning their production and enrichment (Coriat, 2011); (b) the notion that *sharing knowledge freely*, alongside universalism, disinterestedness, organized skepticism, and communism (later changed to communalism) are the foundations of science: "The communism of the scientific ethos," wrote Robert Merton, "is incompatible with the

definition of technology as 'private property' in a capitalistic economy" (1942: 275); (c) the principle of *peer review* which holds that correctness is more likely to emerge if a statement, or an elegant solution to a technical problem (a.k.a. "hack") is submitted to the scrutiny of a community of peers of equal competence; (d) the uniquely contradictory *historical moment during which the Internet was created* (the late 1960s): the Cold War threat of nuclear conflict between the two superpowers led to the US military's request for a distributed network where digital packets could autonomously re-route around destroyed nodes, so intelligence and control were located at the edges, where packets are disassembled and reassembled, not in a central hub as in a traditional phone exchange (Baran, 1964); this was also the height of the counterculture, with its distrust of traditional authorities and emphasis on personal liberation, a belief that would evolve into diverse branches, one of which was the notion that personal computers represented the means to establish free communications and new types of virtual communities (Turner, 2006). Internet technical protocols were established by organizations such as the Internet Engineering Task Force; IETF "hackers" were computer engineers and students influenced by the counterculture, and therefore resistant to hierarchy. The IETF adopted a non-authoritarian methodology to propose improvements to Internet protocols: RFCs (request for comments) were released electronically, so that innovation was not based on the imposition of authority, but on appeals to a community of peers for input (O'Neil, 2009). The founding belief of the IETF was that the legitimate basis for authority was autonomous technical excellence: "We reject kings, presidents and voting. We believe in rough consensus and running code," said Internet pioneer David Clark (Hoffman & Harris, 2009); (e) the UNIX computer system, which Minix and subsequently Linux were modeled on, has a *modular structure*, signifying that as long as common protocols were respected, new components could be added independently; more generally the Internet and the Web are classic examples of "combinatorial innovation" (Varian, 2010); (f) the rise of personal computers as consumer items and attendant *increased proprietary enclosures around software* in the 1980s led MIT computer scientist Richard Stallman to declare that he would "put together a sufficient body of free software so that I will be able to get along without any software that is not free" (Stallman, 1985), and to create the GNU operating system, the General Public License or "copyleft," and the Free Software Foundation. Whilst Stallman's continued advocacy for the right to freely access and modify code in order to improve it has proved an inspiration for many, he sadly demonstrated in 2019 a lamentable insensitivity to sexism (Musil, 2019); (g) finally the emergence of the *mass Internet* in the early 1990s facilitated the rapid global dissemination of computer code (instead of exchanging diskettes by post), leading to the advent of Linux, and in the 2000s of the writeable Internet ("Web 2.0") including wikis, weblogs, and social networking sites.

2 https://guifi.net/
3 https://berlin.freifunk.net/
4 https://briarproject.org/
5 www.memoryoftheworld.org/
6 https://libgen.is/
7 https://monoskop.org/Monoskop
8 https://preciousplastic.com/

References

Aaltonen, A., & Lanzara, G. F. (2015). Building governance capability in online social production: Insights from Wikipedia. *Organization Studies, 36*(12), 1649–1673.

Alderman, J. (2001). *Sonic boom: Napster, mp3, and the new pioneers of music.* New York, NY: Basic Books.

Arazy, O., Lifshitz-Assaf, H., & Balila, A. (2019). Neither a bazaar nor a cathedral: The interplay between structure and agency in Wikipedia's role system. *Journal of the Association for Information Science and Technology, 70*(1), 3–15.

Auray, N. (2005). Le sens du juste dans un noyau d'experts: Debian et le puritanisme civique. In B. Conein, F. Massit-Folléa, & S. Proulx (Eds), *Internet, une utopie limitée: Nouvelles régulations, nouvelles solidarités* (pp. 71–94). Laval, QC: Presses de l'Université Laval.

Baran, P. (1964). *On distributed communication.* Santa Monica, CA: Rand.

Barbrook, R., & Cameron, A. (1996). The Californian ideology. *Science as Culture, 6*(1), 44–72.

Barron, A. (2013). Free software production as critical social practice. *Economy and Society, 42*(4), 597–625.

Bastani, A. (2018). *Fully automated luxury communism: A manifesto.* New York, NY: Verso.

Bauwens, M., Kostakis, V., & Pazaitis, A. (2019). *Peer to peer: The commons manifesto.* London: University of Westminster Press.

Benjamin, R. (2019). *Race after technology: Abolitionist tools for the new Jim Code*. Cambridge: Polity.

Benkler, Y. (2002). Coase's penguin, or, Linux and "The nature of the firm." *The Yale Law Journal, 112*(3), 369–446.

Benkler, Y. (2006). *The wealth of networks: How social production transforms markets and freedom*. New Haven, CT: Yale University Press.

Birkinbine, B. (2020). *Incorporating the digital commons. Corporate involvement in free and open source software*. London: University of Westminster Press.

Bollier, D., & Helfrich, S. (2019). *Free, fair, and alive: The insurgent power of the commons*. Gabriola Island, BC: New Society.

Boltanski, L., Chiapello, E., & Elliott, G. (trans.) (2005). *The new spirit of capitalism*. London: Verso.

Bonaccorsi, A., & Rossi, C. (2004). Altruistic individuals, selfish firms? The structure of motivation in open source software. *First Monday 9* (1). https://doi.org/10.5210/fm.v9i1.1113

Bowker, G., Baker, K., Millerand, F., & Ribes, D. (2010). Toward information infrastructure studies: Ways of knowing in a networked environment. In J. Hunsinger et al. (Eds.), *International handbook of internet research* (pp. 97–117). Dordrecht: Springer.

Broca, S. (2013). *Utopie du logiciel libre: du bricolage informatique à la réinvention sociale*. Lyon: Le passager clandestin.

Burrell, J. (2012). *Invisible users: Youth in the Internet cafés of urban Ghana (Acting with technology)*. Cambridge, MA: MIT Press.

Bruns, A. (2008). *Blogs, Wikipedia, Second Life, and beyond: From production to produsage*. New York, NY: Peter Lang.

Cohen, J. (2019). *Between truth and power: The legal constructions of informational capitalism*. New York, NY: Oxford University Press.

Coleman, E. (2012). *Coding freedom: The ethics and aesthetics of hacking*. Princeton, NJ: Princeton University Press.

Coriat, B. (2011). From Natural-Resource Commons to Knowledge Commons: Common Traits and Differences, *LEM Papers Series 2011/16*, Laboratory of Economics and Management (LEM), Sant' Anna School of Advanced Studies, Pisa, Italy.

Dafermos, G. (2012). Authority in peer production: The emergence of governance in the FreeBSD project. *Journal of Peer Production, 1*, 1–22. Retrieved from http://peerproduction.net/issues/issue-1/peer-reviewed-papers/

Dalla Costa, M., & James, S. (1972). *The power of women and the subversion of the community*. Bristol: Falling Wall Press.

De Angelis, M., & Harvie, D. (2014). The commons. In M. Parker, G. Cheney, V. Fournier, & C. Land (Eds.), *The Routledge companion to alternative organizations* (pp. 280–294). Abingdon: Routledge.

Delfanti, A., & Söderberg, J. (2018). Repurposing the hacker: Three cycles of recuperation in the evolution of hacking and capitalism. *Ephemera, 18*(3), 457–476.

Descy, D. (2006). The wiki: True web democracy, *TechTrends, 50*(1), 4–5.

Eghbal, N. (2016). *Roads and bridges: The unseen labor behind our digital infrastructure*. New York, NY: Ford Foundation.

Euler, J. (2016). Commons-creating society: On the radical German commons discourse. *Review of Radical Political Economics, 48*(1), 93–110.

Fisher, E. (2010). Contemporary technology discourse and the legitimation of capitalism. *European Journal of Social Theory, 13*, 229–252.

Frayssé, O., & O'Neil, M. (2015). Hacked in the USA: Prosumption in the digital age. In M. Frayssé & M. O'Neil (Eds.), *Digital labour and prosumer capitalism* (pp. 1–20). Basingstoke: Palgrave Macmillan.

Freeman, J. (1972). The tyranny of structurelessness. *The Second Wave*. Retrieved from http://struggle.ws/hist_texts/structurelessness.html

González-Barahona, J. M., & Robles, G. (2013). Trends in free, libre, open source software communities: From volunteers to companies. *Information Technology, 55*(5), 179–180.

Haruvy, E., Prasad, A., & Sethi, S. P. (2003). Harvesting altruism in open-source software development. *Journal of Optimization Theory and Applications, 118*, 381–416.

Hoffman, P., & S. Harris. (2009). The Tao of the IETF: a novice's guide to the Internet Engineering Task Force, RFC 4677, 30 November 2009. Retrieved from www.ietf.org/tao.html

Holloway, T., Bozicevic, M., & Börner, K. (2007). Analyzing and visualizing the semantic coverage of Wikipedia and its authors. *Complexity, 12*(3), 30–40.

James, C. L. R. ([1938] 1989). *The Black Jacobins: Toussaint l'Ouverture and the San Domingo revolution*. New York: Vintage.

Jarrett, K. (2016). *Feminism, labour and digital media: The digital housewife*. New York: Routledge.

Jarrett, K. (2019). Through the reproductive lens: Labour and struggle at the intersection of culture and economy. In D. Chandler & C. Fuchs (Eds.), *Digital objects digital subjects: Interdisciplinary perspectives on capitalism, labour and politics in the age of big data* (pp. 103–116). London: University of Westminster Press.

Jasanoff, S., & Kim, S. (Eds.). (2016). *Dreamscapes of modernity: Sociotechnical imaginaries and the fabrication of power*. Chicago: University of Chicago Press.

Jørgensen, N. (2007). Developer autonomy in the FreeBSD Open Source project. *Journal of Management & Governance, 11*(2), 119–128.

Joyce, E. J., Pike, C., & Butler, B. S. (2013). Rules and roles vs. consensus: Self-governed deliberative mass collaboration bureaucracies. *American Behavioral Scientist, 57*, 576–594.

Kaufman, A. S. (1969). Human nature and participatory democracy: Ten years later. In W. E. Connolly (Ed.), *The bias of pluralism* (pp. 178–200). New York, NY: Atherton Press.

Kelty, C. M. (2008). *Two bits: The cultural significance of free software*. Durham, NC: Duke University Press.

Konieczny, P. (2010). Adhocratic governance in the Internet age: A case of Wikipedia, *Journal of Information Technology & Politics, 7*(4), 263–283.

Kreiss D., Finn, M., & Turner, F. (2011). The limits of peer production: Some reminders from Max Weber for the network society. *New Media & Society, 13*, 243–259.

Larkin, B. (2013). The politics and poetics of infrastructure. *Annual Review of Anthropology, 42*(1), 327–343.

Lee, G., & Cole, R. (2003). From a firm-based to a community-based model of knowledge creation: The case of the Linux kernel development. *Organization Science, 14*(6), 633–649.

Lerner J., & Tirole, J. (2002). Some simple economics of open source. *The Journal of Industrial Economics, 50*(2), 197–234.

Lessig, L. (2008). *Remix: Making art and commerce thrive in the hybrid economy*, New York, NY: Penguin.

Levy, S. ([1984] 2010). *Hackers: Heroes of the computer revolution*. Newton: O'Reilly Media.

Lund, A., & Zukerfeld, M. (2020). *Corporate capitalism's use of openness*. Basingstoke: Palgrave Macmillan.

Mansell, R., & Berdou, E. (2010). Political economy, the Internet and FL/OSS development. In J. Hunsinger, M. Allen, & L. Klastrup (Eds.), *International handbook of internet research* (pp. 341–362). Dordrecht: Springer.

Mason, P. (2015). *Postcapitalism: A guide to our future*. London: Allen Lane.

Meretz, S. (2012). Ten patterns developed by the Oekonux movement. *Journal of Peer Production, 1*. Retrieved from http://peerproduction.net/issues/issue-1/debate-societal-transformation/ten-patterns-developed-by-the-oekonux-project/

Merton, R. K. (1942). The normative structure of science. In R. K. Merton, *The sociology of science: Theoretical and empirical investigations* (pp. 267–280). Chicago: University of Chicago Press.

Mindel, V., Mathiassen, L., & Rai, A. (2018). The sustainability of polycentric information commons. *MIS Quarterly, 42*(2), 607–632.

Musil, S. (2019, September 16). Computer scientist Richard Stallman resigns from MIT after Jeffrey Epstein comments. *CNET*. Retrieved from https://www.cnet.com/news/computer-scientist-richard-stallman-resigns-from-mit-after-jeffrey-epstein-comments/

Noble, S. (2018). *Algorithms of oppression: How search engines reinforce racism*. New York: New York University Press.

O'Mahony, S., & Ferraro, F. (2007). The emergence of governance in an open source community. *Academy of Management Journal, 50*, 1079–1106.

O'Neil, M. (2009). *Cyberchiefs. Autonomy and authority in online tribes*. London: Pluto Press.

O'Neil, M. (2014). Hacking Weber: Legitimacy, critique and trust in peer production. *Information, Communication & Society, 17*(7), 872–888.

O'Neil, M. (2015). Labour out of control. The political economy of capitalist and ethical organizations. *Organization Studies, 36*(12), 1627–1647.

O'Neil, M., Muselli, L., Raissi, M., & Zacchiroli, S. (2020a). "Open source has won and lost the war": Justifying commercial-communal hybridization in a FOSS project. *New Media & Society*. https://doi.org/10.1177/1461444820907022.

O'Neil, M., Muselli, L., Cai, X., & Zacchiroli, S. (2020b). Mapping open source capitalism: The firm-volunteer project co-production network and its media representation. ASA *Annual Meeting*, San Francisco, CA.

O'Neil, M., Muselli, L., Cai, X., Pailler, F., & Zacchiroli, S. (2020c). Firm discourses and digital infrastructure projects. *AOIR*, Dublin.

Pentzold, C. (2018). Grounding peer production in practice: Editorial routines and everyday engagement in the "free encyclopedia anyone can edit." *Communication, Culture and Critique, 11*(3), 455–474.

Reagle, J. (2005). A case of mutual aid: Wikipedia, politeness, and perspective taking, *Wikimania 2005*, Frankfurt, Germany.

Riehle, D., Riemer, P., Kolassa, C., & Schmidt, M. (2014). Paid vs. volunteer work in open source. *47th Hawaii International Conference on System Sciences (HICSS)*. Waikoloa, HI (pp. 3286–3295). doi:10.1109/HICSS.2014.407.

Robinson, C. J. ([1980] 2000). *Black Marxism: The making of the Black radical tradition*. Chapel Hill, NC: University of North Carolina Press.

Rothschild-Whitt, J. (1979). The collectivist organization: An alternative to rational–bureaucratic models. *American Sociological Review*, *44*, 509–527.

Schöpf, S. (2015). The commodification of the couch: A dialectical analysis of hospitality exchange platforms. *tripleC*, *13*(1), 11–34.

Söderberg, J., & O'Neil, M. (2014). Introduction. In J. Soderbergh & Maxigas (Eds.), *Book of peer production* (pp. 2–3). Göteborg: NSU Press. Retrieved from http://peerproduction.net/projects/books/book-of-peer-production/

Spek, S., Postma, E., & van den Herik, J. (2006). Wikipedia: Organisation from a bottom-up approach, *WikiSym* 2006, Odense, Denmark (August).

Srnicek, N., & Williams, A. (2016). *Inventing the future: Postcapitalism and a world without work*. London: Verso.

Stallman, R. M. (1985). The GNU manifesto. Retrieved from www.gnu.org/gnu/manifesto.en.html

Terranova, T. (2000). Free labor: Producing culture for the digital economy. *Social Text*, *18*(2), 33–58.

Turner, F. (2006). *From counterculture to cyberculture. Stewart Brand, the Whole Earth Network, and the rise of digital utopianism*. Chicago: University of Chicago Press.

van Dijck, J. (2013). *The culture of connectivity: A critical history of social media*. Oxford: Oxford University Press.

Varian, H. (2010). Computer mediated transactions. *American Economic Review: Papers & Proceedings*, *100*, 1–10.

Verrips, J., & Meyer, B. (2001). Kwaku's car: The struggles and stories of a Ghanaian long-distance taxi driver. In D. Miller (Ed.), *Car cultures* (pp. 153–184). Oxford: Berg.

Wark, M. (2004). *A hacker manifesto*. Cambridge, MA: Harvard University Press.

Weber, M. (1947). *The theory of social and economic organization*. New York, NY: Free Press.

Part II

Concepts: Explaining Peer Production

2

Grammar of Peer Production

Vasilis Kostakis and Michel Bauwens

1 Introduction

In 2005, Michel Bauwens published "The Political Economy of Peer Production." In this seminal article he discussed the principles, characteristics, basic structures, and the future of the then nascent ecosystem of peer production. I (Vasilis) stumbled upon this article, and upon Michel, in 2007 as a postgraduate student at the University of Amsterdam. "The Political Economy of Peer Production," along with Yochai Benkler's work, have had a major impact on my understanding of peer production. Benkler had first observed the seeds of an emerging mode of production in his 2002 article "Coase's Penguin, or, Linux and 'The Nature of the Firm'." He provided the vocabulary for a new field of scholarship. Three years later, building on Benkler's work, Michel developed these insights in his article.

In this chapter, I revisit Michel's 2005 article, adopting, expanding, and refining the operational concepts – or "grammar" – he used to define peer production projects and the institutional ecosystems which sustain them. This chapter is therefore situated, like much of the work of the P2P Foundation which Michel founded in 2005, between the worlds of activism and academic research. After almost 15 years, it is challenging to see what has changed and what has remained the same. In 2005, Wikipedia was four years old, Mozilla Firefox was three years old, the Apache HTTP Server Project was ten years old, and most of the currently prominent open hardware projects did not exist. Michel is cited as a co-author of this chapter because a considerable amount of it is based on his work. He also reviewed the final version. Moreover, this chapter builds upon our latest book published in 2019, co-authored with Alex Pazaitis. However, the truth is that every intellectual product is not the work of one human being, but rather the result of social production; a collective artifact. Nevertheless, for the sake of simplicity, we have compromised and cite two names as authors, though we are indebted to the global peer-to-peer community of practitioners, activists, and scholars.

2 Peer-to-Peer

As capitalism faces a series of structural crises, a new socio-technical dynamic is emerging from its ashes: peer-to-peer (P2P). First, P2P is a type of social relation in human networks, where participants have maximum freedom to connect. Second, P2P is a technological infrastructure that makes the generalization and scaling up of such relations possible. Thus, P2P gives rise to a mode of production, i.e., peer production that includes new governance mechanisms and property frameworks.

P2P is not something new. It has existed since the dawn of humanity and initially was the dominant form of relationship in nomadic hunter-gathering societies. In industrial capitalism (and later in the state-socialist systems), the P2P dynamics were driven to the margins. However, with the affordance of P2P-based technologies, P2P dynamics can now scale up to a global level and create complex artifacts that transcend the possibilities of both state- and market-based models alone.

The Handbook of Peer Production, First Edition. Edited by Mathieu O'Neil, Christian Pentzold, and Sophie Toupin.

So, what are the infrastructural requirements that facilitate the re-emergence of peer production? In his original article, Michel Bauwens (2005) posited that the first requirement is access to the technological infrastructure, i.e., to individual computers that are interconnected and thus enable a universal machine capable of executing any logical task. The Internet, as a point-to-point network, was designed for participation "from the edges" without the use of obligatory hubs such as telephone exchanges for example. Although not fully in the hands of its participants, the Internet is controlled through distributed governance, and outside the complete hegemony of particular private or state actors.

On the one hand, the current Internet might be the product of decisions made four to five decades ago. On the other hand, it has been subsequently shaped by commercial interests after the invention of the Web and the browser, and by governments intent on controlling it. But it has also been taken up by the hacker movements and user communities adapting and changing it to their benefit. The community wi-fi movement, the open spectrum advocacy, and alternative meshwork-based telecommunication infrastructures are exemplary of the latter efforts. However, the future of the Internet, and of the Web that runs on the Internet, is a terrain of struggle, in which different interests are striving for supremacy.

The second requirement, according to Bauwens (2005), is the existence of a "software" infrastructure for autonomous global cooperation. A growing number of collaborative tools, such as wikis, facilitate the creation of trust and social capital, making it possible to create global groups that can create use-value without the intermediary of for-profit enterprises.

The third requirement is a legal infrastructure that enables the creation of use-value and protects it from private appropriation. For example, the General Public License (which prohibits the appropriation of software code) and certain versions of the Creative Commons license fulfill this role. They enable the protection of common use-value and use viral characteristics to spread. The General Public License and related material can only be used in projects that in turn put their source code in the public domain.

The fourth requirement is cultural. The diffusion of mass intellectuality, (i.e., the distribution of human intelligence) and associated changes in ways of feeling and being (ontology), ways of knowing (epistemology), and value constellations (axiology) have been instrumental in creating the type of cooperative autonomy needed to sustain an ethos that can enable peer production (Bauwens, 2005).

3 Principles and Characteristics of Peer Production

According to Bauwens (2005) then, the P2P dynamics have allowed for:

- producing use-value through the free cooperation of producers who have access to distributed capital: this is peer production, different from for profit maximization production or production by state-owned institutions. The product of (commons-based) peer production, as defined by Benkler (2002, 2006), is not exchange value for a market, but use-value for a community of users who may also be the producers.
- community-driven governance mechanisms: this is peer governance, different from market allocation or corporate hierarchy.
- making use-value freely accessible on a universal basis, through new commons-oriented property regimes: this is peer property, different from private property or public (state) property.

Peer production is thus fundamentally different from the incumbent models of value creation under industrial capitalism. In the latter, the owners of the means of production hire workers, direct the work process and sell products for profit maximization. Such production is organized by allocating resources through price signals, or through hierarchical command that takes into account these price signals.

In contrast, peer production is in principle open to anyone with skills (and access to the appropriate technological infrastructure) to contribute to a joint project: the knowledge of every participant is pooled.[1] These participants may be paid, but not necessarily, as this system of production is open to self-motivated contributions. Precisely because peer production projects are open systems in which knowledge can be freely shared and distributed, anyone with the right knowledge and skills can contribute, either paid by companies, clients, or not at all. In these open systems, there are many reasons to contribute beyond or besides that of receiving monetary payment. Peer production allows contributions based on all kinds of motivations, but most importantly on the desire to create something mutually useful to those contributing. Hence, people contribute because they find it meaningful and valuable. Next we review the most important principles and characteristics of peer production.

3.1 Distributed Networks

Peer production occurs in distributed networks. Distributed networks are networks in which autonomous agents can freely determine their behavior and linkages without the intermediary of obligatory hubs. As Alexander Galloway highlights in his book *Protocol: How Control Exists After Decentralization* (2004), distributed networks are not the same as decentralized networks, for which hubs are obligatory. Peer production is based on distributed power and distributed access to resources. In a decentralized network such as the US-based airport system, planes have to go through determined hubs. However, in distributed systems or partly distributed systems, such as the Internet (van Steen & Tanenbaum, 2016) or highway systems, hubs may exist, but are not obligatory and agents may always route around them. One should not forget that a significant amount of the Internet's backbone is wired and thus centralized too (Starosielski, 2015). Nevertheless, although we may now be far from the early vision of the Internet as a highly distributed network, the distributed elements still allow for peer production to transcend some of the restrictions of time and space. Peer production, as we discuss below, is "cosmo-local."

3.2 Commons

Peer production is not a "gift economy" and it is not fully based on reciprocity. Peer production follows the adage: each contributes according to their capacities and willingness, and each takes according to their needs. Thus, any "gifting" is most often non-reciprocal gifting, i.e., the use of peer-produced use-value does not create a reciprocal obligation. Peer production does not usually involve reciprocity between individuals but only between individuals and the collective resource. For example, people are allowed to develop their software based on an existing piece of software distributed under the widely used GNU General Public License, only if their final product is available under the same kind of free and open source license (in this case, GNU General Public License).

Peer production can most easily operate in the sphere of digital goods, where the main inputs are free time and the available surplus of computing resources. Reciprocity-based schemes are necessary in the material sphere where the higher cost of capital intervenes. At present, peer production offers no coherent solution to the material survival of its participants, though as we discuss below, there are some promising solutions to such challenges. Therefore, many people inspired by the egalitarian ethos will resort to cooperative production, the social economy, and other schemes from which they can derive an income, while honoring their values. In this sense, these schemes are complementary. The open cooperative and the platform cooperative movements, however, showcase how peer producers may make a sustainable living (Bauwens et al., 2019; Scholz & Schneider, 2016).

In peer production, people voluntarily and cooperatively construct shareable resources that are governed according to the principles and norms of the productive community, i.e., a commons. The use-value created by peer production projects is generated through free cooperation, without coercion toward the producers, and users have free access to the resulting use-value. The legal infrastructure that we have described above creates digital commons of knowledge, software, design, and culture. These new commons are related to the older form of the commons (most notably the communal lands of the peasantry in the Middle Ages and of the original mutualities of workers in the industrial age), but they also differ from them, mostly through their largely intangible characteristics. The older commons were localized, used, and sometimes regulated by specific communities (Ostrom, 1990); the new commons are available and regulated by global cyber-collectives, usually affinity groups. While the older forms of physical commons (air, water, etc.) increasingly function in the context of scarcity, thus becoming more regulated, the digital commons are non-rival resources enriched through usage (thus they could even be considered "anti-rival").

3.3 Equipotentiality

Michel Bauwens (2005) suggested that P2P processes aim to increase the most widespread participation by equipotential ("equal" + "potential") participants. The processes of peer production are characterized by the adoption of equipotentiality as an organizing principle. This means that everyone can potentially cooperate in a project, and, thus, in principle no authority can pre-judge the ability to cooperate. The quality of cooperation is then judged by the community of peers, i.e., through communal validation. In equipotential projects, participants self-sort themselves into the module to which they feel able to contribute.

A related term, used by Jimmy Wales (2014) of the Wikipedia project, is anti-credentialism, which refers to the fact that no credentials are asked beforehand. This means that there is no *a priori* selection to participation. What matters is the ability to carry out a particular task, not any formal a priori credential. The capacity to cooperate is verified in the process of cooperation itself. Thus, projects are often open to all comers provided they have the necessary skills to contribute to a project. These skills are verified, and communally validated, in the process of production itself. Reputation systems may also be used for communal validation. The filtering is a posteriori, not a priori. Anti-credentialism is therefore to be contrasted to traditional peer review, where credentials are an essential prerequisite to participate. However, the "expert" is not a persona non grata in peer production (O'Neil, 2010). Rather, equipotentiality, through a process that is far from flawless, allows the crowds to engage with experts to produce content, designs, code, and more.

3.4 Holoptism

Another element of the grammar of peer production is that projects are characterized by holoptism, as opposed to panopticism (Foucault, 1977) in which only a centralized power can see the whole. Holoptism, from the Greek όλος ("whole") and όπτικος ("seeing"), is the implied capacity and design of peer production processes that allows participants free access to all the information about the other participants; not in terms of privacy, but in terms of their existence and contributions (i.e., horizontal information) and access to the aims, metrics, and documentation of the project as a whole (i.e., the vertical dimension). This can be contrasted to the panopticism that is characteristic of traditionally hierarchical projects: processes are designed to reserve "total" knowledge for an elite, while participants only have access on a "need to know" basis. In peer production, communication is not top-down and based on strictly defined reporting rules, but feedback is systemic and integrated in the protocol of the cooperative system.

3.5 Stigmergic Cooperation

In peer production some producers may be paid or employed as wage labor, or work for the market as freelancers, but not necessarily. All of them produce a commons. The work is not directed by the corporate hierarchies, but through the mutual coordination mechanisms of the productive community, to which the corporate hierarchies have to defer if they want to participate in this type of production. Peer production is based on open and transparent systems, in which everyone can see the signals of the work done by others, and can, therefore, adapt their contribution to the needs of the system as a whole.

Peer production is often based on "stigmergic," from the Greek στίγμα ("mark, sign") and the έργον ("work"), cooperation. In its most generic formulation, stigmergy is the phenomenon of indirect communication among agents and actions (Marsh & Onof, 2007, p. 1). Think how ants or termites exchange information by laying down pheromones (chemical traces). Through this indirect form of communication, these social insects manage to build complex structures such as trails and nests. An action leaves a trace that stimulates the performance of a next action, by the same or a different agent (ant, termite, or commoner in the case of peer production). Therefore, in the context of peer production, stigmergic collaboration is the "collective, distributed action in which social negotiation is stigmergically mediated by Internet-based technologies" (Elliott, 2006). For example, see how free and open source software code lines and the Wikipedia entries are produced in a distributed and ad hoc manner through contributions by large numbers of people.

3.6 Modularity, Granularity, and Low-Cost of Integration

Other authors have of course made significant contributions to the grammar of peer production. For example, stigmergic collaboration is enabled by three characteristics, identified by Benkler (2006) as embedded in the social design of a peer production project: modularity, granularity, and low cost of integration. To begin with, the project needs to be broken down into smaller components, i.e., into modules. For example, in Wikipedia the content is divided into entries, sections, and paragraphs. People can contribute from one word to thousands of words (or figures). The modules allow for any size of contribution: there are many levels of granularity to match different levels of contributors' motivation and time availability.

So, tasks, products and services are organized as modules, which fit with other modules in a puzzle that is continuously re-assembled. And granularity is a property of creating the smallest possible modules with the aim of lowering the threshold of participation (Benkler, 2006). Further, it should be easy to put the various contributions (the modules) into the final product. Also, effective quality-control mechanisms should be in place to avoid low quality of contributions.

3.7 Heterarchy

That being said, peer production projects do have systems of quality control. These "maintainers" (in free and open source software) or "editors" (in Wikipedia) protect the integrity of the system as a whole and can refuse contributions that endanger the integrity of the system. However, they do not coerce work. Peer production is based on freely engaged and passionate labor. Moreover, it eliminates some costs to capital. Hence, as we discuss next, it can also be embraced by for-profit-maximization forces, which explains the massive growth of peer production as a means to produce software for industry.

Peer production is neither hierarchy-less, nor structure-less (Freeman, 1970; Bauwens in Kostakis, 2010), but usually characterized by flexible and dynamic hierarchies and structures based on merit that are used to enable participation. Heterarchies combine elements of networks and hierarchies. Carole Crumley has offered one of the most common definitions (1979, p. 144), which suggests that heterarchy is "the relation of elements to one another when they are unranked, or when they possess the potential for being ranked in a number of different ways, depending on systemic requirements." Thus, multiple levels exist where the communication among them is crucial in transcending the dysfunctions of traditional, rigid hierarchies.

Peer production introduces a cooperative framework that "includes both ranked and nested structures along with those that are flatter and networked" (Crumley, 2015, p. 9). In peer production, the emergence of dynamic hierarchies empowers a measure of cooperation and autonomy. The sole role of hierarchy is, therefore, the initiation and continuous flowering of autonomous cooperation.

Leadership is also "distributed." Peer production projects are often led by a core of founders, who embody the original aims of the project, and who coordinate the vast number of individuals and micro-teams working on specific patches. Their authority and leadership derive from their input into the constitution of the project (meritocracy), and on their continued engagement. Peer production projects may sometimes involve "benevolent dictatorships" (Kostakis, 2010); however, one must not forget that since the cooperation is entirely voluntary, the continued existence of such projects is based on the consent of the community of producers. One is always free to "fork," i.e., to copy and modify and thus take the project to a different, independent direction; though in reality it could be quite difficult to attract a sufficient number of volunteers to a brand-new project.

3.8 Cosmolocalism

One of the essential features of P2P technologies is the liberation from the limitations of time and space. An ever-larger number of people are not bound to their local circumstances, which includes territory in the virtual sense (e.g., organization or enterprise). This is now possible both for digital and material production. Workers can develop contributory lifestyles, and add and withdraw from paid and unpaid projects throughout their lives.

So, if cosmopolitanism is an ideological reflection of the capitalist mode of production and consumption (Marx & Engels, 1848), cosmolocalism ("cosmopolitanism" + "localism") is an ideological reflection of peer production (Ramos, 2016; Bauwens et al., 2019). Cosmolocalism however comes partly from the understanding of cosmopolitanism from the Enlightenment. In short, cosmopolitanism asserts that all human beings belong to a single community, based on a shared morality and a shared future (Corradetti, 2017; Taylor, 2010). Cosmolocalism reflects the convergence of the global digital commons of knowledge, software, and design with local manufacturing technologies. Such technologies can be found in community-driven places such as makerspaces or fablabs. Put simply, what is light (knowledge) becomes global commons, and what is heavy (machinery) is local and shared. Manufacturing, thus, takes place locally for local communities and specialized purposes. For example, see the production of a wide range of artifacts: from agricultural machines for small-scale farming (Giotitsas, 2019), to low-cost and customized prosthetic arms and off-grid wind and hydroelectric power generators (Kostakis et al., 2018). The shared morality comes through the commons, that is to say, through co-creating and co-managing both globally and locally shared resources (digital and physical).

To recap, peer production is based on open inputs; on a participatory process of coordinating the work; and on shared resources as output. This is in sharp contrast with the capitalist mode of production which is based on labor as a commodity in the input phase, hierarchical command following price signals in the production phase, and products and services for sale in the output phase.

We have presented above some of the constitutive components and operational rules of peer production projects, though we do not claim this list is exhaustive. Below, we discuss some of these components and rules in the context of a triarchy of entities within older and more recent ecosystems of peer production.

4 Entities of a New Commons-Based Ecosystem

Through peer production we observe the emergence of a novel ecosystem of value creation consisting of three entities: productive communities; commons-oriented entrepreneurial coalitions; and for-benefit associations. Our description cannot be all-inclusive because each ecosystem is unique. Moreover, it cannot be definite since we deal with a rapidly evolving mode of production. The aim is to offer a birds-eye-view of the expanding universe of peer production. The following table includes just five of the oldest and better-known peer production ecosystems in the field of intangible production (Figure 2.1).

In addition to the well-documented ecosystems of free and open source software projects as well as Wikipedia, the cases of Enspiral, Sensorica, Wikihouse, and Farm Hack offer new perspectives on the rich tapestry of the increasing number of peer production ecosystems. They all fit within the parameters of our description that build new commons-based ecosystems of value creation (Figure 2.2). These examples also show the shift from purely digital production of software and knowledge, to the emergence of entities producing physical products and sophisticated services. Enspiral creates software; Sensorica is a project to make open source based scientific sources; Wikihouse produces designs to create sustainable housing; the Farm Hack community designs and manufactures open source machinery for small-scale farming. What we see here is a replay of the tripartite institutional structure that we see in digital production. A recent study of the urban commons in Ghent (Bauwens & Onzia, 2017) shows that commons-based urban provisioning systems also exemplify this new structure.

Productive community	Linux	Mozilla	GNU	Wikipedia	Wordpress
Entrepreneurial coalition	e.g, Linux Professional Institute, Canonical	e.g, Mozilla corporation	e.g, Red Hat, Endless, SUSE	e.g, Wikia company	e.g, Automatic company
For-benefit association	Linux Foundation	Mozilla Foundation	Free Software Foundation	Wikimedia Foundation	Wordpress Foundation

Figure 2.1 Five of the longest-running and better-known commons-based peer production ecosystems. Source: Bauwens, M., Kostakis, V. & Pazaitis, A. (2019). *Peer to Peer: The Commons Manifesto.* London: Westminster University Press, p. 16. © 2019 University of Westminster Press.

Productive community	Enspiral	Sensorica	Wikihouse	Farmhack
Entrepreneurial coalition	e.g, Loomio, ActionStation	e.g, Tactus Scientific Inc	e.g, Architecture 00, Momentum Engineering, Space Craft Ltd.	e.g, Open Shops
For-benefit association	Enspiral Foundation	Canadian Association for the knowledge Economy	Wikihouse Foundation	Farmhack nonprofit

Figure 2.2 Four emerging commons-based peer production ecosystems. Source: Kostakis, V. & Bauwens, M. (2019). How to Create a Thriving Global Commons Economy. *The Next System Project*, p. 6.

The current emerging infrastructure (Figure 2.3) consists of the following:

The productive community comprises all the contributors to a project of peer production. The members of this entity may be paid or may volunteer their contributions because of some interest in the use-value of this production. However, all of them produce the shared resource, a commons. The most important characteristic, as compared to systems based on wage labor, is that the system must remain open to contributions (equipotentiality).

The second entity is the commons-oriented entrepreneurial coalition, which attempts to create either profits or livelihoods by generating added value for the market, based on the shared resources. The participating enterprises can pay contributors. The digital commons themselves are most often outside the market, because they are not scarce, and therefore not subject to the laws of supply and demand.

What is crucially important in the relation among the entrepreneurs, the community and the common-pool resource on which they depend, is whether their relationship is generative or extractive. There is a rich literature on the relationship between for-profit enterprises and peer production communities (see, e.g., Dahlander & Magnusson, 2008; Bonacorsi et al., 2006; O'Mahony & Bechky, 2008). Of course, extraction/generation are polarities, and every entity is expected to present a mixture.

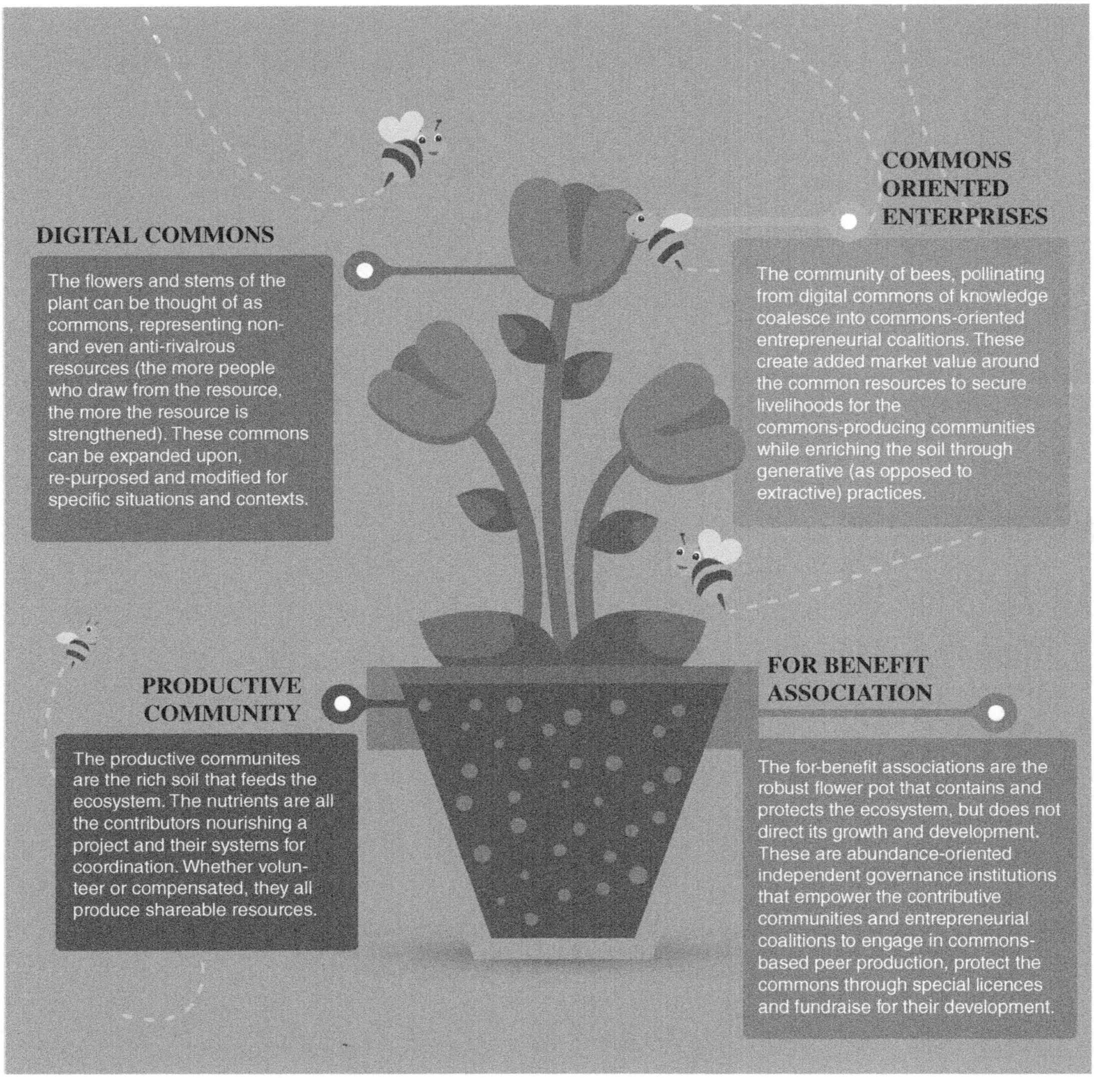

Figure 2.3 The ecosystem of a single commons-based peer production initiative. Source: Bauwens, M., Kostakis, V., Troncoso, S., and Utratel, A. (2017). *Commons Transition and Peer-to-Peer: A Primer.* Amsterdam: Transnational Institute, p.16.

Marjorie Kelly (2012) introduces non-capitalist/generative enterprises, which again comes back to the distinction between markets and capitalism. So, we can have cooperatives or other forms of collectively managed organizations (e.g., nonprofits, NGOs) that have social and environmental goals and use their surplus for these goals, rather than accumulation. To demonstrate the difference between extractive and generative, think of industrial agriculture and permaculture. In the former, the soil becomes iteratively poorer and less healthy, while in the latter case the soil becomes richer and healthier.

In recent years a new type of platform-based extractive entrepreneurs have sought to maximize their profits, and generally do not sufficiently re-invest in the maintenance of the productive communities. Like Facebook, they do not share any profits with the co-creating communities on which they depend for their value creation and realization. Like Uber or AirBnB, they tax exchanges but do not directly contribute to the creation of transport or hospitality infrastructures. So, the problem is that though they develop useful services that rely on underused resources, they do this in an extractive manner. Though they facilitate these services, they also create competitive mentalities which destroy the collaborative and environmental advantages of mutualizing pooled resources. Moreover, extractive enterprises may free-ride on a whole set of social or public infrastructures (e.g., roads as in the case of Uber) and further undermine welfare provisions through the use of fictional worker autonomy to evade taxation and social benefits.

In contrast generative entrepreneurs create additional value around these communities. Seed-forms of commons-oriented entrepreneurial coalitions create added value on top of the commons that they co-produce and upon which they are codependent. In the best cases, the community of entrepreneurs coincides with the productive community. The contributors build their vehicles to create livelihoods while producing the commons. They re-invest the surplus in their well-being and the overall commons system they co-produce.

The third entity is the for-benefit association that can also be seen as the infrastructural organization of the commons, i.e., they manage the infrastructures of commons-based cooperation. Indeed, many peer production ecosystems not only consist of productive communities and entrepreneurial coalitions, but also have independent governance institutions that support the infrastructure for (stigmergic) cooperation. They enable cooperation to take place autonomously and do not command and control the peer production process itself. Behind any commons project, you always find some infrastructural organization, as you cannot have commoning without infrastructure. For example, the Wikimedia Foundation, as the for-benefit association of Wikipedia, does not coerce the production of Wikipedia producers. That is also the case for the free and open source software foundations that often manage the infrastructure and networks of the projects.

By way of contrast, traditional non-governmental and nonprofits organizations operate in a world of "perceived" scarcity. They identify problems, search for resources, and allocate those resources in a directive manner to the solving of the issues they have identified. This approach arguably offers a mirror image to the for-profit model of operating.

For-benefit associations operate for "potential" abundance. They recognize problems and issues but believe that there are enough contributors that desire to assist in solving these issues, most often via holoptism-based stigmergic cooperation. Hence, they maintain an infrastructure of cooperation that allows contributive communities and entrepreneurial coalitions to engage in peer production processes vital for addressing these issues, without directly commanding them. Not only do they protect these commons through licenses, but may also help manage conflicts between participants and stakeholders, fundraise, and assist in the general capacity building necessary for the commons in particular fields of activity (e.g., through education or certification).

Next we discuss the novel aspects of such new interconnected commons-based ecosystems.

5 The Transcendent Aspects of Peer Production

Despite significant differences, peer production and capitalism are highly interconnected. Peer production is dependent on the capitalist market and the capitalist market is dependent on peer production. Most peer producers cannot make a living from peer production, though they derive meaning and value from it, and though it may out-compete the market-based for-profit alternatives in efficiency and productivity terms. Thus, peer production covers only a section of production, while the market provides for many more sections; peer producers are mainly dependent on the income provided by the capitalist market. Peer production has been created within the interstices of the capitalist market.

But a new form of capitalism has been emerging depending on peer production: netarchical capitalism. By "netarchical" we mean the hierarchies within the network that own and control participatory platforms. This version of capitalism is characterized by digital platforms that combine P2P elements, which allow people to interact with each other directly, but they are controlled and monitored by the platform owners. The full centralized control of the rest of the infrastructure is used to extract value from these exchanges.

This new form of capital directly exploits networked social cooperation that often consists of unpaid activities that can be captured and financialized by proprietary "network" platforms. It lives from the positive externalities created through human cooperation and the commons. If previous versions of capitalism were hostile to the commons and tried to destroy it, this new version has learned, at least provisionally, to "tame" the commons. Nevertheless, this also means that it has become parasitic and rent-seeking. Netarchical capitalism is rent-seeking capital that has shifted its control mechanisms to control the whole network itself and functions one step away from real production. For example, social media platforms like Facebook almost exclusively capture the value of their members' social exchange, by monetizing the data and selling the "attention" of their users to advertisers. Crowdsourcing models are based on distributed labor tending to reduce the average income of the producers (for an overview of crowdsourcing's labor markets and the dark side of digital labor in general, see the books edited by Scholz, 2012, and Casilli, 2017). There is no creation of commons by communities, but rather a competition between workers and producers, to get clients on the demand side. Uber, Airbnb, Kickstarter, and TaskRabbit are also examples of the netarchical model.

The support given by major digital economy companies to open source development is another point of reference. The general business model seems to be that business "surfs" on the P2P infrastructure, and creates surplus value through services, which can be packaged for exchange value. The massive use of free and open source software in business, enthusiastically supported by venture capital and large companies such as IBM, is creating new business models. Such business models go "beyond" products and focus instead on services associated with the nominally free and open source software model. Industries are gradually transforming to incorporate user-generated innovation and content. Several knowledge workers are choosing non-corporate paths and becoming mini-entrepreneurs, relying on an increasingly sophisticated participatory infrastructure, a kind of digital corporate commons.

Thus, capitalist forces mostly use partial implementations of peer production. The tactical and instrumental use of P2P infrastructure is only part of the story. Contemporary capitalism's dependence on P2P is systemic. As the whole underlying infrastructure of capitalism becomes distributed, it generates peer production practices and becomes dependent on them.

The for-profit forces that are building and enabling these new platforms of participation represent a new subclass, the "netarchical" (Bauwens, 2009) or "vectoral" (Wark, 2004) class. These new capitalists prosper from the enablement and exploitation of participatory networks. In addition to the examples above, see also Amazon that built itself around user reviews, eBay that lives on a platform of worldwide distributed auctions, and Google that builds on user-generated content.

More broadly, netarchical capitalism is a brand of capital that embraces peer production. It is the force behind the immanence of peer production. Opposed to it, though linked to it in a temporary alliance, are the forces of commoning, those that put their faith in the transcendence of commons-based peer production, in a reform of the political economy beyond the domination of the market.

Indeed, peer production has transcendent aspects that go beyond the limitations set by the for-profit-maximization economy. Historically, though forces of higher productivity may be temporarily embedded in the old productive system, they ultimately lead to deep upheavals and reconstitutions of the political economy. The emergence of capitalist modes within the feudal system is a case in point.

Peer production can become the vehicle of new configurations of production and exchange, no longer dominated by capital and state. This is the "transcendent" aspect of peer production as it creates a new overall system that can subsume the other forms (Bauwens, 2009). One scenario is that capital and state subsume the commons under their direction and domination, leading to a new type of "commons-centric" capitalism. In a second scenario, the commons, its communities, and institutions become dominant and, thus, may adapt state and market forms to their interests.

At a time when the very success of the capitalist mode of production endangers the biosphere and causes increasing psychic (and physical) damage to the population, the emergence of such an alternative is particularly appealing, and corresponds to the new cultural needs of large numbers of the population. It stands as a permanent alternative to the status quo, and the expression of the rising of a new social force: the knowledge workers.

6 Instead of Conclusions: Towards a P2P Theory

The aim of P2P theory is, therefore, to give a theoretical underpinning to the transformative practices of peer production. It aims to understand how a new kind of society, based on the centrality of the commons and within a reformed market and state, is possible. Such a theory has to explain not only the dynamics of peer production, but also their fit with other inter-subjective dynamics. For example, how peer production molds reciprocity modes, market modes, and hierarchy modes; on what ontological, epistemological, and axiological transformations this evolution is resting; and what a possible peer production ethos can be. A crucial element of such a P2P theory would be the development of tactics and strategy for such a transformative practice.

A transformative practice has to acknowledge and address systemic social unfairness and environmental degradation. Yet peer production does not solve many of these problems, especially those involving race and gender. Nor does it directly address the hidden environmental and social costs of digital technologies, which are energy-intensive throughout their life-cycle. Moreover, low-wage laborers (often including children) work under inhumane circumstances so that ever more people in the advanced economies have access to cheap digital technologies. However, P2P theory and practices discuss and introduce new paradigmatic ways of value creation that have the potential to be more radically inclusive and sustainable.

Acknowledgments

Vasilis Kostakis acknowledges financial support from the European Research Council under the European Union's Horizon 2020 research and innovation program (grant agreement No 802512).

Note

1 Although in theory peer production has a high degree of openness, in practice fewer women and people of color (especially from the Global South) are part of peer production. There are several "cultural" barriers that prevent people from joining. Patriarchy and racism are two flagrant examples. For instance, see Reagle (2010) regarding Wikipedia's barriers to genuine openness, and Reagle (2013) for the gender gap within free and open source software.

References

Bauwens, M. (2005). The political economy of peer production. *Ctheory Journal*. www.ctheory.net/articles.aspx?id=499

Bauwens, M. (2009). Class and capital in peer production. *Capital & Class, 33*(1), 121–141.

Bauwens, M., Kostakis, V., & Pazaitis, A. (2019). *Peer to peer: The commons manifesto*. London: Westminster University Press.

Bauwens, M., Kostakis, V., Troncoso, S., & Utratel, A. (2017). *Commons transition and peer-to-peer: A primer*. Amsterdam: Transnational Institute.

Bauwens, M., & Onzia, Y. (2017). *Commons Transitie Plan voor de Stad Gent. In Opdracht van de Stad Gent*. Retrieved from https://tinyurl.com/ybyj5qd4

Benkler, Y. (2002). Coase's penguin, or Linux and the nature of the firm. *Yale Law Journal, 112*(3), 369–446.

Benkler, Y. (2006). *The wealth of networks: How social production transforms markets and freedom*. New Haven, CT: Yale University Press.

Bonaccorsi, A., Giannangeli, S., & Rossi, C. (2006). Entry strategies under competing standards: Hybrid business models in the open source software industry. *Management Science, 52*(7), 1085–1098.

Casilli, A. (2017). Digital labor studies go global: Toward a digital decolonial turn. *International Journal of Communication, 11*, 3934–3954.

Corradetti, C. (2017). Constructivism in cosmopolitan law: Kant's right to visit. *Global Constitutionalism, 6*(3), 412–441.

Crumley, C. L. (1979). Three locational models: An epistemological assessment for anthropology and archaeology. In M. B. Schiffer (Ed.), *Advances in archaeological method and theory* (pp. 141–173). New York, NY: Academic Press.

Crumley, C. L. (2015). Heterachy. In R. Scott & S. Kosslyn (Eds.), *Emerging trends in the social and behavioral sciences* (pp. 1–15). New York, NY: Wiley.

Dahlander, L., & Magnussonn, M. (2008). How do firms make use of open source communities?. *Long Range Planning, 41*(6), 629–649.

Elliott, M. (2006). Stigmergic collaboration: The evolution of group work. *M/C Journal: A Journal of Media and Culture, 9*(2). http://journal.media-culture.org.au/0605/03-elliott.php

Galloway, A. (2004). *Protocol: How control exists after decentralization.* Cambridge, MA: MIT Press.

Giotitsas, C. (2019). *Open source agriculture: Grassroots technology in the digital era.* Basingstoke: Palgrave Macmillan.

Foucault, M. (1977). *Discipline and punish: The birth of the prison.* New York: Pantheon Books.

Freeman, J. (1970). The tyranny of structurelessness. *Berkeley Journal of Sociology, 17,* 151–165.

Kelly, M. (2012). *Owning our future: The emerging ownership revolution – journeys to a generative economy.* Oakland, CA: Berrett-Koehler Publishers.

Kostakis, V. (2010). Identifying and understanding the problems of Wikipedia's peer governance. *First Monday, 15*(3). Retrieved from http://firstmonday.org/htbin/cgiwrap/bin/ojs/index.php/fm/article/view/2613/2479

Kostakis, V., & Bauwens, M. (2019). How to create a thriving global commons economy? *The Next System Project.* Retrieved from https://thenextsystem.org/learn/stories/how-create-thriving-global-commons-economy

Kostakis, V., Latoufis, K., Liarokapis, M., & Bauwens, M. (2018). The convergence of digital commons with local manufacturing from a degrowth perspective: Two illustrative cases. *Journal of Cleaner Production, 197*(2), 1684–1693.

Marsh, L., & Onof, C. (2007). Stigmergic epistemology, stigmergic cognition. *Cognitive Systems Research, 9*(1–2), 136–149.

Marx, K., & Engels, F. (1848). *Manifesto of the communist party.* New York, NY: International.

O' Mahony, S., & Bechky, B. (2008). Boundary organizations: Enabling collaboration among unexpected allies. *Administrative Science Quarterly, 53*(3), 422–459.

O'Neil, M. (2010). Shirky and Sanger, or the costs of crowdsourcing. *Journal of Science Communication, 9*(1). Retrieved from https://jcom.sissa.it/archive/09/01/Jcom0901%282010%29C01/Jcom0901%282010%29C04

Ostrom, E. (1990). *Governing the commons: The evolution of institutions for collective action.* Cambridge: Cambridge University Press.

Ramos, J. (2016). Cosmo-localism and the futures of material production. Retrieved from http://actionforesight.net/cosmo-localism-and-the-futures-of-material-production/

Reagle, J. (2010). *Good faith collaboration: The culture of Wikipedia.* Cambridge, MA: MIT Press.

Reagle, J. (2013). "Free as in sexist?" Free culture and the gender gap. *First Monday, 18*(1). Retrieved from https://firstmonday.org/article/view/4291/3381

Scholz, T. (2012). *Digital labor: The Internet as playground and factory.* New York, NY: Routledge.

Scholz, T., & Schneider, N. (2016). *Ours to hack and to own: The rise of platform cooperativism, a new vision for the future of work and a fairer Internet.* New York: OR books.

Starosielski, N. (2015). *The undersea network.* London: Duke University Press.

Taylor, R. (2010). Kant's political religion: The transparency of perpetual peace and the highest good. *Review of Politics, 72*(1), 1–24.

van Steen, M., & Tanenbaum, A. (2016). A brief introduction to distributed systems. *Computing, 98*(10), 967–1009.

Wales, J. (2014). Transcript: Jimmy Wales. Aljazeera. Retrieved from https://www.aljazeera.com/programmes/headtohead/2014/04/transcript-jimmy-wales-201446143728879415.html

Wark, M. (2004). *A hacker manifesto.* Cambridge, MA: Harvard University Press.

3

Political Economy of Peer Production

Benjamin J. Birkinbine

1 Introduction

Understanding the political economy of peer production requires consideration of at least two interrelated factors: the economic factors that enabled the rise of peer production and the unique cultural factors within peer production communities that have the potential to subvert the prevailing tendencies of global capitalism. The economic factors can be found in the broader structural changes occurring within capitalism throughout the 20th century, particularly after World War II and in the run-up to the Vietnam War era when global geopolitics was being reshaped by a wave of independence movements in the Global South as well as increased war expenditures in the United States. All of these factors set the stage for a response to the economic crisis of 1970s. It was during this time that capital responded to the long downturn by searching for new opportunities for boosting profitability (Brenner, 2006). On the one hand, these changes had dramatic effects on labor processes, commodity supply chains, and manufacturing practices, all of which became increasingly networked to maximize efficiencies and reduce excess capacity.

On the other hand, these changes – and especially the rise of computers and networked communication systems – enabled greater interconnection between people who could collaborate with one another to produce collectively governed resources. As such, there were contradictory forces at play in the increasingly networked global economy. Networking capabilities not only began to transform labor processes across the globe, but simultaneously allowed for production that was no longer driven by market demands. Communities of peers connected by networked communication technologies began producing informational or digital commons, which were made freely available to others. Benkler (2006) framed this type of production as *commons-based peer production*, which he defined as "a new modality of organizing production: radically decentralized, collaborative, and nonproprietary; based on sharing resources and outputs among widely distributed, loosely connected individuals who cooperate with each other without relying on either market signals or managerial commands" (p. 60). For example, the free and open source software movement provides an exemplary case of what commons-based peer production enables, in that it was driven by an ethos of sharing and a strong belief in productive freedom (Coleman, 2013). This community, as well as other communities of commoners around the world, challenged longstanding assumptions in economics about competition, self-interest, and the maximization of individual profit. Indeed, these communities challenged fundamental definitions of *value*, including what it meant and how it circulated within communities.

This chapter considers the political economy of peer production by contextualizing the rise of peer production within capitalism and evaluating to what extent peer production contradicts, or reinforces, these global economic trends. To that end, I begin by tracing the structural changes of capitalism in the 20th century that provide the economic foundation upon which peer production arose. Accompanying these changes was also a shift toward more flexible work routines that privileged creativity, autonomy, and innovation. I argue that the roots of peer production lie in these transformations. Finally, I discuss some more recent scholarship about how value is created and circulates within

The Handbook of Peer Production, First Edition. Edited by Mathieu O'Neil, Christian Pentzold, and Sophie Toupin.

commons-based peer production communities and discuss how these circuits of commons value intersect with capital accumulation circuits. One of the pressing questions running throughout this discussion is whether this emergent value system has the potential to bring about a postcapitalist future, in which the values of care, trust, mutual aid, and conviviality are prioritized rather than capital accumulation.

2 Transformation of Capitalism in the 20th Century

The historical changes of capitalism that enabled the rise of peer production can specifically be found within the 1970s and the response to the economic crisis. In making such an argument, I draw from critical scholars like David Harvey (1989, 2010) and Robert Brenner (2006) by focusing on the structural changes in the global economy and how those changes affected the broader geopolitics of global capitalism and the labor processes. I also draw from Dan Schiller's (1999, 2014) theory of digital capitalism in foregrounding "communication and information as an emerging pivot of the ever-mutating political economy," as it was during this time that global commodity supply chains, financial networks, and military technologies were becoming increasingly networked through information and communication technologies (Schiller, 2014, p. 5). The goal of this broad overview is to identify important historical trajectories, even if this means sacrificing some of the nuance in particular details.

Prior to the 20th century, the early stages of capitalism were marked by industrial growth and increasing complexity in the division of labor. To address this complexity and streamline production, scientific principles were applied to the production process, which simplified the labor process by breaking down various stages involved in the production process into its component parts. These changes are often associated with Taylorism, as it was Frederick Taylor's *The Principles of Scientific Management* (1911) that outlined the inefficiencies of human work and identified ways of optimizing efficiency in production. Once these distinct stages of the production process were identified, those same scientific principles could be used to construct machinery that would supplement or altogether supplant increasingly deskilled human labor. The pressures to increase productivity and the development of assembly-line production are also associated with Fordism, as Henry Ford's assembly line in 1913 made possible the mass production of automobiles.

Long before Ford's assembly-line production, Karl Marx was analyzing the mechanisms of capitalist production, particularly toward the latter years of the Industrial Revolution from roughly the 1840s to the 1880s. His task was a critique of existing political economic thought, while specifically focusing on the consequences of capitalist production for the growing numbers of laborers who were joining the working class. Focusing on industrial factories in the mid-1800s, Marx determined that labor was being exploited in at least two primary ways. First, surplus value was being extracted from labor through the prolongation of the working day, which Marx referred to as *absolute surplus value*. Second, and perhaps most important for the present discussion, was the extraction of what Marx referred to as *relative surplus value*, which drew attention to the ways the "technical processes of labor and the composition of society" were being revolutionized (Marx, 1906, p. 559). In other words, ongoing changes in the technical processes involved in production constantly placed pressure on workers, especially as they were subject to deskilling (i.e., reduction in the skills necessary for performing certain tasks), reskilling (i.e., learning how to operate new technologies), or automating jobs that were previously performed by humans.

These trends would continue into the 20th Century, as global capitalism began to take shape. What emerged was a complex division of labor and an increasingly globalized capitalist economy. These developments also gave rise to a large pool of working-class labor with common interests. As a result, workers in the Global North organized into trade unions to struggle against capital for fair wages, the eight-hour work day, the end of child labor, and other labor rights. As the trade union movement grew, it ensured a certain degree of welfare for workers and created a shared identity among the working class.

As the 20th century wore on, World War II, and particularly its aftermath, significantly reshaped the geopolitical landscape.[1] The United States' competitors in the global market were devastated by World

War II, as their infrastructures were decimated by bombing campaigns, which caused manufacturing to decrease significantly. By comparison, the centers of American manufacturing remained basically untouched, which gave the United States an advantage in the immediate aftermath of the war. As a result, American firms greatly increased their exports, particularly to Western Europe through the Marshall Plan, which was a boon to the American economy. The United States was keenly interested in exporting its products as well as capitalist ideology around the globe during this time, as it was embroiled in the Cold War with the Soviet Union, and the two countries sought to expand their influence around the globe. The United States represented the "first world," which was capitalist and developed, the Soviet Union represented the "second world," which was socialist, and those countries that were not aligned with either capitalism or socialism were referred to as the "third world." While the struggle between the United States and Soviet Union generally frames the dominant narratives of this period, there were also strong political movements within the non-aligned nations (see Prashad, 2007). However, the United States also faced increasing pressure from other "first world" countries during this time frame. In the 1950s and 1960s, Japanese and German manufacturing grew rapidly and, as the global market grew, these competitors began to cut into American market share. The increased production and supply placed downward pressure on prices and, by the 1960s, Japan and Germany were able to undercut American prices, which had higher fixed costs. Exacerbating the pressure felt by the United States was the increased war spending during the Vietnam Era. All of these structural pressures set the stage for the global economic crisis of the 1970s.

3 Response to the Crisis of the 1970s

The crisis of the 1970s placed significant pressure on the profits of US capital. In response, policymakers and shareholders sought measures of restoring profitable capital accumulation. These measures essentially unfolded along two axes. First, policymakers began to cut away at policies designed to protect worker rights. These efforts began in the late 1970s, but really ramped up after leaders like Ronald Reagan, Margaret Thatcher, and Deng Xiaoping took office. These leaders ushered in an era of deregulation and the privatization of national industries, which also cut away at protections for labor (see Harvey, 2005). The upshot of these regulatory changes was a reduction in wages and the outsourcing of jobs, thereby reducing the fixed costs of firms. Second, shareholders and business owners also sought ways of reducing the fixed costs associated with their business operations. This was made possible by new developments in information and communication technologies, most notably software, which allowed owners to track their production processes in a more sophisticated way. The result was a further disciplining of labor and the maximization of efficiency throughout commodity supply chains. Furthermore, commodity supply chains could now be networked across geographic boundaries to take advantage of cost savings wherever they existed. This allowed for what Harvey (1989) refers to as *flexible accumulation*, which is "characterized by the emergence of entirely new sectors of production, new ways of providing financial services, new markets, and, above all, greatly intensified rates of commercial, technological, and organizational innovation" (p. 147). In effect, this new model of production no longer relied on mass production, which was the hallmark of Fordism. Rather, it made possible on-demand production, which reduced stockpiles of mass-produced inventory while simultaneously allowing firms to adapt to ever-changing demands from consumers.

One of the critical elements enabling this type of business model was the development of networked communication technologies, which could be put into the service of capital. It was during this time that information and communication technologies became viewed as a new vehicle for reinvigorating capital accumulation, and the sector saw massive capital investment, especially in the 1990s, while US manufacturing remained stagnant. Indeed, the massive surge in venture capital investment into the so-called "dot-com" companies during the 1990s fueled the financial bubble that eventually burst in the early 2000s (Cassidy, 2002). The investment in digital technologies continues today, as similar investments can be seen in a number of digital services that promise to reinvigorate capital accumulation, whether they are digital platforms like Uber, Lyft, Spotify, or cryptocurrency technologies like Blockchain.

Accompanying these economic changes was also an ideological shift, particularly in the ways in which participation in capitalism was justified. Boltanski and Chiapello (2005) refer to the "spirit of capitalism" as the "ideology that justifies engagement in capitalism" (p. 8). They argue that the spirit of capitalism shifted from the 1970s onwards. During this time, business culture – in the form of management texts – began to abandon the rigidities of hierarchical Fordist production in favor of flexible and network-based forms of organization. In doing so, capitalism effectively assimilated the artistic critique of capitalism of the 1960s and 1970s by touting individual autonomy and emphasizing employee initiative as a path to self-fulfillment, whilst marginalizing its social critique, embodied in trade unions and more concerned with wage inequality. In short, employment was not simply something necessary for earning a living, but it was also a way to achieve creative freedom through self-expression. The espousal of these new forms of freedom and autonomy for workers proved to be an effective way of reducing the fixed costs of corporations. As more workers embraced flexible work routines, they carried the burden of funding their own benefits like insurance, retirement plans, etc. Indeed, this trend can also be viewed today in a number of industries where freelance work has become prevalent, but the trend is particularly pronounced in the media and journalism industries as well as so-called "creative" industries that employ artists, fashion designers, and stylists (see Deuze, 2007; McRobbie, 2016; Cohen, 2016).

4 The Transformation of the Internet and Web-Based Business Models

The structural changes of capitalism throughout the 20th century laid the foundation for the emergence of the Internet as a vehicle for intensifying capital accumulation. Furthermore, this process really began to ramp up toward the end of the 20th century when Internet service provision was privatized. In the wake of its privatization, Internet- and web-based businesses went through a relatively rapid boom-and-bust cycle. While these years were somewhat tumultuous for speculative financial capital, the general tendency to network the world continues, as more and more people, places, and things become interconnected through digital technologies. In what follows, I focus on how the Internet transformed from a publicly funded research and development project to an unprecedented instrument for capital accumulation.

We can begin by looking at the development of the Internet and its various iterations before it was privatized in the 1990s. Early research and development into a distributed network architecture was funded by the Defense Advanced Research Projects Agency (DARPA) in the United States. Established primarily to meet the needs of military demands, the development of DARPANET provided the initial architecture for what would eventually become the Internet in the United States. However, the network passed through various changes, including both ARPANET, MILNET (for military sites), and NSFNET, named after the National Science Foundation, which administered the network until 1995 (see Abbate, 1999). These iterations of the early Internet were funded directly by public money through the Department of Defense and were designed to serve two essential functions. On the one hand, military sites could be linked through MILNET, while ARPANET linked various civilian research communities so they could share information freely with one another. The latter function was achieved through NSFNET's Acceptable Use Policy (AUP), which ensured that the network was used for non-profit research and educational purposes while also prohibiting most commercial use. The policy remained in effect until the early 1990s, at which point commercial Internet service providers were allowed onto the network. In other words, one of the foundational principles of the early Internet was that information on the network ought to be freely available to others. This same principle is also identified as one of the tenets of the "hacker ethic" as defined by Levy (2010). The privatization of the Internet culminated in 1994 after a contentious hearing before the US Congress where the decision was made to relinquish public control of the network to commercial providers.

The privatization of Internet service provision dramatically increased Internet connectivity throughout the 1990s as more and more people migrated online. In response, businesses sought to take advantage of the growing market for Internet services and web-based business models. Companies like America Online thrived in the market for Internet service provision, while the online marketplace eBay was founded in 1995. The introduction that year of the https protocol enabled secure financial dealings so that sellers like Amazon and services like eBay became the dominant business model of the "Web 1.0" era, which was aimed at providing content to consumers on web pages or selling web-based products or services directly to consumers. Meanwhile, Microsoft dominated the market for Internet browsers by negotiating a partnership with IBM, whereby IBM's personal computers would come pre-packaged with Microsoft's Internet Explorer installed on IBM's hardware. Microsoft exemplified the power of companies during this era by developing and selling software either directly to consumers or to original equipment manufacturers (OEMs) like IBM. Since IBM dominated the market for personal computers, as well as computers sold to other enterprises, Microsoft effectively prohibited any competitor from mounting a challenge to its monopolistic position, especially in the market for Internet browsers. Indeed, it was Microsoft's agreement with IBM that eventually led to its conviction for antitrust violations in 2001 after the United States Department of Justice found the company guilty of anticompetitive market behaviors (see *The United States v. Microsoft*, 2001). The Microsoft antitrust conviction coincided with the bursting of the dot-com bubble, which began roughly in early 2000 and lasted through the end of 2002. On the day that the Microsoft antitrust decision was announced, the NASDAQ stock exchange lost approximately 8% of its total value and Microsoft alone lost nearly $80 billion of its market value (Ulick, 2000). From its peak in March, 2000 to its trough in October, 2002, the NASDAQ lost nearly $5 trillion in value (Gaither & Chmielewski, 2006).

Emerging out of the dot-com crash came a reinvention in the way web-based business would be conducted. Companies began to abandon old business models aimed at selling products and services directly to consumers. Instead, the consumers, or data and information about consumers, became the primary products of the emergent "Web 2.0" economy – a term coined by Tim O'Reilly (2005) – which, among other things, was characterized by providing interactive services to users so user-generated content could be monetized by the platforms upon which those users interacted. It was within this context that we saw the rise of Facebook, Google, Wikipedia, and even the transformation of Amazon from simply a book-selling company in the 1990s to one that began offering Amazon Web Services in 2002. This transformation in business models had dramatic consequences for social life, which is increasingly mediated by digital technologies. Ursula Huws (2014) summarizes this succinctly:

> When human sociality is mediated by telecommunications systems, it leaves digital traces wherever it goes, traces that can be mined to generate data that enable advertising to be targeted with ever-greater accuracy. The Internet is thus constituted as a vast virtual shopping mall, with its users bombarded with a constant stream of advertising, preying on their most personal vulnerabilities. (p. 15)

These concerns have been explored within scholarly debates about immaterial labor, free labor, and digital labor (Lazzarato, 1996; Terranova, 2004; Scholz, 2013). The primary concern in these debates has been the nature of work and labor within the information, knowledge, and communication industries with a focus on forms of unpaid labor occurring online (see Andrejevic, 2007, 2012; Fuchs 2012). Particular attention has been paid to users' online behaviors, which are tracked and can be transformed into an audience commodity in the same way that Dallas Smythe (1981) identified with broadcasting. However, whereas Smythe argued that broadcast programs constituted a "free lunch" by producing audiences for advertisers, the parallel to online activity lies in platforms or web sites seeking the attention (or interaction) of users while data is collected about those activities. As we continue to rely on digital devices for an increasing amount of our social lives, the time spent online during both work and non-work time, our digital labor – socially necessary time spent online – offers a more sophisticated form of the audience commodity as browsing data is extracted and transformed into value by service providers and other third-party elements (Fuchs, 2011; McGuigan & Manzerolle, 2013; Turow, 2013).

The tendency of Internet platforms to monitor, measure, collect, and ultimately monetize data about their users has become one of the dominant business models existing online today. The idea is seemingly sound: encourage users to socialize, interact, create, share, or otherwise rely upon your service for connecting with others, and platforms can extract data from those interactions that can be monetized through sales to advertisers or other data companies. In effect, users are generating value *gratis* for these platforms, and there is evidence to suggest that the vast majority of users do not realize this. One recent study showed that 74% of Facebook users did not know the site maintained a list of their interests and traits, and 51% of respondents claimed they were not comfortable with this (Hitlin & Rainie, 2019). However, platforms generally continue to grow rapidly as more people migrate online and begin to use their services. This general trend remains true even if some platforms struggle to become profitable. For example, the music-streaming platform Spotify was unprofitable for its first ten years, and only reported its first profitable quarter at the end of 2018 (see Wang, 2019). Ride-sharing companies, Uber and Lyft, similarly struggle with profitability (see Colley, 2019). Because of the potential for rapid growth and the prospects of future profitability, tech startups continue to draw attention from venture or investment capital. The massive capital investment in these companies, which seem to be valued more for their growth than their profitability, has led some to predict that it is only a matter of time before we experience another tech bubble burst (Colley, 2019).

Despite the seemingly bleak picture presented here, however, there are also contradictory forces within these developments that suggest we may also be witnessing the emergence of a new subjectivity, which is capable of counteracting the incessant drive for capital accumulation. Indeed, it is within these contradictory forces that we can begin to understand the different dynamics at play in the political economy of peer production. At times, these forces exemplify the drive to accelerate capital accumulation, as I have already outlined above, while at other times these forces work to ensure the survival of collectively governed resources.

5 The Commons as Alternatives

The previous sections of this chapter outlined the historical forces that led to the transformation of global capitalism in the 20th century and explained how and why platforms for extracting value from activities taking place online were developed. However, despite the fact that the forces of capitalism (digital, surveillance, or otherwise) seek ways of exploiting the shared experiences of people around the globe, the fact that people are becoming more interconnected also means that they can collaborate in ways that subvert some of the dominant tendencies of global capitalism. Some scholars also see radical potential within these shared experiences and the values inherent in commons-based peer production that may have the potential to bring about an alternative future. In other words, the early parts of this chapter focused on the *external* factors that made possible peer production, but we now need to consider the features *internal* to peer production that make it unique as well as capable of challenging those external forces. In the remaining portions of this chapter, I draw on a growing corpus of scholarship that positions peer production and the commons or "the common" as an alternative to capitalism.

In their examination of "empire," Hardt and Negri (2000, 2004, 2009) argue that the sovereignty of the nation state has been replaced by a decentralized and ever-expanding regime of power that demolishes barriers between what is inside and what is outside of the empire. What emerges from this system of power is an interconnected network, which is capable of managing "hybrid identities, flexible hierarchies, and plural exchanges through modulating networks of command" (Hardt & Negri, 2000, pp. xii–xiii). But whereas these networks enable the nearly seamless flow of global capital, command of commodity supply chains, and increased control and discipline of labor, they also simultaneously give rise to contradictory forces that challenge the prevailing order. For example, the increasing interconnection of peoples, places, cultures, creativity, and so on, makes possible the creation of "the common." By "the common," Hardt and Negri refer not only to "the common wealth of the material world – the air, the water, the fruits of the soil, and all nature's bounty" but also to "those results of social production that are necessary for social interaction and further production, such as knowledges, languages, codes, information, affects, and so on" (Hardt & Negri, 2009, p. vii). The production of the common, particularly as it pertains to peer production online, manifests in the sharing of free and open source software projects, open access book projects, "wikis" dedicated to curating knowledge and information, cultural products in the public domain, and many other projects aimed at creating and preserving shared resources that may be accessed by others.

However, there is more at play in the commons than simply a collectively governed resource. The commons can also be viewed as a "process of becoming," which moves beyond the commons as an identity and orients human activity toward an ethical horizon to which we can aspire (Dyer-Witheford, 2006; Hardt & Negri, 2009; Ryan, 2013; Gutiérrez-Aguilar, 2014; Singh, 2017). For this reason, Linebaugh (2008) prefers the active verb "commoning," which he explains is "embedded in a labor process"; it is collective, and it is "independent of the temporality of the law and the state" (p. 45). Moreover, this framing helps account for the "dual nature" of the commons, which encompasses both the objects of the commons (i.e., a collectively governed resource to which all members of the community claim some association) as well as a subjectivity in which the values of mutual aid, care, trust, and conviviality can be produced and reproduced over time (DeAngelis, 2017). The focus on reproduction is particularly important here, as it corresponds to the critiques from feminist political economy that drew attention to forms of unwaged labor and social reproduction, both of which are instrumental to capitalism but also sites of struggle (Dalla Costa & James, 1975; Jarrett, 2016; Federici, 2012).

Furthermore, positioning the commons as a process of becoming adds critical weight to commoning practices by demonstrating how those activities are capable of bringing about a postcapitalist future. The term "postcapitalism" is generally used to describe a number of different proposals for developing a new economic system to replace capitalism. While there is no unified theory of what postcapitalism is or what it will look like, most thinkers writing about the subject generally agree that the contradictions within capitalism and the recurrent crises of capitalism are unsustainable in the long run. Therefore, they offer proposals for what a postcapitalist future will look like as well as different proposals for how such a system can be achieved. In a sense, these thinkers are revitalizing the tradition of utopian socialism, which offered visions of a future society that was comprised of freely associating individuals living in harmony with one another. Early utopian socialist thinkers, like Henri de Saint-Simon and Charles Fourier influenced Karl Marx, and Marx's work is often situated within a combination of three conceptual areas: political economy, German classical critical philosophy, and utopian socialism (Harvey, 2010). While Marx remains somewhat vague in his specific description of society after the abolition of capitalism, he carried on the tradition of utopian socialism in thinking that another world was possible. In Marx's formulation, such a transition was only possible through the revolutionary struggle of the working class against capital.

More recent accounts of postcapitalism focus on the ways that social relations are beginning to transform in ways that are antithetical to capitalism. For example, one of the earliest uses of the term 'postcapitalism' comes from Peter Drucker's (1994) book, *Post-Capitalist Society.* Noting that the transformation toward more information- or knowledge-based economies in the Global North, and particularly the United States, Drucker argued that such a shift fundamentally changes the social, economic, and political dynamics throughout society. As a management consultant and educator, Drucker's analysis of postcapitalism was not about radically transforming or replacing the capitalist system but recognizing the shifts occurring from within capitalism and how business organizations could adapt their strategies to survive in this new form of capitalism.

Two other recent works have become influential for thinking about postcapitalism. One account comes from the journalist Paul Mason (2015), whose account of postcapitalism does not shy away from a critique of capitalism. Mason argues that the imbalances of power and the increasingly devastating crises at the heart of neoliberal capitalism are ultimately unsustainable. He argues that we need to move beyond capitalism so that it is replaced with an economic system that will be more beneficial and sustainable. To that end, Mason offers five principles for the transition to postcapitalism. The first principle is more of a condition for implementing any project for social change. We need to test all proposals at a small scale then model them many times over before implementing them on a broader scale. Second, ecological sustainability is a necessity for any project in designing the future. Third, the transition to postcapitalism is not simply about economics; it is about human beings, and we cannot privilege economic progress at the expense of social welfare. Fourth, problems must be attacked from all angles, which takes advantage of networked hierarchies. In short, the benefit of networks is that meaningful action can occur at many different levels and it does not need to be centralized. Finally, we should maximize the power of information by learning from the data being generated about social life.

However, the apparatus for data collection would be collaborative and de-centralized rather than under corporate control.

Another influential work about postcapitalism comes from Nick Srnicek and Alex Williams (2016) who provide a series of demands that are necessary for the transition to postcapitalism. First, the authors argue for a fully automated economy, which would free up human beings from the required routines of waged labor. More specifically, they argue for a progressive reduction in work so that necessary labor is reduced as much as possible. Not only would this eliminate some of the social hierarchies between different types of employment, but it would also free up more time for addressing social and community needs. Second, the authors argue for a reduction in the length of the work week with no reduction in pay. Third, the authors argue for a universal basic income that is sufficient for living, given without qualification, and acts as a supplement to the welfare state rather than a replacement. Finally, the authors argue for a diminishment of the work ethic, or the development of a counter-hegemonic approach to work. Such an approach "would overturn existing ideas about the necessity and desirability of work, and the imposition of suffering as a basis for remuneration" (Srnicek & Williams, 2016, p. 125).

At the heart of these proposals is the recognition that commons-based movements and the active process of commoning carry something within them that points to another way of living. In a sense, the proposals offered by Mason (2015) and Srnicek and Williams (2016) take for granted the achievements of commons movements without probing into the specific dynamics of how value is created and circulates within communities of commoners. Massimo DeAngelis (2017), however, has offered such an analysis, particularly of the way that value circulates within the commons and commons-based communities. Figure 3.1 below illustrates one attempt to chart how value circulates in the commons through the circuit of commons value. In this figure, an association of people (A) claim collective ownership of their commonwealth (CW), whether the sources of commonwealth are material, immaterial, commodity (C), or non-commodity (NC). This dual relationship between the association of commoners, as subjects, and their commonwealth, as objects, constitutes the commons (Cs). Through the activity of commoning (cm), the commons are produced and reproduced over time, which ensures their survival.

While this circuit of value differs from capital accumulation circuits, this is not to say that commoners do not interact with capital or the state. Rather, for DeAngelis (2017), circuits of commons value can couple with capital circuits through the commodity form, which is also pictured in the figure. Rather than arguing that these two circuits can or ought to peacefully coexist, DeAngelis simply demonstrates that they do coexist. For example, when commoners must interact with the money form of capital, they do so only as a medium of exchange to gain access to the materials necessary to reproduce the commons and themselves over time. As this relates to the digital commons, free software contributors or users will need access to a computer in order to code the digital commons or access them. In addition, programmers will need to have access to food, water, shelter, and all those things necessary to reproduce their own capacity to code the digital commons over time. These goods may be provided by the welfare state or one's family but, in the absence of such provision, one would need to intersect with capital circuits to obtain them. However, the extent to which commoners engage with capital circuits is left up to the community and will vary depending on the unique situation of each community.

The goal for any commons-based movement, however, is how to ensure that the value created within the community is preserved rather than extracted by capital or the state. Bauwens and Niaros (2017) explore this dilemma through an analysis of value within the commons economy. The authors argue that economic theory is experiencing a "value crisis" in light of the emergent practices of

$$\left.\begin{matrix} & A \\ \left.\begin{matrix}NC\\C\end{matrix}\right\} CW & \end{matrix}\right\} Cs\ldots cm\ldots Cs \left\{\begin{matrix} A & \\ & CW\left\{\begin{matrix}NC\\C\end{matrix}\right.\end{matrix}\right.$$

Figure 3.1 The Circuit of the Commons. Source: DeAngelis, M. (2017). Omnia sunt communia: On the commons and the transformation to post-capitalism. London: Zed Books.

commons-based communities. They argue that whereas value within capitalism is *extractive*, a shift to a *generative* value model would enrich the communities and resources directly involved in production. In response to these concerns, open cooperatives and platform cooperatives (Scholz, 2014) have been developed as organizational forms that seek to directly enrich those involved in production. However, the specific tactics used by open cooperatives to ensure that the value created by their contributors stays within the commons may vary. In other words, rather than imposing a "one-size-fits-all" approach to organizing, the commons framework remains flexible, allowing local communities to make choices about how to govern collective resources. The recognition here is that local communities will have a much more sophisticated understanding of their collective resources, including the institutional, political, cultural, and social contexts within which the resource is situated.

Within commons-based peer production, we also have numerous examples of the ways in which communities negotiate their relationship with capitalist firms. One notable example of how this is done is through the creation of "boundary organizations," which are established to negotiate and establish boundaries between two parties who may have both disparate and shared interests (O'Mahony & Bechky, 2008). Within free and open source software development, communities working on free software projects want to ensure the survival of their software projects and attract other developers to work on the project, and securing corporate sponsorship of a project can be a direct way of attracting more developers to the project (Santos, Kuk, Kon, & Pearson, 2013). However, the community also has an interest in preserving its creative autonomy by not ceding too much influence or power to the corporation. One example of a boundary organization is the Fedora Project Board, which negotiates the boundaries between the Fedora Project as a free software project and Red Hat, Inc. as its corporate sponsor (see Birkinbine, 2017).

These examples illustrate the tension that lies at the heart of peer production and the commons. On the one hand, communities want to ensure creative autonomy for their members as well as the survival of their common resources over time. On the other hand, commons-based peer production provides an intriguing opportunity for capitalist firms looking to extract value from the community's productive activities. Indeed, commons-based peer production has proven to be an efficient and effective model of production, even producing large-scale projects like the Linux kernel, which contains millions of lines of software code (see The Linux Foundation, 2019). When firms want to harness the production power of commons-based communities, they often need to find ways of negotiating access to those communities, particularly because their interests may differ from those of the community.

6 Concluding Remarks

This chapter has provided an overview of the political economy of peer production. I began with an examination of the structural changes within capitalism that enabled the rise of peer production, and I also focused on some of the emergent cultural practices of peer production that subvert prevailing tendencies of capitalism. As such, there is a tension between capital and the commons that is constantly at play in the political economy of peer production. In this sense, it is useful to understand peer production as dialectically situated between capital and the commons, as it can account for the contradictory power of peer production, highlighting the ways that peer production intersects with circuits of capital accumulation but also the ways in which value is created and circulates within peer production communities according to a different logic. On the one hand, the extraction of value from data and information provided *gratis* to corporations has been a boon to capital accumulation schemes – certainly as it pertains to rapid growth and the attraction of speculative financial capital, if not proven formulas for profitability. On the other hand, peer production enables the creation of digital commons that can be shared with an increasingly large online community. The production, maintenance, and reproduction of the commons over time also carries the possibility of ushering in an era where mutual aid, care, trust, conviviality, and the commonwealth are valued more than competition, exploitation, and the maximization of individual or corporate profit.

Note

1 This paragraph provides a basic summary of the argument made by Robert Brenner. For more information, see R. Brenner (2006). *The Economics of Global Turbulence*. London: Verso.

References

Abbate, J. (1999). *Inventing the Internet*. Cambridge, MA: MIT Press.

Andrejevic, M. (2007). *ISpy: Surveillance and power in the interactive era*. Lawrence, KS: University Press of Kansas.

Andrejevic, M. (2012). Exploitation in the data mine. In C. Fuchs, K. Boersma, A. Albrechtslund, & M. Sandoval (Eds.), *Internet and surveillance: The challenges of web 2.0 and social media* (pp. 71–88). New York, NY: Routledge.

Bauwens, M., & Niaros, V. (2017). *Value in the commons economy: Developments in open and contributory value accounting* [PDF file]. Retrieved from www.boell.de/sites/default/files/value_in_the_commons_economy.pdf

Benkler, Y. (2006). *The wealth of networks: How social production transforms markets and freedom*. New Haven, CT: Yale University Press.

Birkinbine, B. J. (2017). From the commons to capital: Red Hat, Inc. and the business of free software. *Journal of Peer Production, 10*. Retrieved from http://peerproduction.net/issues/issue-10-peer-production-and-work/from-the-commons-to-capital/

Boltanski, L., & Chiapello, E. (2005). *The new spirit of capitalism*. London: Verso.

Brenner, R. (2006). *The economics of global turbulence*. London: Verso.

Cassidy, J. (2002). *Dot.con: How America lost its mind and money in the Internet era*. New York, NY: Harper Perennial.

Cohen, N. (2016). *Writer's rights: Freelance journalism in a digital age*. Montreal: McGill-Queen's University Press.

Coleman, E. G. (2013). *Coding freedom: The ethics and aesthetics of hacking*. Princeton, NJ: Princeton University Press.

Colley, J. (2019, January 11). Why 2019 could be the year of another tech bubble crash. *The Conversation*. Retrieved from https://theconversation.com/why-2019-could-be-the-year-of-another-tech-bubble-crash-109468

Dalla Costa, M., & James, S. (1975). *The power of women and the subversion of the community*. Bristol: Falling Wall Press.

DeAngelis, M. (2017). *Omnia sunt communia: On the commons and the transformation to post-capitalism*. London: Zed Books.

Deuze, M. (2007). *Media work*. Cambridge: Polity.

Drucker, P. F. (1994). *Post-capitalist society*. New York, NY: HarperBusiness.

Dyer-Witheford, N. (2006). The Circulation of the Common. Paper presented at Immaterial Labour, Multitudes and New Social Subjects: Class Composition in Cognitive Capitalism conference. University of Cambridge.

Federici, S. (2012). *Revolution at point zero: Housework, reproduction, and feminist struggle*. Oakland, CA: PM Press.

Fuchs, C. (2011, February 1). New media, web 2.0 and surveillance. *Sociology Compass, 5*(2), 134–147.

Fuchs, C. (2012). Critique of the political economy of web 2.0 surveillance. In C. Fuchs, K. Boersma, A. Albrechtslund, and M. Sandoval (Eds.), *Internet and surveillance: The challenges of web 2.0 and social media* (pp. 31–70). New York, NY: Routledge.

Gaither, C., & Chmielewski, D. C. (2006, July 16). Fears of dot-com crash, version 2.0. *Los Angeles Times*. Retrieved from http://articles.latimes.com/2006/jul/16/business/fi-overheat16

Gutiérrez-Aguilar, R. (2014). Beyond the "capacity to veto": Reflections from Latin America on the production and reproduction of the common. *South Atlantic Quarterly, 113*(2), 259–270.

Hardt, M., & Negri, A. (2000). *Empire*. Cambridge, MA: Harvard University Press.

Hardt, M., & Negri, A. (2004). *Multitude: War and democracy in the age of empire*. New York, NY: Penguin.

Hardt, M., & Negri, A. (2009). *Commonwealth*. Cambridge, MA: Belknap.

Harvey, D. (1989). *The condition of postmodernity: An enquiry into the origins of cultural change*. Oxford: Blackwell.

Harvey, D. (2005). *A brief history of neoliberalism*. Oxford: Oxford University Press.

Harvey, D. (2010). *A companion to Marx's capital*. New York, NY: Verso.

Hitlin, P., & Rainie, L. (2019, January 16). Facebook algorithms and personal data. *Pew Research Center*. Retrieved from www.pewinternet.org/2019/01/16/facebook-algorithms-and-personal-data/

Huws, U. (2014). *Labor in the global digital economy: The cyberteriat comes of age*. New York, NY: Monthly Review Press.
Jarrett, K. (2016). *Feminism, labour and digital media: The digital housewife*. New York, NY: Routledge.
Lazzarato, M. (1996). Immaterial labor. In P. Virno and M. Hardt (Eds.), *Radical thought in Italy: A Potential Politics* (pp. 133–150). Minneapolis, MN: University of Minnesota Press.
Levy, S. (2010). *Hackers: Heroes of the computer revolution*. Sebastopol, CA: O'Reilly Media.
Linebaugh, P. (2008). *The magna carta manifesto*. Berkeley, CA: University of California Press.
Marx, K. (1906). *Capital: A critique of political economy, Vol. 1*. New York, NY: Modern Library.
Mason, P. (2015). *Postcapitalism: A guide to our future*. New York, NY: Farrar, Straus and Giroux.
McGuigan, L., & Manzerolle, V. (Eds.) (2013). *The audience commodity in the digital age: Revisiting a critical theory of commercial media*. New York, NY: Peter Lang.
McRobbie, A. (2016). *Be creative: Making a living in the new culture industries*. Cambridge: Polity.
O'Mahony, S., & Bechky, B. A. (2008). Boundary organizations: Enabling collaboration among unexpected allies. *Administrative Science Quarterly*, *53*(3), 422–459.
O'Reilly, T. (2005, September 30). What is web 2.0. *O'Reilly.com*. Retrieved from www.oreilly.com/pub/a/web2/archive/what-is-web-20.html
Prashad, V. (2007). *The darker nations: A people's history of the third world*. New York, NY: The New Press.
Ryan, A. B. (2013). The transformative capacity of the commons and commoning. *Irish Journal of Sociology*, *21*(2), 90–102.
Santos, C., Kuk, G., Kon, F., & Pearson, J. (2013). The attraction of contributors in free and open source software projects. *Journal of Strategic Information Systems*, *22*(1), 26–45.
Schiller, D. (1999). *Digital capitalism: Networking the global market system*. Cambridge, MA: MIT Press.
Schiller, D. (2014). *Digital depression: Information technology and economic crisis*. Urbana, IL: University of Illinois Press.
Scholz, T. (Ed.). (2013). *Digital labor: The Internet as playground and factory*. New York, NY: Routledge.
Scholz, T. (2014, December 5). Platform cooperativism vs. the sharing economy. Medium. Retrieved from https://medium.com/@trebors/platform-cooperativism-vs-the-sharing-economy-2ea737f1b5ad
Singh, N. (2017). Becoming a commoner: The commons as sites for affective socio-nature encounters and co-becomings. *Ephemera: Theory & Politics in Organization*, *17*(4), 751–776.
Smythe, D. W. (1981). *Dependency road: Communications, capitalism, consciousness, and Canada*. Norwood, NJ: Ablex.
Srnicek, N., & Williams, A. (2016). *Inventing the future: Postcapitalism and a world without work*. New York, NY: Verso.
Terranova, T. (2004). *Network culture: Politics for the information age*. Ann Arbor, MI: Pluto.
The Linux Foundation. (2019). 2017 State of Linux Kernel Development. *The Linux Foundation*. Retrieved from www.linuxfoundation.org/2017-linux-kernel-report-landing-page/
Turow, J. (2013). *The daily you: How the new advertising industry is defining your identity and your worth*. New Haven, CT: Yale University Press.
Ulick, J. (2000, April 3). Nasdaq sinks 350 points. CNN Money. Retrieved from https://money.cnn.com/2000/04/03/markets/markets_newyork/
United States v. Microsoft, 253 F.3d 34 (D.C. Cir. 2001).
Wang, A. X. (2019, February 6). Spotify turns a profit for the first time – and reveals its bigger ambitions. *Rolling Stone*. Retrieved from www.rollingstone.com/music/music-news/spotify-fourth-quarter-profit-790725/

4

Social Norms and Rules in Peer Production

Christian Pentzold

1 Introduction

The regulation of peer production projects is usually achieved by the users themselves. Ideally, this self-organization and self-management depend on shared social norms and rules. Some of these institutional arrangements characterize the larger population of peer production projects, whilst others are the attribute of particular projects. The chapter provides an overview and comparison of peer production's institutions, traces their origins, and describes their implications for cooperation and governance.

Peer production is often advertised as a truly alternative form of providing information goods and services, one that is less constraining than hierarchical firms or contractual markets. Hence, voluntary participation and freely available outcomes are taken to epitomize a novel form of libertarian cooperation unfettered by rules and restrictions. This kind of ideal setting of "organizing without organizations," as Clay Shirky (2008) described it, should be characterized by egalitarianism, communal evaluation, and flat, fluid, or even absent hierarchies so as to maximize individual freedom and autonomy (Bruns, 2008; Stevenson, this volume). Peer production does, therefore, not only rest on the freedom of membership and commons goods but is believed to also be freedom-producing (Tkacz, 2015; see also Borschke, this volume; Dulong de Rosnay, this volume). It thus produces free information and services, but also ought to enhance the practice and experience of generosity, altruism, and comradery (Benkler & Nissenbaum, this volume; Firer-Blaess & Fuchs, 2014).

Ideas of independence, voluntarism, and equality, deliberation and self-amending procedures are deeply ingrained in the rhetoric and culture of peer production (Fish et al., 2011; Jemielniak, 2016; see also Bauwens & Kostakis, this volume; Dafermos, this volume). When considering the actual interactions in projects, these claims seem vastly exaggerated: Free and open source software initiatives like Debian have, for instance, devised extensive ranks and procedural schemes to steer contributors. In Wikipedia, editors complain about cliques and the power plays of an oligarchy of users and they struggle with bureaucracy and formalized rules.

Addressing this discrepancy, it would be too easy to just dismiss the discourse around the virtues of peer production as an ill-founded pipedream that never was in touch with its reality. Instead of decrying the apparent gap between the rationale of peer production as an alternative mode of cooperation and its imperfect practical accomplishment, it is more productive to study the processes through which organizational forms and institutions come into being. That way, the consolidation of norms, codes of conduct, and contributor positions does not arise as an inadvertent setback but can be approached as an inevitable and even requisite development. On these terms, Kreiss, Finn, and Turner (2011) refer to Max Weber's classical sociological thought in order to dismiss the idea that peer production must stand against bureaucracy. Indeed, bureaucracy can protect against arbitrary rule. Loose hierarchies or the absence of rules are, therefore, not per se more egalitarian or democratic because "regulations hold the whims of individual members in check," Kreiss et al. (2011, p. 251) argue. The authority of rules and formalized workflows do not automatically preclude peer production or stifle creativity but are

The Handbook of Peer Production, First Edition. Edited by Mathieu O'Neil, Christian Pentzold, and Sophie Toupin.

necessary to make it happen. Anarchy has, it seems, never been an option, even though the ordering of peer production projects must not necessarily take the shape of a formalized bureaucracy (O'Neil, 2014). In fact, a key characteristic of peer production institutions is that they are openly debated by participants on mailing lists in Debian or on wiki pages as in Wikipedia.

Institutions enable the conversion of the freedom gained from a number of legal and managerial restrictions into a freedom achieved for productive participation (Berlin, 1969). This chapter charts and discusses rules and norms as potentially conducive "forms of closure" (Tkacz, 2016, p. 33) in peer production. This sounds oxymoronic, yet it points to a fundamental friction between, on the one hand, the ambition to open forms of cooperation that are less restricted by property rights or hierarchical structures and, on the other, the need to establish rules that close activities set to violate this openness.

The overview offered in this chapter starts from the idea that voluntary, commons-based cooperation needs some kind of ordering and that institutions can help to organize the freedom of action found within peer production. Following the social constructivism of Berger and Luckmann (1966), the chapter assumes that institutionalization happens whenever people interact in a more habitual manner. Institutions such as norms and rules thus hint at the typification, habitualization, and ordering of social action. De-institutionalization, in turn, occurs through the transition of rules or norms into new contexts or when their original setting changes. Regulations can also be altered by strategic efforts. Furthermore, institutions might be transformed due to conflicting rulings or dissenting normative expectations (Oliver, 1992; Zucker, 1977). For instance, in her study of the Debian community of programmers, Coleman (2013) finds what she calls moments of "punctuated crisis" (p. 124). There, the conflicting modes of governance of democratic majoritarian rule, a guildlike meritocracy, and an ad hoc process of rough consensus clash and prompt negotiations and new rulings.

The chapter is organized along a set of questions. The second part asks why peer production needs any sort of rules and norms. Based on this discussion of institutional conditions, the third part will map what sorts of rules and norms prevail in peer production. It delineates the three institutional levels of policies, guidelines, and basic normative understandings that are geared towards the products and the processes of peer production. Next, the fourth part examines how rules and norms come into existence and are made to function. Finally, the chapter reflects to what end peer production's institutions congeal into governance regimes, bureaucracy, and hierarchies.

The chapter mostly refers to Wikipedia and free and open source software projects such as Linux and Debian (see also Couture, this volume; Haider & Sundin, this volume). These paragons showcase the power and potential of peer production. They are taken as reference cases since these mature initiatives provide rich information about the long-term processes of building and transforming institutions for voluntary and free cooperation.

2 Why Rules and Norms? Institutional Conditions for Peer Production

Rules and social norms play an important part in organizing peer production. Given the anonymity, ease of entry, and limited social cues which characterize peer projects, this can be a challenging mission (Resnick & Kraut, 2011). In that respect, institutions can be seen as an answer to the question of "how to steer the integration of dispersed knowledge resources and how to coordinate such activities to the purpose of creating common value" (Aaltonen & Lanzara, 2015, p. 1650). They provide for the collective capability of participants separated in space and time to bundle together the piecemeal contributions and direct them towards the joint production of a valuable outcome. In order to continuously create and ensure collective agency, rules and social norms can rarely be one-time solutions. They have to be adapted to the dynamics of a particular project, for instance, in terms of new user cohorts, content growth, or a changing technological and legal environment.

Scott (2001) defines institutions as "multifaceted, durable, social structures, made up of symbolic elements, social activities, and material resources" (p. 49). Institutions are created because of "the development, recognition, and naming of a recurrent problem to which no existing institution provides a satisfactory repertoire of responses" (Scott, 2001, p. 96). In peer production institutions take the shape of formal regulations, informal social norms, as well as shared actions and evaluations. Furthermore, institutions are encoded requirements and courses of activity embedded in software and hardware (de Laat, 2007; Markus, 2007; O'Mahony & Ferraro, 2007; Schroeder & Wagner, 2012). The institutional ecology of cooperation is established in order to protect the integrity of projects.

Most notably, the idea of freedom is enshrined in licenses such as the General Public License or the Creative Commons License (Lakhani & von Hippel, 2003; Lessig, 1999; Raymond, 1999; Stewart & Gosain, 2006; Vieira & de Filippi, 2014; see also Dulong de Rosnay, this volume). In projects such as the Linux kernel operating system, they guarantee some of the pivots of peer production and safeguard its essential modes of operation that rest on sharing, copying, adapting, and disseminating incremental contributions. This backbone is charged with a normative impetus, namely, that proprietary software "is antisocial, that it is unethical, that it is simply wrong," as Richard Stallman (2004) exclaimed. This again rests on the moral idea of a hacker culture where all information should be free and decentralization ought to be promoted so as to change life for the better (Levy, 1984).

Institutions are an ill-defined category. Following Scott (2001) we can distinguish "three pillars of institutions" (p. 51). They encompass regulations, norms, as well as conventional orders of knowledge and action (DiMaggio & Powell, 1991). Regulative rules come as provisions, statutes, laws or decrees that specify how an activity must be executed. They imply monitoring and enforcement through sanctions or gratifications that either penalize deviant behavior or reward compliance. In peer production projects, rules are vital resources to draw on in order to justify as well as to challenge a decision or action (Bryant, Forte, & Bruckman, 2005; Lakhani & Von Hippel, 2003; Pentzold, 2017, 2018; Viègas, Wattenberg, & Dave, 2004; Viègas, Wattenberg, & McKeon, 2007). Rules crystallize informal conventions or implicit standards in tangible written form and prompt what Giddens (1984) has called an instrumental behavior. It can mean conformity, yet other responses such as prevention, defiance, or forms of gaming the system might also pose viable options.

The application of formal regulations can be backed up by normative demands. Norms can be seen, according to Scott (2001), as "conceptions of the preferred or the desirable, together with the construction of standards of which existing structures or behavior can be compared and assessed" (p. 54f.). They usually take shape in maxims, sayings, or moral doctrines like the dictum that "All information should be free," as documented by Levy (1984, p. 40) in his account of the hacker ethic. Because peer production projects only have a limited potential to enforce rules or prosecute wrongdoers due to their voluntary and open nature, they very much rely on informal norms (Raymond, 1999; Stewart & Gosain, 2006). Norms ideally do not need external sanctions because they encompass an obligatory moral request that takes effect by way of internalized commitment. In Max Weber's (1978) terms, norms feature the "prestige of being considered binding, or, as it may be expressed, of 'legitimacy'" (p. 31). Thus, while rules request an instrumental logic of individual interests and alternative ways to act, some of which are permitted while others are not, norms engender a logic of acceptability that is oriented towards an accordance of action and expectations (March & Olsen, 1989).

Both rules and norms rest on collective forms of knowledge and shared mindsets which endow them with sense and meaning. These cultural frameworks of wider belief systems as well as repetitive patterns of action often remain tacit. As such, they "not only constrain options: they establish the very criteria by which people discover their preferences" (DiMaggio and Powell, 1991, p. 11). Similarly, Hall and Taylor (1996) suggested that "institutions influence behavior not simply by specifying what one should do but also by specifying what one can imagine oneself doing in a given context" (p. 948). This horizon of thought and agency cannot be reflected upon as a whole. The "way we do things" escapes instrumental efforts to efface, add, or change particular elements as it unfurls in long-term processes of habitualization. This happens, as Berger and Luckmann (1966) stated, when an action is repeatedly performed. It becomes instituted as an expected pattern which is reproduced in further activity and as such then becomes the matter of codification, instruction, and reinforcement. The Code of Conduct adopted by the Debian project in 2014 demonstrates this dynamic. It sets down a number of basic principles. For instance, it includes the request to be respectful, cooperative, and concise.

As a social imaginary, suggests Charles Taylor (2002), basic normative understandings are "not a set of ideas; rather it is what enables, through making sense of, the practices of a society" (p. 91). Compliance does, therefore, occur because other kinds of thinking or acting are quite inconceivable (Scott, 2001; Zucker, 1977). Consequently, even when peer production projects do not feature a written code of conduct, accepted contributions will still follow an implicit cultural code. With a distinction made by John Searle (1997) we can say that work in peer production might, at times, not be guided by regulatory rules of commands and interdictions, but they are anchored by constitutive rules of sensemaking and evaluation.

The three institutional pillars can only be separated for analytical purposes. In practice, they are entangled: prescriptive rules go along with normative creeds which again only make sense against a background of taken-for-granted understandings. In peer production, institutions define what are considered to be correct forms of participation. They justify different positions among contributors and their stratified rights and obligations to execute decisions. They also establish authority among the stratified project participants and they account for their license to employ or manipulate technological levers.

It is important to note that the software and hardware of peer production enclose an institutional dimension *sui generis* (Lessig, 1999). As virtually all projects rest on a technological infrastructure, design implications are essential instruments that configure the agency of contributors. The programs and algorithmic procedures set down in code are used in order to materialize regulations and social norms by way of organizing access to technical features as well as to the programming facilities themselves (Kesan & Shah, 2005).

Studying the institutional conditions in Wikipedia, Butler, Joyce, and Pike (2008) distinguished a set of perspectives among editors. One way of defining institutions was as rational efforts to achieve consistent and reliable decisions and to codify role positions and duties. From a different perspective, they were taken to represent evolving, competing entities which propagated themselves: rules generate more rules. Another view accentuated the construction of meaning and identity that defines the character and ambition of the project. Institutions were furthermore framed as external signals that indicated to audiences or users not actively participating that the project attends to problems, and finds ways to address them. They were also regarded as being internal signals that raise awareness for topics or perspectives and draw boundaries of a project's inside and outside. Wikipedia's set of rules, norms, and basic understandings was specified in terms of negotiated settlements and trophies that mark the end of conflict, or to signal binding consensus. Finally, institutions also served as control mechanisms set in place to ensure appropriate action.

Overall, institutions in peer production help to order the dispersed engagement of volunteers and to bring together meaningful, valuable outcomes. They should facilitate, not suppress productive engagement. In open projects with unsolicited membership and the constant exit option, we often find a plastic interpretation of institutions, not a strict enforcement though there are also institutionalized forms of sanctioning and ostracizing users. The flexible handling of rules and social norms pertains to the interests and agendas of the users involved. Hence, institutions in peer production are sites of conflict and specification and thus form part of editorial power plays (Kriplean et al., 2007). Besides, they are a matter of socialization and instruction (Viègas et al., 2007). In this regard, Gabriella Coleman (2013) referred to an "ethical enculturation" (p. 124) she encountered among the contributors to the free and open source software project Debian. In order to stay relevant and mirror the requirements and concerns of the users, the institutions had to be actively practiced and passed on to incoming participants.

3 What Rules and Norms? Policies, Guidelines, and Basic Understandings in Peer Production

Institutions in peer production can be classified into those oriented towards products and those centering on work processes. Thus, they encompass, on the one hand, content standards about the form and quality of the generated and delivered goods and services. On the other, they include interactional standards and procedural standards that arrange the cooperation among project members. Reflecting the institutional register of regulations, norms, and basic understandings, peer production projects have formulated cognate distinctions with different degrees of authority, from axiomatic principles and enforced rules to advice or cues. They either prescribe, explain, or suggest correct forms of conduct and valid contributions to the project, respectively. In sum, they define a more or less strict scope of activity. This incorporates, along a decreasing level of exigency and validity, actionable rules, moral tenets, and non-binding musings.

Somewhat exemplary for a number of other peer production efforts, the English-language Wikipedia features three levels of policies, guidelines, and essays. The online encyclopedia rests on a core set of obligatory "five pillars" that originate from the beginnings of the project in 2001. One precept determines "What Wikipedia is not," and thus the content scope of the articles. A second specifies the so-called "neutral point of view." It demands authors to represent all significant views fairly, proportionality, and without bias. The third principle, "Wikipedia is free," states the copyright status of the project that allows anyone to edit, use, modify, and distribute. "Civility" as the fourth axiom reminds the contributors to respect each other. This includes the policy to "Assume Good Faith," which requires editors to treat and think of others well. Hence, this catalyst for cooperation works, Reagle (2010) explained, thanks to the "dovetailing of an open perspective on knowledge claims (epistemic) and other contributors (intersubjective)" (p. 161). To this end, the authors are framed as being cooperative, goodwill contributors striving towards productive joint work. In the English Wikipedia, the set of fundamental ideas is completed with the call: "Be bold." That way, the authors hope to account for the evolving character of their trade. Participants should first of all aim to improve Wikipedia which might then also mean to scrutinize, adapt, or suspend existing policies and guidelines.

Although these five rules are marketed as being central and unchangeable, they nevertheless vary to some extent from one language version to another. For instance, in the German edition, the fifth maxim is missing. Further core institutions like the wikiquette, which has not yet congealed in a Wikipedia code of conduct as in Debian, attended to the conduct among users and therefore to the desirable manners and forms of social interaction. They required Wikipedians to be nice to each other, to be honest, and to abstain from legal threats. Other policies, guidelines, and essays dealt with, for example, the resolution of disputes, the organization of editing, and the handling of vandalism. They rest on regulatory practices that date back to the time of early Usenet applications and mailing lists (Baym, 1996; Sternberg, 2012).

In sum, they form the "Wikipedia policy environment" (Morgan & Zachry, 2010, p. 165), that has evolved in reaction to the editors' need to handle emerging contingencies. The majority of policies tended to focus on process and legal issues whereas guidelines often dealt with content matters, and essays were mostly dedicated to user behavior. Like in other peer production ventures, Wikipedia's "genre ecology" (Morgan & Zachry, 2010, p. 165) of policies, guidelines, and essays is mushrooming in character (Butler, Joyce, & Pike, 2008; Halfaker et al., 2013). In January 2008, Morgan and Zachry (2010) sampled 47 policies, 232 guidelines, and 404 essays. In his survey of March 2009, Reagle (2010a) found 686 pages in page categories relevant for organizing collaboration with 104 proper rule pages. Five years later, Jemielniak (2016) collected more than 1,200 regulatory documents in the English-language edition and counted 150,000 words in the 50 main policies.

The three groups of Wikipedia institutions provide no clear-cut and exclusive order but rather are a "helter-skelter hodgepodge" (Kostakis, 2010) with different levels of validation, integrity, irreversibility, and legitimacy. The extensive and somewhat haphazard corpus of provisions, conventions, and personal advice or opinions seeks to stipulate aspects of working on and in Wikipedia (Tkacz, 2016). Some purvey concrete descriptions of procedures and rationales for making decisions, others instead are composed of formulaic mottoes without tangible directives. They are complemented by a repository of software utilities created for aligning the content and patrolling contributions according to the standards in place. A vast number of robots or bots are, for example, maintained in order to handle menial tasks like fixing broken links or correcting typos. In Wikipedia's "sophisticated technomanagerial system" (Niederer & van Dijck, 2010, p. 1373), these tools, in combination with semi-autonomous editing interfaces, increasingly support the enforcement of rules, for instance, when they facilitate the detection and reversion of vandalism (Geiger & Ribes, 2010). In this way, technological assistants assume an increasing share of the contributions and also sustain the implementation of institutions. In July 2017, bots made about 20 per cent of all edits (Geiger, 2017). Referring to the technical protocols that steer the activity of bots, Müller-Birn, Dobusch, and Herbsleb (2013) spoke of an "algorithmic governance" (p. 80) sustaining peer production.

4 How to Create Rules and Norms? Institutional Work in Peer Production

There is no single way of establishing institutions in peer production. Given the participatory and emerging nature of its endeavors, users stress the idea that regulations and normative stances have to reflect established consensus and routines: practice reifies in policy. In effect, institutions should be linked back to changing consensus about the aims and scope of a project, conventional ways to contribute, or shifting values. Rules also ought to respond to urgent issues that require regulation. In this respect, a Wikipedia manual for example declared that "policies and guidelines are typically altered to reflect changing practice on the site or to solve a problem that has arisen" (Ayers, Matthews, & Yates, 2008, p. 369). Rules and social norms codify the status quo of practice and sensemaking in peer production and provide them with an obligatory force.

These purposive actions of creating and implementing institutions can be understood as a kind of "institutional work," that is, in the words of Lawrence, Suddaby, and Leca (2009), "practical actions through which institutions are created, maintained, and disrupted" (p. 1). In peer production, the construction of rules, the formulation of social norms, and the disposition of more widespread institutional understandings should be an ongoing, inclusive, and open-ended process. It runs through different stages from informal, local, and ad hoc rulings to more long-term and holistic determinations (Kriplean et al., 2007). In ideal form, institutions are created in proposals which are then discussed and modulated until a consensus can be reached among the contributors. If commonly accepted, they become codified both in texts as well as in software which then direct their interpretation and enforcement. In this regard, institutions usually are "socially constructed, routine-reproduced," as Jepperson (1991, p. 149) held, despite instrumental schemes to install particular regulations or to exact a ruling deemed official and binding. In principle, the formulation and adaptation of institutions is conceived of as an incremental bottom-up process.

Rules in peer production typically arise from many individual contributions. Their repute stems from the alleged broad inclusion of perspectives, the user acceptance of the process in which the institutions are configured and implemented, and the concrete provisions of what to do in terms of promoting productive cooperation and social interaction. In written form, they are shared and can be inspected and revised if need be. Yet rightly because the process and its outcomes are designed to be open and integrative, there is no linear development from heuristics to negotiated formal rules. Instead, the contributors uphold an enduring debate. The contentious struggles about the adequate form, scope, and application of institutions originates in the fluctuating user base with veterans leaving projects while new participants enter. These shifting populations, in connection with the fluid formation of the projects, propel continuous discussions around the organization of cooperation and the inspection of project products.

In order to compensate for this kind of friction, peer production collectives have proceduralized and decentralized the creation and enforcement of institutions (Beschastnikh, Kriplean, & McDonald, 2008; Jemielniak, 2016). For instance, in Wikipedia, Forte, Larco, and Bruckman (2009) discovered, on the one hand, a formalization of rule-making and, on the other, tendencies to decentralize their implementation. As the project matured, the formation of governance mechanisms was refined and assigned to institutionalized bodies. Hence, in the early Wikipedia conflicts among users were usually resolved according to fiat with the project founder Jimmy Wales acting as ultimate "benevolent dictator." Over time, a special arbitration committee assumed many of these competencies. It follows a set course of evidence and deliberation in order to make decisions. These procedures were reproduced in separate projects and sub-projects of Wikipedia. Therefore, the effort to apply rules was decentralized and spread across a broader set of contributors. Decentralization, therefore, helped to reflect community concerns and to involve a growing editor base in these tasks.

However, the increasing standardization of activities implies the decelerating creation of new policies whereas the interpretation of existing rules rests with senior members, rendering them less open to reversion (Kriplean et al., 2007; Morgan et al., 2012). "Calcification of policy is disproportionately felt by newer editors, who see their policy edits rejected at a higher rate," hence concluded Halfaker et al. (2013, p. 683). In the English-language version of Wikipedia, the growth of policies began to slow down in 2006 (Forte et al., 2009). In 2005, 217 new policies were proposed, 12 of which were accepted. In 2011, this number had dwindled to only 16 new proposals while the acceptance rate of 7.5 per cent remained stable (Halfaker et al., 2013). In contrast, institutional innovation has switched to essay format. These less restricted and binding documents responded to an increasing bureaucratization of the overall policy-making procedures and supported a kind of soft regulatory mechanism. So the number of essays had increased from 69 in 2005 to more than 185 in 2006 and in the following years (Morgan & Zachry, 2010).

5 Whither Rules and Norms? Governance, Hierarchies, and Bureaucracy in Peer Production

Rules, social norms, as well as basic understandings and values constitute an essential element in the governance of peer production. Governance can be understood, following Markus (2007), as "the means of achieving the direction, control, and coordination of wholly or partially autonomous individuals" (p. 152). In peer production, governance does not preclude autonomy, but presupposes free will and a non-contractual nature of cooperation. On this note, O'Mahony (2007) defined it as a "community-managed governance model" (p. 144). Governance manifests in the independence from commercial organizations, a pluralism of interests and solutions, decentralized decision-making, and the prospect of autonomous choice.

For free and open source software, governance includes the free licensing of the software provided and usually of the technological infrastructure used for collaboration. It also encompasses the signaling and allocation of tasks to be completed, the inspection and selection of contributions, and the control of release versions (Markus, 2007). Besides written rules and normative creeds, governance rests on code features such as the concurrent versions system (CVS) used to keep track of changes to a file. The source code management system therefore helps to control software revisions. Developers also check the file transfer protocol (FTP) for the transmission of files between clients and servers on a computer network. Governance also covers the communication among project members in chats, via mailing lists, or forums. It deals, for instance, with bug reports, votes, or the communicative sanctioning of troublemakers like flaming, shunning, or kill-filing, that is, automatically discarding posts from particular users.

A core concern of peer production governance are hierarchies. Despite their egalitarian ambitions, structures of authority and influence have loomed in most projects (Dahlander & O'Mahony, 2011; de Laat, 2007; O'Neil, 2009; Weber, 2004). They frequently rest on seniority and thus on the length of continuous participation, as well as on the volume of engagement. Along these criteria, many ventures exhibit a loose onion-like segmentation into core contributors, regular or occasional contributors, and users or beta-testers (Raymond, 1999). Peer production relies on voluntary support and personal commitment so that "heavy-handed control can deter participation" (Shah, 2006, p. 108). In other words, leadership does not rest on formal command positions. Rather, authority is a matter of meritocratic achievement and acceptance (von Krogh, Spaeth, & Lakhani, 2003; Stewart, 2005). It presupposes a number of qualities such as communicative skills, the ability to recognize valuable contributions as well as the competence to set goals, control tasks, and motivate fellow users.

The role model for this type of "benevolent dictator" (Weber, 2004, p. 89f.) is Linus Torvalds who founded Linux. In the project, he assumed a gatekeeper position from which he was able to monitor the release of official versions of the software. Being aware that his authority depended on the goodwill and appreciation of participants, he "never orders anyone to do anything and even his suggestions are mild-mannered" (Moon & Sproull, 2000). Such attitude and diligence are, however, precarious and nothing that can be taken for granted. Hence, Torvalds himself has been criticized for departing from this ideal by becoming more restrictive and imperious over time.

In their survey of governance styles in free and open source software, Shaikh and Henfridsson (2017) call this kind of governance autocratic clearing. "Autocratic clearing," they explicate with the example of the Linux kernel, "is a system of management with singular coordinating points that obliges other actors to channel all work and decisions through a central 'clearing house' before accomplishment" (p. 124). Other modes of ordering they examined involved a more stratified regime of decision-making and peer review of segments contributed to the Linux code. Next to such "oligarchic recursion" (p. 125) they found evidence for a more autonomous form of self-governance where branches broke away from the main project. Yet another type of governance process embodied the negotiation of mutually agreed decisions considering divergent viewpoints, what Shaikh & Henfridsson (2017) have named "meritocratic idea-testing" (p. 127).

The governance arrangements organizing cooperation in peer production and its outcomes are not fixed but evolve over a project's life-cycle. This change reflects the influx of new users and the growth of the project. It also indicates transformations in the institutional environment of a project and in the relations between contributors (Mateos-García & Steinmueller, 2008). For Debian, O'Mahony and Ferraro (2007) describe four phases of governance. Initially, the collective enterprise was organized by de facto governance. It was based on autocratic rule without any consistent involvement of the contributors. In the next phase of designing governance, a hierarchy of positions was devised that should exist apart from the actual people assuming a certain role. This plan was then set to work in the phase of implementing governance. It demanded that the execution of positional powers had to rest on the commission of fellow developers. The final stage of a stabilizing governance was achieved when elections were successfully held in order to transfer a mandate from one user to another (Schweik & Englisch, 2012).

Along these steps, Debian as a mature free and open source software project has established a "Social Contract" that codifies key principles of cooperation and the normative expectations undergirding the users' collaboration. The responsibilities and the structural relationships of project governance are furthermore defined in a Debian Constitution and there are elections for project leadership. The many consecutive versions of this document show that it is not locked up but is an element of ongoing debate, and subject to amendments (Sadowski et al., 2008).

In Wikipedia, what has been devised as a fluid and transparent process risks petrifying into quite rigid bureaucratic configurations with counter-productive consequences. Instead of facilitating cooperation, standardization and hierarchization block new users and hinder user retention. They complicate decision-making procedures, promote elitism, and lead to an organizational deadlock that defies the ideas of openness and inclusivity. These implications of bureaucratization as more of a burden than a driver of peer production have especially been studied for Wikipedia (Butler et al., 2008; Carr, 2011; Halfaker et al., 2013). Despite early accounts of an ad hoc form of non-hierarchical governance (Konieczny, 2009, 2010), in many language versions hierarchies have tended to become more pronounced (Shaw & Hill, 2014). In line with this finding, the number of links to policy pages in Wikipedia discussions has increased, in particular in reference to the rules about signing individual posts, the use of reliable published sources, and the neutral point of view (Beschastnikh et al., 2008). While the citation of policies has risen, their creation slowed down (Forte & Bruckman, 2008). The attention towards existing rules thus went hand-in-hand with the diminishing flexibility to change these rules and declining revision activity (Keegan & Fiesler, 2017).

6 Conclusion

The organization and governance of peer production projects has to balance the independence of individual contributors and the interdependence of a collective endeavor. In-between the promise of autonomy and the implementation of due process, peer production projects struggle with the suitable level of provision and tolerance (Kostakis, 2010). Initiatives have fed on principles of free choice and increasing individual agency. Yet they are challenged by the reality of an increasingly vast and pervasive texture of rules and requirements which should secure the accuracy and value of the resource and respond to societal expectations.

In this ongoing transformation, standards in Wikipedia have, for instance, changed from "folksy 'description of how we do things'" to "punitive 'the way one must do things because otherwise they will punish me'," Lih (2009, p. 224) claimed. Being aware of these tendencies of increasing formalization, the editors of the online encyclopedia have also devised the somewhat paradoxical rule to "Ignore all rules." It states that if a rule hinders users from improving Wikipedia it should be willfully ignored. The call to be bold and edit thus seeks to balance mushrooming specifications with the vital commitment to openness. Interestingly, this principle is not found in all Wikipedia language versions, which points to the heterogeneity of the arrangements prevalent among the myriad endeavors that make up peer production. There are forms of isomorphism where different projects follow connate social norms and rules as well as distinct institutional settings peculiar for an individual enterprise.

A blueprint for the future design and implementation of peer governance is arguably provided by Elinor Ostrom's (1990) principles of successful common-property regimes. They are a key reference for analyzing the provision and allocation of shared resources in "settings in which the individuals involved have exercised considerable control over institutional arrangements" (p. 61). Sustainable, long-term settings rest on the congruence between institutions and local conditions. They are marked by collective choice arrangements where people affected by rulings should also participate in their formulation and enforcement. This kind of reciprocal responsibility also ought to foster the monitoring and conflict resolution mechanisms. Graduated sanctions allow community members to adapt penalties in accordance with the severity of wrongdoings. Ostrom furthermore points to the need for local enforcement where communal jurisdiction is recognized by external authorities, and she underscores the importance of multiple layers of organization and governance. In effect, commons-based projects form a type of nested enterprises where productive participation is arranged at different levels that reach from small circles to the entire community.

With these directions for successful mutual and inclusive governance beyond markets or firms, Ostrom's work on property-rights regimes seems to be a natural fit. Indeed, many peer production projects have sought to emulate some of her advice (Aaltonen & Lanzara, 2015; Forte & Bruckman, 2008; Keegan & Fiesler, 2017; Kiesler et al., 2011; Pentzold, 2011; Viègas et al., 2007). Yet while Ostrom deals with the allocation and provision of scarce natural resources, peer production mostly centers on information goods whose use is nonrival (Hess & Ostrom, 2011). In consequence, urgent issues revolve around the creation of common resources, not their consumption. This is unfinished business. In order to expand and reinforce the collective capability of contributors, peer production's institutions have to be continuously aligned with the dynamically unfolding conditions and requirements of its projects.

References

Aaltonen, A., & Lanzara, G. F. (2015). Building governance capability in online social production. *Organization Studies, 36*(12), 1649–1673.

Ayers, P., Matthews, C., & Yates, B. (2008). *How Wikipedia works. And how you can be part of it.* San Francisco, CA: No Starch Press.

Baym, N. (1996). The emergence of community in computer-mediated communication. In S. G. Jones (Ed.), *Cybersociety* (pp. 138–163). London: Sage.

Berger, P., & Luckmann, T. (1966). *The social construction of reality.* New York, NY: Penguin.

Berlin, I. (1969). *Four essays on liberty.* Oxford: Oxford University Press.

Beschastnikh, I., Kriplean, T., & McDonald, D. W. (2008). Wikipedia self-governance inaction. *Proceedings International Conference on Weblogs and Social Media*, Seattle, WA (pp. 1–9).

Bruns, A. (2008). *Blogs, Wikipedia, Second life, and beyond.* New York, NY: Peter Lang.

Bryant, S. L., Forte, A., & Bruckman, A. (2005). Becoming Wikipedian. *Proceedings 2005 International ACM SIGGROUP Conference on Supporting Group Work*, Sanibel Island, FL (pp. 1–10).

Butler, B., Joyce, E., & Pike, J. (2008). Don't look now, but we've created a bureaucracy. *Proceedings SIGCHI Conference on Human Factors in Computing Systems*, Florence, Italy (pp. 1101–1110).

Carr, N. (2011). *The shallows.* New York, NY: W. W. Norton.

Coleman, G. (2013). *Coding freedom: The ethics and aesthetics of hacking.* Princeton, NJ: Princeton University Press.

Dahlander, L., & O'Mahony, S. (2011). Progressing to the center. *Organization Science, 22*(4), 961–979.

de Laat, P. (2007). Governance of open source software. *Journal of Management & Governance, 11*(2), 165–177.

DiMaggio, P. J., & Powell, W. W. (1991). Introduction. In W. W. Powell & P. J. DiMaggio (Eds.), *The New Institutionalism in Organizational Analysis* (pp. 1–38). Chicago, IL: University of Chicago Press.

Firer-Blaess, S., & Fuchs, C. (2014). Wikipedia: An info-communist manifesto. *Television & New Media, 15*(2), 87–103.

Fish, A., Murillo, L., Nguyen, L., Panofsky, A., & Kelty, C. (2011). Birds of the internet. *Journal of Cultural Economy 4*(2), 157–187.

Forte, A., & Bruckman, A. (2008). Scaling consensus: Increasing decentralization in Wikipedia governance. *Proceedings of Hawaiian International Conference of Systems Sciences (HICSS)*, Waikoloa, HI (pp. 1–10).

Forte, A., Larco, V., & Bruckman, A. (2009). Decentralization in Wikipedia governance. *Journal of Management Information Systems, 26*(1), 49–72.

Geiger, S. R. (2017). Beyond opening the black box. *Big Data & Society*, July–December 2017, 1–14. doi:10.1177/2053951717730735.

Geiger, S. R., & Ribes, D. (2010). The work of sustaining order in Wikipedia. *Proceedings ACM Conference on Computer Supported Cooperative Work*, Savannah, GA (pp. 117–126).

Giddens, A. (1984). *The constitution of society*. Cambridge: Polity.

Halfaker, A., Geiger, S. R., Morgan, J., & Riedl, J. (2013). The rise and decline of an open collaboration system. *American Behavioral Scientist, 57*(5), 664–688.

Hall, P. A., & Taylor, R. C. R. (1996). Political science and the three new institutionalisms. *Political Studies, 44*(5), 936–957.

Hess, C., & Ostrom, E. (2011). An overview of the knowledge commons. In C. Hess & E. Ostrom (Eds.), *Understanding knowledge as a commons* (pp. 3–26). Cambridge, MA: MIT Press.

Jemielniak, D. (2016). Wikimedia movement governance. *Journal of Organizational Change Management, 29*(3), 361–378.

Jepperson, R. (1991). Institutions, institutional effects, and institutionalization. In W. W. Powell & P. J. DiMaggio (Eds.), *The new institutionalism in organizational analysis* (pp. 143–163). Chicago, IL: University of Chicago Press.

Keegan, B., & Fiesler, C. (2017). The evolution and consequences of peer producing Wikipedia's rules. *Proceedings of the 11th International AAAI Conference on Web and Social Media*, Montréal, Canada (pp. 112–121).

Kesan, J. P., & Shah, R.C. (2005). Shaping code. *Harvard Journal of Law & Technology, 18*, 319–400.

Kiesler, S., Kraut, R. E., Resnick, P., & Kittur, A. (2011). Regulating behavior in online communities. In R. E. Kraut & P. Resnick (Eds.), *Building successful online communities* (pp. 125–178). Cambridge, MA: MIT Press.

Konieczny, P. (2009). Governance, organization, and democracy on the *Internet. Sociological Forum, 24(*1), 162–192.

Konieczny, P. (2010). Adhocratic governance in the Internet age. *Journal of Information Technology & Politics, 7*(4), 263–283.

Kostakis, V. (2010). Peer governance and Wikipedia. *First Monday, 15*(3). Retrieved from http://firstmonday.org/ojs/index.php/fm/article/view/2613/2479

Kreiss, D., Finn, M., & Turner, F. (2011). The limits of peer production. *New Media & Society, 13*(2), 243–259.

Kriplean, T., Beschastnikh, I., McDonald, D. W., & Golder, S. (2007). Community, consensus, coercion, control. *Proceedings ACM Conference on Supporting Group Work*, Sanibel Island, FL (pp. 167–176).

Lakhani, K. R., & Von Hippel, E. (2003). How open source software works: "Free" user-to-user assistance. *Research Policy, 32*(6), 923–943.

Lawrence, T., Suddaby, R., & Leca, B. (2009). Introduction: Theorizing and studying Institutional work. In T. Lawrence, R. Suddaby, & B. Leca (Eds.), *Institutional work* (pp. 1–27). Cambridge: Cambridge University Press.

Lessig, L. (1999). *Code and other laws of cyberspace*. New York, NY: Basic Books.

Levy, S. (1984). *Hackers. Heroes of the computer revolution*. New York, NY: Penguin.

Lih, A. (2009). *The Wikipedia revolution*. New York, NY: Hyperion.

March, J. G., & Olsen, J. P. (1989). *Rediscovering institutions*. New York, NY: Free Press.

Markus, M. (2007). The governance of free/open source software projects. *Journal of Management & Governance, 11*(2), 151–163.

Mateos-García, J., & Steinmueller, W. (2008). The institutions of open source software. *Information, Economics and Policy, 20*(4), 333–344.

Moon, J. Y., & Sproull. L. (2000). Essence of distributed work. *First Monday, 5*(11). Retrieved from http://firstmonday.org/ojs/index.php/fm/article/view/801/710

Morgan, J. T., Mason, R. M., & Nahon, K. (2012). Negotiating cultural values in social media: The case of Wikipedia. *Proceedings 45th Hawaii International Conference on System Sciences*, Waikoloa, HI (pp. 3490–3499).

Morgan, J. T., & Zachry, M. (2010). Negotiating with angry mastodons. *Proceedings 16th ACM International Conference on Supporting Group Work*, Sanibel Island, FL (pp. 165–168).

Müller-Birn, C., Dobusch, L., & Herbsleb, J. (2013). Work-to-rule. The emergence of algorithmic governance in Wikipedia. *Proceedings 6th International Conference on Communities and Technologies (C&T'13)*, Munich (pp. 80–89).

Niederer, S., & van Dijck, J. (2010). Wisdom of the crowd or technicity of content? Wikipedia as a sociotechnical system. *New Media & Society, 12*(8), 1368–1387.

Oliver, C. (1992). The antecedents of deinstitutionalization. *Organization Studies, 13*(4), 563–588.

O'Mahony, S. (2007). The governance of open source initiatives. *Journal of Management & Governance, 11*(2), 139–150.

O'Mahony, S., & F. Ferraro (2007). The emergence of governance in an open source community. *Academy of Management Journal, 50*(5), 1079–1106.

O'Neil, M. (2009). *Cyberchiefs. Autonomy and authority in online tribes*. London: Pluto.

O'Neil, M. (2014). Hacking Weber: Legitimacy, critique, and trust in peer production. *Information, Communication & Society, 17*(7), 872–888.

Ostrom, E. (1990). *Governing the commons: The evolution of institutions for collective action*. Cambridge: Cambridge University Press.

Pentzold, C. (2011). Imagining the Wikipedia community. *New Media & Society, 13*(5), 704–721.

Pentzold, C. (2017). Editorial surveillance and the management of visibility in peer production. *International Journal of Communication, 11*, 2462–2481. doi:1932-8036/20170005.

Pentzold, C. (2018). Grounding peer production in practice: Editorial routines and everyday engagement in the "free encyclopedia anyone can edit". *Communication, Culture and Critique, 11*(3), 455–474.

Raymond, E. S. (1999). *The cathedral and the bazaar*. Sebastopol: O'Reilly Media.

Reagle, J. (2010a). *Good faith collaboration. The culture of Wikipedia*. Cambridge, MA: MIT Press.

Reagle, J. (2010b). Be nice. Wikipedia norms for supportive communication. *New Review of Hypermedia and Multimedia 16*(1–2), 161–180.

Resnick, P., & Kraut, R. E. (2011). Introduction. In R. E. Kraut & P. Resnick (Eds.), *Building successful online communities* (pp. 1–20). Cambridge, MA: MIT Press.

Sadowski, B. M., Sadowski-Rasters, G., & Duysters, G. (2008). Transitions of governance in a mature open software source community. *Information Economics and Policy, 20*(4), 323–332.

Schroeder, A., & Wagner, C. (2012). Governance of open content creation. *Journal of the American Society for Information Science and Technology, 63*(10), 1947–1959.

Schweik, C. M., & Englisch, R. C. (2012). *Internet success*. Cambridge, MA: MIT Press.

Scott, W. R. (2001). *Institutions and organizations*. Thousand Oaks, CA: Sage.

Searle, J. (1997). *The construction of social reality*. New York, NY: Free Press.

Shah, S. K. (2006). Motivation, governance and the viability of hybrid forms in opensource software development. *Management Science, 52*(7), 1000–1014.

Shaikh, M., & Henfridsson, O. (2017). Governing open source software through coordination processes. *Information and Organization, 27*, 116–135.

Shaw, A., & Hill, B. M. (2014). Laboratories of oligarchy? *Journal of Communication,64*(2), 215–238.

Shirky, C. (2008). *Here comes everybody*. New York, NY: Penguin.

Stallman, R. (2004). The GNU Project. Retrieved from www.gnu.org/gnu/thegnuproject.en.html

Sternberg, J. (2012). *Misbehavior in cyber places*. Lanham, MD: University Press of America.

Stewart, D. (2005). Social status in an open-source community. *American Sociological Review, 70*(5), 823–842.

Stewart, K. J., & Gosain, S. (2006). The impact of ideology on effectiveness in open source software development teams. *MIS Quarterly, 30*(2), 291–314.

Taylor, C. (2002). Modern social imaginaries. *Public Culture, 14*(1), 91–124.

Tkcaz, N. (2015). *Wikipedia and the politics of openness*. Chicago: University of Chicago Press.

Vìègas, F., Wattenberg, M., & Dave, K. (2004). Studying cooperation and conflict between authors with history flow visualizations. *Proceedings SIGCHI Conference on Human Factors in Computing Systems*, Vienna (pp. 575–582).

Vìègas, F., Wattenberg, M., & McKeon, M. M. (2007). The hidden order of Wikipedia. *Proceedings 2nd International Conference on Online Communities and Social Computing*, Beijing, China (pp. 445–454).

Vieira, M. S. & de Filippi, P. (2014). Between copyleft and copyfarleft: Advanced reciprocity for the commons. *Journal of Peer Production, 4: Value and currency.* Retrieved from: https://ssrn.com/abstract=2468731

Von Krogh, G., Spaeth, S., & Lakhani, K. R. (2003). Community, joining, and specialization in open source software innovation. *Research Policy, 32*(7), 1217–1241.

Weber, M. (1978). *Economy and society.* Berkeley, CA: University of California Press. (original publication 1922)

Weber, S. (2004). *The success of open source.* Cambridge, MA: Harvard University Press.

Zucker, L. G. (1977). The role of institutionalization in cultural persistence. *American Sociological Review, 42*(5), 726–743.

5

Cultures of Peer Production

Michael Stevenson

1 Introduction: Peer Production as Cultural Production

How can we make sense of cultures of peer production, which exist in diverse national, cultural and language contexts, span several industries and domains, and comprise a range of different organizational structures?[1] Peer production is commonly defined as a mode of production – that is, a social and material structure in which labor takes place. Specifically it is defined as a decentralized mode of production in which a group of participants collaborate on a common project (say, an operating system or an encyclopedia), and in which individuals self-select for specific parts of an overall project and generally volunteer their time (Benkler, 2002). Given this definition of peer production as a mode of production, it makes little sense to talk about a single culture – defined as a more-or-less shared set of values, expressed in a common identity and shared practices – much like it makes little sense to speak of the culture of factories or the culture of firms. In practice, however, there are certainly comparisons to be drawn between the values and practices commonly found in peer production projects, in particular in the well-known examples of Wikipedia and FOSS production.

To set the groundwork for understanding cultures of peer production, this chapter argues that it is necessary to see that peer production is, by and large, a form of *cultural production*. Cultural production refers to aesthetic and intellectual forms of production, from literature and music to journalism and academic scholarship. It may seem strange at first to compare the production of open source software or Wikipedia to these fields, since for example we often think of traditional cultural production as an individual pursuit (ignoring how such practice is always collaborative (Becker, 1984)). In the following, I show how this analogy and corresponding theory helps us to see how (1) peer production projects are clearly embedded in existing cultural fields, and often represent an autonomous form of production that seeks to resist certain economic and political pressures in favor of core values such as meritocracy and openness and (2) such autonomy is achieved through the enactment of those core values, which are in turn related to the social hierarchies, forms of exclusion, and other limitations that characterize these projects and the groups of people who populate them.

2 Overview and Limitations

The chapter synthesizes existing research to produce a framework for understanding cultures of peer production. I argue that the cultures of peer production projects will tend to be characterized by *autonomy*, *meritocracy*, *openness*, and *inspiration*, with each of these concepts corresponding to one of the following sections. In Section 3, I discuss the *autonomy* of peer production projects and their position within larger fields of software and media production, or what might be described as a larger field of "tech" or the field of new media production (Stevenson, 2016). The distinction of peer production projects is closely related to core values and assumptions, in particular *meritocracy*, and Section 4 examines this value and the cultural practices associated with it. Section 5 discusses *openness* as a key

The Handbook of Peer Production, First Edition. Edited by Mathieu O'Neil, Christian Pentzold, and Sophie Toupin.

value and disposition within peer production cultures, elaborating on this with examples such as Wikipedia's "good faith collaboration" (Reagle, 2012). Section 6 asks what kind of moral "order of worth" (Boltanski & Thévenot, 2006) justifies commitment to peer production projects, and argues that we should especially consider the role of *inspiration* and the importance of the affective dimension of contributing to these projects.

In this introduction to the cultures of peer production, I have made various decisions that should be noted. First, this chapter draws its framework largely from Bourdieu's field theory (1993) and Boltanski and Thévenot's sociology of critique (2006). In addition to the fact that a number of other sociological theories have been used to make sense of cultures of peer production, it should be noted that the theories I have chosen contain significantly different assumptions: critics argue that Bourdieu's theory does not account for the agency of individuals, something that Boltanski and Thévenot specifically aim to address by arguing that individuals reflect on their actions and seek to justify them through appeals to various kinds of valuation, or "orders of worth" (see, e.g., O'Neil, 2011 for a discussion of this difference). Despite this internal inconsistency, I have chosen this combination as it helps to highlight (1) the relational forms of structure that give meaning to peer production projects and actions within them, for example through oppositions that actors draw between autonomy and commercial or political influence, and (2) how participants justify their commitment to these projects and specific actions within them. Second, although some definitions of peer production include entities such as Google's search engine (which analyzes the hyperlinks produced by a decentralized group of actors in order to produce better search results), this chapter draws on a specific subset of commons-based peer production based on intentional collective action, specifically Wikipedia and free and open source software (FOSS) production. In addition to their intentional nature, in Bourdieu's terminology I am limiting my examples to "autonomous" examples of peer production, at the expense of more commercial forms. For example, one can study Reddit's platform for news and discussion or Amazon's platform for consumer reviews as peer production, as these contain many of the same principles for the decentralized creation of value. Likewise, there are examples of open source production that are perceived as more "closed," such as the Android OS (Currie et al., 2013). I would argue that more commercial or closed forms of peer production lack the degree of autonomy found in Wikipedia and many open source communities, as discussed in the next section. My normative position is that these autonomous communities should be studied with the intent of serious critique and engagement that can ultimately be put to use in making them stronger and helping them live up to the lofty ideals they set for themselves. That said, it must be noted that the values and practices I identify can and often will be present in commercial forms of peer production, if not to the same degrees or articulated in the same ways. Finally, my emphasis here is on describing a framework rather than providing exhaustive detail about specific communities or cultures – my hope is that supplying a "bigger picture" at the expense of rigorous description will help readers to compare and contrast the different cultures of peer production they will encounter in this *Handbook* and elsewhere.

3 Autonomy

Field theory helps us to see how peer production projects are situated in broader structures of power. With this framework, Bourdieu (1993) outlines how society consists of fields, semi-autonomous domains of social activity. In addition to fields of power (i.e., the fields of politics, economics, and law) there are cultural fields that Bourdieu defines as fields of aesthetic and intellectual production: art, journalism, and academia would fit this description. All fields are semi-autonomous: on the one hand they have an independence demonstrated through internal consistency – shared "rules," values, practices and so on; on the other hand, they continually impact one another to differing degrees. The journalistic field certainly impacts the fields of political and economic power, for example, but not to the same degree that it is in turn shaped by them. Internally, fields are shaped hierarchically, meaning actors have different levels and kinds of power. According to Bourdieu actors are also positioned in relation to two poles: towards the field's autonomous pole actors demonstrate a capacity to resist pressures from outside fields (in particular political and market pressures), whereas towards the field's heteronomous pole they are less likely to demonstrate a commitment to the field's autonomous values. In the sense used here, autonomy refers to the autonomy of a peer production project rather than the autonomy of individuals working on that project (although that is also a key part of the cultures of peer production).

To make this more concrete, think of how FOSS production relates to (1) mainstream and proprietary software production and (2) market pressures. FOSS is often understood as having a dichotomous relationship with proprietary software, for example Linux vs. Apple/Microsoft. In this sense Linux distributions like Debian and Ubuntu seek to distinguish themselves as autonomous actors in comparison to the heteronomous positions of large companies who are more likely to bow to market demands and political restrictions. They do this not just through the fact that they make their source code freely available, but also by enacting sets of values that they perceive to be lacking in mainstream commercial development, such as a commitment to open technological standards, improving access to computing, or allocating resources to projects or solutions they deem technically superior. They thus actively pursue autonomy as they perceive it. They oppose market pressures in different ways, not just through their opposition to intellectual property regimes but also to certain judgments of quality (e.g., Debian prides itself on being an operating system geared towards other software developers). Importantly, Bourdieu consciously uses the term "semi-autonomy": pressures from the market, state, and other fields are always present, but it is a matter of degree.

Using field theory, we can thus consider how a peer production project enacts a form of relative autonomy in relation to other actors in a field of production – open source vs. proprietary software companies, Wikipedia vs. commercial reference media, etc. Again, making the connection to older cultural fields, think of how the example of the Debian Linux distribution is similar to how a reputable newspaper or weekly political magazine distinguishes itself from more commercial or mainstream actors. Where Debian's developers seek to maintain a commitment to technical elegance and to serving other developers in ways that distinguish their product from more proprietary or mass-market operating systems, quality news organizations similarly seek to uphold their own field-specific definitions of quality and distinguish themselves from popular journalism such as tabloid newspapers – such journalism might offer a less entertaining, denser product that is appreciated by a smaller audience, while their peers in tabloid news are more likely to adjust content to expectations of what the audience would like to read and subject themselves to criticisms of superficiality, sensationalism, and the like. At the same time, note that in the context of "semi-autonomy" this is a matter of degree, and it hardly means that quality newspapers ignore their audience or advertising revenue, or that Debian software developers never accommodate the needs of proprietary applications. Likewise, this does not mean that tabloid journalists feel their work is not valuable, as they for example can point out that they engage more readers (see e.g. Deuze, 2005a). What it does mean is that actors who take up an autonomous position within a field will demonstrate a commitment to certain shared values and seek to uphold what they perceive as a more "pure" or "quality" form of cultural production.

Autonomy is closely tied to another key concept in Bourdieu's field theory, namely symbolic capital. Symbolic capital can be thought of as a field-specific form of reputation, and it is gained informally through recognition from peers and critics and formally through prestigious rewards (e.g., particular roles, jobs, awards, mentions in the trade press, etc.). Think of this, again, in the context of journalism: there is symbolic capital in writing for *The New York Times*, in receiving a Pulitzer prize, when peers cite your work in other journalistic venues, and so on. Symbolic capital is related to autonomy such that it is often at odds with economic capital. As many of us will recognize from the experience of watching a favorite music artist "sell out," there is a reputational cost that may come with commercial success. Gaining symbolic capital increases one's status in a particular field, and combines with other forms of capital (economic, cultural, and social) to generate hierarchies.

Much of the interest around peer production is focused on the "puzzle" of individuals collectively organizing their labor largely independent of firms (e.g., managers with the authority to hire and fire) and price-based mechanisms, such as wages (Benkler, 2002). Bourdieu would likely argue that such non-market based production has long existed in semi-autonomous fields, and is only a puzzle because we remain blind to (1) the ways in which all cultural production is distributed and (2) the different forms of capital that exist beyond economic capital. Wikipedia, open source, and other forms of peer production are certainly new, but they also extend the logics of existing cultural fields.

If autonomy is defined as the capacity to act in ways that resist pressures from other fields, then what comes in the place of that pressure? What guides action within, say, the art world, if it is not the market, state, or other external power? Field theory assumes that the competition for symbolic capital is what motivates actors, and that this circulation of symbolic capital is in turn geared toward reinforcing core values and principles that, while not immutable, are at the heart of what is considered good practice or purpose within a field, and thus what must be protected from outside influence. These values can be as diverse as creative expression and aesthetic innovation (summarized as "art for art's sake") in the art world (Bourdieu, 1993) to serving the public interest in the field of journalism (Benson & Neveu, 2005; Deuze, 2005b).

Here, too, we can look to Wikipedia and open source software to see how this interplay of autonomy, values, and symbolic capital is present in peer production projects. For Wikipedia, notions of universal access and the unfettered pursuit of knowledge would be considered important shared values, and participants are recognized (and recognize each other) for their efforts in support of them. Among Wikipedia participants, symbolic capital is gained through "barn stars" and other forms of peer recognition, as well as through appointments to positions within Wikipedia's bureaucratic hierarchy. Less formally, symbolic capital is gained by being a visible contributor to the overall project, whether through providing expertise for certain topics, productively engaging in backend discussion, building software tools to help editors work more efficiently, and so on. As participants gain symbolic capital, this esteem will give them a better chance of mobilizing peers when it comes to conflicts and key decisions relating to the project.

The values enacted by Wikipedia and open source software communities are notable for how they are related to process as much as end product. As Kelty (2008) argues, open source software communities can be understood as "recursive publics." That is, the identity of these communities are built on a kind of meta-level: if publics normally form around a shared interest (e.g., fans of the same musician or citizens engaging with a particular political issue), what is remarkable in the geek cultures of Wikipedia and open source software production is that identity is largely tied to values and practices related to *how* communication takes place (including in the technical sense of which tools are used), rather than the content of that communication. As Kelty writes, such publics are "vitally concerned with the material and practical maintenance and modification of the technical, legal, practical, and conceptual means of [their] own existence" (2008, p. 3). For example, what ties the identity of Wikipedians together is less the content of the encyclopedia, and more their shared commitment to their desire for open and transparent processes for writing articles and settling disputes.

Another point that may be extrapolated from the discussion of autonomy above is that the most famous peer production projects can be seen in the context of what Kelty calls geeks' "reformist" identity. Kelty notes that these communities of software makers often narrate their own identities through implicit and explicit analogies to the Protestant Reformation, where the control of technology and media (religion) is taken away from corporations (the Church) and restored to its original purpose by true believers (the geeks). For Kelty, the implicit analogy can be sensed in how notions of good and evil are deployed by these communities. For example, an open source software producer may find it "evil" when a large software company ignores an open standard and instead pushes a proprietary one onto its users. Likewise, a Wikipedian dedicated to the universal access of information may find it "evil" for companies to copyright or otherwise prevent access to useful knowledge, such as when academic publishers put scholarship behind expensive paywalls. In this way, we can see Wikipedia and open source as relatively autonomous forms of technical and media production, and that this position is narrated and articulated through stories (or "myths") that echo the story structure of the Reformation (and these stories may be reproduced socially).

In sum, Bourdieu's concepts of field, semi-autonomy, and symbolic capital help us to see common threads in (1) how peer production projects are situated within a larger context of cultural production (where for example open source communities provide alternatives to proprietary computing software), (2) how this positioning within a larger field can be seen as enacting a kind of semi-autonomy, meaning these projects demonstrate some capacity to resist pressures from other fields and in particular market and state power, and (3) how such semi-autonomy is related to the establishment and protection of core values, of which many are related to the *process* by which projects are governed, (4) how symbolic capital, or field- and project-specific reputation, is accrued through demonstrating commitment to these core values and is intertwined with formal and informal hierarchies within peer production projects. In the next two sections, I discuss two overarching values that typify peer production: meritocratic rule and openness. Importantly, these values relate to the processes of collaboration and governance rather than the end products.

4 Meritocracy

Although peer production projects are generally open to participation by anyone, there are certainly hierarchical tendencies within them and with this comes forms of closure and exclusion. On the one hand, there are formal hierarchies – project leaders, maintainers, committees, Wikipedia editors, and so on, and these various positions entail different rights and privileges. In Wikipedia's case, for example, the user class "bots" has more rights than anonymous users, reminding us that such large-scale peer production also requires socio-technical forms of government (Niederer & van Dijck, 2010). Proponents of peer production argue that these hierarchies are flexible arrangements that serve a larger goal of ensuring equal participation, rather than serving to sustain themselves (Bauwens, 2005). On the other hand, there are informal hierarchies that can be understood as accumulations of symbolic capital – think for example of a Wikipedia admin who holds particular esteem and is deferred to by other editors and admins. In idealized form, these hierarchies in peer production arise from *merit* as opposed to *credentials*, meaning there is no prior separation of experts and non-experts, qualified persons and unqualified persons. Instead, the aim is that individuals earn their position through demonstrations of their ability and commitment to the project. In practice, both Wikipedia and open source communities display strong beliefs in the importance of rewarding people based on individual merit, and in the *possibility* of a functioning meritocratic system. I emphasize that the belief also relates to the possibility of meritocracy because this is a contentious position to hold. Before moving to critiques of the belief in meritocracy in peer production and related online communities, however, this section first develops the concept of meritocracy in peer production further.

4.1 Liberalism

The desire for a meritocratic form of hierarchy, in which individual talent and dedication is rewarded accordingly, is linked to a broader commitment to the political philosophy of liberalism. As Coleman (who uses the term "hackers" to refer to the collective identity of open source software makers) explains, hackers working on open source projects display a commitment to liberalism in the sense of:

> historical as well as present-day moral and political commitments and sensibilities that should be familiar to most readers: protecting property and civil liberties, promoting individual autonomy and tolerance, securing a free press, ruling through limited government and universal law, and preserving a commitment to equal opportunity and meritocracy. (Coleman, 2013, p. 2)

This commitment to liberalism is not straightforward, but rather hackers enact a "liberal critique from within liberalism" (Coleman, 2013, p. 3). They critique a liberal belief in the value of intellectual property by promoting another liberal value, that of free speech. Hackers, in other words, critique proprietary software on the basis that it limits their "productive freedom," or freedom to "autonomously improve on their peers' work, refine their technical skills, and extend craftlike engineering traditions" (Coleman, 2013, p. 3).

How is meritocracy enacted? As Coleman argues, hackers' commitment to meritocracy gives open source software production a contradictory character. On the one hand, hackers emphasize that individuals must "pull themselves up by their own bootstraps," meaning they should apply themselves to learning their craft and earning any respect or rewards through long hours of dedication to programming and engineering. On the other hand, to create the conditions for this, hackers must provide not only the software they build and its source code, but also the documentation and explanation necessary for others to assess, evaluate, and learn from their work (Coleman, 2013). Beyond that, since the work of programming is often far more complex than implementing a well-defined plan and involves a host of social and cultural competencies in addition to technical ones, hackers rely on each other for all kinds of knowledge that help them to do their jobs or contribute to open source projects. In other words, where meritocracy suggests self-reliant, individual feats of technical skill, here it also implies providing the necessary resources that create the equal opportunity in which a meritocracy works.

Hacker culture, in particular in the online discussion spaces for discussing programming problems, reflects this contradictory emphasis on self-reliance and helping others: on the one hand, the experienced programmers in these spaces can be notoriously hostile to beginners asking questions that have been answered numerous times before, leading to the response "RTFM" ("read the fucking manual") and similar (Coleman, 2013, p. 107). On the other hand, the fact that these spaces exist and that so many programmers volunteer time to maintain them suggests programmers are fully aware of their reliance on the help of others and the need for this circle of support to remain (Coleman, 2013). Meanwhile, technically challenging questions invite a great number of responses, not just because they represent an intellectual challenge but also because it gives hackers an opportunity to demonstrate their skills and intelligence to an audience of peers.

4.2 Humor and Merit

To outsiders it may seem strange, but anyone who has spent time in open source communities or editing Wikipedia will know that humor plays an important role in peer production. Although such humor is significant for many reasons (including smoothing over conflicts or acting as a kind of "communal gift"), it has a specific relationship to the value of meritocracy (Coleman, 2013, p. 105). As Coleman points out, humor and craftiness in general allows hackers to draw attention to their technical competence in a socially acceptable way (Coleman, 2013). To write a script or piece of documentation that is funny to another expert coder, one has to play with form (conventions and corresponding expectations), and thus one must demonstrate a deep familiarity with the programming language, programming styles, algorithmic techniques, and general approaches to solving problems. Similarly, to make jokes that draw on or highlight Wikipedia's complex set of policies and guidelines or key events from its history one must be steeped within them. In Bourdieu's terminology, we could say that such clever software developers or Wikipedians are signaling their *cultural capital*, or command of the various skills, knowledge, and competences that will allow them to succeed within a particular field or project.

4.3 Critiques of Meritocracy in Geek Cultures

In theory, a meritocracy works when individuals have access to the same material and intellectual resources, and thus an equal opportunity from which to develop themselves and display their individual talent. In practice, access to these resources will always be unequal, since it is impossible for everyone to have the same time, financial resources, prior knowledge, and so on.

In particular, Alice Marwick (2013) and Marie Hicks (2016) have each criticized the belief in meritocracy in computer cultures in ways that can be extended to analyzing peer production. Although Marwick's analysis is focused on startup culture (which in Bourdieu's terms is a relatively "heteronomous" form of software production opposite the "autonomous" mode of FOSS production), her criticism of the "myth of meritocracy" is likely to be highly relevant to many peer production projects. As she argues, this belief is not only generally false but has disproportionate negative effects on women and people of different races, nationalities, ethnic and economic backgrounds. Beyond more obvious cases of sexism and racism, there will always be issues of unconscious bias that impact judgments of others and their contributions – Marwick notes in her interviews with both male and female Silicon Valley workers, for example, that female entrepreneurs and tech workers are often judged by their appearance rather than technical or business competence (Marwick, 2013, p. 264). The fact that Silicon Valley and workers in the US and European tech industries generally are perceived as favoring progressive politics hardly means that ethnocentrism and erroneous assumptions about others disappear (and rather that this ethnocentrism takes on new forms, see English-Lueck, 2002). By upholding an ideology of meritocracy, open source communities and Wikipedia are less likely to see how participation and achievement are dependent on being able to fit in with a group that is overwhelmingly male and white. As Ford and Wajcman (2017) note, for example, participation in Wikipedia is often adversarial, unwelcoming to newcomers and suited to people with high degrees of technical skill, all of which favors its existing (mostly male) community and discourages women from taking part. Likewise, Nafus (2012) demonstrates the strong link between the meritocratic ideal in open source communities (present in claims that technical contributions "don't have gender") and the hostile attacks women often face in these communities when raising issues of sexism.

Marie Hicks's work deals with the history of computing, but is equally relevant to the present topic. Hicks (2016) notes how highlighting the contributions of "genius" women in the history of computing – while noble in its intentions – strengthens the belief that success is a function of individual talent, obscuring inequalities of access and the fact that any individual contribution is dependent on a great deal of "hidden work" by countless others. This kind of critique raises the question of whether we should aim to improve ostensibly meritocratic systems favored in peer production and geek culture, or replace them with something else entirely.

In sum, this section has taken a closer look at meritocracy as a core value of FOSS and Wikipedia communities and the wider technological and media fields they operate within. Meritocracy is steeped in a liberal worldview that prizes individual achievement and equal opportunity, and as such we can see in peer production cultures an emphasis on both individual merit as well as the need to assist others. The meritocracies enacted by these communities can be found in crystallized form in various reputation systems and governance structures. However, meritocracy will always remain an ideal, or myth, given that power always ensures inequalities of access and various forms of inequality such as sexism and racism persist even in relatively progressive geek communities.

5 Openness

In Benkler's famous formulation, the success of peer production is dependent on several factors related to the envisioned product, namely that there is low uncertainty about what constitutes quality and that the product is modular and granular, meaning it can be split into independent parts and allows for both smaller and greater sized contributions. This means the project is suited for decentralized, self-selecting producers of different levels of expertise and differing levels of commitment). What may go missing in such an economic analysis is that peer production requires more than a set of affordances, rules and procedures, and is arguably dependent on a kind of moral education of its participants. As Pentzold (2011) argues in the context of Wikipedia, participants understand the glue tying their community together not as the bureaucratic governance structure (consisting of wiki software and all of the rules and regulations guiding decision-making on Wikipedia), but rather as shared values. "[U]sers primarily understand their collective as an *ethos-action community* tying community membership not to admission procedures but to the personal acceptance of a set of moral obligations and rules of conduct" (Pentzold, 2011, p. 716). Logically, such personal acceptance will arise not just from existing preferences and previously held beliefs, but also through processes of socialization and education. An example of such moral education in service of peer production is Debian's New Member Protocol, a form of apprenticeship that also requires an individual essay presenting one's views on the value of open source (Coleman, 2013, p. 124)

The moral education involved in peer production will be geared towards several shared values, obligations, and rules of conduct. Perhaps most importantly, however, is instilling a disposition of openness, understood as tendencies toward humility and empathy that we associate with an "open mind." Here it is important to see that openness is not a straightforward concept, and can refer to different things including the free access of information and technology (as in "open source code"), or a liberal political philosophy (Tkacz, 2014). The "open-mindedness" I focus on here is a crucial part of what Reagle defines as the larger meaning of openness in Wikipedia and similar communities, namely as "an *accessible and flexible* type of *collaboration* whose result may be *widely shared*" (Reagle, 2012, p. 75, emphasis in original). As Reagle notes, Wikipedia's famous policy of "neutral point of view" (NPOV) is often misunderstood as a kind of positivist notion of truth. However, in practice it is pursued "rather as a stance of dispassionate open-mindedness" (Reagle, 2012, p. 56). In other words, neutrality and openness are articulated as an attitude or disposition.

On Wikipedia, such openness-as-disposition is encoded and enacted in different ways. Perhaps most important among these is the colloquial Wikipedia norm of "assuming good faith" ("Wikipedia," 2019). Assuming good faith means taking opposing views seriously and a willingness to entertain the view that one's own perspective is incomplete and flawed. Although prolonged flame wars get the most attention, as Reagle points out these are outweighed by civil interactions where compromise is sought and good faith is assumed on all sides. A related practice that Wikipedia encourages is "writing for the opponent" ("Wikipedia," 2018). This challenges editors to spend more time writing and polishing points of view they themselves do not hold. Here, the guideline can be read in two ways: on the one hand it encourages openness through exposure to difference; on the other it suggests this is also the best way to convince others of your own arguments (which of course somewhat violates the principle of good faith). Regardless, it remains a practice that requires some degree of empathy and respect (Reagle, 2012, p. 58).

Importantly, while openness is often understood as an organizational goal or requirement in peer production, it must in fact be a quality of the culture. Source code that anyone can see, an encyclopedia that anyone can edit, this form of openness is relatively easy to implement and sustain through licensing. But openness in terms of remaining open to new, diverse, and challenging ideas, and to change more generally – this will never be easy, due to the hierarchical structures that emerge in any social organization. There is also evidence that growth itself in peer production projects leads to conservatism: Halfaker (Halfaker et al., 2013) show that Wikipedia's solutions to its enormous growth over the years – formalizing policies and guidelines, using software to do mass reverts of suspect contributions, etc. – have the unintended consequence of rejecting the contributions of desirable participants, and thus discouraging them from investing more time in the project. Therefore, peer production projects must find ways to foster and maintain openness through various formal and informal means of education and socialization.

5.1 Humility and Openness in Open Source Cultures

Hacker culture and FOSS communities tend not to be associated with humility, at least in the popular consciousness. There is also a kind of romantic heroism attributed to visionary and brazen hackers who push forward with their single-minded projects, beating the odds and showing up naysayers (Levy, 1994). A lack of humility can also be seen in the overzealous factions of different open source communities who use humor and mockery to disparage competing projects. Such in-crowd humor builds camaraderie within one group while closing the group off to others (Coleman, 2013).

However, technical work itself is also extremely humbling. As a programmer you must continually seek support from others, and deal constantly with failure in the form of bugs and broken code. In such a context, being open to new technologies, methods and solutions, and recognizing their strengths is necessary. Openness and humility can therefore also be seen as intrinsic to technical work and in particular open source software production, which preaches releasing "early and often" and being open to the improvements suggested by others. It should also be noted such intrinsic openness extends beyond seeking optimal technical solutions, and in some ways open source culture was founded on the difficult work of inclusivity. Although the history of the rapid rise of open source software in the 1990s is sometimes presented as a cooptation of the free software movement (Morozov, 2013), in practice it required finding ways to balance the wishes of civic- and market-oriented factions of the community and ensure their collaboration (Stevenson, 2018).

5.2 Gender, Identity, and Failures of Openness

To note that such openness is present in these communities is not to deny that there is a crisis of openness currently facing them. Throughout many projects and venues for open source culture, at issue now is the exclusivity that is largely associated with geek masculinity, and most notably manifest in abrasive and abusive behaviors both on- and offline towards women. As Reagle (2013) notes, the values of meritocracy and openness in free culture and open source communities make them vulnerable to difficult personalities, and these values can also be rhetorically deployed in ways to excuse or downplay inequality and uncivil or hostile behavior.

To combat this tendency a large number of open source communities have in recent years adopted a "code of conduct" that give more clarity about which behaviors are unacceptable and give leadership more leverage in policing behaviors (Finley, 2018). How these open communities continue to address this issue will vary in both approach and success, however it seems obvious that their fates will hinge on their ability to recognize their own biases and maximize good faith collaboration while also policing these communities. In this sense, continued interrogation of the concept of openness and its role in these communities seems like fruitful ground for both practitioners and researchers.

In sum, this section has discussed openness as a disposition that benefits peer production. Through various forms of moral education, Wikipedians and open source programmers ideally internalize a tendency towards humility and empathy, seeing competing views not as attacks to be defended against but rather as questions to be taken seriously and integrated into improving a project as a whole.

6 Inspiration

If openness defines a disposition that aids peer production through conflicts and key decisions, how is such a seemingly counterintuitive attitude justified by opinionated hackers and Wikipedians? What is it that motivates participants to put aside individual differences, and leaders to allow their projects to grow in unexpected ways? In short, the answer is a dedication to a higher purpose, and in some sense a belief that a project ultimately leads a life of its own. Here again the analogy with cultural production can be drawn: successful peer production projects appear to draw on a mix of passion, reflexive contemplation, self-importance, and charismatic leadership similar to that found in literary, artistic, and journalistic movements. This passion for the *integrity* of a larger whole can be seen as a unifying theme when considering participation in Wikipedia and open source projects. As O'Neil writes, dedication to a project can be understood as a critique of contemporary alienation:

> [P]articipants in these projects seek a feeling of unity between their identities as consumers and producers, between their status as experts and amateurs, between their roles as leaders and followers, between their activities of work and play, and between themselves and their fellow participants in the project – a project which they see, more often than not, as a cause to defend. (O'Neil, 2011; see also Turner's (2009) discussion of the psychic wholeness experienced by tech workers at Burning Man)

This seeking of unity is enacted through talk of values, traditions, and identity. For example, any major change in the organizational structure of Wikipedia will undoubtedly be discussed by proponents and opponents in terms of the "purpose" of Wikipedia, making such purpose the object of debate and potentially leading to edits of Wikipedia's policies and guidelines (for the most famous example of this, see discussions of the "Spanish fork," where Spanish Wikipedians led resistance to the display of advertising on Wikipedia: e.g. Tkacz, 2014).

It may sound strange and unscientific to talk about passion in the context of Wikipedia or open source software, but this affective dimension is certainly an important factor. Here again, the analogy with cultural production is instructive: think of how often we talk about the passion or zeal of great artists and journalists. This is very similar to how Wikipedians and open source developers view senior members of their communities. Meanwhile intense feelings of community at open source conferences (Coleman, 2013) can be compared to a similar "sense of community" felt by those in participatory cultural forms such as folk music (Bohlman, 1988).

These comparisons help us to see not only the importance of emotion, but also the specificity of the "moral economy" of peer production. Here, moral economy essentially means the kinds of moral justifications underlying the specific behaviors, decisions, rules, and other forms of social activity that characterize a project. In this light, it is useful to draw here on Boltanski and Thévenot's framework (Table 5.1) of justificatory regimes or "orders of worth."

Table 5.1 Boltanski and Thévenot's "orders of worth" (1999, p. 368).

	Inspired	*Domestic*	*Civic*	*Opinion*	*Market*	*Industrial*
Mode of evaluation (worth)	Grace, nonconformity, creativeness	Esteem, reputation	Collective interest	Renown	Price	Productivity, efficiency
Format of relevant information	Emotional	Oral, exemplary, anecdotal	Formal, official	Semiotic	Monetary	Measurable: criteria, statistics
Elementary relation	Passion	Trust	Solidarity	Recognition	Exchange	Functional link
Human qualification	Creativity, ingenuity	Authority	Equality	Celebrity	Desire, purchasing power	Professional competency, expertise

Source: Boltanski, L., & Thévenot, L. (1999). The Sociology of Critical Capacity. European Journal of Social Theory, 2(3), 359–377. https://doi.org/10.1177/136843199002003010.

In the Table 5.1, the authors outline six "orders of worth," each of which represents an important source of justifications for specific actions in many different social situations. For example, let's say a project team at company X is deciding whether their software product should include a particular feature. Arguments will be put forward that appeal to the overarching value of the market – i.e., whether it will be profitable to include the feature ("the market order"). Arguments will also be made for what is most technologically appealing or efficient ("the industrial order"). And there may also be arguments for a kind of civility ("the civic order"), for example in discussion of whether the product will be compatible with open standards or competing products.

When analyzing cultures of peer production, it is important to see the role of the inspirational mode of justification, from the value placed on charisma as a quality of leadership to the emphasis placed on growing or maintaining levels of passion among project members. In my previous research (Stevenson, 2018) I outlined the importance of this moral economy in the historical development of the Perl programming language, a scripting language that was a key technology for the early World Wide Web and far outgrew its humble beginnings as a UNIX tool created for system administrators. Support for Perl and the many volunteer hours spent by developers on increasing its value could be justified in many ways – for a typical Perl developer it could be justified economically (becoming more proficient with Perl increased their job prospects), technologically (through a dedication to producing the best possible software), as well as in a civic mode (contributing to a commons that would benefit everyone). What stands out, though, is where the inspirational mode of justification was drawn upon to smooth over conflicts and provide grounds for consensus and action when this seemed to be most difficult: most notably, in a case where Perl's grassroots volunteers and a commercial company collaborated on the Windows version of Perl. Perl's creator Larry Wall had a keen sense of the need for the Perl community to feel inspired, and for a shared sense of Perl evolving organically rather than according to any single person's or community's interests. Feeling inspired, volunteers would be willing to devote significant time and energy to a project they felt carried significance beyond meeting particular technological, economic, or even civic needs. Believing that the project was bigger than any single individual's preferences and should be allowed to evolve in unexpected ways, Perl supporters were more likely to accept when decisions they did not agree with in the first instance were made.

The case of Perl and Larry Wall confirms what O'Neil previously argued in the context of open source software communities and other "online tribes" (O'Neil, 2009). Building on Weber's (1978) classic categorization of forms of legitimate authority, O'Neil argues that what characterizes the organization of online tribes is the infusion of charismatic authority within what are largely bureaucracies, or organizations defined by rational-legal authority, though these bureaucracies are more transparent and democratic than the classic type defined by Weber (O'Neil, 2011). (It should be noted that this strange mashup is not limited to peer production, and Boltanski and Chiapello (2005) note that increasing appeals to the inspirational mode of justification and charismatic authority are one of the elements of what they call the "new spirit of capitalism.") Mixing these forms of authority means these tribes appear to have it both ways: on the one hand, a project like Perl preaches inclusivity and has a host of committees, appointed maintainers, project groups and so on, and thus various formal and informal means of ensuring voices are heard; on the other hand, if Perl founder Larry Wall decides to act against the wishes of the majority and implement a particular major change, his charismatic authority will likely ensure support and acceptance. This contradictory stance is epitomized by the common description of open source project maintainers as "benevolent dictators." As Kreiss et al. (2011) argue, this means that although peer production projects pride themselves on their opposition to "closed" bureaucracies, they in fact introduce other kinds of closure and opacity, potentially bringing about the kinds of inequality they seek to avoid.

This ambiguity of authority is reflected in cultures of peer production. Much discussion will express a distrust of elite power within the organization, often jokingly called "the cabal," suggesting a secretive group exercising power with impunity. This use of humor can be seen to work through and playfully challenge the competing forms of authority within these projects (Coleman, 2013, p. 122). As individual projects grow and become more established, the limits of charismatic authority are often tested. As a project's complexity, diversity, and wider significance increases, so will calls for more transparency and democracy. However, even in Wikipedia and established open source projects, the transition to rational-legal authority and more democratic forms of governance does not mean charismatic authority disappears. As Dafermos (2012) notes in the context of the FreeBSD project transitioning from an informal selection process to democratic elections for its core team members, the significance of charisma remains: "The persuasive authority of core team members is legitimized mainly through the recognition of the authenticity of their technical charisma by committers."

The importance of the inspirational mode of justification, and thus the value placed on passion and the justification of charismatic authority, is arguably linked to a key issue facing Wikipedia and several open source projects today, namely the prospects of declining participation and (relatedly) increasing pressures on project members that do stay involved. What happens when the inspiration dries up, or when what once felt like a revolutionary product begins to feel routine? In the case of Perl, an instructive moment in this regard occurred in 1999, when Larry Wall and other senior project members saw that the programming language was losing market share, momentum, and key contributors to competing languages. While some supported developing a charter and thus sought to improve Perl's prospects by transitioning to a more formalized governance structure, Larry Wall and others rejected this idea, arguing that what was needed was to inspire the community again. Wall's way forward was chosen, and this resulted in the launch of the Perl 6 project, an entirely new language written from scratch. Once again Wall inspired many others to volunteer their time to a large-scale, visionary collaborative project with him at the helm, however this time the ambition perhaps outgrew the community's capacity, and Perl 6 remained in development for years, still requiring a great deal of work before it will stand a chance of widespread adoption. The story of Perl 6 and other examples of mixed success or even failure in the world of open source software reveal the tenuousness of peer production, where inspiration is both a boon and a requirement, and thus may work like a double-edged sword.

7 Conclusion

This chapter has sought to synthesize existing research with sociological theory in order to shed light on the cultures of peer production. Building on Bourdieu's field theory, I have argued that attention to peer production as cultural production helps us to understand several key aspects of Wikipedia and open source communities. First, such projects do not exist in a vacuum, and their cultures are tied to notions of autonomy within a broader field of software and media production. Wikipedians enact a form of knowledge production that stands in contrast to certain bureaucratic and commercial practices that they perceive as unsatisfactory, if not "evil." Likewise, Linux communities and other FOSS actors enact a more autonomous form of software production than, say, the commercial and proprietary operating systems produced by Apple and Microsoft. Second, the core values these projects embrace are related to the forms of reputation or symbolic capital that circulate within them, and this in turn creates hierarchies. Third, these values are enacted and maintained through various forms of socialization and a range of cultural practices from sharing code to sharing jokes. Meanwhile enactment of these values does not come without social costs and blind spots: in particular the belief in meritocracy and openness of these communities can lead to a blindness for inequalities that do exist. Fourth, drawing also on the work of Boltanski and Thévenot, as forms of cultural production these projects are notable for the importance of an affective charge: autonomous peer production projects require that participants feel connected to a larger common purpose.

Although this chapter has limited its focus to peer production in its more autonomous (and well-known) forms of Wikipedia and open source software, the aim is that this framework will help readers map and analyze the cultures of other peer production projects. What larger cultural field are they situated in? To what degree do they appear to be independent or autonomous, and what kinds of companies or other actors would they distinguish themselves from? In what ways are hierarchies created among participants? What shared values do they have and how are they enacted? Most importantly, how do these various elements common to cultural production – a sense of autonomy, shared social norms, and hierarchies of symbolic power – interrelate?

In addition to benefiting analysis and research, understanding peer production through the framework of cultural production may benefit practitioners. In particular, it may help practitioners think through the problem of gender and racial inequality that endanger open source communities and Wikipedia by alienating new participants, in particular women. As many critics have argued and various researchers have shown, these projects often fail when it comes to living up to their professed liberal value of equality: seen in the framework offered here, what is happening is a classic dynamic of fields, namely a conservatism that favors existing power structures, including existing class and gender inequalities. As Bourdieu argues, determinations of cultural capital (i.e., what constitutes valuable traits, knowledge, or skills within a particular social context) and symbolic capital (i.e., acts of recognizing valuable work or contributions) are never neutral, and often serve to reproduce the power of a dominant class. This can be seen at the level of peer production projects as well, where for example Wikipedia's culture – wrapped up as it is in the social norm of fiery debate, the celebration of technical skill and the lack of patience for new participants who don't know or understand all the rules and policies in place – is incredibly unwelcoming for newcomers and in particular women.

Much the same can be said of many FOSS communities. These tendencies were notably nurtured in the early, male-dominated development of these projects, and by naturalizing such "culture" and demanding that new volunteers adapt to it, these projects are now justifiably facing harsh criticism. The most critical voices would likely argue that the cultures of open source software and Wikipedia are rotten at the core, and this may be true. Despite this, seen in the framework of field theory, I would argue that we should also see some ambivalence here: peer production projects tend to have an inspired, passionate group who codify their shared identity and sense of autonomy in various cultural norms, guidelines, and practices. These projects, like any other form of social action, are not immune to hierarchy and the cultures of these projects will tend to favor those who are in power. However, the conservatism of upholding a project's original values can arguably also be progressive, such as when Spanish Wikipedians forked the project in response to co-founder Larry Sanger suggesting that Wikipedia would take on advertising (Tkacz, 2014). In doing so, these Wikipedians demonstrated how such a move would destroy the integrity of the project (both in the sense of its authenticity and unity), and thus successfully argued against it in line with a shared sense of the project's purpose. Perhaps the question facing these communities is not how to replace their liberal values, social structures, and norms wholesale, but how to renew their sense of project autonomy as well as their corresponding cultures in line with an ethic of inclusivity and open-mindedness.

Acknowledgments

Research for this chapter was supported by the Netherlands Organization for Scientific Research (NWO) as part of the research program Innovational Research Incentives Scheme Veni in connection with the project "The Web that Was" (275-45-006).

References

Bauwens, M. (2005). The political economy of peer production. *CTheory*, December 1. ctheory.net/articles.aspx?id=499.

Becker, H. S. (1984). *Art worlds*. Berkeley, CA: University of California Press.

Benkler, Y. (2002). Coase's Penguin, or, Linux and the nature of the firm. *Yale Law Journal, 112*(3), 367–445.

Benson, R., & Neveu, E. (Eds.). (2005). *Bourdieu and the journalistic field*. Cambridge: Polity.

Bohlman, P. V. (1988). *The study of folk music in the modern world*. Bloomington, IN: Indiana University Press.

Boltanski, L., & Chiapello, E. (2005). The new spirit of capitalism. *International Journal of Politics, Culture, and Society, 18*(3), 161–188.

Boltanski, L., & Thévenot, L. (2006). *On justification: Economies of worth* (C. Porter, Trans.). Princeton, NJ: Princeton University Press.

Bourdieu, P. (1993). *The field of cultural production* (R. Johnson, ed., 1st ed.). New York, NY: Columbia University Press.

Coleman, E. G. (2013). *Coding freedom: The ethics and aesthetics of hacking*. Princeton, NJ: Princeton University Press.

Currie, M., Kelty, C., & Murillo, L. F. R. (2013). Free software trajectories: From organized publics to formal social enterprises? *Journal of Peer Production, 3*. Retrieved from http://peerproduction.net/issues/issue-3-free-software-epistemics/peer-reviewed-papers/free-software-trajectories-from-organized-publics-to-formal-social-enterprises/

Dafermos, G. (2012). Authority in peer production: The emergence of governance in the FreeBSD project. *Journal of Peer Production*, 1(1). Retrieved from http://peerproduction/issues/issue-1/peer-reviewed-papers/authority-in-peer-production/

Deuze, M. (2005a). Popular journalism and professional ideology: Tabloid reporters and editors speak out. *Media, Culture & Society, 27*(6), 861–882. https://doi.org/10.1177/0163443705057674

Deuze, M. (2005b). What is journalism? Professional identity and ideology of journalists reconsidered. *Journalism, 6*(4), 442–464. https://doi.org/10.1177/1464884905056815

English-Lueck, J. A. (2002). *Cultures@SiliconValley*. Stanford, CA: Stanford University Press. http://www.sup.org/books/title/?id=3738

Finley, K. (2018, September 26). The woman bringing civility to open source projects. *Wired*. Retrieved from www.wired.com/story/woman-bringing-civility-to-open-source-projects/

Ford, H., & Wajcman, J. (2017). "Anyone can edit," not everyone does: Wikipedia's infrastructure and the gender gap. *Social Studies of Science, 47*(4), 511–527. https://doi.org/10.1177/0306312717692172

Halfaker, A., Geiger, R. S., Morgan, J. T., & Riedl, J. (2013). The rise and decline of an open collaboration system: How Wikipedia's reaction to popularity is causing its decline. *American Behavioral Scientist, 57*(5), 664–688. https://doi.org/10.1177/0002764212469365

Hicks, M. (2016). Against meritocracy in the history of computing. *CORE: The Magazine of the Computer History Museum, 20*, 28–33.

Kelty, C. (2008). *Two bits: The cultural significance of free software*. Durham, NC: Duke University Press.

Kreiss, D., Finn, M., & Turner, F. (2011). The limits of peer production: Some reminders from Max Weber for the network society. *New Media & Society, 13*(2), 243–259. https://doi.org/10.1177/1461444810370951

Levy, S. (1994). *Hackers: Heroes of the computer revolution*. New York, NY: Dell Pub.

Marwick, A. E. (2013). *Status update: Celebrity, publicity, and branding in the social media age*. New Haven, CT: Yale University Press.

Morozov, E. (2013). The meme hustler. *The Baffler, 22*. Retrieved from www.thebaffler.com/salvos/the-meme-hustler

Nafus, D. (2012). "Patches don't have gender": What is not open in open source software. *New Media & Society, 14*(4), 669–683. https://doi.org/10.1177/1461444811422887

Niederer, S., & van Dijck, J. (2010). Wisdom of the crowd or technicity of content? Wikipedia as a sociotechnical system. *New Media & Society, 12*(8), 1368–1387.

O'Neil, M. (2009). *Cyberchiefs: Autonomy and authority in online tribes* (1st ed.). New York, NY: Pluto.

O'Neil, M. (2011). The sociology of critique in Wikipedia. *Journal of Peer Production, 0.* Retrieved from http://peerproduction.net/issues/issue-0/peer-reviewed-papers/sociology-of-critique/

Pentzold, C. (2011). Imagining the Wikipedia community: What do Wikipedia authors mean when they write about their "community"? *New Media & Society, 13*(5), 704–721. https://doi.org/10.1177/1461444810378364

Reagle, J. (2012). *Good faith collaboration: The culture of Wikipedia.* Cambridge, MA: MIT Press.

Reagle, J. (2013). "Free as in sexist?" Free culture and the gender gap. *First Monday, 18*(1). https://doi.org/10.5210/fm.v18i1.4291

Stevenson, M. (2016). The cybercultural moment and the new media field. *New Media & Society, 18*(7), 1088–1102. https://doi.org/10.1177/1461444816643789

Stevenson, M. (2018). Having it both ways: Larry Wall, Perl and the technology and culture of the early web. *Internet Histories, 2*(3–4), 264–280. https://doi.org/10.1080/24701475.2018.1495810

Tkacz, N. (2014). *Wikipedia and the politics of openness.* Chicago, IL: University Of Chicago Press.

Turner, F. (2009). Burning man at Google: A cultural infrastructure for new media production. *New Media & Society, 11*(1–2), 73–94.

Weber, M. (1978). *Economy and society: An outline of interpretive sociology.* Berkeley, CA: University of California Press.

Wikipedia:Writing for the opponent. (2018). In Wikipedia. https://en.wikipedia.org/w/index.php?title=Wikipedia:Writing_for_the_opponent&oldid=851817754

Wikipedia:Assume good faith. (2019). In Wikipedia. https://en.wikipedia.org/w/index.php?title=Wikipedia:Assume_good_faith&oldid=895290482

6

Commons-Based Peer Production and Virtue (reprint)

Yochai Benkler and Helen Nissenbaum*

Commons-based peer production is a socio-economic system of production that is emerging in the digitally networked environment. Facilitated by the technical infrastructure of the Internet, the hallmark of this socio-technical system is collaboration among large groups of individuals, sometimes in the order of tens or even hundreds of thousands, who cooperate effectively to provide information, knowledge or cultural goods without relying on either market pricing or managerial hierarchies to coordinate their common enterprise.[1] While there are many practical reasons to try to understand a novel system of production that has produced some of the finest software, the fastest supercomputer and some of the best web-based directories and news sites, here we focus on the ethical, rather than the functional dimension. What does it mean in ethical terms that many individuals can find themselves cooperating productively with strangers and acquaintances on a scope never before seen? How might it affect, or at least enable, human action and affection, and how would these effects or possibilities affect our capacities to be virtuous human beings? We suggest that the emergence of peer production offers an opportunity for more people to engage in practices that permit them to exhibit and experience virtuous behavior. We posit: (a) that a society that provides opportunities for virtuous behavior is one that is more conducive to virtuous individuals; and (b) that the practice of effective virtuous behavior may lead to more people adopting virtues as their own, or as attributes of what they see as their self-definition. The central thesis of this paper is that socio-technical systems of commons-based peer production offer not only a remarkable medium of production for various kinds of information goods but serve as a context for positive character formation. Exploring and substantiating these claims will be our quest, but we begin with a brief tour through this strange and exciting new landscape of commons-based peer production and conclude with recommendations for public policy.

I Commons-Based Peer Production – Examples

The best-known examples of commons-based peer production are the tens of thousands of successful free software projects that have come to occupy the software development market. Free or open source software development is an approach to developing software that resembles nothing so much as an

*We have benefited from the generous help and wisdom of others, particularly Julia Driver, Francis Grodzinsky, Gilbert Harman, George Kateb, participants at the Conference on Computer Ethics: Philosophical Enquiry 2003 and the valuable research assistance of Daniel J. Bloch. Arguments were improved by the careful and astute comments of reviewers for this Journal.

Source: From Yochai Benkler, Yale Law School and Helen Nissenbaum, Culture & Communication, New York University. *The Journal of Political Philosophy*: Volume 14, Number 4, 2006, pp. 394–419 © Blackwell Publishing.

The Handbook of Peer Production, First Edition. Edited by Mathieu O'Neil, Christian Pentzold, and Sophie Toupin.

idealized barn raising – a collective effort of individuals contributing towards a common goal in a more-or-less informal and loosely structured way. No single entity "owns" the product or manages its direction. Instead, it emerges from the collaboration of groups of developers, ranging from a few individuals up to many thousands. Many of the participants are volunteers working in their spare time. Some are paid by corporations that do not themselves claim ownership in the product, but benefit from its development by selling services or equipment associated with the software. The flagship products of free or open source software development – the GNU/Linux operating system, the Apache web server, Perl and BIND – are the most famous. But at any given moment there are tens of thousands of free software development projects, and hundreds of thousands of software developers collaborate on them in various forms to produce some of the world's best software.[2] As Moglen pointed out, free software gains its salience from its functionality. One can compare the products of free software development communities with those of corporations, like Microsoft's. There is a technical answer to the question: is this software better or worse?[3] It is this measurable quality that has forced businesses and governments to take notice of free software. It is what caused the President's Technology Advisory Committee in 2000 to recommend U.S. adoption of open source software as a strategy for supplying mission critical software.[4] Measurable contributions to its machines and services caused IBM to invest over a billion dollars to support development of the Linux kernel and Apache Web Server software, without seeking ownership in the product.

While its functional success forces observers to take free software seriously as a sustainable form of production, what makes free software interesting from a social or moral perspective is its social and human structure. No one "owns" a free software project, though individuals own – in a formal sense – the software they contribute. Its touchstone is that all these individual contributors agree that none of them shall exclude anyone else from using it – whether they contributed to the development or not. No one is a formal manager who tells different people what they must do so that the project can succeed. Though leadership is present in many projects, it is based on no formal power to limit discussion, prevent subgroups from branching off if they are unhappy with a leadership decision, and in any event never involves the assignment of projects – no one can require or prohibit action by anyone. The effort is sustained by a combination of volunteerism and good will, technology, some law – mostly licensing like the GNU General Public License that governs most free software development – and a good bit of self-serving participation. But all these factors result in a model of production that avoids traditional price mechanisms or firm managers in organizing production or motivating its participants.

While the measurable efficacy of free software has captured wide attention, free software does not exhaust the universe of instances where one sees this emerging phenomenon of "barn raising"-like production on the Net. As one begins to look at information, knowledge and cultural production on the Internet, it becomes clear that free software is but one, particularly salient, instance of a more general phenomenon, the phenomenon of commons-based peer production. To provide something of a sense of this phenomenon and its human characteristics, we offer a few more examples. The first two capture the potential efficacy of widespread volunteer effort. The latter begin to give texture to the claim that these efforts offer a platform for qualitatively attractive human behavior.

The simplest example of large-scale volunteer production is distributed computing. Take SETI@home for example. The project is a scientific experiment that uses Internet-connected computers in a Search for Extraterrestrial Intelligence (SETI). The data sets collected from large radio telescope observations are immense. The project was organized to harness the computer processing cycles of millions of volunteers with computers connected to the Internet to process these vast data sets. Participants download a small free program that functions as a screen saver when they are not using their computers. At that point, it downloads and analyzes radio telescope data. According to statistics maintained on the SETI@home website, as of August, 2003, the project had absorbed over 4.5 million users from 226 countries, and provided an average computation speed almost twice that of the fastest "supercomputer" then in operation in the world. The approach, called distributed computing, has been similarly harnessed to simulate the process of protein folding (Folding@home), to model the evolution of drug resistance and design anti-HIV drugs (Fightaidsathome), and a host of other scientific and publicly minded projects.

A step up in human participation, but still fairly mechanical, was the NASA Clickworkers experiment. In this project, tens of thousands of individual volunteers collaborated in five-minute increments to map and classify Mars's craters, performing tasks that would normally require full-time PhDs working for months on end, freeing those scientists for more analytic tasks. In the first six months of the project's operation, over 85,000 users visited the site with many contributing to the effort, making over 1.9 million entries (including redundant entries of the same craters, used to average out errors). An analysis of the quality of markings up to that point showed "that the automatically computed consensus of a large number of Clickworkers is virtually indistinguishable from the inputs of a geologist with years of experience in identifying Mars craters."[5]

Both Clickworkers and distributed computing on the model of SETI@home offer examples that are easy to comprehend and measure, but involve relatively mundane and small-scale contributions. They require relatively little of their participants. To outline the type of behaviors that one sees in these collaborations, we turn to three richer examples of large-scale collaboration, where contributions are larger and require more of the knowledge of the participants and their willingness to participate in a cohesive social process.

The first such project is the Wikipedia project, which involves some 30,000 volunteers who collaborate to write an encyclopedia. While they have not been able to generate a complete encyclopedia in their roughly three years of operation, they have made substantial progress, producing about 250,000 articles in English and many more articles in other languages. Readers are invited to test their own evaluation of the quality, but we would venture that Wikipedia holds its own by comparison to all other online encyclopedias, excluding, perhaps, Britannica. What Wikipedia provides, then, is a rich example of a medium sized collection of individuals, who collaborate to produce an information product of mid-brow quality and who are reasonably successful.

The Wikipedia project runs on a free software collaborative authorship tool, Wiki, which is a markup language similar in concept to HTML but optimized to permit multiple users to edit a single document and interlocking documents while generating archives of the changes made to each. Unlike the projects we will describe in the following few paragraphs, Wikipedia does not include elaborate software-controlled access and editing capabilities. On the contrary, its most interesting characteristic is the self-conscious use of open discourse, usually aimed at consensus, and heavy reliance on social norms and user-run quasi-formal mediation and arbitration, rather than on mechanical control of behavior. It begins with a statement of community intent – to produce an "encyclopedia" – rather than a series of opinion pieces. It continues with a technical architecture that allows anyone to contribute, edit and review the history of any document easily. These two characteristics account for the vast majority of document development. Someone starts a definition. Others update, contributing substance and editorial improvements. Occasionally, disagreements will arise. These are usually dealt with in a "Talk" page associated with every definition. On relatively rare occasions when disagreement persists, there are mechanisms for mediation, and ultimately arbitration, run by participants in the community who are chosen by other users for their sustained commitment to the project. For more decisions, such as determining the policies of Wikipedia, consensus rather than majority vote is the practice. In the cases of vandalism, which occurs on Wikipedia occasionally, the first line of defense is provided by the editors themselves, who have simple means of reverting to an earlier, clean version. The last line of defense is for system operators to block a user.

The important point is that Wikipedia requires much more than mere mechanical cooperation among participants. It requires a commitment to a particular approach to conceiving of one's task, and a style of writing and describing concepts, that are far from intuitive or natural. It requires self-discipline. It enforces the behavior it requires primarily through appeal to the common enterprise in which the participants are engaged, coupled with a thoroughly transparent platform that faithfully records and renders all individual interventions in the common project and facilitates discourse among participants about how their contributions do, or do not, contribute to this common enterprise. This combination of an explicit statement of common purpose, transparency, discourse and the ability of participants to identify each other's actions and counteract them – that is, edit out "bad" or "faithless" definitions – seems to have succeeded in keeping this community from devolving into inefficacy or worse.

What is surprising from the perspective of established conceptions of social cooperation is that this success occurs not in a tightly knit community with many social relations to reinforce the sense of common purpose and the social norms embodying it, but in a large and geographically dispersed group of otherwise unrelated participants: there are as we write about 25,000 participants in the English language Wikipedia, and another 30,000 participants contributing to younger Wikipedia projects in many other languages.

Perhaps the most visible collective commentary project on the Internet as of the mid-2000s is Slashdot, a collaboration platform used by between 250,000 and 500,000 users. Users post links to technology stories they come across, together with comments on them. Others then join in a conversation about the technology-related events, with comments on the underlying stories as well as comments on comments. Comments are in turn "moderated" by other readers in small increments, for quality and relevance: To "moderate" in this system means to grade a comment – to mark whether it is relevant or not, high or low quality, etc. Users who are registered, rather than anonymous, and who have posted for a while, are given by the system limited moderation privileges. Each moderator has a single vote, positive or negative, on any given comment. Out of the collective judgment of the users who chose to moderate a given comment, a collective judgment is computed. The comment is then associated with a certain value, ranging from -1 to 5, indicating its quality and relevance to the topic of conversation, as judged by the moderators in the aggregate. Users can then set their browsers to read only comments above a certain threshold they choose to use, or they can organize their reading of the comments based on the quality judgments of their peers. Out of these mechanisms a newsletter emerges that is widely read as a highly informative source of information about computer software in particular, and information and communications technology more generally.

The relative roles of technology and social norms in Slashdot and Wikipedia are very different. The Slashdot software platform, Slash, is given a very active role in moderating the discussion and the peer review process. Rather than relying on self-discipline and a sense of common purpose, the software builds in limits on use that are designed to constrain anti-social behavior. For example, each user receives only five moderation points in any three day period. This severely limits the amount of influence any one person can have on the collective judgment of a group of hundreds of thousands of users. Users cannot post constantly and automatically – their submissions will be rejected if they try to post comments too often in any minute or hour. But the system also relies on collective judgment and mutual review. Every person who moderates comments is subject to peer review. Users who agree to perform this peer review, or "metamoderation," receive a series of anonymous moderations produced by other participants. They rank these moderations as fair or unfair. A moderator whose judgments are consistently considered by others to be unfair will no longer be permitted by the system to moderate comments. Because there is no ultimate single document, the conversation facilitated by Slashdot does not require the kinds of dispute resolution mechanisms that Wikipedia has employed. Disagreements persist in the record of the conversations, to be resolved by subsequent readers.

There are many other peer-production projects on the Net. Somewhat similar to Slashdot is Kuro5hin, designed to enable its users to share more substantial writing efforts than the usually brief comments produced for Slashdot. Scoop, the software that runs Kuro5hin, functions in many senses similarly to Slash. The most interesting difference is that in Kuro5hin, there is a substantial degree of pre-publication peer review in a "submissions" area, whereas in Slashdot all the discussion is public. Kuro5hin also involves editorial commentary, rather than simply substantive discussion and an up-or-down moderation. A more distinct, but no less impressive exercise, is the Open Directory Project. That site relies on tens of thousands of volunteer editors to determine which links should be included in a human-edited directory similar to Yahoo. Acceptance as a volunteer requires application. Not all are accepted, and quality relies on a peer review process based substantially on seniority and engagement as a volunteer. The site is hosted and administered by Netscape, which pays for server space and a small number of employees to administer the site and set up the initial guidelines, but licenses the database freely. The volunteers are not affiliated with Netscape. Out of the joy of doing so, or for other internal or external motivations, they spend time in relatively small increments selecting sites for inclusion in the directory. The result has been perhaps the most comprehensive, highest quality human-edited directory of the Web – competing with, and often outperforming, Yahoo in this category.

II Commons-Based Peer Production – Principles

The phenomenon of large- and medium-scale collaborations among individuals, organized without markets or managerial hierarchies, is emerging everywhere in the information and cultural production system. Elsewhere, Benkler has provided a detailed analysis of the economics of this emerging phenomenon;[6] here we briefly recapitulate this analysis, with a particular focus on characteristics that are relevant to the specific arguments of this paper.

At its core, peer production is a model of social production, emerging alongside contract- and market-based, managerial-firm based and state-based production. These forms of production are typified by two core characteristics. The first is decentralization. Authority to act resides with individual agents faced with opportunities for action, rather than in the hands of a central organizer, like the manager of a firm or a bureaucrat. The second is that they use social cues and motivations, rather than prices or commands, to motivate and coordinate the action of participating agents. As a descriptive matter, the phenomenon is a product of the emergence of digital networks and the rising importance of information and cultural production. The wide distribution of low-cost processors, coupled with increasingly ubiquitous computation, changes the capital structure of information production. Physical capital is widely distributed and owned by those individuals who also are capable of contributing the other major input into information and cultural production – human effort and creativity. Because it obviates the need for centralized capital investment, this capital structure makes possible – though does not require – the reorganization of at least some information and cultural production along decentralized lines. In this technical-economic context, peer-production enterprises appear to be emerging as newly feasible social and technical systems that motivate and organize human collective contributions by means other than contracts and monetary compensation for the use of physical capital.

Commons-based peer-production relations regularly exhibit three structural attributes. First, the potential objects of peer production must be modular. That is, they must be divisible into components, or modules, each of which can be produced independently of the production of the others. This enables production to be incremental and asynchronous, pooling the individual discrete efforts of different people, with different capabilities, who are available at different times. Second, the granularity of the modules is important. Granularity refers to the sizes of the project's modules, and in order for a peer-production process successfully to pool a relatively large pool of contributors the modules should be predominantly fine-grained, or small in size. This allows the project to capture contributions from large numbers of contributors whose motivation level will not sustain anything more than quite small efforts towards the project. Novels, for example, at least those that look like our current conception of a novel, are likely to prove resistant to peer production. But as we have already suggested, encyclopedia entries, judgments about the worth of one or another website and components of software programs are commonly and effectively produced in this fashion. In addition, a project will likely be more efficient if it can accommodate variously sized contributions. Heterogeneous granularity will allow people with different levels of motivation to collaborate by contributing smaller or larger grained contributions, consistent with their level of motivation.

Finally, a successful peer-production enterprise must have low-cost integration – the mechanism by which the modules are integrated into a whole end product. Integration must include both quality controls over the modules and a mechanism for integrating the contributions into the finished product. First, the project must include an established, low-cost way of defending itself against both incompetent and malicious contributions. Given that peer production is dependent on self-identification of people for projects, each community must have a way of weeding out contributions from those who misidentify their talents. Second, the project must include a mechanism for integrating the competent modules into a finished product at sufficiently low cost. As one observes actual peer-production communities, a number of robust methods have emerged. First, one sees automated integration and iterative peer production of integration. For example, the use of free software mechanically to integrate modules of some other information good is a primary mechanism by which particular peer-production projects like Slashdot, Kuro5hin and Wikipedia have lowered the cost of integration to the point where they can succeed and sustain themselves. Second, one sees peer-production enterprises using a variety of approaches towards solving collective action problems that are relatively familiar from the offline commons literature.[7] These include various formal rules, like the GNU GPL (General Public License) that prevents defection from many free software projects, including most prominently the GNU/Linux operating system. They also include technical constraints that prevent or limit the effect of defection, as in the case of the limited voting power that Slash or Scoop give each individual editor on Slashdot or Kuro5hin, respectively. Social norms too play a role in sustaining some of these collaborations, both where there are small groups, and where there are larger groups and the platform allows for good monitoring and repair when individuals defect. This approach is particularly salient in the Wikipedia project. Finally, the NASA Clickworkers project suggests that the sheer size of some of these projects enables the collaboration platform to correct for defection by using redundancy of contributions and averaging out the contributions of outliers – be they malicious or incompetent.

When they succeed in motivating and organizing collaborations, peer production enterprises have two primary advantages from a purely economic perspective over both markets and firm hierarchies. The first is an information gain. Because individuals have widely variable creativity, experience, insight, motivation and availability, human capital tends to be hard to specify for efficient contracting or formal organizational assignment. Firms and markets therefore simplify decision making by losing a lot of information about the tremendous variability of human creativity and motivation over time and context. Peer production, by contrast, allows individuals to self-identify for tasks that attract them and for which they are suited. As long as a peer-production enterprise institutes mechanisms for peer review of some sort to weed out mistakes, peer-production enterprises generate more textured and dynamically updated information about the capabilities and availability of agents for actions. Second, the variability in fit of people to projects and existing information resources is great. This leads to substantial increasing returns to scale to the number of people, resources and projects that may be pursued without need for a contract or other transaction permitting the use of the resource for a project. The larger the number of people who can potentially work on projects, the larger the number of resources with which they can work to pursue projects; and the larger the number of projects they can initiate and imagine, the higher the probability that the best set of persons will be able to work on the set of resources with which they would be most productive, towards the project most suitable from that combination. Peer production, by making information resources freely available to potentially huge collections of individuals, maximizes the effect.

Before turning to an analysis of the relationship between the emergence of peer production and virtue, it is important to underscore one further central characteristic of peer production. By definition, peer-production enterprises are non-price based, that is, they are devoid of marginal payments to contributors for contributions. While some contributors contribute because of an expectation of learning and earning a reputation that could translate into a job in the future, most of the participation cannot easily be explained by a relatively mechanistic reliance on economic incentives. Rather, it seems that peer-production enterprises thrive on, and give opportunity for, relatively large scale and effective scope for volunteerism, or behavior motivated by, and oriented towards, positive social relations. People contribute for a variety of reasons, ranging from the pure pleasure of creation, to a particular sense of purpose, through to the companionship and social relations that grow around a common enterprise. What makes peer-production enterprises work best has been the capacity to harness many people, with many and diverse motivations, towards common goals in concerted effort. While understudied and difficult to predict and manage by comparison to a more simple picture of human motivation as driven by personal wealth maximization, peer production begins to offer a rich texture in which to study the much more varied and multifarious nature of human motivation and effective human action.

The economic potential of the phenomenon of commons-based peer production may or may not be sufficient reason to support its growth and justify attention to factors or conditions needed for its flourishing. It certainly suggests the potential staying power and sustainability of this mode of production in an economy and society heavily attentive to economic performance. Here, however, we are interested in considerations beyond economic efficiency. Taking a moral perspective, we argue that the remarkable social and technical phenomenon of commons-based peer production fosters virtue by creating a context or setting that is conducive to virtuous engagement and practice, thereby offering a medium for inducing virtue itself in its participants.

III Commons-Based Peer Production and Virtue

Before we are able to develop our claim about virtue and commons-based peer production, we need to say something about the notion of virtue that we will be using. Those of us educated in a Western, analytic school of philosophy tend to think of virtue as essentially an Aristotelian creation, passed down and elaborated along the way by just about every historically significant moral philosopher, including Kant, Nietzsche and Hume, falling into disfavor in the mainstream until roughly two decades ago and benefiting from a revival of interest through persuasive interventions of contemporary moral philosophers such as Philippa Foot, Alasdair MacIntyre, Martha Nussbaum, Bernard Williams, Michael Slote, Rosalind Hursthouse and others. Yet the idea of virtue is far richer than even that pedigree would suggest, spanning centuries as well as cultures and religions. Robust notions of virtue have been recorded in commentaries on ancient Confucian philosophers, such as those by Xunzi, born circa 310 B.C.E. Not only has it been theorized by countless philosophers and theologians throughout ages and across continents, but it has a robust meaning in natural languages, regularly uttered in everyday speech.

We have tried not to tie our discussion to a specific theory or doctrine of the virtues and virtue ethics. By staying as close as possible to an intuitively plausible sense of virtue, remaining neutral on many of the most controversial theoretical questions, and plumbing only the most robust insights of scholarship where they are relevant to our arena of application, we have aimed for as broadly appealing a conceptual foundation as possible. This neutrality applies to three broad areas: We make no commitment to the specific number or particular catalog of the virtues but assume the existence and character of those that have enjoyed broad recognition. We are agnostic on the foundations of virtue, whether in ideals of human flourishing and the good life (Aristotle, MacIntyre), naturalism (Foot and Hursthouse) or utilitarianism (Hume, Driver).

Finally, we take no position on the meta-ethical question of whether virtue ethics is best understood as a rival to deontological and Utilitarian accounts of "the problems and phenomena of ethics."[8] As a working position, we prefer David Wiggins' ecumenical line: What a grown-up moral philosophy might attempt is an account of morality that embraces the full gamut of moral predications, seeing them as mutually irreducible and mutually indispensable, allowing no primacy to character traits or practices or states of affairs – or allowing primacy to all at once.[9] Ideally, this stance will allow the primary focus of our discussion, and inevitable controversy, to settle on concrete claims rather than these much debated elements of the conceptual landscape.

At a minimum, however, we take virtue ethics to be an important approach to moral evaluation, which offers a framework for appraising people over time, in contrast with other dominant approaches that appraise actions, atomistically, in terms of consequences or compatibility with deontological rules. In other words, where the basic unit of moral evaluation for rival frameworks is individual actions (or action-types), the basic unit of moral evaluation for virtue ethics is the person (or soul or character), an entity persisting over time. Accordingly, Rosalind Hursthouse writes, "If you have the virtues of, say, generosity, honesty, and justice, generous, honest, and just is the sort of person you are."[10] Most contemporary virtue theorists and virtue ethicists consider the character to be the bearer of virtue, "an admirable character trait;"[11] that is, "those qualities of character the possession and exercise of which make human beings flourish."[12] Finally, most virtue theories allow that character, the bearer of virtues, is not immutable and its virtue may ebb and flow.

By most accounts, virtues are dispositional properties, though particularly with many of the most important virtues such as, honesty, justice, courage and benevolence they are not simple dispositions to act in narrowly specified ways. Being honest, for example, does not merely involve telling the truth, or never telling a lie, but a complex pattern of beliefs, desires, emotions, preferences, sensitivities, opinions and broadly related actions and practices, even as general as how one rears one's children. Equally clear is that particular instances of beliefs, actions, emotions and so forth need not imply the virtue of honesty, as, for example, the conman who happens to tell the truth in a court of law. This point, thoroughly discussed in contemporary writings such as those of Hursthouse and others, is also recognized in ancient works, such as in those of Xunzi, who holds that virtues are dispositions involving the faculties of choice, judgments, desire, emotion and action. Further, virtues can be manifested in a great variety of ways, depending on circumstances.[13]

Later in the paper, we will refer to other general features of virtues but, first, it will be useful to plunge directly into an analysis of systematic associations between particular virtues and characteristics inherent to commons-based peer production. To avoid quibbling over differences that may have more to do with contingencies of language than with substance, it makes sense to put forward these associations in terms of clusters of virtues associated with socio-technical systems of commons-based peer production rather than individually named virtues.

A Cluster I: Autonomy, Independence, Liberation

As noted above, an essential feature of commons-based peer production is volunteerism and self-selection. In the first place, individuals have chosen freely to participate and are free to continue or cease to participate as they please. Usually, they are able to contribute when and how much they want, and can select aspects of production according to their own criteria. In the typically decentralized, non-hierarchical settings, even if participants seek to please and impress peers, they need not cower to a boss or any other such authority. As volunteers, they exercise independence of will, initiative, even self-reliance, discretion and free-spiritedness. No matter what other demands constrain their lives, participation in peer production constitutes an arena of autonomy, an arena where they are free to act according to self-articulated goals and principles. In this arena, they manifest, in Charles Taylor's terms, the virtue of "liberation," manifest in bearers "directing their own lives, . . . deciding for themselves the conditions of their own existence, as against falling prey to the domination of others, and to impersonal, natural, or social mechanisms which they fail to understand, and therefore cannot control or transform."[14]

Note that when one is speaking of autonomy as a virtue, as an attribute of character, the ambiguity between autonomy as an instantiated property versus a human potentiality recedes as it is unlikely we would value in another the merely presumed capacity to be free. It is actual liberation, actual self-direction, that is admirable in a character. In the context of commons-based peer production, this may mean independence from the wide-ranging commercial entities influencing our actions and choices as well as from the typical array of institutional entities, whether employers, banks, agents of government, or whoever.

B Cluster II: Creativity, Productivity, Industry

Through their involvement in commons-based peer projects, participants are able to reach beyond the humdrum routines many of us experience in our workdays, including those of us privileged to live materially comfortable lives in industrialized and wealthy nations. Even our recreational choices tend to be passive and limiting, such as selecting among TV channels, watching movies and shopping in malls. While the industrially organized, mass market economy largely structures so many of our choices as comparatively passive consumption choices, or comparatively regimented production choices in fairly controlled work environments, peer production opens up new avenues for creative, productive practices. Few of us will write novels, create encyclopedias or works of art, or produce effective computer programs. Fewer still will do so in their spare time. But peer production offers a medium for contributing our thoughts, our knowledge, our know-how, or merely the spare cycles of our PCs toward a meaningful product.

Peer production offers the possibility of engagement in what MacIntyre terms a "practice," namely, a "socially established human activity through which goods internal to that form of activity are realized in the course of trying to achieve those standards of excellence which are appropriate to, and partially derivative of, that form of activity, with the result that human powers to achieve excellence, and human conceptions of the ends and goods involved, are systematically extended."[15]

In commenting on other forms of peer-based engagement, a broader category that includes commons-based peer production, others have highlighted their potential for active rather than passive intellectual and social participation online. Andy Oram, for example, has asserted that, "peer-to-peer technologies return the Internet to its original vision, in which everyone creates as well as consumes."[16] Those engaged in peer-to-peer activities "are active participants, not just passive 'browsers.'"[17] They are writing code, collaborating in community networks, commenting on the news, and so on.

C Cluster III: Benevolence, Charity, Generosity, Altruism

If the previous two virtue clusters can be considered "self-regarding" virtues,[18] this cluster and the next can be considered "other-regarding," or in Hume's terms, "social virtues." Although each element of this cluster has a distinctive character and is favored in varying measure by theorists and scholars of virtue, their common core is the disposition to benefit others, "to seek the good of others as an end in itself, and in circumstances in which it is not required of us."[19] Among all the virtues, these are central to almost all theories of the virtues and virtue ethics, serving as in David Hume's work, as one of the bedrocks of a general moral theory. Benevolence and generosity require not only that the good of others be furthered, but that a cost be borne by the generous individual alone.

Participants in commons-based peer production benefit others by contributing time and effort that could, in principle, be spent in more directly self-serving pursuits. In helping others, in small ways such as donating spare cycles, or larger ways such as creating carefully researched encyclopedia entries without receiving conventional, tangible payments or favors in return, peers exercise kindness, benevolence, charity and generosity. In the specific case of free and open source software, the literature is ambiguous on the centrality of the role that this cluster of virtues plays. Some, like Richard Stallman, founder of the free software movement, seem more animated by the linkage with virtues in the autonomy and self-reliance cluster, though they also highlight themes of helping friends and neighbors through the sharing of software.[20] Some proponents of open source, trying to bring it into the business world's mainstream, have sought to depoliticize free software by explaining the motivations of participants in terms palatable to believers in *homo economicus*.[21] And yet, many who have worked hard to make open source tractable to economists and business people have lauded its underlying "gift culture."[22] Although it is entirely possible that the persistent and pervasive practice of spending time and effort producing something of value and giving it freely to be used by others for no compensation can be explained as self-serving behavior in pursuit of, say, reputation, a more efficient and direct explanation in many, if not most cases, is the pleasure or satisfaction of giving – generosity, kindness, benevolence.[23]

D Cluster IV: Sociability, Camaraderie, Friendship, Cooperation, Civic Virtue

This cluster of virtues is thematically related to Cluster III, but not identical to it. Cluster III virtues involve giving to others, sometimes needy others, to benefit them – and if Hunt's thesis is to be believed – at a cost to the giver. In this cluster, the virtues also imply giving, but the open-hearted contribution is to a commons, a community, a public, a mission, or a fellowship of which the giver is a part, and the giving dimension might be only one aspect of it. Its core is a conception of the self as part of a collective and of one's efforts as a part of a collective effort, whether the collective or common search for extra-terrestrial life, the quest for a free encyclopedia for all, or for a balanced, popular vision of advanced technologies in society. The giving, therefore, does not merely involve agents parting with something of value, but agents working in cooperation with others to give or produce something of value to all.

Whereas generosity, benevolence, and so on are universally present in the explicit lists of virtues proposed by ethicists from Aristotle to Rosalind Hursthouse, from Xunzi to Martha Nussbaum, the virtues within this cluster are rarely encountered in explicit terms. Although, arguably, they are implicitly present in, for example, David Hume's list of the virtues of "humanity, benevolence, friendship, public spirit, and other social virtues of that stamp,"[24] Hume does not develop the notion of public spirit, nor say much about its relation to benevolence. MacIntyre, too, seems interested in the social contribution of virtues: "The catalogue of the virtues will therefore include the virtues required to sustain the kind of households and the kind of political communities in which men and women can seek for the good together and the virtues necessary to enable us to understand what more and what else the good life for man is."[25] But in extended discussion, he does not develop in detail the relation between political and other communities and specific virtues.

Among political theorists, however, these social, or civic virtues have had greater salience. Defenders of liberalism, for example, have sought to characterize "liberal virtues," including among them the dispositions to engage in voluntary associations and to promote common ends in social cooperation.[26] Michael Sandel, a renowned contemporary proponent of Republicanism, traces the commitment to these civic virtues in the United States back to the founders, who sought to nourish "public virtue" in the citizenry. A complex virtue, public virtue involved a commitment to the public good, or, in John Adams' words, "a positive passion for the public, the public interest, honour, power and glory."[27] According to Sandel, the Revolutionaries valued "public good" second only to liberty, meaning by public good, "more than the sum of individual interests. Accordingly, the practice of politics should be aimed not merely at brokering a compromise among competing interests but to transcend these interests by seeking the good of the community as a whole."[28] In Sandel's account, the founders believed "civic virtue" to be the bedrock of liberal democracy.[29]

In a similar way, participants in a commons-based peer effort cooperate, build upon the work of others, contribute time, effort and expertise to create and enhance a public good. The self-reliance, vaunted by Richard Stallman and other proponents of free software, is not in tension with fellowship – rather, it is only in tension with the incapacity to make do for oneself, and hence reliance on the commercial other to make do for one. The act of creating for oneself and one's fellows is an act both of self-reliance and of fellowship – like barn-raising or establishing a community watch.

IV From Structure to Virtue

So far, our analysis has established a structural connection between key defining properties of commons-based peer production and the possibility of engagement in creative, autonomous, benevolent and public-spirited undertakings. This is a less ambitious claim than the one with which we set out, namely, that the practice of commons-based peer production is connected to virtue. The remaining sections of the paper will argue that a plausible causal connection can be drawn in two directions: first, that virtue leads people to participate in commons-based peer projects, and second, that participation may give rise to virtue. Based upon these observations, we conclude the paper with some prescriptions for public policy and design.

Even if participation in peer production yields benefits to others, contributes to the common good, is a setting for cooperative activity, has the capacity to engender autonomy and so forth, the claim of virtue requires that participants are, in fact, acting generously, exercising autonomy and so forth. Supporting this claim requires more than pointing out that the behaviors benefit others, promote the public good and so forth, but also that they are performed with the right kinds of attitudes and motives. (There is also the theoretical question of the role of motivation in defining particular virtues, which lies outside the scope of this paper.) This issue is not new for many admirers of open source and free software, who continuously seek to account for the motivations of project participants.[30] Establishing the strongest version of the claim, that *all* participants are motivated by benevolence, good-will, fellowship and so forth, is impossible and also unnecessary. Without a doubt, there are many reasons people participate. It will suffice if we are able to show that virtuous motivations are at least a substantial part of the picture.

Laudable actions, in general, may arise out of a variety of motives. We know, for example, that people give gifts for many reasons besides sheer generosity, including a wish to reciprocate, to win favor, to impress onlookers or recipients or to place recipients in the giver's debt.[31] People behave fairly or even-handedly as much out of fear of disapproval or selfish ambition – say, to promote their own professional status – as out of the virtue of justice. And likewise with other virtues and related behaviors. Moreover, it is likely that participants in the projects described above and others like them are motivated by many factors, which may also vary systematically across projects and people and within individuals, over time. Some may be lonely and seeking company, others may wish to promote their chances of a good job, still others may seek the benefits of learning the craft through participation in one of the peer-production projects. None of these reasons is morally reprehensible. But for purposes of drawing the connection between commons-based peer production and virtue, it is crucial that we discover a substantial set of participants whose motivations implicate the four clusters of virtue. Such knowledge is of empirical fact and not purely a matter of analysis.

Despite considerable interest in the issue of motivation, particularly regarding free and open source projects, we found no empirical studies that could rigorously confirm the causal connection between virtue and participation in commons-based systems of peer production. There are, nevertheless, suggestive findings – more than mere anecdotes – that lend systematic support to our thesis. One datum comes from the SETI@home website, where volunteers are asked by organizers to participate in a poll which includes a question about motivation: "What is your main reason for running SETI@home?" Participants are offered a set of multiple choice answers, including an option of "Other," which allows for free form answers as well. With a number of 117,894 participants when the site was last visited the results were:

Find ET for the good of humanity 58.33%
Find ET to become famous 3.08%
Keep my computer productive 16.92%
Get my name on a top 100 list on the web site 2.29%
Other 19.37%[32]

The free-form responses stimulated by the "Other" category were even more telling. As illustrated below, respondents indicated a clear attraction to the opportunities SETI offered for contributing to the public good, for promoting welfare by helping with scientific research, and for the opportunity to be part of an interesting, possibly momentous project. For example:

- "Because the SETI is one of the greatest science programs running and I like to participate in this great search and, of course, for humanity."
- "Support a worthwhile cause. Participate in the largest parallel processing effort."
- "Find ET and it is just plain cool to help with the research."
- "Find ET and to be part of an exceptional distributed computing project."
- "Find ET for good of humanity and prove the net power."
- "Find ET for humanity. Keep my computer productive, and to just be a part of this great project!"
- "Helping out the Scientific Community."

In the context of free software and open source, the rhetoric of movement leaders like Richard Stallman, Linus Torvalds and Eric Raymond clearly endorses the relevance of values such as autonomy, self-reliance, gift-giving, collaboration, active participation, liberation and creativity in motivating participation.[33] These ideological and anecdotal accounts resonate with findings of the Free/Libre and Open Source Software (FLOSS): Survey and Study, the first large-scale study of the role and importance of open source and free software worldwide. Funded by the European Commission and carried out by Berlecon Research and the International Institute of Infonomics at the University of Maastricht, the FLOSS survey and study not only generated primary data on usage and development, indicators of value dissemination and distribution, business models and economic and regulatory implications. It also included a survey of 2,784 developers worldwide on a variety of topics with a section devoted to reasons (or motivations) for participating. A few of the questions drew noteworthy answers.

In commenting generally on the Open Source/Free Software (OS/FS) scene, the greatest percentage agreed that it enabled more freedom in software development. In significant numbers they also cited new forms of cooperation, opportunities to create more varieties of software and innovative breakthroughs. When asked what they thought other OS/FS developers expected from them, the majority said "share my knowledge and skill," a large percentage said "help in realizing ideas for software projects," and in relation to these and other responses a tiny fraction said "provide better job opportunities" and "make money." As to why they began, the highest proportion answered "to learn and develop new skills," closely followed by "to share my knowledge and skills," "to participate in new forms of cooperation" and "because I think that software should not be a proprietary product." As to why they continue, the first is still dominant, though with a smaller margin, while the other three, namely sharing, cooperating and objecting to proprietary control over software, all rise significantly as motivating forces. It is also striking that although a number of other non-moral reasons achieved mid-range scores, the ones that consistently earned lowest were, "to make money" and "to get a reputation in the OS/FS scene."

We admit that neither SETI@home nor FLOSS study results are conclusive. For one thing, they both suffer from the possibility of a self-reporting bias, people wanting to portray themselves as more altruistic than they in fact are. With SETI@home there is the additional problem of self-selected respondents. Indeed, no matter how many participants we survey, or how closely we observe them, we may never prove to the satisfaction of some (diehards) that these benevolent, independent, civic-minded and productive behaviors were performed with the right types of attitudes to qualify for virtue. We share this plight with a host of others, who somehow must prove to skeptics that true altruism or sympathy or generosity exists. Although the burden of so doing strikes us as absurd, it lies outside the scope of this paper to enter this larger debate. A limited conclusion, however, seems credible: insofar as any other-regarding action is possible, there is good reason to hold that a sizable proportion of peer participation is prosocial, or morally praiseworthy in the ways discussed.

The general question of what warrants the move from observed behaviors to assertions about virtue is a core component of most theories of virtue, but lies outside the scope of the paper. It is normally assumed that to assert the presence of an underlying disposition requires not only that the relevant actions are performed, or performed on a one-off basis, but that they are performed habitually. Pitching in, for example, is praiseworthy, but only one who pitches in regularly is seen as instantiating the virtue of good-fellowship. The projects described earlier and many more provide precisely this type of opportunity for repeated engagement in relevant actions over an extended period of time, so that we can say that those who participate are the kinds of people who give to others with no prospect of direct payoff or punitive sanctions. Such engagement may reasonably be seen by others, *as well as by participants themselves*, as expressions of the associated virtues.

Assuming we have shown that a significant number of participants in peer production are acting in ways that are morally and politically praiseworthy, not only producing utility but doing so for the right reasons and with the right motives, we would like to take our account one step further, suggesting not only that in many instances reliable and long-term participation constitutes evidence of virtue but that participation may also lead to virtue. This causal connection from behavior to virtue is arguably the more interesting one.

For those of us accustomed to conventional philosophical thinking about action as the upshot of beliefs, desires, emotions and so forth, the reverse direction of causation may seem odd. Nevertheless this is widely embraced by virtue theorists and ethicists, who agree that significant sources of virtue include good habits and practice. Grounded also in common sense ideas about the power of good habits, this idea is evident in Leon Festinger's notion of cognitive dissonance, which captures the paradoxical (but ultimately compelling) idea that people's attitudes and beliefs are frequently formed as a consequence of their actual choices and actions and not vice versa. Aristotle endorses the importance of practice in his account of the three sources of character formation: nature, explicit teaching and – most central to our discussion – habit. Virtuous action performed habitually could induce or contribute to the attainment of virtue itself. Because of this, Aristotle recommends that children, in addition to being taught explicitly about the virtues, should be trained to adopt certain habits because "it is our actions that determine our dispositions."[34] Moreover, "[i]t is the repeated performance of just and temperate actions that produces virtue. . .. It is therefore quite fair to say that a man becomes just by the performance of just actions, and temperate by the performance of temperate actions."[35] Interpreters of Xunzi, the ancient Confucian philosopher, have attributed to him a similar account of the sources of virtuous character. Accordingly, virtuous practice, alongside learning and training in various rituals, is a crucial element contributing toward development of full virtue. Virtue is thus a product of correct practice; and the capacity to act correctly is a step toward virtue itself. Immanuel Kant affirms this principle: "Helping others to achieve their ends is a duty. If a man practices it often and succeeds in realizing his purpose, he eventually comes to feel love for those he has helped. Hence the saying: you ought to love your neighbor . . . means do good to your fellow-man, and this will give rise to love of man in you."[36]

The Aristotelian ideal of a dedicated mentor to guide the character development and education of each child is no longer feasible in the contemporary landscape. We rely on a far more diverse and less systematic set of offerings, including the mass media, public education and other social institutions (such as museums, religious institutions and a myriad of others), which serve, directly and indirectly, as sources of learning, training and even character formation. If, as we have suggested, participation in commons-based peer production is an instance of an activity that not only enables the expression of virtuous character but serves as a training ground for virtue, it holds the potential to add to the stock of opportunities for pro-social engagement. With this in mind, we now turn to consider some implications of what we have said for public policy.

V Public Policy

In chapter fifteen of *After Virtue*, Alasdair MacIntyre remarks: "Only in fantasy do we live what story we please. In life, as Aristotle and Engels noted, we are always under certain constraints. We enter upon a stage which we did not design and we find ourselves part of an action that was not of our making."[37] MacIntyre means something quite general about the ways people's lives reflect their distinctive narratives mingled together with those of others, constrained by the stories and experiences of families, tribes, traditions, communities, social, political and cultural institutions and historical circumstances, which together shape choices and possibilities by exposing possibilities and defining good and bad behavior, vice and virtue.

A practical corollary of the observation that social and institutional arrangements have the power to shape human behavior and disposition is to strive for positively valued behaviors and dispositions through purposive interventions in these arrangements. This ambition is not unprecedented either in the actions of political leaders or in evaluations of leaders, policies and institutions by members of the academy.[38] Sandel cites noteworthy historical instances in which stances on critical policy issues, emerging during formative periods of the United States, turned on the potential effects on human character of the alternatives. As an example, Sandel cites George Mason's vehement opposition to the Port Bill to promote development of large commercial cities, which was grounded in the belief that development would undermine the moral virtue of residents, diminishing their "frugality, probity and strictness of morals."[39] According to Sandel, such arguments would not have been out of place for the environment in which there was general agreement regarding the principle that: "The public life of a republic must serve a formative role, aimed at cultivating citizens of a certain kind." The government was seen to have "a stake in cultivating citizens of a certain kind." The Constitution of 1787 was another vehicle designed to save "American republicanism from the deadly effects of [the] private pursuits of happiness," and "from the acquisitive preoccupations that so absorbed Americans and distracted them from the public good." In addition to policy and the Constitution, there would be other public institutions aimed at "improving" moral and civic character, from "education, to religion, and more broadly, to the social and economic arrangements that would define the character of the new nation."[40] It goes without saying that the argument supporting the role of social and political institutions in shaping character is separable from substantive beliefs about what counts as virtue.

In *Character and Culture*, the philosopher Lester Hunt shares insights into ways the design of social institutions may influence the development of virtues. He writes:

> [T]here are some virtues that it is comparatively easy to acquire and instill in the context of American institutions. At any rate, parents who are trying to get their children to respect the property of others or keep their promises probably do not so often have the feeling that they have the whole world working against them. Our institutions do seem to be arranged so that they facilitate the acquisition of some good traits of character. . .. We may find, for instance, that some of the limits to the powers that moral instructors possess are not due to the immutable facts of human nature but to institutions that we have the power to change. Perhaps people are no more generous or just than they are, no less envious and vengeful than they are, because of the institutions that influence their behavior, and not because the guardians of virtue – whoever they may be – have failed to be sufficiently vigilant or skillful.[41]

The institutions Hunt has in mind may be loosely construed as cultural rituals like gift-giving, or explicit regulatory vehicles as in the historical cases Sandel discusses, or contemporary counterparts such as tax-deductions for charitable gift-giving. As Hunt observes, the virtue-enhancing properties of any given social institution are complex and subject to a variety of contingencies:

> A theory of the origin of a trait of character does not state that from a specific concrete situation, in all its complexity and with all the features of it that individuals might perceive and to which they might respond, one specific result must emerge. Rather, it picks out certain features of many actual situations – as, for instance, that the people in them are taught a certain type of rule – and shows that these features support the formation of certain traits of character.[42]

Accordingly, the efficacy of the virtue-enhancing properties of commons-based peer production will, likewise, be tempered by contingencies of the highly variable background conditions in which participation takes place.

The past few decades of the philosophical and social study of technology (STS) has shown, however, that it is possible to give a more concrete interpretation of MacIntyre's remark by considering the stage not only as a social context but as the material context, designed by others, into which we must enter. For the philosophers and social scientists who study technology, this metaphor draws attention to a world in which we are constrained not only through the narratives and expectations of the self and other social agents and institutions, but by the material world which is constituted in increasing measure by technology. From Lewis Mumford's authoritarian and democratic technics to Marshall McLuhan's medium as a shaper of content, to Langdon Winner's artifacts with politics, to Bruno Latour's inscription of morality in machines – each of these expresses the common idea that technical systems and devices, in virtue of their properties, architecture or functionality, have the capacity both to limit and to facilitate what individuals and collectivities are able do.[43] Rejecting the view of technology as neutral, producing outcomes only as a result of the uses and applications chosen by people, these theorists of technology, and others, hold that technology embodies values. Values may be "built into" technical design characteristics of technologies, which, in interaction with the social, political, economic and cultural characteristics of the contexts in which they are embedded, produce outcomes skewed in one way or another. In drawing attention to the ways technologies enhance or suppress social, political and moral values, these philosophers, legal scholars and social scientists frequently see their work as continuous with social, political and moral commentary. Brian Pfaffenberger observes:

> All around us today are artifacts that were generated in the technological dramas of their time: railways, canals, aviation artifacts, radios, and more. And yet their meaning, together with their location in what was formerly a deeply felt grammar of political action, is utterly lost; in their place is what appears to be nothing more than a material record of "technological progress." What was once the conscious product of human cultural and political action, passionate and meaningful, is now a silent material reality within which we lead our daily lives, mutely acting out patterns of behavior that once had obvious connections to the root paradigms of our culture. . .. To become fully aware of the political circumstances of their lives, new generations of students, at every level of education, must be trained (as Hughes suggests) to 'fathom the depth of the technological society, to identify currents running more deeply than those conventionally associated with politics and economics.' Because STS offers a way to recontextualize technological artifacts, it is therefore the political philosophy of our time, and it deserves to stand at the center of any curriculum that teaches political awareness and civic responsibility.[44]

In other words, technical systems and devices are as much a part of political and moral life as practices, laws, regulations, institutions and norms that are more commonly seen as vehicles for moral and political values.

The political aim of this paper is similar to those pointing to social or political consequences of specific technical systems, for example, those warning of privacy threats posed by pervasive video and data surveillance and biometric measurement technologies, diminishment of autonomy and accountability due to automated command and control systems and alterations in hierarchies of power and authority due to design features of certain network infrastructures. Unlike many political analyses of technologies, however, ours does not warn of a direct threat of harm. Rather, it warns of a threat of omission. We might miss the chance to benefit from a distinctive socio-technical system that promotes not only cultural and intellectual production but constitutes a venue for human character development.

Some might challenge the underlying premise, arguing that "low-cost" contributions to the public good, such as those afforded by commons-based peer production, are not capable of training virtue. Because an inherent feature of these production schema is to facilitate contribution at relatively low personal cost, its virtue-enhancing capacities are questionable. This characterization misses the point. As noted above, commons-based peer production generates new modes of contributing to the public good by facilitating the collaborative engagement of thousands of ordinary individuals in the voluntary, creative, communal, regular, non-commercial production of intellectual and cultural goods, for a wide variety of reasons and motives. There are also many different *types* of projects, demanding highly variable degrees of effort. Whereas some, such as SETI@home, call for relatively low levels of engagement and effort, others such as Wikipedia[45] call for significant commitment, work, time, patience, dedication, fairness and civic-mindedness.

Effort (or cost) may be a requirement for moral action but it is surely not boundless; effort may be required, but not supererogation. Where to draw the line is not a question we need settle here, except to observe that moral action lies somewhere between the extremes of no-effort and supererogation, and is surely also a function of the good it engenders. There is no reason to appraise participation in peer production any differently. Accordingly, although what critics say may apply to a fraction of the cases, it certainly does not apply to all when we take into consideration the effort involved, the good produced and the widely shared intentionality and self-understanding as providing that public good. When a society endorses social policies that, for example, offer protection to whistle-blowers, provide tax-credits for charitable donations, or generate institutional safeguards against corruption, it is not shrinking the moral sphere, but structuring the environment to lessen the burden of valued practices. In a similar way, commons-based peer production opens a path previously restricted by economic cost and industrial organization to small numbers of professional producers of information, knowledge and culture to large numbers of ordinary people, enabling them to contribute to the public good in a particular domain. The path does not bypass virtuous action, but generates new opportunities for it.

Despite the positive potential of commons-based peer production, there is cause for concern over its future flourishing. The primary sources of resistance are rent-seeking behavior by incumbent firms of the industrial information economy and well-meaning, but ill-informed policies and judicial decisions. The former stems from the fact that commons-based peer production presents an alternative to, and therefore a form of competition with, the incumbents. The latter arises because policy makers and judges have been habituated by the twentieth-century economics of information production to think that market production, in particular established firms, are critical to growth, innovation and creativity. Relatively few have adjusted to the new economics of networks, and recognized the potential and value of peer production.

The most visible conflict between commons-based peer production and incumbent firms is in the area of free or open source software. Here, FLOSS is seen as a potential alternative, and is actively supported and used by major corporations and many governments. This has led Microsoft, in particular, to try to resist these developments.[46] More subtle is the way in which the incumbents of the cultural industries, in particular Hollywood and the recording industry, have successfully lobbied Congress and litigated in courts to expand copyright and to introduce new related technical requirements – like the Digital Millennium Copyright Act of 1998 and the persistent efforts to pass hardware regulation. While these efforts have mostly succeeded in allowing these firms to extract higher rents from their inventories, they impose higher entry barriers than is necessary to the public domain. Because all information production requires access to existing information inputs, these constraints on the public domain create barriers to effective pursuit of peer production. While not always aimed directly at peer production, these efforts raise the costs of maintaining information production activities and dampen the development of some technologies that can be useful to peer production, as well as to other peer-to-peer uses that are more legally controversial – like music file-sharing.

VI Conclusion

We have argued that participation in commons-based peer production fosters important moral and political virtues. We have not made the case that it is therefore incumbent upon the state to support peer production. That would require a greater commitment to a perfectionist state agenda than we have stated or defended here, or are willing to defend. Nonetheless, we have offered new reasons to find peer production to be a morally attractive set of social, cultural and economic practices. There is a growing literature on the relative efficiency of peer production in many domains of information production, and some exploration of its attractiveness from the perspective of a variety of liberal commitments: to democracy, autonomy and social justice.[47] Here we have contributed additional reason to think that peer production is normatively attractive. For those who hold one of a broad range of conceptions of virtue, peer production can be said to provide a social context in which to act out, and a set of social practices through which to inculcate and develop, some quite basic human, social and political virtues. It is in light of the whole picture that we recommend vigorous support for this exceptional socio-technical phenomenon that serves not only as the source of knowledge and information but as a platform for virtuous practices and the development of virtue in its participants.

Notes

1 Yochai Benkler, "Coase's penguin, or Linux and the nature of the firm," *Yale Law Journal*, 112 (2002), 369–446; Benkler, "Sharing nicely: on sharable goods and the emergence of sharing as a modality of economic production," *Yale Law Journal*, 114 (2004), 273–358.
2 Josh Lerner and Jean Tirole, "Some simple economics of open source," *Journal of Industrial Economics*, 50 (2002), 197–234.
3 Eben Moglen, "Anarchism triumphant: free software and the death of copyright," *First Monday*, 4 (August 1999); Available at http://www.firstmonday.org/issues/issue4_8_mogden (accessed April 25, 2005).
4 President's Information Technology Advisory Committee, *Developing Open Source Software to Advance High End Computing* (Washington, D.C.: Government Printing Office, 2000).
5 "Clickworkers results: crater marking activity," July 3, 2001; available at http://clickworkers.arc.nasa.gov (accessed April 25, 2005).
6 Benkler, "Coase's penguin."
7 Charlotte Hess and Elinor Ostrom, "Artifacts, facilities, and content: information as a commonpool resource," *Journal of Law & Contemporary Problems*, 63 (2003), 111–145. See also Ostrom, *Governing the Commons: The Evolution of Institutions for Collective Action* (Cambridge, Mass: Cambridge University Press, 1992).
8 Michael Slote, *From Morality to Virtue* (New York: Oxford University Press, 1987). For further information on "human flourishing and the good life," see Aristotle, "Nichomachean ethics," *Moral Philosophy: Selected Readings*, ed. George Sher (San Diego: Harcourt Brace Jovanovich, 1987) and Alasdair McIntyre, *After Virtue* (Notre Dame, Ind.: University of Notre Dame Press, 1984). On "naturalism," see Philippa Foot, *Virtues and Vices and Other Essays in Moral Philosophy* (Berkeley: University of California Press, 1978) and Rosalind Hursthouse, *On Virtue Ethics* (Oxford: Oxford University Press, 1999). On "utilitarianism," see Julia Driver, *Uneasy Virtue* (Cambridge: Cambridge University Press, 2001) and David Hume, *Enquiry Concerning the Principles of Morals*, ed. P.H. Niddich, 3rd edn (Oxford: Clarendon Press, 1975; originally published 1751).
9 David Wiggins, "Natural and artificial virtues: a vindication of Hume's scheme," *How One Should Live: Essays on Virtue*, ed. Roger Crisp (Oxford: Clarendon Press, 1996), pp. 131–40.
10 Hursthouse, *On Virtue Ethics*, p. 11.
11 Slote, *From Morality to Virtue*, p. 10.
12 Peter Simpson, "Contemporary virtue ethics and Aristotle," *Virtue Ethics: A Critical Reader*, ed. Daniel Statman (Washington D.C.: Georgetown University Press), pp. 245–59 at p. 246.
13 Lester Hunt, *Character and Culture* (New York: Rowman & Littlefield, 1997).
14 Charles Taylor, "The diversity of goods," *Moral Philosophy: Selected Readings*, ed. George Sher, 2nd edn (New York: Harcourt Brace, 1996), pp. 581–93 at p. 589. Taylor adds, "I have integrity to the degree to which my actions and statements are true expressions of what is really important to me" (p. 589).
15 McIntyre, *After Virtue*, p. 175.
16 Andy Oram, *Peer-to-Peer: Harnessing the Power of Disruptive Technologies* (Sebastopol: O'Reilly & Associates, 2001), p. ix.
17 ibid., p. 51.
18 See Slote, *From Morality to Virtue*.
19 Hunt, *Character and Culture*, p. 63, for an extensive discussion of generosity and benevolence.
20 Richard Stallman, "Philosophy of the GNU project"; available at http://www.gnu.org/philosophy/ (accessed April 25, 2005).

21 Eric Raymond, "Homesteading the noosphere," *First Monday*, 3 (April 1998); available at http://www.firstmonday.org/issues/issue3_10_raymond (accessed April 25, 2005). See also Lerner and Tirole, "Some simple economics of open source."
22 Eric Raymond, *The Cathedral and the Bazaar: Musings on Linux and Open Source by an Accidental Revolutionary* (Cambridge, Mass.: O'Reilly Associates, 1999): "It is quite clear that the society of open source hackers is in fact a gift culture" (p. 81).
23 See Hunt, *Character and Culture*, p. 192 and fn 5, for interesting points on Aristotle and Nietzsche on productive work and gift giving as valuable both to us and to others.
24 Hume, *Enquiry Concerning the Principles of Morals*, Sec. I.ii, p. 204.
25 McIntyre, *After Virtue*, p. 543.
26 See, for example, William Galston, *Liberal Purposes: Goods, Virtues, and Diversity in the Liberal State* (New York: Cambridge University Press, 1991). In addition, see Stephen Macedo, *Liberal Virtues* (Oxford: Clarendon Press, 1990).
27 John Adams, "Letter to Mercy Warren, April 16, 1776," *Warren-Adams Letters*, ed. Worthington C. Ford (Boston: Massachusetts Historical Society, 1917), vol. 1, p. 222, as quoted by Michael J. Sandel, *Democracy's Discontent: America in Search of a Public Philosophy* (Cambridge, Mass.: Harvard University Press, 1996), p. 126.
28 Sandel, *Democracy's Discontent*, p. 127.
29 See discussion in ibid., p. 126.
30 See Lerner and Tirole, "Some simple economics of open source." See also Benkler, "Coase's penguin" and "Sharing Nicely."
31 See such key works on gift giving as: Bronislaw Malinowski, *Argonauts of the Western Pacific* (New York: Dutton, 1950; originally published 1922); Marcel Mauss, *The Gift: Forms and Functions of Exchange in Archaic Societies*, trans. Ian Cunnison (New York: Norton, 1967; originally published 1925); Maurice Godelier, *The Enigma of the Gift*, trans. Nora Scott (Chicago: University of Chicago Press, 1999; originally published 1997).
32 "Why people are running SETI@home;" available at http://setiathome.ssl.berkeley.edu/ motivation.html (accessed December 20, 2002).
33 Pekka Himanem, *The Hacker Ethic, and the Spirit of the Information Age* (New York: Random House, 2001).
34 Sher, *Moral Philosophy*, p. 478.
35 ibid., p. 480.
36 Immanuel Kant, *The Doctrine of Virtue: Part II of the Metaphysic of Morals*, trans. Mary Gregor (New York: Harper & Row, 1969), as quoted by J.B. Schneewind, "The misfortune of virtue," *Ethics*, 101 (1990), 42–63 at p. 60.
37 McIntyre, *After Virtue*, p. 213.
38 Contemporary theorists who have explored links between political institutions and virtue include William Galston and Stephen Macedo, as well as Michael Sandel.
39 Sandel, *Democracy's Discontent*, pp. 125–126.
40 ibid., quotations from respectively, pp. 127, 131, 129 and 133.
41 Hunt, *Character and Culture*, p. 150.
42 ibid., p. 186.
43 Significant works that explicate related ideas include: Langdon Winner, "Do artifacts have politics?" in *The Whale and the Reactor* (Chicago: University of Chicago Press, 1986); Lawrence Lessig, *Codes and Other Laws of Cyberspace* (New York: Basic Books, 2000); Joel Reidenberg, "Lex informatica: the formulation of information policy rules through technology," *Texas Law Review*, 76 (1998), 553–93; Philip Brey, "Disclosive computer ethics," *ACM SIGCAS Computers and Society*, 30 (Dec. 2000), 10–16; Batya Friedman and Helen Nissenbaum, "Bias in computer systems," *Human Values and the Design of Computer Technology*, ed. Batya Friedman (New York: Cambridge University Press, 1997), pp. 21–40; Lucas Introna and Helen Nissenbaum, "Shaping the web: why the politics of search engines matters," *The Information Society*, 16 (2000), 1–17; Yochai Benkler, "Communications infrastructure regulation and the distribution of control over content," *Telecommunications Policy*, 22 (1998), 183–96; Bruno Latour, "Where are the missing masses: the sociology of a few mundane artifacts," *Shaping Technology/Building Society*. Eds. W. Bijker and J. Law (Cambridge, Mass.: MIT Press, 1992), pp. 225–58; Lewis Mumford, *Technics and Civilization* (New York: Harcourt, Brace and World, 1934); Marshall McLuhan, *Understanding Media: The Extensions of Man* (New York: McGraw-Hill, 1964); Neil Postman, *Amusing Ourselves to Death* (New York: Viking, 1985).
44 Brian Pfaffenberger, "Technological dramas," *Science, Technology and Human Values*, 17 (Summer 1992), 282–312, in part quoting Thomas P. *Hughes, American Genesis: A Century of Innovation and Technological Enthusiasm, 1870–1970* (New York: Penguin, 1989), p. 4.
45 Joseph M. Reagle Jr. "A case of mutual aid: Wikipedia, politeness, and perspective taking;" available at http://reagle.org.joseph/2004/agree/wikip - agree.html (accessed April 25, 2005).
46 Joe Wilcox and Stephen Shankland, "Why Microsoft is wary of open source," News.com, June 18, 2001; available at http://news.com.com/2100 - 1001-268520.html?legacy=cnet (accessed April 25, 2005).
47 Yochai Benkler, "Freedom in the commons: towards a political economy of information," *Duke Law Journal*, 52 (2003), 1245–76.

Part III

Conditions: Enabling Peer Production

7

Prophets and Advocates of Peer Production

George Dafermos

1 Introduction: A New Specter is Haunting the World

Since the beginning of the 21st century, peer production has been theorized by various thinkers as an alternative model of production, governance and property which can antagonize and subvert capitalism. And without doubt it has come a long way since then: from being a theory that only a handful of academics and intellectuals were interested in 20 years ago, it has managed to find mainstream appeal, becoming a material force for social, economic, and political change. It is important to realize that this development did not occur automatically, by itself. Various actors played a decisive role: scholars such as Yochai Benkler helped to popularize the notion in the academic world, while intellectuals-cum-activists such as Michel Bauwens were pivotal in framing peer production in ways that appealed to activists, progressive entrepreneurs, and policymakers around the world. If one were to write a history of peer production theory, these actors would figure prominently in it as the key advocates of peer production in the beginning of the 21st century and as the "prophets," so to speak, of a new mode of production founded on its principles.

This chapter of the *Handbook* retraces the history of these attempts at theorizing and disseminating peer production. It looks at how the theory spread through the advocacy work of its leading proponents and discusses the differences in their approaches. In parallel, it explores the effect they have had so far on the academic, economic, and political world.

The remainder of this chapter is structured as follows: in the next section, we discuss the enabling conditions for the emergence of the phenomenon of peer production in the 1990s, as typified by the development of online projects such as Linux, in connection with the theoretical works that prepared the ground for the development of peer production theory in the next decade. Then we look at how peer production was theorized in the 2000s through the work of its main proponents in this period: Yochai Benkler, Michel Bauwens' P2P Foundation, and the Oekonux Project. Following on from this historical background, we discuss the wider effect that peer theorists have had in the current decade on the community of scientists, policymakers, and collaborative economy actors.

2 Before 2000: A Period of Gestation

Although the concept of peer production had not yet made its appearance, the 1990s were critical for the development of the theory. To begin with, the 1990s established the necessary material conditions for its emergence. This period made personal computers and Internet connectivity accessible to a large part of the population, which meant that for the first time in human history, a critical mass of people

The Handbook of Peer Production, First Edition. Edited by Mathieu O'Neil, Christian Pentzold, and Sophie Toupin.

could take advantage of the Internet as a medium that makes it possible to communicate and collaborate virtually with others around the world. The result of this democratization of access to technology was the development of a new generation of online community projects, such as Linux, which exemplified what could be achieved by unlocking the potential of the Internet for distributed collaboration. The development of a computer operating system like Linux is a rather complex undertaking that cannot be performed by just a single individual. It requires teamwork, that is, the collaboration of a group of people, who need to coordinate somehow what they do. Historically speaking, only large organizations could mobilize that kind of resources. On the one hand, they have the financial capacity to put a large number of people to work on a daily basis on something as complex and time-demanding as the development of an operating system. On the other hand, they have managers and hierarchies for coordinating the work of these people. That is why the development of Linux from 1991 onwards was so important: because it showed the world that sophisticated technology production does not necessitate the financial resources or the managerial hierarchies of large organizations. In other words, the Linux project was paradigmatic of a mode of production that contrasted sharply with the dominant model in the software industry.

First of all, Linux was not just another product coming out of a company; instead, it was developed by an online community of hobbyists and technology aficionados sharing code improvements and "bug-fixes" over the Net. Strictly speaking, Linux was not a commodity: it was not produced for sale. On the contrary, Linux was developed by a global community of volunteer technologists on account of its *use value*: a tool they were developing with the aim of using it themselves for their own needs, "scratching their own itch," so to speak. Its production did not rely on the wage labor of professionals, but on the volunteer contributions of a multitude of self-motivated software programmers. And that was not all. Linux was special in yet another important respect: anyone could download the software and use it any way they liked. Unlike proprietary software with its usual copyright restrictions, Linux was distributed under the GNU General Public License (GPL),[1] which gives users the freedom to run the software, to study it, to modify it, and to redistribute it, thereby ensuring its character as a *digital commons*. By contrast to proprietary software, which places restrictions on what end users can do with it, Linux treated its community of users as a base of potential co-developers, empowering them to participate in its development process by contributing modifications and bug-fixes to the project. It should come as no surprise then, given the character of its development model, that Linux seemed to suggest that distributed networks and the Internet render corporate hierarchies obsolete.

Of course, Linux was not the only software project of that kind. An entire ecosystem of software had been using the GPL since the mid-1980s when that license was created by Richard Stallman. We should not forget that until the early 1970s, sharing software was a common practice among computer programmers (von Hippel & von Krogh, 2003; Levy, 1984). That gradually changed with the emergence of a rapidly growing market for software: to put it simply, it was bad for business, as Bill Gates (1976) argued in his famous article, which criticized code sharing as an obstacle to the development of the software industry. The development of the software industry in the 1980s was synonymous with the expansion of companies like Microsoft, whose software was distributed with restrictive licenses that legally forbade users from copying it and giving it to others. In addition to employing such legal measures to control the use and re-distribution of their products, they also made it technically impossible to tinker with them: their software was distributed in binary copies without the source code, that is, the human-readable instructions that programmers need in order to understand how a computer program works "under the hood." Many technologists took issue with these restrictions. For Stallman, who was a well-known member of the hacker community at MIT,[2] this type of software was a serious threat against the freedom of computer users. Driven by the conviction that this threat should be actively resisted, in 1985 he set up the Free Software Foundation, devoting the rest of his life to the advocacy of free software (Moody, 2001, pp. 14–30; Williams, 2011). In the context of this objective, a few years later (1989) he created the GNU General Public License (GPL): unlike conventional software licenses, which are used to constrain what users can do with a given computer program, the GPL was designed with the aim of granting users the right to use, modify, and share the software *freely*. Stallman was a prolific software developer in this period, who wrote some very popular programs for coders. He also authored many influential texts, such as the *GNU Manifesto*, in which he calls on the hacker community to join him in the development of a free operating system (Stallman, 1985). His crusade resonated with the feelings of many members of the hacker community in the 1980s and 1990s, who distributed the programs they wrote under the GPL (Williams, 2011). As a result, the common pool of GPL-licensed software grew dramatically over time: by the early 2000s, hundreds of GPL-licensed programs were publicly available on software repositories like SourceForge (Krishnamurthy, 2002).

Neither was Linux the only software project that leveraged the Internet for distributed collaboration in the 1990s. Yet, it was certainly one of the most emblematic, with hundreds of developers and a rapidly growing community of users across the globe. So, its success meant that a lot of people started to pay attention. Up until the end of the 1990s, software released under the GPL was known as free software. But then in 1998, with the aim of making free software more attractive to businesses, a group of prominent figures from the free software community, who were emboldened by the success of Linux, came up with the term "open source" and began to evangelize its adoption (Moody, 2001, pp. 166–169). The same year they formed the Open Source Initiative to support the promotion of the new brand name. In the context of propagandizing their ideas, the development of a new discourse, which would emphasize the business potential of free software, was critically important (Raymond, 1998b, 1999b).

An integral part of that strategy was a fresh theorization of the phenomenon of distributed software development by online communities. The prime mover of that effort was Eric S. Raymond, one of the main proponents of the term "open source." In his 1997 essay *The Cathedral and the Bazaar*, which had a lasting influence, Raymond (1998a) argued passionately that Linux represents an alternative software development model that has the potential to dominate the industry. As its metaphorical title implies, the essay makes a distinction between two rival models of software development: the "cathedral" and the "bazaar." "The cathedral model has been the default for almost all commercial software development since the invention of the computer" (Jones, 2003, p. 390): the development process is hierarchical and largely closed to actors outside the core developers' group. By contrast, Linux epitomizes the "bazaar," which is open and participative: the software is developed and shared over the Internet in full public view and anyone can participate, in accordance with their interests and skills. At the same time, the "bazaar" of Linux is distributed, making use of the Internet to harness the potential of a global community of developers and users. Raymond's comparison of the two models comes to the conclusion that Linux's open source development model is set to become a leading force in the software industry: its ability to mobilize a worldwide community of developers and users through the Internet means that it can be far more productive and innovative than any closed group of "cathedral-builders."

The influence of Raymond's "manifesto" was enormous (Behrenshausen, 2016; Hamerly et al., 1999; Jones, 2003, p. 390; Moody, 2001, p. 152; Weber, 2004. p. 113; Wiley, 2014). Pregnant with meaning, the story of *The Cathedral and the Bazaar* captivated the imagination of an entire generation of aspiring high-tech entrepreneurs, who were fascinated by the prospect of tapping into the creative contributions of a global community of developers and users.[3] The most widely known case is that of Netscape, which, under the influence of Raymond's work (Hamerly et al., 1999; Jones, 2003, p. 390; Raymond, 1998a; Suárez-Potts, 2001; Williams, 2011, Chapter 11), decided in 1998 to "open source" its flagship product (that is, the predecessor of today's popular Mozilla Firefox web browser) by releasing its source code on the Internet under an open source license. Raymond (1999a) followed up on this work with a series of essays, which further develop the narrative of the open source development model and the theories put forward in *The Cathedral and the Bazaar*.[4] Should anyone wish to trace the historical origins of peer production theory, Raymond's writings in the late 1990s should be the starting-point. Although he does not use the term "peer production," he theorizes the model of Linux and open source software development as a commons-based mode of production, with its own distinctive mode of governance and property, which has the potential to dominate the software industry and even expand beyond it into other spheres of economic activity. But that is fundamentally the same analysis and interpretation of the phenomenon of distributed software development that is encapsulated in later theories of peer production. For that reason, Raymond's work constitutes an embryonic form, so to speak, of the theory of peer production. Raymond's colleagues from the "open source camp" were also active in propagandizing the view that Linux represents an alternative model of software production, and the publishing house of Tim O'Reilly, who was a key figure in that group, played an important role in its diffusion.[5] In the space of just a few years, the work of Raymond, along with the contributions of his colleagues, formed a discourse which had a great influence (Behrenshausen, 2016; Weber, 2004, p. 113). And so, it is certain that it functioned as a catalyst for the development of the theory of peer production in the coming years.

In short, the wide availability of affordable PCs and Internet connections in the 1990s was the enabling condition for the phenomenon of distributed software development by online communities over the Internet, with Linux being a shining example. This phenomenon, in turn, spurred the development of new theories on the "open source development model," which carry within themselves the seeds for the development of the concept of peer production, as theorized by thinkers in the next decade.

3 2000s: Emergence of Peer Production Theory

The theory of peer production emerged in the new millennium. Israeli-American law scholar Yochai Benkler has been commonly credited with coining the term, which is first introduced in a paper he published in the *Yale Law Journal* in 2002. In that paper, titled "Coase's Penguin or Linux and the Nature of the Firm," Benkler detects in the development model of Linux the emergence of a "third mode of production" on the Internet, which is distinct from both markets and firms, as neither the motivation of participants, nor the coordination of their work, is achieved through "market prices or managerial commands." Here, Benkler argues for the first time that the development of free and open source software (FOSS) projects like Linux cannot be accounted for by financial incentives: the majority of participants are volunteers, who are mobilized, above all, by intrinsic motivations such as creativity and self-fulfillment at work. The other main point that he makes in this article concerns the anti-hierarchical organizational structure of these projects. For Benkler, the reason why Linux is paradigmatic of the model of "peer production" is precisely because its developers self-select the tasks they perform and coordinate their work without bureaucrats and bosses.[6] Benkler continued to work on these ideas and four years later he published his magnum opus, *The Wealth of Networks* (2006), in which he expanded on his definition of peer production, clarifying its characteristics and analyzing its various types and forms.[7] The influence of the *Wealth of Networks* was enormous: it remains to this day the most influential writing on peer production.[8] It established Benkler as a leading theorist of commons-oriented peer production and played an important role in establishing peer production, distributed networks, and the digital commons as a promising research field in the social sciences, thereby giving many academic researchers an incentive to engage with the subject.

However, Benkler was not the only prominent figure in the early years of the development of the theory. Equally influential was the work of Belgian intellectual Michel Bauwens on the other side of the Atlantic, who had already from the early 2000s began to theorize "peer-to-peer" as a new template for society, economy, and politics (see, for example, Bauwens, 2002). The views of those two thinkers had much in common. Bauwens' theories were based on a similar analysis of Linux and FOSS as an alternative mode of production, governance, and property (Bauwens, 2002, 2005). Like Benkler, Bauwens argued that peer production is a mode of production that is neither directed by market exchange, nor governed by bureaucrats and managers. He, too, emphasized that it is inseparable from a commons-based property regime. The only substantial point on which his views differed from Benkler's was with regard to peer production's transcendent character: they shared the conviction that the phenomenon of peer production was bound to spread beyond the confines of the software industry, but Bauwens went even further, arguing that peer production has "the potential to succeed capitalism as the core value and organizational model of a post-capitalist society" (Bauwens, 2012). Aside from this crucial point, the differences between their theories do not so much lie in their content, as in their "targeting" and mode of diffusion. Benkler's work was very well-received by the academic community, but Bauwens' discourse appealed to a different audience. As an Internet-age intellectual, Bauwens was an active participant in online mailing lists and discussion forums related to peer production theory. His passionate and provocative contributions to these debates influenced many "netizens" much more than Benkler's dense academic prose.

In 2005, Bauwens founded the P2P Foundation (P2PF). In the beginning, this basically consisted in an online mailing list devoted to the discussion of peer production. It became very popular and attracted a lot of sympathizers and followers from all over the world. Within a short time, Bauwens added a wiki, which he and his numerous online collaborators began to develop into an all-encompassing online repository of knowledge on the topic of peer production. Bauwens' collaborative style and his open source approach towards the documentation and development of peer production theory appealed to many researchers and thinkers who formed, in a sense, an online research group around him and the P2PF. Much of the work of that group had an activist bent, attracting many radicals who were interested in exploring peer production as a weapon in the struggle against the capitalist system. Reflecting the aspirations of the actors in its network, the P2PF soon began to evolve into a think-tank for the theory of peer production, advocating social, economic, and political change. In the context of its advocacy, it also began to get involved in the organization of events and conferences. The first one was in 2007 when Bauwens co-organized a workshop with Andreas Wittel,[9] out of which came the inspiration for a special issue of *Capital & Class* titled "Parallel Visions of Peer Production." Based on the contributions of the participants of this workshop, it was published in 2009 (Moore & Karatzogianni, 2009). This was the first time that an established scientific periodical had devoted an entire issue to peer production.

The P2PF was not alone in theorizing peer production from a radical perspective at the time. A project that was heavily involved throughout this period in the development of peer production theory was Oekonux. The project was launched in 1999 in Germany by Stefan Merten and a small group of activists and intellectuals who were interested in exploring the subversive potential of this model. Its main thesis was that the development of Linux and FOSS is prefigurative of a mode of production without the alienation that is characteristic of wage labor, which can transcend capitalism, leading to a free "society beyond labor, money, exchange" (Merten interviewed in Richardson, 2001). In this "GPL society," there would be no coercion and people would engage in productive projects out of intrinsic motivation: this would allow their "self-unfolding," while benefiting society as a whole. Based on an analysis of peer production's transcendent potential that was largely influenced by the work of Karl Marx, Oekonux theorized peer production as a "germ form," thus hypothesizing that the peer production model could gradually become hegemonic, superseding capitalism (for a more extensive discussion of Oekonux theories, see Euler, 2016; Meretz, 2012; Merten, 2000, 2009; Merten & Meretz, 2009; Richardson, 2001).

Oekonux members interacted mainly through two mailing lists: one for German speakers and another for discussions in English. In the beginning, most of the members of the group were from Germany, but because of the open and outward-looking character of the project, that soon changed. From 2001 until 2009, Oekonux organized four important international conferences: in 2001 in Dortmund, in 2002 in Berlin, in 2004 in Vienna and in 2009 in Manchester.[10] These conferences were unique in that they were the first (and only) ones to focus exclusively on the exploration of FOSS and peer production as a mode of production for the transition to post-capitalism. Their unique character attracted thinkers from all over the world, such as Graham Seaman (2003, 2004), Christian Siefkes (2007, 2009), Johan Söderberg (2008), and Raoul Victor (2003, 2004, 2009) who made significant contributions to the project. Michel Bauwens was also involved in the project and as a long-time member of the mailing list, he played an active role in the Oekonux debates. In fact, the ideas of Oekonux resonated so well with his own that in 2009 he co-organized the fourth and final Oekonux conference on "Free Software and Beyond: The World of Peer Production" in Manchester, highlighting the affinity between the theories and aspirations of the P2PF and Oekonux. This synergy between the two projects attracted the interest of many people from the network of the P2PF, thereby helping the propagation of Oekonux theories.

4 Post-2010: Peer Production Theory Moves into the Mainstream

Until 2010, the number of academics focusing on peer production was quite limited. That changed in the 2010s with the development of increasingly more research groups in various universities around the world. Benkler and his colleagues from the Berkman Klein Center for Internet and Society (which Benkler co-directs) at Harvard have been very active[11] and helped to establish peer production as a serious research field in the social sciences. Equally extensive and influential has been the work of the P2P Lab of researchers led by Vasilis Kostakis at the Tallinn University of Technology[12] and of the research group on the digital commons ("Dimmons") led by Mayo Fuster Morell at the Open University of Catalonia.[13]

From a more general point of view, the 2010s was the decade in which the theory of peer production matured and began to have a wider impact. The P2PF played a crucial role in this process. Bauwens had been very actively writing and publishing articles, organizing events and conferences and promoting the P2PF as a think-tank, to which commons-friendly policymakers, ethical entrepreneurs and activist organizations can turn for advice. In parallel with the P2PF, he also contributed to the development of like-minded projects that shared the P2PF's objective to agitate for the commons and peer production, such as the Commons Strategies Group (CSG), which he founded in 2009, with the American commons theorist David Bollier and Silke Helfrich, a German author and activist. CSG organized two large international conferences in Berlin in 2010 and 2013, with 200 speakers from all over the world, and published two important anthologies in 2013 and 2015 (Bollier & Helfrich, 2013, 2015).

Bauwens' influence was also instrumental in the development of the P2P Lab, which constitutes the research branch of the P2PF (Bauwens & Pantazis, 2018). This was founded in 2012 by Vasilis Kostakis, a core member of the P2PF and one of the most active academic researchers in its network, who has been Bauwens' main theoretical collaborator since then. The P2P Lab is a research collective made up of Kostakis and his postgraduate and PhD students from Tallinn University of Technology. The group has been a rich source of publications in academic journals, helping to spread the theories of the P2PF to the community of social scientists. At the same time, it has been P2PF's main vehicle for participating in various scientific research projects (such as P2P Value,[14] Phygital,[15] and Cosmolocalism[16]) which function as a platform for the advocacy of peer production.

During this period, the work of Bauwens and the P2PF began to have a strong influence on public policy actors. In 2013, Bauwens was invited to become the research director of the FLOK Society Project,[17] a government-supported activist-research project in Ecuador, with the aim of developing a set of policy proposals for the transformation of the country through peer production and the commons. Bauwens recruited the core team of researchers, with whom he worked closely for about a year in Ecuador. In the summer of 2014, FLOK Society organized a large international conference in Quito, with hundreds of participants from all over the world, which attracted a lot of media attention. Funded by three ministries of the Ecuadorian government, it was the first time that peer production theorists had ever worked in such close proximity to political decision-makers: that in itself was a strong signal that the theory was becoming a force to be reckoned with in the world of politics.[18]

That actually seemed to be true in some parts of the world in the mid-2010s, as in the case of Greece where Bauwens' theories and the example of FLOK Society had caught the attention of some Syriza party members and work-groups, who invited Bauwens and FLOK researchers George Dafermos and John Restakis to Athens in the autumn of 2014 for a series of seminars (Bauwens, 2014). This marked the beginning of a period of close collaboration, which continued for several months, between Syriza's work-groups and Bauwens' team of colleagues from FLOK Society and the P2PF. It is indicative of its legacy that Syriza's official "Government Program," which was released in 2015, refers to commons-based peer production as a pillar of the productive transformation of the Greek economy (Syriza, 2015). With a view to reinforcing their advocacy of the commons and peer production, in 2014 Bauwens and his colleagues launched Commons Transitions as a think-tank focused on research and consulting.[19] In 2017, Bauwens followed up on this work with another FLOK-like project. This time he went to the city of Ghent in Belgium, where he spent three months with his colleague from the P2P Lab, Vasilis Niaros, in order to lead a research project, which had the support of the mayor and the political coalition of the city council, on the "commons city of the future" (Bauwens & Onzia, 2017a, 2017b). The emphasis on the *city* as a locus of policy intervention has a strategic significance in the work of Bauwens and his P2PF colleagues since the mid-2010s, reflecting the growing political momentum of new municipalist movements in various European cities (such as the Barcelona en Comú citizen platform that governed Barcelona from 2015 to 2019), which aspire to bring about radical change by taking control of their local government (Utratel & Troncoso, 2017). Most importantly, reaching out to these new political forces has been quite a fruitful endeavor, judging by the fact that the "cornerstone of new municipalism," as some participants and researchers of Barcelona en Comú have remarked, "is a reinvigorated notion of the commons, as proposed by... Yochai Benkler, ... Michel Bauwens and organisations like Commons Transition" (Calafati & McInroy, 2017).

One of the main concepts that the above projects emphasized in their advocacy of peer production to public policy actors was that of the so-called "Partner State," which is a proposal for the development of a commons-friendly government (at both local and national levels) that enables community organizations to participate in the management and provision of public goods. Practically speaking, the concept denotes a local or national government that is supportive of cooperative organizations by developing policies and regulatory frameworks that enable them to play an important role in the economy.[20] Through their advocacy of the "Partner State," research projects like FLOK Society infused peer production theory with a theory of the State, in the context of which cooperatives are actively involved in the management and provision of public goods and services. At the same time, the emphasis of their work on cooperatives brought Bauwens and his colleagues in touch with actors from the *new* cooperative movement, who were receptive to the idea that their goals synergize well with those of commoners and peer producers. From a theoretical point of view, this dialogue between peer theorists and cooperators led to the development of the concept of "open cooperativism" (Bauwens & Kostakis, 2014, 2015; Conaty & Bollier, 2014; Pazaitis et al., 2017): the concept, which is basically a proposal for cooperatives to become more actively engaged in the production of the commons, has been at the epicenter of Bauwens' recent work (e.g., Bauwens & Pantazis, 2018). At a more practical level, it resulted in influencing cooperative projects, like the famous Catalan Integral Cooperative (CIC) in Catalonia, to become vocal proponents of the commons and peer production (see, for example, Catalan Integral Cooperative, 2015, Dafermos, 2017, and Duran interviewed in Bauwens et al., 2014).[21]

Unlike the P2PF, which evolved into the most important hub for the theory of peer production, Oekonux did not fare well in this decade. For no particular reason, discussion on its mailing lists had dwindled since the end of the 2000s. And so, by 2013, the project was officially over.[22] The radical perspective of its work, however, continued to have a strong influence on researchers and activists engaged in the development of the theory, playing a very important role in the development of other activist-research projects, such as the online *Journal of Peer Production (JoPP)* which released its inaugural issue in 2011.[23] The initiative was proposed at the fourth Oekonux conference in 2009 in Manchester by newcomer Mathieu O'Neil, who was inspired by the conference, and was supported by Athina Karatzogianni, Michel Bauwens, George Dafermos, Stefan Merten, Christian Siefkes, and Johan Söderberg (later joined by Nathaniel Tkacz and Maurizio Teli). This group of people had met each other through Oekonux and were, to various degrees, attuned to its theories (O'Neil, 2012a). This was a key event in the propagation of peer production as an object of interest: since 2011 when the inaugural issue came out, no other scientific periodical has featured as many theoretical and empirical investigations of peer production as the *JoPP*.[24] In the beginning, the journal was closely associated to Oekonux, with the journal website and mailing list hosted on the Oekonux server.

However, in 2012, "a series of serious disagreements" between lead editor Mathieu O'Neil and Stefan Merten, who administered the Oekonux infrastructure, "about the way the journal should operate" led O'Neil and some core members of the journal's editorial board to the decision to "fork" the project (O'Neil, 2012b). In the context of this process, the journal "migrated" to a server run by the P2PF and, in parallel, its name was changed from *Critical Studies in Peer Production* (*CSPP*) to the *Journal of Peer Production* (O'Neil, 2012b). This move was received positively for the most part by the rest of the members of the journal's editorial group, who gave their support to the new project. A few months later, the *Journal of Peer Production* (JoPP) released its first issue, and the *CSPP* release was rebranded as *JoPP* #0. Since then, the *JoPP* has published 14 other issues that span a wide range of subjects related to peer production. What is more, in contrast to traditional academic journals, the project has been organized in accordance with commons-based peer production principles: *JoPP* articles are freely shareable,[25] and the original submissions and reviews of scientific articles are also made public. Moreover, the *JoPP* exemplifies a truly democratic and transparent governance model, based on decision-making through dialogue on the project's publicly archived mailing list, to which anyone can subscribe.[26]

To recap: in this period, peer production theory had a visible impact on academia, leading to the development of various research projects and groups around the world, such as the P2P Lab. Influenced by the radical core of the theory, many of these projects (like the *JoPP*) had a decidedly activist bent. Moreover, the theory began to have a wider effect outside academia. Through projects like FLOK Society which combined research and advocacy work, peer theorists embarked on a dialogue with public policy actors in various countries like Ecuador, Greece, and Belgium. At the same time, by reaching out to the new cooperative movement, peer theorists managed to successfully create a collaborative link between peer producers and cooperators.

5 Conclusion: What is the Impact of Peer Production Theory Today?

Peer production theory has undoubtedly come a long way since its emergence in the early 2000s. That is most easily discernible in the academic world, where it has had an important research impact, leading to the development of increasingly more research groups at various universities around the world. The three research groups (at the Berkman Klein Center for Internet and Society at Harvard University, the Open University of Catalonia and Tallinn University of Technology, respectively) mentioned in the previous section are merely the most influential, but in fact there are more, such as the Heteropolitics research project[27] led by Alexandros Kioupkiolis at the Aristotle University of Thessaloniki in Greece or the P2P Models research project at the Complutense University of Madrid.[28] The effect of the practice is also apparent in the growth of academic literature on the subject: the development of journals like the *Journal of Peer Production* and the publication of academic *Handbooks* such as the one you are now holding in your hands attest to its growing influence. Another indication of its far-reaching effect is the development of several university courses related to peer production across Europe and the United States.[29]

Aside from its academic impact, an important accomplishment of peer production theory is that it has managed to connect with the collaborative economy actors par excellence, that is, cooperatives. This *cooperative turn* in peer production theory is largely due to the advocacy work of Bauwens and his fellow theorists from the P2PF, who transmitted the theory to the new generation of cooperators, thereby attracting them to commons-oriented production models. However, through their preoccupation with the new cooperative movement, these theorists soon realized that the cooperative legal form can be a valuable tool for peer producers who wish to formalize their entrepreneurial activity. That is, they realized that the cooperative legal form can reinforce the democratic governance of a peer production project, ensuring the democratic participation and equality of its members in the decision-making process. Thus, it did not take them long to start propagating the "cooperative idea" through their network of contacts with peer producers. In other words, their propaganda was "bi-directional": on the one hand, they turned cooperators to the commons and peer production as a potential ally in the struggle for the development of a truly collaborative economy that is capable of antagonizing and subverting capitalism. On the other hand, they oriented peer producers towards the cooperative legal form, familiarizing them with its advantages. In this way, peer theorists contributed decisively to the creation of a productive bond between cooperators and peer producers, which is very likely to have a significant impact in the coming years.

In addition to its influence on the cooperative economy, the contemporary effect of peer production theory extends into the world of politics. To a large extent, that is due to projects like FLOK Society in Ecuador, which were instrumental in establishing peer production in the eyes of policymakers as a credible alternative to capitalism. By familiarizing public policy actors with the main tenets of the theory of peer production, these projects raised awareness of its strategic benefits. Their advocacy highlighted two main advantages of peer production compared to capitalism: first, it emphasized its productivity, pointing out that it is a model of producing quality products, which can out-compete capitalism in terms of efficiency. Second, it underscored its democratizing effect on the governance of productive projects in the economy as well as on the relationship between citizens and the State. These are ideas that various progressive political forces around the world seem to be receptive to, as they form a weapon for challenging the ideological hegemony of the capitalist system (as the most productive and democratic of all existing economic systems). What remains to be seen is whether these progressive political forces can further "weaponize" the theory.[30] If they manage to sharpen its subversive edge, that will make the theory of peer production cut like a well-honed knife, opening up an incredible range of potentialities for a radical break with the whole organizational framework of social, economic and political life in its present form.

6 Acknowledgments

This publication has received funding from the European Research Council (ERC) under the European Union's Horizon 2020 Research and Innovation Program No. 724693 (Heteropolitics, ERC COG 2016). Furthermore, I would like to thank Michel Bauwens, Alexandros Kioupkiolis, Vasilis Kostakis, Stefan Meretz, Stefan Merten, Mathieu O'Neil, Christian Pentzold, Sophie Toupin, and Nicholas Anastasopoulos for their feedback on the draft version of this chapter.

Notes

1 www.gnu.org/licenses/old - licenses/gpl-1.0.html
2 In 2019 Stallman received harsh criticism on account of some comments he made on an MIT mailing list, which were considered sexist and utterly unacceptable. As a result of the outcry, Stallman was forced to resign from both MIT and the FSF.
3 Characteristically, Jimmy Wales, the founder of Wikipedia, mentioned in an interview that Raymond's book "opened my eyes to the possibility of mass collaboration" (Wales quoted in Schiff, 2006, July).
4 Many of these essays, including *The Cathedral and the Bazaar*, were published in book form in 1999 by O'Reilly.
5 For those who wish to delve more deeply into the subject, the anthology *Open Sources: Voices from the Open Source Revolution* published by O'Reilly in 1999 includes many key contributions to open-source discourse.
6 For the sake of clarity, Benkler nowhere argues that peer production is entirely devoid of authority: rather, his thesis is that the authority that project leaders, like Linus Torvalds, do exercise in the context of peer production is certainly not of the "command-and-control" type encountered in hierarchical organizations. As he writes: "Torvalds' authority is persuasive, not legal or technical and certainly not determinative" (Benkler, 2006, p. 105).
7 As one of the reviewers of this chapter pointed out, early peer production theory is characterized by the omission of any discussion regarding the variants of peer production commonly encountered in Asia, Africa, and South America, in which people engage for livelihood purposes. This is certainly true and can be largely explained by the fact that early peer production theory was primarily focused on "pure" models, that is, modes of production which cannot be accounted for by financial incentives, as they are geared for use, rather than market exchange. However, as the theory began to attract wider interest, various researchers and theorists drew attention to a plethora of market-oriented "hybrid" models of peer production in the Global North as well as the Global South, which are geared towards the production of commodities (see, for example, Arvidsson's (2019) study of commons-based petty production in countries outside the Global North).
8 Indicatively, according to Google Scholar, *The Wealth of Networks* has been cited more than 10,000 times.
9 http://wiki.p2pfoundation.net/Nottingham_Peer_Production_Workshop
10 http://second.oekonux - conference.org, http://third.oekonux - conference.org, http://www.oekonux - conference.org
11 See, for example, the list of publications at Benkler's homepage at Harvard at https://hls.harvard.edu/faculty/directory/10071/Benkler/publications
12 www.p2plab.gr/en/
13 http://dimmons.net
14 https://p2pvalue.eu
15 https://phygitalproject.eu
16 www.cosmolocalism.eu
17 https://floksociety.org
18 For a more extensive discussion of the FLOK Society Project, see Chapter 7 in Schneider (2018).
19 http://commonstransition.org/about - commons-transition-2/
20 For a discussion of the "Partner State" concept, see Bauwens & Kostakis (2015), Kostakis (2011a, 2011b), Kostakis & Bauwens (2014), Orsi (2009) and Restakis (2015).
21 As CIC's founder Enric Duran says characteristically: "Commons-oriented peer production has proven to be highly successful in initiatives such as LINUX, Wikipedia and many others. We feel we're part of this, and it has inspired many of our approaches. It's clearly the best method for producing collective knowledge and information, as it combines functionality and participation without hierarchy" (Duran interviewed in Bauwens et al., 2014).
22 Merten's announcement of the decision to discontinue the project: http://oekonux.org/list - en/archive/msg06188.html

23 http://peerproduction.net
24 That is not to say that other academic journals have not shown an interest in the subject. From 2010 onwards, papers on peer production have been published in journals such as *Fibreculture*, *First Monday*, *Triple-C* and *Information, Communication & Society*. But as the first and only journal in which peer production theory occupied center-stage, the contribution of the *JoPP* has been decisive.
25 *JoPP* articles are distributed under the Creative Commons Attribution-ShareAlike 4.0 International (CC BY-SA 4.0) license.
26 https://lists.ourproject.org/cgi - bin/mailman/listinfo/jopp-public
27 www.heteropolitics.net
28 https://p2pmodels.eu/
29 For example, the University of Colorado has a class on "Peer Production and Crowdsourcing," Tallinn University of Technology has a course on "Peer Production and the Theory of the Commons" and the University of Texas runs a class on "Social Media, Peer Production and Web 2.0."
30 Obviously, I do not claim that this will be an easy task. In fact, several researchers have raised the concern that the discourse of peer theorists could be assimilated to capitalist ideology. Their critique focuses on two main points. First, they reproach peer theorists for the emphasis they put on conceptual categories, such as that of ethical entrepreneurs or ethical markets, which are anything but disruptive to the hegemony of bourgeois ideology (e.g., Rigi, 2014). As Kioupkiolis writes, "the lexicon of entrepreneurship is the essential vocabulary of neoliberalism. . .to advocate for the commons by mobilizing time and again the discursive 'apparatus of capture' of neoliberalism bespeaks political shallowness, if not a commons-washing of capitalist business as usual" (Kioupkiolis, 2017, p. 74; also, see Rigi, 2014, p. 394). Second, the critics argue that the theories of the leading proponents of peer production encapsulate a deterministic and technocratic vision of system change that downplays the crucial role of social and political struggles in the process of historical transformation (Kioupkiolis, 2017, pp. 52–84; Rigi, 2014, pp. 401–403). Without going into detail (as an extensive discussion of the question of recuperation is, for the sake of brevity, beyond the scope of this chapter), these critiques admittedly raise questions that peer production advocates have to seriously engage with in order to harness the subversive potential of the theory.

References

Arvidsson, A. (2019). Capitalism and the commons. *Theory, Culture & Society*. Retrieved from https://journals.sagepub.com/doi/abs/10.1177/0263276419868838

Bauwens, M. (2002, November). Peer-to-peer: From technology to politics to a new civilisation? Talk presented at 2nd Oekonux Conference, Berlin, Germany. Retrieved from http://second.oekonux-conference.org/documentation/texts/Bauwens.html

Bauwens, M. (2005). The political economy of peer production. *C-Theory*. Retrieved from https://journals.uvic.ca/index.php/ctheory/article/view/14464/5306

Bauwens, M. (2012). From the theory of peer production to the production of peer production theory. *Journal of Peer Production*, 1. Retrieved from http://peerproduction.net/issues/issue-1/invited-comments/from-the-theory-of-peer-production-to-the-production-of-peer-production-theory/

Bauwens, M. (2014). What the P2P Foundation Did in 2014. *P2P Foundation Blog*. Retrieved from https://blog.p2pfoundation.net/what-the-p2p-foundation-did-in-2014-1-founders-report/2014/12/30

Bauwens, M., Gorenflo, N., & Restakis, J. (2014). Integral revolution: An interview with Enric Duran about CIC. Retrieved from https://cooperativa.cat/en/integral-revolution/

Bauwens. M., & Kostakis, V. (2014). From the communism of capital to capital for the commons: Towards an open co-operativism. *Triple-C: Communication, Capitalism & Critique*, 12(1). Retrieved from www.triple-c.at/index.php/tripleC/article/view/561

Bauwens, M., & Kostakis, V. (2015). Towards a new reconfiguration among the state, civil society and the market. *Journal of Peer Production*, 7. Retrieved from http://peerproduction.net/editsuite/issues/issue-7-policies-for-the-commons/peer-reviewed-papers/towards-a-new-reconfiguration-among-the-state-civil-society-and-the-market/

Bauwens, M., & Onzia, Y. (2017a). *Commons transition plan for the city of Ghent*. Ghent: City of Ghent and P2P Foundation. Retrieved from https://stad.gent/sites/default/files/page/documents/Commons%20Transition%20Plan%20-%20under%20revision.pdf

Bauwens, M., & Onzia, Y. (2017b). Special report: A commons transition plan for the city of Ghent. *Commons Transition*. Retrieved from http://commonstransition.org/commons-transition-plan-city-ghent/

Bauwens, M., & Pantazis, A. (2018). The ecosystem of commons-based peer production and its transformative dynamics. *Sociological Review*, 66(2), 302–319.

Behrenshausen, B. (2016, May) 19 years later, the cathedral and the bazaar still moves us. *Opensource.com*. Retrieved from https://opensource.com/life/16/5/19-years-later-cathedral-and-bazaar-still-moves-us
Benkler, Y. (2002). Coase's penguin, or, Linux and the nature of the firm. *Yale Law Journal*, 112(3), 369–446.
Benkler, Y. (2006). *The wealth of networks*. New Haven, CT: Yale University Press.
Bollier, D., & Helfrich, S. (Eds.) (2013). *The wealth of the commons: A world beyond market and state*. Commons Strategies Group.
Bollier, D. & Helfrich, S. (Eds.) (2015). *Patterns of commoning*. Commons Strategies Group.
Calafati, L., & McInroy, N. (2017). Local government and the commons: The time has come. *CLES*. Retrieved from https://cles.org.uk/blog/local-government-the-commons-the-time-has-come/
Catalan Integral Cooperative (2015). The P2PF/CIC strategic plan will be presented in AureaSocial on Tuesday, May5th.CatalanIntegralCooperativeWebsite.Retrievedfromhttps://cooperativa.cat/en/the-p2pfcic-strategic-plan-will-be-presented-the-nextat-aureasocial/
Conaty, P., & Bollier, D. (2014). Toward an open co-operativism: A new social economy based on open platforms, co-operative models and the commons. *Commons Transition*. Retrieved from http://commonstransition.org/toward-an-open-co-operativism/
Dafermos, G. (2017). The Catalan integral cooperative: An organizational study of a post-capitalist cooperative. P2P Foundation/Robin Hood Coop. Retrieved from http://commonstransition.org/the-catalan-integral-cooperative-an-organizational-study-of-a-post-capitalist-cooperative/
Euler, J. (2016). Commons-creating society: On the radical German commons discourse. *Review of Radical Political Economics*, 48(1), 93–110.
Gates, B. (1976). Open letter to hobbyists. *MITS Computer Notes*, 1(9), 3.
Hamerly, J., Paquin, T., & Walton, S. (1999). Freeing the source: The story of Mozilla. In S. Ockman & C. DiBona (Eds.), *Open sources: Voices from the open source revolution*. Sebastopol, CA: O'Reilly. Retrieved from https://www.oreilly.com/openbook/opensources/book/netrev.html_original
Jones, S. (2003). *Encyclopedia of new media: An essential reference to communication and technology*. Thousand Oaks, CA: Sage.
Kostakis, V. (2011a). *The political economy of information production in the social web: Towards a "partner state approach"* (Doctoral dissertation). Retrieved from https://digi.lib.ttu.ee/i/?610
Kostakis, V. (2011b). Commons-based peer production and the neo-Weberian state: synergies and interdependencies. *Halduskultuur – Administrative Culture*, 12(2), 146–161. Retrieved from http://halduskultuur.eu/journal/index.php/HKAC/article/download/45/54
Kostakis, V., & Bauwens, M. (2014). *Network society and future scenarios for a collaborative economy*. Basingstoke: Palgrave Macmillan.
Kioupkiolis, A. (2017). *Heteropolitics: Refiguring the common and the political* (Unpublished manuscript, on file with author).
Krishnamurthy, S. (2002). Cave or community? An empirical examination of 100 mature open source projects. *First Monday*, 7(6). Retrieved from www.firstmonday.org/ojs/index.php/fm/article/view/960/881
Levy, S. (1984). *Hackers: Heroes of the computer revolution*. New York: NY: Anchor Press/Doubleday.
Meretz, S. (2012). Ten patterns developed by the Oekonux Project. *Journal of Peer Production*, 1. Retrieved from http://peerproduction.net/issues/issue-1/debate-societal-transformation/ten-patterns-developed-by-the-oekonux-project/
Merten, S. (2000). GNU/Linux-Milestone on the way to the GPL society. *Oekonux Project*. Retrieved from http://oekonux.org/texts/meilenstein/english.html
Merten, S. (2009). The (open source) world in 2020 – a history based look into the future. *Oekonux Project*. Retrieved from www.oekonux.org/texts/TheWorldIn2020.html
Merten, S., & Meretz, S. (2009). Germ form theory: Peer production in a historical perspective. *Oekonux Project*. Retrieved from www.oekonux.org/texts/GermFormTheory.html
Moody, G. (2001). *Rebel code. Inside Linux and the Open Source revolution*. Cambridge, MA: Perseus Press.
Moore, P., & Karatzogianni, A. (2009). Parallel visions of peer production. *Capital & Class*, 33(1), 7–11.
O'Neil, M. (2012a). Editorial notes: Productive negation. *Journal of Peer Production*, 1. Retrieved from http://peerproduction.net/issues/issue-1/editorial-notes/
O'Neil, M. (2012b). Announcing fork. Email sent to the mailing list of Critical Studies in Peer Production. Retrieved from www.oekonux.org/journal/list/archive/msg00820.html
Orsi, C. (2009). Knowledge-based society, peer production and the common good. *Capital & Class*, 33(1), 31–51.
Pazaitis, A., Kostakis, V., & Bauwens, M. (2017). Digital economy and the rise of open cooperativism: The case of the enspiral network. *Transfer: European Review of Labour and Research*, 23(2), 177–192. Retrieved from http://journals.sagepub.com/eprint/cQtJrUauKHrIGGYmMZtq/full
Raymond, S. (1998a). The cathedral and the bazaar. *First Monday*, 3(3). Retrieved from https://firstmonday.org/article/view/578/499

Raymond, S. (1998b). Goodbye, "free software"; hello, "open source." Retrieved from http://www.catb.org/~esr/open-source.html

Raymond, S. (1999a). *The cathedral and the bazaar: Musings on Linux and open source by an accidental revolutionary*. Sebastopol, CA: O'Reilly.

Raymond, S. (1999b). The revenge of the hackers. In S. Ockman & C. DiBona (Eds.), *open sources: Voices from the open source revolution*. Sebastopol, CA: O'Reilly. Retrieved from www.oreilly.com/openbook/opensources/book/raymond2.html

Restakis, J. (2015). Public policy for a social economy. *Journal of Peer Production*, 7. Retrieved from http://peerproduction.net/issues/issue-7-policies-for-the-commons/peer-reviewed-papers/policy-for-a-social-economy/

Richardson, J. (2001). *Free software and GPL society: Stefan Merten of Oekonux interviewed by Joanne Richardson*. Retrieved from http://subsol.c3.hu/subsol_2/contributors0/mertentext.html

Rigi, J. (2014). The coming revolution of peer production and revolutionary cooperatives. A Response to Michel Bauwens, Vasilis Kostakis and Stefan Meretz. *TripleC: Communication, Capitalism & Critique*, 12(1). Retrieved from www.triple-c.at/index.php/tripleC/article/view/486

Schneider, N. (2018). *Everything for everyone: The radical tradition that is shaping the next economy*. New York, NY: Nation Books.

Schiff, S. (2006, July). Know it all: Can Wikipedia conquer expertise? *New Yorker*. Retrieved from www.newyorker.com/magazine/2006/07/31/know-it-all

Seaman, G. (2003). The two economies; or: Why the washing machine question is the wrong question. *Oekonux Project*. Retrieved from www.oekonux.org/list-en/archive/msg01030.html

Seaman, G. (2004, May). Technology and Society. Talk presented at 3rd Oekonux Conference, Vienna, Austria. Retrieved from http://third.oekonux-conference.org/documentation/texts/Seaman.html

Siefkes, C. (2007). *From exchange to contributions. Generalizing peer production into the physical world*. Berlin: Edition C. Siefkes (self-published). Retrieved from http://peerconomy.org/text/peer-economy.pdf

Siefkes, C. (2009, March). Peer production everywhere: How can we do it and where can We wtart? Talk presented at 4th Oekonux Conference, Manchester, UK. Retrieved from http://oekonux-conference.org/program/events/19.en.html

Söderberg, J. (2008). *Hacking capitalism: The free and open source software movement*. New York, NY: Routledge.

Stallman, R. (1985). The GNU manifesto. Retrieved from www.gnu.org/gnu/manifesto.en.html

Suárez-Potts, L. (2001). Interview with Frank Hecker. Retrieved from www.openoffice.org/editorial/ec1May.html

Syriza (2005). *Plan of government program*. Athens: Syriza. Retrieved from www.dikaiologitika.gr/images/archive/sxedio-syriza.pdf

Utratel, A. M., & Troncoso, S. (2017). Promoting the commons in the time of monsters. *Common Dreams*. Retrieved from www.commondreams.org/views/2017/06/03/promoting-commons-time-monsters

Victor, R. (2003). Free software and market relations. *Oekonux Project*. Retrieved from www.oekonux.org/texts/marketrelations.html

Victor, R. (2004, May). Marxism and free-software. Talk presented at 3rd Oekonux Conference, Vienna, Austria. Retrieved from http://third.oekonux-conference.org/documentation/texts/Victor.html

Victor, R. (2009, March). Money and Peer Production. *Talk presented at 4th Oekonux Conference*, Manchester, UK. Retrieved from http://oekonux-conference.org/program/events/29.en.html

Von Hippel. E., & Von Krogh, G. (2003). Open source software and the "private-collective" innovation model: Issues for organization science. *Organization Science*, 14(2), 209–223.

Weber, S. (2004). *The success of open source*. Cambridge, MA: Harvard University Press.

Wiley, D. (2014). Eric Raymond, "The cathedral and the bazaar." In D. Wiley (Ed.), *An open education reader*. Retrieved from https://openedreader.org/chapter/the-cathedral-and-the-bazaar/

Williams, S. (2011). *Free as in freedom: Richard Stallman's crusade for free software*. Sebastopol, CA: O'Reilly.

8

Virtue, Efficiency, and the Sharing Economy

Margie Borschke

1 Introduction

Peer production is often assumed to be a virtuous and public-spirited undertaking, a networked socio-economic system of production, that is highly efficient, promotes individual agency, harnesses collective knowledge, creates robust technologies and information, and contributes to sustaining the public domain in the Internet era. As an organizational innovation enabled by network technologies, peer production is also closely associated with the rise of social networking technologies, practices, and platforms in the 2000s and an ethos of participation, sharing, and remix. The centrality of virtue and pragmatism to the academic conceptualization of peer production has also informed how adjacent practices associated with participatory media and the platforms of the sharing economy[1] are imagined in everyday network culture and in the metaphors used to describe and explain these new forms of communicative and market exchange. This connection is not coincidental – if we look closely at the cultural history of metaphors such as remix and sharing, we find an explicit connection to the literature of commons-based peer production, in particular the work of Yochai Benkler, whose theorization of information commons and non-market production was expanded to apply to other realms of cultural practice. Peer production may have helped to describe and explain new forms of online collaboration but there lurks a tension at the heart of theorizations of peer production and participatory media: that is the tension between virtue and pragmatism, between a belief that particular kinds of networked spaces and practices can enable the development of personal and social virtues and also be more efficient than other forms of production. This tension becomes more visible in adjacent practices and the platforms of the sharing economy where ethical debates about agency, property, privacy, and collective rights abound and where utopian rhetoric can appear to act as a cover for the corporate drive for efficiency over ethical concerns. This chapter considers how this tension can be understood by examining how the ideals of peer production were shaped by its own cultural and intellectual history and how this plays out in broader conversations about participatory media, online collaboration, and what has come to be called the "sharing economy."

2 Commons-Based Peer Production and Adjacent Practices

Yochai Benkler (2002a) coined the term "commons-based peer production" to describe new collaborative practices that were enabled by digital network technologies, promoted decentralized organization, harnessed diverse motivations for participation, and separated "governance and management from property and contract" (Benkler, 2017, p. 265). Exemplary practices include free and open

The Handbook of Peer Production, First Edition. Edited by Mathieu O'Neil, Christian Pentzold, and Sophie Toupin.

source software production (e.g., Linux) and the open access production of online information resources such as Wikipedia. Participants on such projects were understood to be self-selecting and empowered agents, choosing and executing their contributions to collaborative projects outside of the influence of market incentives and hierarchical organizational structures. The fruits of their collective labors were held in common rather than privately owned. Although the free software approach was originated by programmers who objected to the increasingly proprietary nature of software in the mid-1980s after two decades of open and collaborative work (Stallman, 2002), it was the success of free and open source software like Linux and online projects like Wikipedia in the 2000s that prompted industry and academics to take notice. Peer production did not only produce successful results, it also did so with surprising efficiency, desired outcomes that challenged economic and managerial orthodoxies about centralized top-down decision-making and market incentives. Wrote Benkler (2016), "Peer production is the most significant organizational innovation that has emerged from Internet-mediated social practice" (p. 91). It seemed nothing short of revolutionary.

Commons-based peer production, however, was more than just an efficient innovation. It was theorized to be a more ethical mode of production, one of the ways in which the new economy would not be evil. This approach to organizing talent and labor efforts using online media and communications technologies, not only produced better results, but it was suggested by Benkler and Nissenbaum (2006; also in this volume) that decentralized networked organization could promote the cultivation of virtues. It could do so by enabling a person to dedicate themselves to more personally rewarding activities, the promotion of individual liberty and innovation in the service of the common good. According to Benkler and Nissenbaum (2006), "Socio-technical systems of commons-based peer production offer not only a remarkable medium of production for various kinds of information goods but serve as a context for positive character formation" (pp. 394–395). The contrast with neoliberal political economic assumptions about self-interested actors pursuing private goods in an iron cage of bureaucracy where actors were structurally stripped of meaningful agency, was sharp and clear.

Benkler and Nissenbaum (2006) brought together a rising interest in virtue ethics in political philosophy with work in science and technology studies on the material and technical constraints and affordances of networks, arguing that the organizational structures of peer production could foster many moral and political virtues including autonomy, independence, liberation, creativity, productivity, industry, benevolence, charity, generosity and altruism. Peer production practices, they argued, created a context for virtuous engagement and practice. Benkler and Nissenbaum claim (2006), "Peer production can be said to provide a social context in which to act out, and a set of social practices through which to inculcate and develop, some quite basic human, social and political virtues" (p. 419). By developing rules for engagement and a space in which the individual had access to easy, low-stakes opportunities to engage in virtuous practice (e.g., working towards a common good for its own sake) they were more apt to engage and be virtuous. Subsequently, the success of these virtuous actions would feedback and in turn promote the exercise of new virtue. Virtuous acts made the system of production more efficient but this efficiency in turn promoted the exercise of more virtue.

These early academic theories about peer production, also came to be associated in the academic literature (e.g. Jenkins, 2005; Lessig, 2008) and in everyday discourse with other disruptive technologies, platforms, and practices including those we might call adjacent practices such as crowd sourcing (e.g., Amazon's Mechanical Turk and ReCAPTCHA), online markets for labor and services (e.g., Uber, Airbnb) and user-generated content (e.g., social media platforms and practices). Although Benkler (2016) explicitly distinguishes these adjacent practices from his theorization of commons-based peer production, academics interested in theorizing these related socio-technical platforms and industries interested in marketing their innovations to a wider public, often cited and built upon Benkler's early theorizations of peer production (e.g., 2002a, 2002b, 2006), tapping into common rhetoric about the virtue and efficiency of non-hierarchical organization, and the value of participation and sharing in the networked environment (e.g., Jenkins, 2005; Lessig, 2008). Collaboration and decentralization were celebrated. Mass participation would disrupt the worst sins of corporate bureaucracy and transform the conditions of labor into something less alienating, more personally and socially rewarding; remix culture and the sharing economy were born.

Yet these adjacent practices are also the site where in recent years undesirable outcomes and ethical concerns have become more evident. The sharing economy seems to have launched new, quasi-legal forms of exploitative labor practices, contributing to problems associated with precarious work and widespread inequality, and sidestepped regulations intended to protect citizens, workers, and consumers. Popular media now regularly asks if Uber is taking its workers for a ride, questioning why a multi-billion dollar company is allowed to avoid paying legislated benefits such as sick leave and holiday pay and its independent contractors have no power to negotiate the rate they are paid. Meanwhile many cities ponder whether short-term rental platforms such as Airbnb are compromising the availability of affordable housing and changing the character of local communities. How could it be that structures and approaches that were thought to promote the development of social and personal virtues could also produce such perverse and ethically questionable outcomes? In the next section I highlight the rhetorical dimensions of network culture to show how particular ideals about the value of participation and collaboration as positive behaviors are embedded within the metaphors we use to imagine Internet culture. I also consider scholarly critiques of peer production and its claim that participation is always an unquestionable good.

3 Participation, Collaboration and the Rhetorical Dimensions of Network Culture

Both "remix" and "sharing" are now common metaphors used to describe online communication and distribution practices, but this was not always the case. Recent scholarship (Borschke, 2017; John, 2017) on the origins, history, and rhetorical significance of both terms, tracked the rise of their metaphorical usage to the first decade of the 21st century, coinciding with and, as I will highlight, drawing upon the development of scholarly theories of peer production practices. The shared historical context for these concepts is the 2000s, a decade that ushered in the era of widespread broadband access to the Internet in much of the world and witnessed the rise and adoption of peer-to-peer technologies and social media. It was also a decade of conflict between everyday users and the media and entertainment industries over intellectual property, amidst new user expectations about access and the development of a culture of downloading and archiving. These conflicts were an important backdrop for the development of remix and sharing as metaphors that could be invoked to describe, explain and, as I will argue, defend new technologically-enabled social and cultural practices as well as promote the rising valuing of participation and collaboration.

Metaphors work by analogy: they help turn the unfamiliar into the understandable, by describing something new in terms of something familiar (cf. Borschke, 2017, p. 33–40; Herman, 2008; Lakoff & Johnson, 1980; Loughlan, 2006). In the online space, we rely on many metaphors of remediation without much thought – from your desktop to the cloud, much of our life online and off is animated by metaphors. But no metaphor arrives fully formed and the cultural history of everyday network metaphors provides an opportunity to consider how the values, assumptions, and social contexts that underpin them developed. As Nicholas John (2017) observed in his book *The Age of Sharing*, "[T]he cultural and discursive association between the Internet and the values of sharing is not a natural given. Tarleton Gillespie's observations concerning the term 'platform' are equally apt here: 'A term like 'platform' does not drop from the sky, or emerge in some organic, unfettered way from public discussion' (Gillespie, 2010, p. 359). This, of course raises the question of where it does come from" (p. 45).

Remix is a metaphor that came to us from the recording studio, the DJ booth, and the dance floor (Borschke, 2017, pp. 33–112) and whose meaning has broadened, used to describe media made from extant media and many new expressive practices and forms enabled by digital reproduction technologies and networks. In the context of debates about intellectual property in the network environment, "remix" also has ideological heavy lifting to do, serving to defend and normalize everyday copying practices (e.g., posting a YouTube video on Facebook or producing a meme as a form of political comment) and new forms of collaborative authorship from intellectual property regimes and efforts to expand their scope. Similarly, John (2017) examines how "sharing," a word with both distributive and communicative meanings, offers a versatile rhetorical framework for making sense of a wide range of participatory practices enabled and encouraged by social media platforms (e.g., sharing your opinions, your photos, links to stories, etc.) and other digital and network technologies and media. Both "remix" and "sharing" perform imaginative work in the rhetorical construction of everyday network culture; in particular, they help to make sense of the digital reorganization of communication and distribution, storage, and transmission, in the network sphere. Both metaphors also tend to have positive normative connotations including openness, fairness, and equality, opening up a utopian space for describing new practices and imagining new possible futures. But in the context of unwanted unethical outcomes, these same positive connotations can serve different rhetorical purposes. The virtuous connotations of sharing – being open, considerate, selfless (or at least not selfish), working for the common good – can also offer a cover from scrutiny from negative outcomes, sidestepping ethical questions about labor exploitation and the monetization of personal data by adopting and promoting the language of sharing.

It would be easy to write off undesirable uses of these metaphors as buzzwords or corporate spin, but the same metaphors favored by the technology industries were also adopted and promoted by scholars of social media and network culture. Both "remix" and "sharing" offered useful rhetorical frameworks that helped to describe and explain the blurring of distinctions between communication and distribution in the online space and to consider the possibilities of personal empowerment and public engagement afforded by then new technologies. Remix was a form and practice with origins in the creative use of analog media formats and technology was redefined to be quintessentially digital; alongside neologisms such as "produsage" (Bruns, 2008), remix aimed to signify a perceived shift from the so-called passive consumption of the 20th century to the participatory engagement characteristic of user-generated content in the 21st; it is also often presented as a tactic of resistance to the corporatization of culture, a practice that could transform copying into sharing (Borschke, 2017). Sharing was another constitutive network activity and to share in the online space was to participate in its construction, a mass collaboration that was thought to empower users – to turn them into netizens – and lead towards a more egalitarian future.

The utopian character of these metaphors and their rise in the 2000s was also deeply informed by scholarship on peer production and its celebration of personal liberty and the common good, in parallel with innovation and efficiency. Peer production's combination of virtue and pragmatism proved to be an attractive model for explaining modes of socio-technical change and new business models that promoted efficient production and innovative outcomes. The whole world could work smarter not harder if participation and collaboration were promoted as desirable network actions. Participation and collaboration, however, are not inherently positive or always ethical. Many of the ethical questions that arise with regards to peer production and adjacent practices relate to this ambiguity. Kreiss, Finn, and Turner (2011) critiqued the utopian character of many scholarly claims about peer production, questioning the claim that it was an inherently positive force of change and challenging the optimism of Web 2.0's early theorists (e.g., Benkler, 2002a, 2002b; Castells, 2007; Jenkins, 2006; Lessig, 2004; Shirky, 2008). "We do not believe that networked information exchange necessarily levels that social playing field, or that networked modes of social action are replacing their industrial antecedents," they wrote (Kreiss, Finn, & Turner, 2011, p. 244). They questioned academic orthodoxies and claims that peer production necessarily promoted agency, ethical and egalitarian relationships, adaptability and efficiency. Even the claim that peer production was outside of markets and property regimes, they charged, might not only be wrong, but it might work to "obscure the ways in which technologically enabled peer production may be changing our lives for the worse" (Kreiss et al., 2011, p. 244). The unbridled optimism, they charged, might be less about the present than it was a tacit critique of industrial bureaucracy in the recent past.

> If peer production offers flexibility and egalitarian interpersonal relations, bureaucracies seem to many to confront workers with rigid, hierarchical social relations. If bureaucracies hoard wealth and power, peer production distributes it. If bureaucracies breed psychologically divided "organization men," peer production supports the whole person. Above all, peer production promises to liberate the individual subjects of mass consumer culture by fashioning them into active producers. (Kreiss et al., 2011, p. 244)

Yet, as Kreiss et al. argue, these hopes do not necessarily come to fruition. As the widespread use of open source software has shown, the practice is compatible with commerce and bureaucracy and free labor may benefit private firms and bureaucracies more than individuals. Incentives to participate in the online sphere also contribute to the erosion of boundaries between private selves and public activities, an experience that may have many undesirable consequences. If status and reputation come from making free contributions to mass projects, those with paid positions that encourage their participation have advantages that independent actors do not. Participation and collaboration are more ambiguous than their idealizations. Peer production might not always produce ethical outcomes, let alone promote virtuous character traits.

4 Utopian Thinking and Copyright Conflicts

The expectation that virtue and efficiency coincide and are promoted by the practices of peer production influenced how we think about and idealize other forms of networked participation and collaboration; this expectation is baked into the metaphors we have adopted to animate our online lives. A tacit critique of the recent past and a response to pressing debates and conflicts around intellectual property is another characteristic of this shared lineage. In this final section I draw attention to how conflicts over copying and recirculation shaped how we reimagined participation and collaboration in the context of digital and network technologies and why these actions tend to be positively associated with the promotion of ethical outcomes and virtuous character attributes.

A key tenet of commons-based peer production is a commitment to the perpetuation of an information commons. It is important to note that this is a commitment to weak protection of intellectual property rather than no protection at all – share and share alike licensing, for instance, aims to protect the commons from enclosure even while it protects users' liberties. In order to understand the significance of such protections, we should consider the special status of copies and copying in the Free and Open Source Software movement and why its role is considered so crucial in promoting and guaranteeing freedom, cooperation, and creativity.

Almost three decades before the term peer production was coined, programmer and activist Richard Stallman began thinking about how to protect the collaborative intellectual culture of software development and users from the rise of proprietary software in the nascent software industry. He called this approach "free software" and codified his position in a manifesto associated with the development of a free operating system GNU. In this manifesto Stallman (2002) states:

> Free software means software that respects users' freedom and community. Roughly, it means that the users have the freedom to run, copy, distribute, study, change and improve the software. Thus, "free software" is a matter of liberty, not price. To understand the concept, you should think of "free" as in "free speech," not as in "free beer." We sometimes call it "libre software" to show we do not mean it is gratis. (p. 3)

This manifesto demanded four essential freedoms including the freedom to use software without conditions, the freedom to read the source code and to modify or tinker with it; it also recognized the important role that copies played, explicitly demanding the freedom to "redistribute copies so you can help others" and to distribute copies of your modifications and implicitly acknowledging the role of copying in reading, studying, and changing or improving code. Stallman was arguing for a freedom from constraints that copyright and patent laws imposed on users, but he also implied that these negative liberties would guarantee certain kinds individual and social virtues – liberty, agency, helping others, benefiting the whole community. Autonomy was not incompatible with generosity and fellowship, and it was through copying as a mode of sharing that self-reliance could be reconciled with community mindedness (Benkler & Nissenbaum, 2006, pp. 407–408.)

Free culture advocates in the 21st century demanded similar rights for users, including the freedom to modify and distribute other kinds of cultural works (e.g., music, text, images, video) without consent or compensation. Cheap digital technologies and online networks took reproduction and distribution out of the hands of industry, making these actions accessible to anyone with a computer and an Internet connection; they even potentially transformed what were previously seen as exclusively industrial economic activities into potential forms of participation and collaboration online. Distribution could be a form of communication; communication could result in distribution; this was the affordance of the digital copy.

Simultaneously, industries which benefited from existing copyright regimes first sought to protect (via enforcement, punitive actions, education campaigns, etc.) and expand the scope of copyright protection. Later they began to develop ways of monetizing this everyday social production and exchange of copies. It is in the context of this struggle between users and industry that the rhetoric of remix and sharing comes into being.

Lawrence Lessig's books *Free Culture* (2004) and *Remix* (2008) entered into these debates between the mainstream media and entertainment industries which benefited from existing copyright regimes and networked users who wanted on-demand access to their favorite music, films, television, and so on, and also wanted to use and adapt copyrighted media in the social networks where they were spending increasing amounts of time. To defend these new forms of technologically-enabled social participation and expression, Lessig invoked the metaphors of remix and sharing to describe, explain, and defend the place of copying practice in network culture. John (2017, p. 74) attributes the popularization of the term the "sharing economy" to Lessig and highlights Lessig's reliance on Benkler's theorizations of peer production, and the new value of voluntary unremunerated participation over other kind of labor. Writes John (2017), "In Lessig's discussion. The 'sharing economy' refers to all interactions that do not involve money . . . his paradigm of the sharing economy . . . is Wikipedia, because 'Its contributors are motivated not by money, but the fun or joy in what they do'" (Lessig, 2008, p. 74, p. 162). Of course, Lessig was also a key promoter of remix as a descriptive and explanatory metaphor and his book aimed to defend users by showing that these new practices of reproduction and circulation were a kind of speech – an explicit echo of Stallman's argument about free software as a kind of speech – and a potentially creative and innovative type of speech at that. If the intention of copyright was to promote knowledge and innovation, then copyright laws had not caught up with what new technologies enabled. Copyright was not just an obstacle to user's liberties; it was double damned for being inefficient and an obstacle to innovation.

5 Conclusion

My aim, in tracing the genealogy of these metaphors and their relationship with peer production and its intellectual antecedents, is to highlight that many of the assumptions that underpin remix and sharing as favored modes of participation and collaboration were in a sense reactive – they aimed to defend. That said, I acknowledge that they were reactions to real threats to freedom and agency and their utopian character offered possible solutions. Embedded in the history of these ideas are fears but also hopes and possibilities. My efforts to disentangle description from normative assumptions is not an effort to dislodge words and concepts from their current uses and aims but to use their cultural histories to unpack sources of tension, to help us to understand how actions that promote virtues might also promote inequalities.

The 2000s were an era of blue sky and utopian thinking about the possibilities created by the Internet and new forms of networked social organization, communication, and distribution. These ways of imagining network technologies came about in the context of a number of key socio-technical changes and innovations including the widespread adoption of broadband connectivity, the growth of social media, and the proliferation of peer-to-peer distribution technologies. Commons-based peer production, as exemplified by free and open source software production and online knowledge projects such as Wikipedia, offered a description and explanation of a specific set of practices and products that were not easily described or explained by traditional economic or organizational theories. Peer production with its efficient, non-hierarchical practices and projects seemed to defy classic economic explanations of self-interest and market rewards, and instead offered the possibility of personal liberty and agency through an ethos of participation, collaboration, the common good, and collective rights.

These utopian possibilities were attractive prospects to both social media entrepreneurs and the academics who aimed to defend these new networked practices and innovations against intellectual property regimes that limited their use. Drawing and expanding upon the ideology of free software developed in the 1980s and its particular interest in the regulation of copies and copying, critics pointed out that to limit a person's right to copy in a networked digital environment, was tantamount to limiting their right to speak, to comment and, in the case of software, to read, to learn, and to tinker. The context of debates on intellectual property in the 2000s, was a perfect storm in which technological disruption and user innovations butted up against existing copyright, patent, and trademark laws and a media and entertainment industry clamoring for longer, stronger, and more expansive intellectual property regimes. There was much to be critiqued about this struggle, both from the standpoint of most artists for whom copyright was a poor form of remuneration for labor and from users who were in a new position of having their rights to expression, access, and use trammeled. There were little guys to defend and big fish to fry. However, as my earlier work on remix suggests, there are problems for scholars of culture when the metaphor used to describe a phenomenon and critique its recent past also obscures that same history (Borschke, 2017). Similarly, Kreiss, Finn, and Turner (2011) noted that peer production's ethical ideals were less a description of the present than they were a critique of the recent past. Efforts to extend the ideals of peer production into other parts of culture such as music, video, and art, were also more defensive than they were descriptive or explanatory. Remix would defend the meme makers and video satirists; but sharing could also provide a virtuous cover for new forms of commercial exploitation.

Although scholars like Benkler are careful to distinguish commons-based peer production from its adjacent practices, it is here in the extensions of these ideals where the cracks in the façade are more evident and pronounced. The "sharing economy" is not merely a misnomer, the ethical problems regarding new forms of labor exploitation and related concerns about agency, property, privacy, and collective rights in the context of a drive towards precarious employment and the monetization of personal data exploit a tension between virtue and efficiency in theories of peer production. The drive to defend ethical ideals and practices with instrumental thinking about efficiency is problematic. This is not to suggest that peer production is inherently unethical – it is not – or that we should not consider how our systems of production can serve both individuals and the public good, rather it encourages us to think deeply about how our hopes ***and*** our fears are embedded in utopian thinking and to consider what we might learn from these tensions. Appeals to the efficiency of virtue not only render the idea of moral excellence as instrumental, but they also offer an instrumental defense of both unethical intentions and outcomes.

Note

1 Adjacent practices are those practices that are commonly associated with peer production such as crowd-sourcing, user-generated content, online labor market platforms, etc., but do not meet specialized academic definitions. See Benkler (2016) for a discussion of distinctions.

References

Benkler, Y. (2002a). Coase's penguin, or, Linux and "the nature of the firm". *Yale Law Journal, 11*(3), 369–446.

Benkler, Y. (2002b). Freedom in the commons: Towards a political economy of information. *Duke Law Journal, 52*, 1245–1276.

Benkler, Y. (2006). *The wealth of networks: How social production transforms markets and freedom.* New Haven, CT: Yale University Press.

Benkler, Y. (2016). Peer production and cooperation. In J. M. Bauer & M. Latzer (Eds.), *Handbook on the economics of the Internet* (pp. 91–119). Cheltenham: Edward Elgar Publishing.

Benkler, Y. (2017). Peer production, the commons, and the future of the firm. *Strategic Organization, 15*(2), 264–274.

Benkler, Y., & Nissenbaum, H. (2006). Commons-based peer production and virtue. *Journal of Political Philosophy, 14*(4), 394–419.

Borschke, M. (2017). *This is not a remix: Piracy, authenticity and popular music.* London: Bloomsbury Academic.

Bruns, A. (2008). *Blogs, Wikipedia, Second Life, and beyond: From production to produsage.* New York, NY: Peter Lang.

Castells, M. (2007). Communication, power and counter-power in the network society. *International Journal of Communication*, *1*, 238–266.
Gillespie, T. (2010). The politics of "platforms." *New Media and Society*, *12*(3), 347–364.
Herman, B. D. (2008). Breaking and entering my own computer: The contest of copyright metaphors. *Communication Law & Policy*, *13*(2) 231–274.
Jenkins, H. (2005). *Convergence culture*. New York, NY: New York University Press.
John, N. A. (2017). *The age of sharing*. Cambridge: Polity.
Kreiss, D., Finn, M., & Turner, F. (2011). The limits of peer production: Some reminders from Max Weber for the network society. *New Media and Society*, *13*(2), 243–259.
Lakoff, G., & Johnson M. (1980). *Metaphors we live by*. Chicago, IL: University of Chicago Press.
Lessig, L. (2004). *Free culture*. New York, NY: Penguin Press.
Lessig, L. (2008). *Remix: Making art and commerce thrive in the hybrid economy*. New York, NY: Penguin.
Loughlan, P. L. (2006). Pirates, parasites, sowers, fruits, foxes. The metaphors of intellectual property. *Sydney Law Review*, *28*(2), 211–226.
Shirky, C. (2008). *Here comes everybody: The power of organizing without organizations*. New York, NY: Penguin.
Stallman, R. (2002). *Free software, free society: Selected essays of Richard M. Stallman*. Boston, MA: Free Software Foundation.

9

Open Licensing Peer Production

Mélanie Dulong de Rosnay

1 Introduction

This chapter traces the evolution of legal conditions meant to support the production and flourishing of "commons-based peer production" (Benkler, 2006) in a diversity of fields covered by copyright, mostly in the digital realm. From software to creative works, including scientific articles, cultural heritage, public sector information, and open data, a wealth of digital, knowledge, intellectual or information commons can be peer produced. The rules which guarantee that they can remain in the commons, under open conditions, have been the subject of heated debates about the politics of technology (Berry, 2008) and heavy legal fine-tuning along the years, opposing different definitions and nuances in openness reflecting underlying philosophies within the peer production political economy, such as liberal and commons-based approaches.

According to the Open Knowledge Definition,[1] a standard that proposes minimum criteria to define openness, an open license grants to anyone "permission to access, reuse and redistribute a work, with few or no restrictions," namely the requirements to attribute authors (BY), to apply the same freedoms to derivative works (Share Alike or SA), or to reserve the right to modify the work (Non Derivative or ND), the right of reserving commercial use (NC for no commercial use) being excluded from these possible restrictions.

Having well-tailored legal documents setting the rules to access and reuse peer products is key to maintaining their availability in time and allowing their collaborative development. They allow for works to be shared, improved, and built upon. Without them, copyright law would apply by default to grant exclusive rights to authors, preventing peers from accessing, reusing or building on these works without prior written authorization, for each use and user. They help to waive such rights, so as to legally accompany the development of peer production.

Refining them to the specific needs of different formats of works, as well to different ideologies, led to improving them, and to an ecology of licenses. Some are better suited to specific formats, some require only attribution, some require to share alike. Different legal options and conditions set up by peer production authors or platforms to potential users can ensure different levels of openness, leading to the construction of informational, cultural, knowledge, or digital commons.

Roman law set up three categories of property rights: *usus*, the right to use a good; *fructus*, the right to grow or rent it; and *abusus*, the right to dispose exclusively of, destroy, or resell it to others. These rights can belong to a single owner, or to different users, identified or unidentified, actual or potential. They are reflected in open licenses (Table 9.1). Tangible commons resources are also managed following a bundle of rights, a related concept of US legal realism, where rights of access, exploitation, management, governance (who takes the decision on who may exercise which of the other rights), exclusion (deciding who will

The Handbook of Peer Production, First Edition. Edited by Mathieu O'Neil, Christian Pentzold, and Sophie Toupin.

Table 9.1 Open Licenses (derived from https://opendefinition.org/licenses/).

License	*Type or format*	*BY*	*SA*	*Comment*
GNU-GPL	Software	Yes	Yes	The first
GNU Free Documentation License	Software documentation	Yes	Yes	Before Creative Commons
Free software licenses	Software	Yes	Yes	www.gnu.org/licenses/license-list.html
Free Art License	Non-software works	Yes	Yes	Equivalent to GNU GPL
Open source software licenses	Software	Yes	No	https://opensource.org/licenses
Public domain	Any	No	No	A legal state (not a license) characterized by the absence of copyright
CC0	Any	No	No	Public domain dedication
Public Domain Mark	Works	No	No	Public domain recognition
Creative Commons Attribution	Non-software works	Yes	No	Similar to open source
Creative Commons Attribution Share Alike	Non-software works	Yes	Yes	Similar to GNU GPL
CC BY NC SA CC BY NC CC BY ND CC BY NC ND	Non-software works	Yes	No	Not suitable for free cultural works because they allow commercial use, modifications, or both
Open Database License ODbL	Databases	Yes	Yes	Equivalent to GNU GPL and CC BY SA
Peer production and copyfair license	Any	Yes	Yes	Introducing reciprocity
CERN and TAPR open hardware licenses	Hardware	Yes	Yes	For hardware
Pico peering agreement	Internet service providers	No	No	Non-copyright based, but similar values
Open source seeds	Seeds	Yes	Yes	For seeds

Source: Derived from "Conformant Licenses", https://opendefinition.org/licenses/. Licensed under CC BY 4.0.

have access rights) and alienation (right to sell or transfer other rights) can be segmented (Ostrom, 1990). In the same way, peer production also follows the legal model of a bundle of rights, where different rights under copyright can be segmented: the right of access only will be the first level of rights (corresponding to *usus*), followed by the right of *fructus*, the right of commercial exploitation (authorization to perform any use including commercial, or reserve those, NC for no commercial use), or modification (authorization to prepare derivative or reserve those, or ND for no derivatives), and the right of *abusus*, exclusion, and alienation and the prohibition thereof (further transmission of freedoms to derivative works).

Originally, free software or "copyleft" licenses such as the GNU General Public License (GPL) were standard contracts between the producers of commons, who could trust that other users would not abuse their work, and these other users who were granted the right to modify and distribute this work (provided they also granted these same rights to others). In order to achieve that objective, a legal hack was developed, "copyleft," or Share Alike (SA), ensuring that copyright applicable by default would be exercised differently, in order to grant rights to users rather than exclude them. The introduction of open source software licensing was then intended to make peer production more compatible with firms.

In the early 2000s, free and open licenses have expanded their range of options in order to be fit for purpose to the wider cultural field (including Creative Commons options accepting the reservation of commercial and derivative rights). They also intersected with the growing preoccupation with the access to knowledge or "digital rights" movement, as an alternative to the expansion of intellectual property. The late 2000s and the 2010s have been characterized by a focus on rules which would be more suitable to data, as well as databases, with the development of licenses or clauses aiming to tend to the special needs of functional works, also oscillating between full open access and certain restrictions inspired by the copyleft ethos.

Section 1 will briefly present the legal hack of copyleft and its first application for software licensing, as well as open source licenses. Section 2 will describe the evolution of open licensing to cover cultural works. Section 3 will address open licensing in the open science, open data, and open hardware or tangible environments. These fields came into being to answer the needs of the communities and the development of specific licenses tailored to formats.

Section 4 will introduce current efforts to develop reciprocity and expand the range of open licensing options and models to emerging needs.

2 Free, Libre, and Open Source Software Licenses

Free, libre, and open source software licenses are the first open licenses in history, allowing the development of software as a key area of peer production. They have been analyzed so many times in the literature (Carver, 2005; Coleman, 2013; Kelty, 2008; Liang, 2005) that this part will remain short. They exemplify the political philosophy (Berry, 2008) and political economy (Farchy, 2009) debate between free (open source, no downstream restriction on derivatives, Section 2.2) and libre (Share Alike, or copyleft, Section 2.1) in the free, libre, open or information commons community.

2.1 Copyleft

MIT computer scientist Richard Stallman and lawyer Eben Moglen created the first legal document enabling copyrighted creative work to be placed into the commons and neutralize the privatization of derivatives. Framed as a legal hack, the GNU General Public License (GPL) embeds four freedoms: the freedom to access the software code, the freedom to copy the software, the freedom to modify it, and the freedom to release modified versions under the same conditions. This was in contrast to proprietary software, where the code is inaccessible to the majority of users, and cannot be improved for other functionalities; or modified for any reasons such as format; or verified for bugs, or security issues, or Trojan horses; or adapted to special needs. This revolution in both the world of software and of copyright, begun as a subversion of licensing agreements, and became very popular, since most software is now reusable, that is to say released under free or open source conditions.

2.2 Open Source

A multitude of other licenses were then drafted for free and then open source software (FLOSS), removing the copyleft requirement to offer open source software on open access conditions without requiring modified versions to be released.

Some controversies accompanied the development of the movement, since this legal innovation. A legal controversy was whether these texts would be enforceable (Guadamuz, 2004, 2009, 2013; Kreutzer, 2011). It became widespread practice for judges to enforce such licenses which thus passed the test of legal resistance and plasticity, meaning they can be applied to a range of situations and objects. Another issue deriving from the multiplicity of different licenses in the FOSS ecosystem is the compatibility issue, since it can be difficult to merge software licensed under different (even if very close) conditions, prompting the dominance of the GNU GPL on the market. Several versions were developed, in the spirit of software, fixing legal bugs, or enforcing new user requirements, leading to stability, meaning they do not need more versioning.

Free software and open licensing prompted open collaboration (O'Neil et al., 2016) and peer production to a certain extent, since most projects hosted on GitHub, the largest repository of open source software, have a limited number of contributors and therefore do not engage with a large community of peers ("65% had only one committer, 83% two or less, and 90% three or less"; Kalliamvakou et al., 2014). But there are success stories in terms of peer production collaboration, quality and wide adoption: the Linux kernel, as well as most Internet servers are free software. In the age of online surveillance, with "good data" practices as an alternative (Daly et al., 2019), free software became a guarantee of transparency, allowing privacy activists to check, under peer production conditions, whether the users' personal data are protected or not, while closed code would not allow such checking to be collaborative. The ethos of access to code and transparency is also related to access to information storage, where governments hold and process public data.

3 Open Licenses for Cultural Works

3.1 First Developments

The GNU GFDL, a software documentation license, has been used for non-software works such as Wikipedia, before the encyclopedia switched to the Creative Commons copyleft option in 2009 after a vote by the community of authors. One of the inconveniences of using the GFDL (GNU Free Documentation License) was the requirement to include a list detailing all contributions, which could be very long and cumbersome, and already achieved in wiki history.

Expanding intangible commons to non-software works inspired the drafting of the very first artistic licenses, such as the Free Art License, first written in French and translated into English and other languages without a legal adaptation porting the process to other national laws.

3.2 Creative Commons

The Creative Commons organization (CC) was created in 2001 by a group of scholars and activists around Lawrence Lessig based on the Eric Eldred case, as copyright law duration was expanded in US law and no works would be included in the public domain to be freely available for this reason, in a legal context of enclosure of the public domain (Boyle, 2010). Therefore, CC licenses have initially been designed in order to facilitate the dedication, if not to the public domain, at least to a semi-commons (Pallas-Loren, 2007), a state with various flavors of openness through private means (Chen, 2009; Dusollier, 2006, 2007; Elkin-Koren, 2005) by living authors.

This section is significantly longer than the other sections because of the instrumental role Creative Commons played in the development of a legal standard for peer production, free culture, remix culture, and sharing culture (Aigrain, 2012; Lessig, 2004, 2008;) and the open movement (Cheliotis, 2009). The section will focus on some of the legal questions raised by the licenses, which exist in different flavors combining various licensing elements (Attribution, Commercial rights, Derivative Rights, and Share Alike, the equivalent of copyleft) and have led to several versions, following the model of free software development which collectively corrects "bugs," including translations, legal adaptations, or porting to a large number of national legislations by chapters worldwide.

This diversity of legal rights created policy issues, because of the level of freedoms, but also technical legal issues: interoperability, namely the legal incompatibility among works licensed under the numerous Creative Commons licenses led to executive decisions by CC headquarters. The licenses requiring only Attribution and possibly Share Alike carried a badge of honor, "Approved for Free Cultural Works,"[2] designed to signal those licenses which fulfill the criteria of FOSS activists. Legal porting operations were interrupted, with the most recent version drafted so as to be implemented worldwide without reference to national definitions.

Creative Commons also developed a range of other licenses and licensing instruments, with various degrees of success. The sampling and developing nations licenses, aiming at authorizing different levels of sampling, and at only granting rights to users located in countries classified as such, did not take off, and were promptly withdrawn. To the contrary, both the Public Domain Mark and the CC0

Dedication to the Public Domain were maintained as more suitable for data as analyzed in the section dedicated to open licenses for functional works.

The "Share Alike" option allows derivative works to be licensed under the same license as the original work, but also under a later version or a compatible version of the same license. Versions 2.0, 2.5, and 3.0 allowed the use of translated versions that had been transposed into the law of another jurisdiction after validation by the Creative Commons organization. Versions 3.0 and 4.0 provided for compatibility of licenses that have been approved by their home organizations as equivalent in effect and recognized as compatible: currently the "Free Art License" in version 1.3 provides for a similar mutual compatibility clause, with both Copyleft Attitude and Creative Commons organizations having recognized both licenses as compatible and equivalent in effect. The definition of "Share Alike" according to the Creative Commons and the compatibility method have been adopted in the same way by the Open Database License (ODbL) v1.0 which applies copyleft to databases.

Conversely, the GNU Free Documentation License (GNU GFDL) provided in version 1.3 a unilateral temporary migration clause to the "CC BY SA 3.0" for massively multi-author collaborative sites; the objective was to allow the Wikimedia Foundation, the organization that structures part of the Wikipedia encyclopedia projects, to change its license.

Share Alike applies the copyleft principle to non-software works and is used by different types of authors of literary and artistic works. First, producers of utilitarian or functional resources, intended to be assembled, combined and derived, such as databases or wikis and Wikipedia, are emblematic examples of creations whose production is based on free software, peer production, and licensed under free conditions. There are also public institutions and activists in the Free Culture movement. On the other hand, the combination of the options "Attribution and Share Alike" (CC BY SA), although validated by the label of Free Works ("Approved for Free Culture Work") as including the freedoms recognized as essential, is not recommended by the open data, open education, and open science movements which privilege the sole attribution (CC BY), or even the voluntary public domain (CC0).

A justification for the Share Alike option is saving transaction costs in collaborative projects, which this license is supposed to protect from private appropriation and enclosure of the commons. The Share Alike option allows forking or bifurcation of projects, as communities can create forks and use the content generated in the previous project to continue in other directions. Thus, when Wikitravel was sold to a company that added ads to the site, the German and Italian versions were able to leave with their content in 2006 and 2007, by creating the WikiVoyage project, before joining the Wikimedia Foundation with other languages in 2012.

However, the Share Alike clause causes some issues of legal incompatibilities among national laws due to the porting strategy (Dulong de Rosnay, 2010). Also, the multiplicity of options results in transaction and information costs (Elkin-Koren, 2005) for the licensor and the licensee of the license and blurring of the message of what constitutes a free or open work. Indeed, not all Creative Commons licenses, including all those with the "Share Alike" option, are considered free in the sense of free software if they reserve commercial rights (NC), not allowing commercial companies or projects such as those of the Wikimedia Foundation to reuse the work without additional authorization. This perspective can be interpreted as a restriction of freedom (in the sense of free software) and a fragmentation of commons, in the sense that the work cannot be reused by all. On the contrary, another perspective is to consider the NC option as a protection of original authors against commercial appropriations of their contribution that could take place without offering them remuneration.

The Share Alike option can lead to discouraging the exploitation of free software for commercial purposes. Indeed, if identical reproduction is free for all, it will be difficult to monetize copies, as the example of Wikipedia shows. However, without the NC option, Share Alike theoretically allows the scraping of content, a sort of capture that could strengthen commercial sites that value traffic and the exploitation of personal data more than access to the work itself. It should be noted that the Share Alike clause organizes the identical sharing of derivative works (under the same conditions) The peer-to-peer licenses under development (see last section of this chapter) intend to develop more reciprocity than the Share Alike clause, while avoiding the pitfalls of the "No Commercial Use" NC clause.

The lack of interoperability between works under different Creative Commons licenses has been analyzed by the legal doctrine (Dulong de Rosnay, 2010; Dusollier, 2007; Elkin-Koren, 2005; Giannopoulou, 2014; Katz, 2005). If works under free license cannot be mixed together without authorization and a table must be consulted to find out under which license to distribute the derivatives, this leads to a fragmentation of the commons, which has been observed for free software since content licensed under one license can technically not be remixed under other similar though different licenses if they have not been formally recognized as compatible (Dulong de Rosnay, 2010; Dusollier, 2007; Elkin-Koren, 2005; Katz, 2005).

The Share Alike clause, by providing for compatibility between equivalent licenses, introduces another risk of legal incompatibility, which results from the lack of harmonization between national laws, as the adapted, ported, or transposed licenses do not therefore cover exactly the same spectrum of rights. However, if licenses are declared compatible with each other, authors are supposed to consent to future adaptations of their works being licensed under unidentified conditions, which may pose a problem of internal validity and consent under contract law (Dulong de Rosnay, 2010).

Creative Commons licenses, up to version 4.0, were adapted into national laws. These legal adaptations reflected actual legal differences between national laws, and were supposed to ensure compatibility with domestic legal systems, interpretation by the judges, and direct applicability by the courts (Maracke, 2010). Several court decisions have successfully applied the licenses. National transpositions have also led to the creation of a network of Creative Commons affiliates, lawyers, professionals, and information commons activists. Nevertheless, the practice of transposition was abandoned in 2014 after version 4.0, in order to simplify licenses and avoid compatibility problems caused by the "Share Alike" clause, which recognized as downstream compatible licenses that had been transposed into national laws and therefore included substantial differences in the scope of the rights and the definition of the works concerned, besides not being accessible in all languages. Despite the end of the national transposition process, which is replaced by a simple translation, there is still a risk of a difference between language versions in the scope of limitation of liability and the absence of a guarantee by the licensor that the work does not constitute an infringement of copyright, a generic clause of free licenses.

In addition, the licenses still include such a clause which provides that the conditions of the license which would contravene national law are not applicable. The scope of the guarantees that can be waived by contract varies according to the place of production of the original and the modification, as this branch of the law is not harmonized internationally. A contractual limitation of the author's liability as it exists in version 4.0 may be void under French or Italian law. The Share Alike clause can therefore lead to the spread of legal risks for licensees, a source of pollution of the commons, and of legal risk and weakening of open licenses if the author does not guarantee that the work offered does not contain any infringement, and a licensor builds upon it in good faith (Dulong de Rosnay, 2013). Can more rights be reasonably transferred than one thinks one holds? This contractual problem does not arise with other copyleft licenses deemed compatible or in the process of being compatible (Free Art License and GNU GFDL) which remain silent on the issue of the guarantee of non-infringement by the licensor, leaving the choice of jurisdiction outside of the license and therefore determined by applicable law.

3.3 The Access to Knowledge (A2K), Free Culture, and Public Domain Advocacy Movements Supporting Digital Commons

The promotion of Creative Commons licenses in the 2000s has been ensured, among other supporters, by Free Culture collectives of students within universities. Also, an important tool for local translation, legal adaptation but also legitimation of the digital commons and legal information to digital commons authors and institutions has been provided by CC jurisdictions affiliates, a mix of academics (law, computer science, economy, communications), and digital rights non-profits.

CC supporters' and affiliates' policy purposes overlap with several broader advocacy communities: digital rights (Postigo, 2012) activists promoting users' rights, privacy, freedom of speech; Access to Knowledge (A2K), a movement addressing not only digital commons, but also Access to Medicine, or Intellectual Property for the Public Interest, a community structured around the yearly eponymous Global Congress on Intellectual Property for the Public Interest. Starting with support for more balanced copyright reform, with broader exceptions and limitations to exclusive rights, the demands of open licensing advocates, if fulfilled, would lead to a generalization of the contractual effect of open licensing in the law: more free access, more transformative rights for artists, authors, researchers, and teachers, more exceptions to exclusive rights especially in case of non-commercial uses such as libraries, educational institutions, and archives, or associations working with visually impaired persons. If the copyright reform agenda was achieved, open licensing would become less necessary, because copyright by default would allow a broader range of freedoms and rights to users.

However, the US tax status of the CC organization made it difficult to perform advocacy tasks, and prevented the take-off of CC as one of the most powerful lobbying organization representing the digital commons, a space which has been occupied by other actors such as the Wikimedia Foundation, the main user of open licensing worldwide, together with dedicated digital rights nonprofit organizations.

Several organizations focusing on digital commons are active in Geneva, where copyright international treaties and policies are designed in multilateral institutions. As the most prominent open content licensing actor, CC applied to get an observer seat at WIPO, the UN World Intellectual Property Organization, and occasionally sends a representative at SCCR (Standing Committee on Copyright and Related Rights) and CDIP (Committee on Development and Intellectual Property) diplomatic conferences in order to promote the inclusion of alternative views of intellectual property – that is, in addition to views of IP supporting more exclusive rights, a vision of intellectual property supporting digital commons as embodied in open licenses. Several WIPO reports have been drafted by open licensing legal experts on topics such as the preservation of the public domain, the legal feasibility of voluntary dedications to the public domain, applications of open licensing for public sector information, and open education.

In 2008, CC joined the Communia International Association for the Digital Public Domain, an association which was incorporated into Europe after a research project (2007–2011) which gathered initially most European chapters together, aiming at performing not only research and promotion about open licensing, but also about the public domain at large, including copyright reform, exceptions, and limitations. CC, through its policy and legal staff participation to the work of the association, has been an active contributor to Communia management, policy briefs, responses to consultations, and advocacy work for copyright reform in Europe.

4 Open Licenses for Functional Works

This section will examine the specificities of functional works, such as scientific articles, educational resources, databases (of scientific data, public sector information, or digitized public domain works), and digital commons or Internet-related artifacts, namely computer-generated open hardware, and community networks allowing connections to the Internet or to local area networks. All of them have been relying on specific flavors of open licensing, sometimes adapted to suit their particular needs. All of them have been accompanied by policy guidelines and manifestos advocating for openness and the best ways to achieve such a result in the respective fields.

4.1 Open Access, Open Science, and Open Education

The movement for open access in science began in 2001 with the Budapest Open Access Declaration, which was followed by the Bethesda and Berlin declarations. They have established a number of principles to define and achieve this openness, including a difference between the simple financial freedom to read an article and legal open access, guaranteeing rights to authors and users. Two types of licenses have been validated by the authors of these declarations to ensure open access to publications: those requiring attribution only (such as CC) and those adding a copyleft clause (such as CC BY SA).

Science Commons is the former branch of Creative Commons dedicated to the development of openness and licensing for science. In terms of publications, Science Commons created, together with other nonprofit organizations advocating open science such as SPARC, addendums for academics to amend standard copyright licensing agreements where authors have to relinquish all of their rights to commercial publishers, and may not have the right to self-archive their own publications, nor to access them if their library does not have, or interrupts its subscription to the commercial publishers' journal bundle.

When it comes to scientific data, Science Commons promoted the use of CC0 in order to reach the closest possible conditions to the public domain and limit liability risks in case of breach of license, in the case of a missing attribution when mining hundreds of databases.

Finally, in the case of biological data, Science Commons produced a number of material transfer agreements adapting the core principles of open licensing.

Openness for academic publishing can follow three different models. Green open access consists in being able to self-archive a version of one's paper, sometimes the preprint, before publisher layout and copy-editing, in an institutional or a disciplinary repository. Gold open access is a commercial practice used by some publishers surfing on the fashion of openness and charging a hefty fee to authors choosing to apply a CC license to their paper to make it open access. Platinum open access or freemium are hybrid models, where authors do not pay to have their article freely available. The funding is collected by publishers through library subscriptions based on the number of downloads and in exchange for some services (formatting, statistics), and revenues are shared between platforms and journals. Freemium has been adopted by the Open Edition platform for Social and Human Science Journals.

Funders or government mandates are useful instruments to require the distribution of the results of the research they support under open access conditions equal to open licensing.

The open education resources sector has also been employing CC Attribution licenses to distribute textbooks which can be used and adapted by teachers and students for free.

4.2 Open Data

Closely related to the movement for open science reclaiming the reproducibility of scientific results, the movement for open data in the field of science, but also for public sector information produced by governments or public bodies (see next section), draws its inspiration from the potential unleashed by information technologies to access and process more data, and the expectation it might produce more results for innovation and society.

Open Data refers to openly available digital data, or more broadly the techno-political movement at the origin of the opening of data produced mainly by the public and academic sectors, but also increasingly by the private sector. To be recognized as open, data must be made available in an open standard format and allow free and open reproduction and reuse by all. Government agencies, associations, and non-governmental organizations are involved in defining the concept and producing processing platforms and tools (e.g., data mining or visualization) for the reuse and manipulation of this data. According to policy declarations, only a minimum of conditions is acceptable for data to be recognized as open, including obligations to assign a source, to Share Alike, and to respect integrity or avoid derivatives (ND). This mainly concerns public data and data resulting from scientific research, considered as information commons in which case open access can be set up either by mandatory law or through voluntary institutional policies and open licensing.

The benefits of open data for knowledge and democracy are not immediate, the translation of data into policy, business, economic, or scientific derivative leading to positive results for society being neither automatic nor easy. The open data movement's efforts to make data available and reusable are a good stimulus for democracy and the economy. However, care should not be taken only to improve the degree of openness of data, but also to curate them, and incentivize peer production practices based on reuse. Otherwise there is a risk of ending up with portals that are empty shells. Finally, the impact of big data should be considered, as peer produced information requires more than the processing of data sets made possible by their openness, a processing that is not accessible to all and can lead to increased inequality.

4.3 Public Sector Information and Open Government

The movement for Open Government and Open Data applied to public sector information is based on – in addition to arguments supporting open data for science – the rationale that State transparency, accountability, and freedom of information are clear benefits. Taxpayers have already paid for the production of the data, which could in addition be reused by other services, including other governmental agencies, without legal frictions or copyright royalties.

In addition to fundamental rights and the democratic imperative, there are economic and political justifications (Uhlir, 2004), backed by a technicist and neoliberal discourse: transparency promotes citizen participation in public decision-making and the accountability of governments; the taxpayer financing the production of public and scientific data should be able to access the results; innovation by companies, associations and citizens, and thus economic growth and employment, will be stimulated by the availability of open data, on the basis of which services and applications can be developed; finally, public policies and citizens' decisions should be based on scientific data and evidence, the advancement of scientific knowledge and discoveries being accelerated and facilitated by cross-referencing and searching databases, which must therefore be legally and technically available and open.

Public data cover information contained in administrative documents held by public agencies: this includes the databases they produce or receive, such as general interest data and reference data. The open data produced by public institutions are diverse and varied: statistics and data on transport, the environment, land registry, housing, weather, education, employment, culture, trade, public expenditure, and so on. Geographic data are generally the most open part of public sector information, with other services building applications upon them. Open data can also be commons-based and peer produced: this is the case with the geographical data of the Open Street Map project, or the scientific data of the human genome (Personal Genome Project).

Open Data actors are government agencies, researchers, and associations that will promote the release and reuse of data by facilitating open legal and technical availability, with open formats and standards to link, structure, and enrich data. They produce recommendations, criteria, and definitions of what constitutes open data. According to the Open Definition already mentioned in the introduction, data will be open if it can be used, modified, and shared freely by anyone for any purpose and with only the attribution, integrity, identical sharing, license notice, or source format as acceptable conditions.

Several policy instruments are available to open public sector data: legal implementation can be achieved through copyright laws, public sector information laws, and open licenses. Some countries are directly releasing public sector information in the public domain, while others need open licensing to reach the same objective, in order to facilitate distribution on dedicated platforms, under conditions ranging from contractual dedication in a voluntary public domain, to the use of copyleft. Several associations, including the Open Knowledge Foundation, have developed open licenses specifically for the *sui generis* rights of database producers (the Open Database License, or ODbL).

Government platforms such as data.gov distribute public data under different types of open licenses, which will allow reuse, even for commercial purposes under certain conditions. Australia, the United States, Brazil or Burkina Faso use a Creative Commons Attribution license, while the Dutch government releases its data under CC0, the Creative Commons protocol for voluntary and early dedication to the public domain. France, Great Britain, and Italy have chosen to develop their own open government licenses, while declaring them compatible with Creative Commons Attribution licenses in order to allow interoperability with other data sets.

The use of standard licenses, as opposed to the drafting of vanity open licenses by governments, reduces information and transaction costs, but imposes a fairly heavy contractual obligation to reusers and does not allow the insertion of specific clauses, such as the offer of guarantees or the prohibition to alter data.

4.4 Cultural Heritage

Cultural heritage data and digitalized public domain works held by libraries and museums are also part of a broad open licensing movement (Evens, 2016). Europeana is a public sector platform for creative works, but also metadata produced by libraries and museums. An example of peer production by the GLAM community (galleries, libraries, archives, and museums) are the projects developed within Wikimedia Commons by volunteers working with public institutions to provide the high resolution reproductions of public domain works and to make them available with their metadata. Commons-based peer production plays a dual role here, first by providing access to digital collections of public domain works and thus avoiding the enclosure of the commons managed by libraries and museums. Second, the Wikimedia GLAM community has been vocal in several instances of copyright reform. One example is the fight to defend "freedom of panorama", the right to upload reproductions of works which are on the physical public domain (Dulong de Rosnay & Langlais, 2017).

A counter example to these best practices is the commodification of the public domain and public domain digitized archives by institutions who practice "copyfraud" (Mazzone, 2011) by applying restrictive copyright licensing conditions. Licensing standards recommended in such cases are the Public Domain Mark (PDM), another instrument developed by Creative Commons, where institutions can certify that a work is in the public domain, meaning that it is no longer covered by copyright because its term of protection expired, typically 70 years after the death of the authors. Metadata added by librarians and documentation professionals to describe and retrieve the work should be licensed under CC0, in the vein of open data recommendations.

4.5 Open Hardware and Open Infrastructure

This section presents licenses developed to accompany the distribution and creation process of other forms of peer produced resources, located between tangible and intangible commons.

As for open design, two open hardware licenses (Pearce, 2016; Powell, 2012) have been developed to address specific issues which were not covered by FLOSS licenses, since many hardware designs are patentable rather than copyrightable.

Open licenses adaptations are also required to fit the specificities of 3D printing generated works (Jee & Sohn, 2018). The hardware can be released under a FLOSS license, and the content files have been experimenting with either software licenses, or CC licenses, some being attracted by the reservation of commercial rights option. The field has not reached mature standardization and trademark, and patent or contract issues with third party rights may still arise with complex works gathering several types of intellectual property.

The spirit of openness and peer production has also been exported to open biology and open seeds, to counter the effect of patenting and regenerating a commons through copyleft applied to the generation of seeds extraction of genetic information and enhancement of varieties.[3]

All of these creations and innovations would otherwise have been submitted to exclusive rights under other subparts of intellectual property than copyright: design, patents, the *sui generis* intellectual property system of plant variety (Deibel, 2013) breeders' rights.

Between open hardware and open infrastructure, community networks, alternative Internet service providers of peer produced connectivity, rely on local activists' volunteer work, and modifiable hardware. In addition to free software and open source routers, based on FLOSS licenses, community networking prompted the drafting of specific legal texts akin to open licenses. The PICO peering agreement,[4] an open license for community networks, is not subverting intellectual property, since the body of regulation which is touched upon is not a resource. It is dealing with telecommunications law, network neutrality, and quality of service.

Such legal texts can still be included in a broad definition of open licensing for peer production, because they aim at guaranteeing openness, defined as freedom for users, and at securing the rights and responsibilities of peer contributors, exactly like traditional open licenses.

5 Developments in Open Licensing

Finally, beyond the need to legally tailor the ethics of open source and peer production to different outputs, some recent open licensing developments are driven by the reinforcement of some of the original political principles of the commons.

5.1 Copyfarleft, Copyfair, and Commons Reciprocity Options

According to Dmitry Kleiner (2007),

> copyleft cannot allow workers to accumulate wealth beyond subsistence, copyleft alone cannot change the distribution of productive assets, which is what any revolutionary strategy must seek to do (…). A copyfarleft license should make it possible for producers to share freely and to retain the value of their labour

product, in other words it must be possible for workers to make money by applying their own labour to mutual property, but impossible for owners of private property to make money using wage labour.

The Peer Production license[5] embeds these anticapitalist principles, allowing commercial commons-based reuse by cooperatives, and nonprofit actors, but preventing "commercial entities intent on making profit through the commons without explicit reciprocity."

The Copyfair principle requires reciprocity in commercial activities. It "aims to subject commercialization of any such knowledge commons to some form of contribution to that commons."[6] For instance, open access content could not be scrapped by a predatory commercial website whose business model would rely on the exploitation of user traffic and other personal data.

These new licensing prototypes have to solve many implementation issues. A reciprocity clause "that restricts commercial usage according to how much the user has contributed to the common pool" (Vieira & De Filippi, 2014) requires monitoring the existence of actual and past contribution by identified peers, instead of the copyleft provision loose model allowing open-ended possibilities for future, unidentified peers. Such a reciprocity model would rely on the identification of contributors and the definition of metrics to assess individual contributions and their ethical nature.

5.2 Data Commons

The last evolution observed in open licensing is led by practical projects involving personal data. At first, there is a contradiction between open access and personal data, since privacy prohibits sharing or reusing data without consent. But models can be imagined where the wealth of information gathered from personal actions could be re-injected into the commons while maintaining a level of user control rather than privatizing and enclosing it for profit. Many privacy policies inspired by Creative Commons have been started[7] but none has become a standard yet.

New directions to constitute personal data commons for data (Bassi et al., 2019) generated by users and citizens in social media platforms, in the Internet of Things, or in cities, are being considered, in order to develop alternatives to the smart and surveillance models centralizing and commodifying users' data. They would be exercising their consent in a different way from agreeing to restrictive terms of use and the commodified processing of their personal data, and opt in for ways to rather dedicate them to a commons.

Data commons can follow the model of scientific data pools, data trusts, or data collaboratives[8] such as the projects analyzed by the Governance Laboratory (GovLab) at New York University (Verhulst et al., 2019). While policy guidelines and ethical principles for the management of personal data are numerous, unfortunately, end-user licensing agreements for these projects are not public, beyond standard terms and conditions for reuse. Reasons for this may be because lawyers prefer to tailor them for each project and market them, or because they have not come to fruition yet, due to the complexity of crossing copyright and personal data laws in order to open them and the lack of international harmonization of personal data legislations.

Data commons can also use decentralized technologies for collection, analysis, and storage platforms. Governance principles for commons-oriented data management practices can be found in the European project Decentralized Citizen Owned Data Ecosystems or Decode (Fuster Morell et al., 2017) and are being implemented in Amsterdam and Barcelona. Such principles are framed around consent and security and are to be transposed into ontologies for rights expression languages and licensing terms merging principles of openness and anonymization to enable data to be re-injected into the commons without enclosure nor risks for citizens' privacy.

6 Conclusion

Openness and licensing in peer production have been oscillating between full open access and some restrictions aimed at promoting informational commons and avoiding their enclosure. After a take-off in software, and culture, supporting activists of fairer copyright laws and users' rights, the open licensing movement addressed more functional works and created specialized licenses such as open data and open hardware licenses. The broad acceptance of open licensing by institutions such as libraries and museums for their digitized collections, as well as by governments for public sector information, led to raising awareness of peer production. On the one hand, openness points to a neoliberal vision rather than a social-democratic, commons-based (Broumas, 2017) or Marxist vision (Rigi, 2013), and possibly open- or commons-washing[9] can lead to a dilution of the original values of the commons, the ethos of sharing with a community for the common good. On the other hand, the open licensing scene has been trying to develop options refining copyleft, with the copyfarleft and commons reciprocity licensing models proposals aiming at better translating the commons-based peer production ethos into open licensing conditions. They might find applications in fields such as personal or scientific data pools, data generated by tangible applications in cities, or connected homes. The question of the appropriation and co-optation of production platforms and access infrastructure, as well as of user-generated and/or personal data, requires designing further hacks to resist peer production commodification and enclosure.

Notes

1 http://opendefinition.org/od/2.0/en/
2 https://creativecommons.org/share - your-work/public-domain/freeworks/
3 www.opensourceseeds.org/en/licence
4 http://picopeer.net/
5 http://wiki.p2pfoundation.net/Peer_Production_License
6 http://wiki.p2pfoundation.net/CopyFair_License
7 https://wiki.creativecommons.org/wiki/CC - inspired_projects_for_Terms_of_Service_and_Privacy_policies
8 http://datacollaboratives.org/explorer.html?#data - pooling
9 https://www.netcommons.eu/?q=content/commonswashing-information-technologies-and-online-platforms-semantic-appropriation-commons

References

Aigrain, P. (2012). *Sharing: Culture and the economy in the Internet age*. Amsterdam: Amsterdam University Press.

Bassi, E., Ciurcina, M., De Martin, J. C., & Fenoglietto, S. (2019). Licensing of digital commons including personal data – update. *DECODE European project deliverable 1.9*. https://decodeproject.eu/publications/licensing-digital-commons-including-personal-data-update

Benkler, Y. (2006). *The wealth of networks: How social production transforms markets and freedom*. New Haven, CT: Yale University Press.

Berry, D. (2008). *Copy, rip, burn: The politics of copyleft and open source*. London: Pluto.

Boyle, J. (2010). *The public domain: Enclosing the commons of the mind*. New Haven, CT: Yale University Press.

Broumas, A. (2017). Social democratic and critical theories of the intellectual commons: A critical analysis. *TripleC: Communication, Capitalism & Critique. Open Access Journal for a Global Sustainable Information Society*, *15*(1), 100–126.

Carver, B. (2005). Share and Share Alike: Understanding and enforcing open source and free software licenses. *Berkeley Technology Law Journal*, *20*(1), 443–481.

Cheliotis, G. (2009). From open source to open content: Organization, licensing and decision processes in open cultural production. *Decision Support Systems*, *47*(3), 229–244.

Chen, S.-L. (2009). To surpass or to conform – what are public licenses for? *University of Illinois Journal of Law, Technology & Policy*, *1*, 107–139.

Coleman, E. G. (2013). *Coding freedom: The ethics and aesthetics of hacking*. Princeton, NJ: Princeton University Press.

Daly, A., Devitt, S. K., & Mann, M. (2019). *Good data. Theory on demand*. Amsterdam: Institute of Network Cultures.

Deibel, E. (2013). Open variety rights: Rethinking the commodification of plants. *Journal of Agrarian Change*, *13*(2), 282–309.

Dulong de Rosnay, M. (2010). *Creative Commons licenses legal pitfalls: Incompatibilities and Solutions.* Amsterdam: Institute for Information Law of the University of Amsterdam. Retrieved from https://halshs.archives-ouvertes.fr/halshs-00671622

Dulong de Rosnay, M. (2013). Open content licenses without representation: Can you give away more rights than you have? *European Journal of Law and Technology (EJLT), 4*(3).

Dulong de Rosnay, M., & Langlais, P.-C. (2017). Public artworks and the freedom of panorama controversy: A case of Wikimedia influence. *Internet Policy Review, 6*(1).

Dusollier, S. (2006). The master's tools v. the master's house: Creative Commons v. copyright. *Columbia Journal of Law & Arts, 29*, 271–293.

Dusollier, S. (2007). Sharing access to intellectual property through private ordering. *ChicagoKent Law Review, 82*, 1391–1435.

Elkin Koren, N. (2005). What contracts cannot do: The limits of private ordering in facilitating a creative commons. *Fordham Law Review, 74*, 375–422.

Evens, T. (2016). Creative commons licences in cultural heritage institutions in Flanders. *Journal of Librarianship and Information Science, 48*(2), 209–217.

Farchy, J. (2009). Are free licences suitable for cultural works? *European Intellectual Property Review, 31*(5), 255–263.

Fuster Morell, M., Carballa Smichowski, B., Smorto, G., Espelt, R., Imperatore, P., Rebordosa, M., Rocas, M., Rodríguez, N., Senabre, E., & Ciurcina, M. (2017). Multidisciplinary framework on commons collaborative economy? *DECODE European Project Deliverable 2.1.* Retrieved from www.decodeproject.eu/publications/multidisciplinary-framework-commons-collaborative-economy

Giannopoulou, A. (2014). The Creative Commons licences through moral rights provisions in French law. *International Review of Law, Computers & Technology, 28*(1), 60–80.

Guadamuz, A. (2004). Viral contracts or unenforceable documents? Contractual validity of copyleft licenses. *European Intellectual Property Review, 26*(8), 331–339.

Guadamuz, A. (2009). The license/contract dichotomy in open licenses: a comparative analysis. *University of La Verne Law Review, 30*(2), 101–116.

Guadamuz, A. (2013). Comparative analysis of national approaches on voluntary copyright relinquishment. World Intellectual Property Organization, CDIP/13/INF/6. Retrieved from www.wipo.int/meetings/en/doc_details.jsp?doc_id=272263

Jee, S. J., & Sohn, S. Y. (2018). License choice in open 3D printing content community: Are current license options sufficient? *Telematics and Informatics, 35*(8), 2242–2253.

Kalliamvakou, E., Gousios, G., Blincoe, K., Singer, L., German, D. M., & Damian, D. (2014). The promises and perils of mining GitHub. *Proceedings of the 11th Working Conference on Mining Software Repositories – MSR 2014* (pp. 92–101).

Katz, Z. (2005). Pitfalls of open licensing: An analysis of Creative Commons licensing. *IDEA - The Intellectual Property Law Review, 46*(3), 391–413.

Kelty, C. M. (2008). *Two bits: The cultural significance of free software* (ed. M. M. J. Fischer & J. Dumit). Durham, NC: Duke University Press.

Kleiner, D. (2007). Copyfarleft and copyjustright. Retrieved from www.metamute.org/editorial/articles/copyfarleft-and-copyjustright

Kreutzer, T. (2011). Validity of the Creative Commons zero 1.0 universal public domain dedication and its usability for bibliographic metadata from the perspective of German copyright law. Retrieved from www.rd-alliance.org/sites/default/files/cc0-analysis-kreuzer.pdf

Lessig, L. (2004). *Free culture: The nature and future of creativity.* New York, NY: Penguin.

Lessig, L. (2008). *Remix: Making art and commerce thrive in the hybrid economy.* New York, NY: Penguin.

Liang, L. (2005). *Guide to open content licenses v1.2.* Rotterdam: Piet Zwart Institute, Institute for Postgraduate Studies and Research, Willem de Kooning Academy.

Postigo, H. (2012). *The digital rights movement: The role of technology in subverting digital copyright.* Cambridge, MA: MIT Press.

Maracke, C. (2010). Creative Commons international. The international license porting project – origins, experiences, and challenges. In D. Bourcier, P. Casanovas, M. Dulong de Rosnay, & C. Maracke (Eds), *Intelligent multimedia. Sharing creative works in a digital world* (pp. 67–88). Series in Legal Information and Communication Technologies, Vol. *8*. Florence: European Press Academic Publishing.

Mazzone, J. (2011). *Copyfraud and other abuses of intellectual property law.* Stanford, CA: Stanford Law Books.

O'Neil, M., Gorbatai, A., & Jemielniak, D. (2016). Management and the future of open collaboration. *Journal of Organizational Change Management, 29*(3).

Ostrom, E. (1990). *Governing the commons: The evolution of institutions for collective action.* Cambridge: Cambridge University Press.

Pallas-Loren, L. (2007). Building a reliable semicommons of creative works: Enforcement of Creative Commons licenses and limited abandonment of copyright. *George Mason Law Review*, *14*, 271.

Pearce, J. (2016). Return on investment for open source scientific hardware development. *Science and Public Policy*, *43*(2), 192–195.

Powell, A. (2012) Democratizing production through open source knowledge: From open software to open hardware. *Media, Culture and Society*, *34*(6), 691–708.

Rigi, J. (2013). Peer production and Marxian communism: Contours of a new emerging mode of production. *Capital & Class*, *37*(3), 397–416.

Uhlir, P. (2004). *Policy guidelines for the development and promotion of governmental public domain information*, UNESCO, Paris. Retrieved from https://fas.org/sgp/library/unesco_govinfo.pdf

Verhulst, S., Young, A., Winowatan, M., & Zahuranec, A. (2019). Data collaboratives. Leveraging private data for public good. Retrieved from https://datacollaboratives.org/existing-practices.html

Vieira M. S., & De Filippi P. (2014). Between copyleft and copyfarleft: Advanced reciprocity for the commons. *Journal of Peer Production*, *4*. Retrieved from http://peerproduction.net/issues/issue-4-value-and-currency/invited-comments/between-copyleft-and-copyfarleft-advance-reciprocity-for-the-commons/

10

User Motivations in Peer Production

Sebastian Spaeth and Sven Niederhöfer

1 Introduction

The advantages of "commons-based peer production" (Benkler, 2002, p. 375) on a national or firm level seem obvious: (1) it is better at identifying and assigning human resources to production processes, (2) increasing returns allow huge clusters of potential contributors to interact with huge clusters of information resources and (3) it achieves lower transaction costs by "not relying on property and contract as the organizing principles of collaboration" (Benkler, 2002, p. 376). By reducing transaction costs through peer self-selection, Benkler argues an economic case for a commons-based peer production process. This does not explain why individuals would volunteer to contribute to such a system, though. Olson famously stated: "Indeed, unless the number of individuals in a group is quite small, or unless there is coercion or some other special device to make individuals act in their common interest, ***rational, self-interested individuals will not act to achieve their common or group interest***" (Olson, 1965, p. 2 emphasis in original). In the case of public goods, this is the classic consequence of under-provisioning and overuse: the tragedy of the commons (Hardin, 1968).

Yet, various forms of peer production obviously thrive, ranging from the obscure (such as proofreading of scanned classic texts) to industry-wide dominant solutions (e.g., Wikipedia). Thousands of volunteers, both unpaid and paid, act in and contribute to their respective projects. Here, we examine what motivates these individuals to contribute on an individual level.

In 2000, the economists Lerner & Tirole called these peers' behavior "startling" and succinctly condensed their dilemma in the oft-cited question "Why should thousands of top-notch programmers contribute freely to the provision of a public good?" (Lerner & Tirole, 2000, p. 2). Similar questions have been asked about other forms of peer production besides free and open source software creation, such as Wikipedia contributors (Schroer & Hertel, 2009) as well as participants in electronic networks of practice in the legal professions (who are not famous for their altruism) (Wasko & Faraj, 2005).

In this chapter, we summarize current research on user and peer motivations in peer production systems, focusing on the following: (1) individual motivations to participate, (2) selection of tasks, and (3) participation in peer production as a social practice, which influences motivations and highlights the critical role of institutions in enabling peer production. Generally, we focus on motivation at the individual level (that is, not addressing the question of why organizations choose to partake in peer production).

The Handbook of Peer Production, First Edition. Edited by Mathieu O'Neil, Christian Pentzold, and Sophie Toupin.

2 Individual Motivation

Most early studies on user motivations in peer production settings have adopted a form of self-determination theory (Deci & Ryan, 1985), which rests on the assumption that individuals exhibit certain behaviors in order to satisfy three basic needs, namely competency, social relatedness, and autonomy (Deci & Ryan, 2000, pp. 233–235). The theory distinguishes between intrinsic and extrinsic motivation. While intrinsic motivation rests on the inherently interesting and enjoyable act of performing the task itself, extrinsic motivation requires an outcome that is distinct from the task itself (Deci & Ryan, 1985, 2000). Thus, extrinsically motivated individuals satisfy their basic needs indirectly, e.g., by being monetarily rewarded (Osterloh & Frey, 2000). Some studies have extended the basic concept and have added internalized extrinsic motivation (e.g., Chandler & Connell, 1987; Deci & Ryan, 2000). Internalization of extrinsic motivation refers to "an active, natural process in which individuals attempt to transform socially sanctioned mores or requests into personally endorsed values and self-regulations" (Deci & Ryan, 2000, pp. 235–236). Thus, internalized motivation is by definition extrinsic but may be internalized by the individual and, accordingly, be perceived as self-regulating behavior rather than externally imposed (Deci & Ryan, 2000; Roberts, Hann, & Slaughter, 2006).

Much of the early research was conducted in the context of open source software programming (von Krogh, Haefliger, Spaeth, & Wallin, 2012) and online community contributions (such as Wikipedia contributors). Generally, surveys have identified a diverse set of motivations for starting and continuing engagement in peer production (Benkler, 2017) spanning intrinsic, internalized extrinsic and extrinsic motivations. Table 10.1 provides an overview of identified individual-level motivations, based on von Krogh et al. (2012).

2.1 Intrinsic Motivation

2.1.1 Fun

One major intrinsic motivation discussed is enjoyment (Moilanen, 2012; Torvalds & Ghosh, 1998). It drives efforts due to the joy, creativity, and challenges associated with performing the task (Shah, 2006). Fun deriving from participating, creating objects, or "hacking" software has been consistently cited as important drivers of effort (Hausberg & Spaeth, 2020; Lakhani & Wolf, 2005; Moilanen, 2012). This holds true even for contracted and paid contributors (Hars & Ou, 2002).

While these studies focus on the quantitative impact of fun and enjoyment on contributions, Shah (2006) further argued that they play an important role for sustained long-term contribution.

2.1.2 Ideology

Besides enjoyment and fun, many report that ideology-based intrinsic aspects are important factors driving contributions (David & Shapiro, 2008; Ghosh, Glott, Krieger, & Robles, 2002; Lakhani & Wolf, 2005). Ideology is typically based on certain norms, values, and beliefs (Stewart, Ammeter, & Maruping, 2006). Especially requirements such as freedom of use and modification as well as publication of source code or the overall conviction that open source is the best way to develop software are typically strongly supported (David & Shapiro, 2008).

Empirical findings, however, suggest that the impact of ideology-driven motivation may differ between peer production systems. For instance, Hertel et al. (2003) find a positive and significant

Table 10.1 Overview of individual-level motivations in open source software production.

Intrinsic	*Internalized extrinsic*	*Extrinsic*
Enjoyment and fun ideology	Own-use value	Pay
Kinship amity	Learning	Career
Altruism	Reciprocity	
	Reputation	

Source: Based on von Krogh, G., Haefliger, S., Spaeth, S., & Wallin, M. W. (2012). Carrots and rainbows: Motivation and social practice in open source software development. MIS Quarterly, 36(2), 649–676.

relationship between social and political motives and code contributions within the Linux developer community. In contrast, Xu and Li (2015) do not find a significant relationship between ideology-based motivation and participation levels for Wikipedia. Thus, they argue that individuals may choose to participate in multiple encyclopedic communities, diminishing the effect of strong ideological beliefs on the overall contribution level.

2.1.3 Kinship

Similarly to ideological motives, kinship amity has been identified as an intrinsic inducement. Kinship amity is associated with the desire to belong to a certain group or community where members are treated as "kin" (Hars & Ou, 2002). Thus, they exhibit an altruistic behavior as they do not expect something in return for their contributions. However, it differs from altruism in that it is limited to a specific community (von Krogh et al., 2012). Some surveys document kinship motives as respondents report a desire to interact with like-minded people (David & Shapiro, 2008; Hertel et al., 2003; Moilanen, 2012) This driver can be quite strong. As such, self-identification as a Linux developer appears to be the strongest predictor for the average hours spent per week on Linux-related activities (Hertel et al., 2003).

2.1.4 Altruism

Some literature also discusses the role of altruism as an intrinsic motivation (e.g., Hars & Ou, 2002; Ke & Zhang, 2008). Altruism refers to the behavior where peers seek to increase the welfare of others at their own expense without necessarily expecting something in return (Hars & Ou, 2002). Thus, it is often considered in the context of open source projects since participants contribute to a public good while investing their own time or bearing opportunity costs (Osterloh & Rota, 2007; von Krogh & von Hippel, 2006). While there is some empirical support for altruism for a minority of survey participants, especially students and hobby programmers (Hars & Ou, 2002; Hemetsberger, 2004), it seems that altruism is too simple an explanation to account for the occurrence of peer production.

2.2 Internalized Extrinsic Motivation

2.2.1 Own-use value

Between intrinsic and extrinsic inducements, research also identified a set of internalized extrinsic motivations. For instance, Raymond suggested that individuals might be motivated to contribute to public goods in pursuit of "scratching a personal itch" (Raymond, 1999), that is, they have a personal need that needs solving. The sometimes termed "own-use value" has been widely documented in surveys (Hars & Ou, 2002; Hemetsberger, 2004; Lakhani & Wolf, 2005). While most studies find a strong positive relationship between personal need and successful participation (Hars & Ou, 2002; Hertel et al., 2003), a few studies identify a negative relationship between personal need and the amount of one's contribution (Hausberg & Spaeth, 2020; Roberts et al., 2006). These studies argue that a personal need may only serve as a temporary driver to contribute, with peers stopping their contributions as soon as their need is sufficiently addressed. Consequently, they might leave the community after a short period.

It is interesting to note, though, that motivations are dynamic. Shah (2006) reminds us that peers' motivations evolve over time, moving from utility driven to more intrinsic motives.

2.2.2 Learning

An often identified second internalized intrinsic motivation is learning (David & Shapiro, 2008; Shah, 2006; Ye & Kishida, 2003). It typically includes aspects such as the acquisition or improvement of new skills or knowledge and is associated with feedback on peers' work (Ye & Kishida, 2003).

Several studies find support for learning motivations. Some show a positive impact on individual contribution levels, such as hours per week spent on open source software or hardware contribution (Hars & Ou, 2002; Hausberg & Spaeth, 2020) or lines of code written (Roberts et al., 2006). Other studies show significant relations between learning motivations and a higher intention to participate in the future (Spaeth, Haeflinger, Von Krogh, & Renzl, 2008; Wu, Gerlach, & Young, 2007).

Ye & Kishida (2003) argue that learning is one of the major forces that motivate developers to contribute as it may create the intrinsic satisfaction and the opportunity to earn a higher rank within the community. Some scholars, moreover, link learning with future job opportunities as it may increase human capital (Hars & Ou, 2002; Xu, Jones, & Shao, 2009).

2.2.3 *Reciprocity*

Besides own use and learning, the literature frequently discussed reciprocity as internalized extrinsic motivation. Reciprocity is based on a "give and get mentality" of individuals (Alexy & Leitner, 2012), which is embedded in a gift culture (Bergquist & Ljungberg, 2001; Zeitlyn, 2003). As such, individuals are motivated to contribute as they have already received something and feel the obligation to give something back to the community or they expect to receive something in return for their contribution (Lakhani & von Hippel, 2003). Moderate support for reciprocity as a motive was reported in several studies covering hackerspaces (Moilanen, 2012), virtual consumer communities (Hemetsberger, 2004), and open source projects (Lakhani & Wolf, 2005). In a similar vein, some studies suggest that individuals who received help from other contributors are more inclined to return help once they have gained more experience and knowledge (Ghosh et al., 2002; Hertel et al., 2003).

Further, Shah (2006) finds that individual contributions associated with obligation-based activities were rather small and limited to a shorter term. Specifically, she finds that individuals driven by this motivation tended to contribute in pursuit of releasing themselves from the obligation, and then retreated. Thus, reciprocity – similar to need-based motivation – might not necessarily lead to a sustaining motivation to contribute.

2.2.4 *Reputation*

Further, some scholars investigated individual reputation as driver for peer production. In this regard, von Krogh et al. (2012) identify two types of reputation discussed in the literature, namely peer reputation and outside reputation. Peer reputation is directed toward peers within the community. It is associated with demonstrating skills and capabilities to other contributors by making contributions of high quality and earning a corresponding status (Raymond, 1999; Roberts et al., 2006; Stewart, 2005). It thus serves as an indicator of community membership in the absence of hard and formal membership criteria (von Krogh, Spaeth, & Lakhani, 2003) and may be linked to learning motivations (Ye & Kishida, 2003). Peer reputation is also what drives other famous peer production systems, such as the production of academic knowledge (von Krogh & Spaeth, 2007).

Outside reputation is concerned with the anticipated reactions of related people outside the community and the resulting prestige (Shah, 2006; von Krogh et al., 2012). Research on open source software production has frequently found a link between peers' motivation and outside reputation (Hars & Ou, 2002; Hemetsberger, 2004; Hertel et al., 2003; Roberts et al., 2006). Outside reputation also overlaps with the extrinsic motivation of potential career benefits as elaborated in the following.

2.3 Extrinsic Motivation

In addition to intrinsic motivations, individuals can be motivated by extrinsic motives such as direct compensations or expected career benefits (e.g., Hars & Ou, 2002; Lerner & Tirole, 2002; Wu et al., 2007).

2.3.1 *Pay*

Indeed, contributors in peer production systems such as open source communities do not necessarily work for free. Rather, for some, contributing is part of their job, or at least their supervisor knows about and tolerates such engagements during working hours (Hertel et al., 2003; Lakhani & Wolf, 2005). Regarding direct compensation, some scholars find that developers motivated by payment on average spent more time working on open source software than volunteer contributors (Hertel et al., 2003; Lakhani & Wolf, 2005; Roberts et al., 2006). However, Alexy & Leitner (2012) only found an overall positive effect of payment on total motivation in cases where contributors did not show a strong norm against payment.

2.3.2 Career

Apart from financial incentives, signaling theory (Spence, 1976) provides an additional framework for extrinsic motivation, which was taken up by Lerner & Tirole (2002). Their original argument that peers' behaviors may be driven by anticipated career benefits in an attempt to signal their skills and abilities to the labor market has been validated in numerous studies. As such, positive relations have been reported with respect to contribution intensity (Hars & Ou, 2002) and participation intention (Hertel et al., 2003).

Complementary findings show that signaling indeed may lead to a higher average wage (Hann, Roberts, & Slaughter, 2013). In their study, they report a positive and significant relation between the developers' rank within the Apache foundation and their average wage and argue that employers may use a developer's rank as a measure of productive capabilities.

However, one question remains open as to the causality of the observed links: Do peers attempt to become employed so they can continue to do what they love to do, or do they participate in peer production as part of a rational signal to the labor market?

To sum up the above discussion, one may note that the general picture shows a range of user motivations (Benkler, 2017; David & Shapiro, 2008), whereas individual motivation is not based on a single motive but rather on a dynamic mix of several interacting inducements. Accordingly, some scholars started to research the links and mutual influences among different motivations within what Roberts et al. (2006) call the "system of motivations."

2.4 Crowding Out

One popular effect often considered in peer motivation is "crowding out" (Frey, 1997), i.e., the undermining of intrinsic motivations through the introduction of extrinsic incentives. An early meta-analysis by Deci et al. (1999), considering 128 studies in the field of psychology, finds evidence for crowding out intrinsic motivation, especially in those cases where individuals affected by external interventions perceived them as controlling. In contrast, external factors perceived as supportive can lead to the opposite effect of crowding in (Frey & Jegen, 2001).

2.4.1 Crowding effects caused by extrinsic motives

For open source software, Osterloh & Rota (2007) had suggested that extrinsic rewards may impede voluntary sharing of software and knowledge. However, much of the empirical research finds no evidence of a crowding out Roberts et al. (2006). Studies rather identify a positive effect of pecuniary rewards on overall contribution levels in various open source communities (Alexy & Leitner, 2012; Krishnamurthy, Ou, & Tripathi, 2014; Lakhani & Wolf, 2005; Roberts et al., 2006). Thus, financial (extrinsic) rewards seem to crowd in intrinsic motivation if they are perceived as supportive rather than controlling (Frey & Jegen, 2001; Krishnamurthy et al., 2014). Crowding in denotes a positive effect of external interventions (such as pay) on intrinsic motivation (Frey & Jegen, 2001).

As these results appear counterintuitive, Alexy and Leitner (2012) examine the role of payment norms and find that crowding in is related to a community's and an individual's norms against payment and commercial involvement. In the face of strong anti-commercial and anti-payment norms, intrinsic motivation tended to be crowded out. Indeed, the results of a study by Krishnamurthy et al. (2014) examining the acceptance of monetary rewards in open source software development suggests that even intrinsically motivated participants might be willing to accept financial rewards without loss of intrinsic motivation. Individuals exhibiting a higher kinship-based motivation who are thus more exposed to social norms have a lower propensity to accept financial rewards. These findings reflect the contingency associated with crowding effects.

2.4.2 Crowding effects caused by internalized extrinsic motives

Further studies dealing with crowding effects consider the impact of internalized extrinsic motives on intrinsic motivation. A few studies find a crowding-in effect of reputation on intrinsic motivation (Gallus, 2016; Roberts et al., 2006). Gallus explains this effect as the result of an increase in the identification with the community, which can be caused by symbolic reward systems as, for example, in the case of Wikipedia (Gallus, 2016). Further studies find a crowding in effect of learning on enjoyment-based motivation (Hausberg & Spaeth, 2020).

The effects are not clear-cut though: some studies in this field also find negative effects arising from internalized extrinsic inducements. Hausberg and Spaeth (2020) find in their study on user motivations in the open source hardware context a crowding out of enjoyment-based intrinsic motivation when makers were highly motivated by reputational benefits. The study of Roberts et al. (2006) unveils a negative effect of own-use value on contribution levels in open source software development.

2.5 Summary of Self-Determination Theory

Overall, we can say that contributors to commons-based peer production systems are motivated not by a single motive, but by a whole range of intrinsic, internalized extrinsic and extrinsic motives. Further, motivation is not static: there is instead an evolution of motives over time (Shah, 2006).

Empirical support is provided for almost all the discussed motives. However, differences regarding their magnitude can be observed. While kinship amity is found to be a strong driver of intrinsic motivation, the effects of ideology vary significantly between different peer production systems. Furthermore, altruism appears a too simple explanation in the context of peer production.

Similarly, in the case of internalized extrinsic motives, own-use value and reciprocity are widely documented (Hars & Ou, 2002; Hemetsberger, 2004; Lakhani & Wolf, 2005), but do not necessarily lead to more contributions as some studies find a short-term pursuit of quickly solving personal needs (Hausberg & Spaeth, 2020; Roberts et al., 2006) or fulfilling obligations of giving something back to the community (Shah, 2006). Learning and reputation appear as sustainable drivers (Hars & Ou, 2002; Roberts et al., 2006; Shah, 2006; Ye & Kishida, 2003).

Regarding extrinsic motives, payment can lead to an increased overall motivation, especially when the individual and the community are not principally against payments (Alexy & Leitner, 2012). Moreover, signaling effects are discussed in the literature as a further motive, which is linked to anticipated career benefits (Hars & Ou, 2002; Lerner & Tirole, 2002).

As motivations may interact (Roberts et al., 2006), several scholars suggested there were crowding out effects of extrinsic factors in intrinsic motivation (Haruvy, Prasad, & Sethi, 2003; Osterloh & Rota, 2007). However, empirical investigations cannot confirm the assumption, but find instead the opposite effect of crowding in (Alexy & Leitner, 2012; Krishnamurthy et al., 2014; Lakhani & Wolf, 2005; Roberts et al., 2006). For internalized extrinsic motives, the effects vary significantly.

3 Choosing a Task to Work On

Peer production systems, such as open source software or hardware communities, typically cover plenty of different tasks (Hausberg & Spaeth, 2020; Moilanen, 2012), which can coarsely be categorized into creational tasks and community-related tasks (Xu & Li, 2015). While creational tasks include direct contributions to the public good, such as writing source code of software or editing content (e.g., Wikipedia articles), community-related tasks cover duties such as administration issues or technical user support (Lakhani & von Hippel, 2003; Xu & Li, 2015).

As the individual's motivation depends on the characteristics inherent in the task (e.g., creative or challenging), variations might also be reflected by differences in the tasks' attractiveness and selection (Lakhani & von Hippel, 2003). Benkler (2002) postulates that task self-allocation in peer production systems is based on more or less well-informed assessments of the individual's own fit to the respective task or job.

3.1 "Sexy" Tasks

The reasons for individuals to take on attractive – "sexy" – creational tasks are obvious. Individuals performing such tasks may receive direct or indirect rewards such as fun and enjoyment (Hausberg & Spaeth, 2020), learning benefits (Roberts et al., 2006), the enhancement of reputation (Lakhani & von

Hippel, 2003), or the satisfaction of a personal need (Shah, 2006) as elaborated above. In addition, this self-selection might lead to neglecting less attractive but vital tasks, because peer production systems rely heavily on the self-selection of tasks by volunteers, potentially jeopardizing the survival of individual peer production systems (Benkler, 2002; Lakhani & von Hippel, 2003; von Krogh et al., 2012).

While specifically creational tasks exhibit high popularity among peers (Moilanen, 2012), covered by a variety of motivations (Lakhani & von Hippel, 2003; Roberts et al., 2006; Shah, 2006), some scholars are especially concerned with community-related tasks as they may be mundane or tedious, without providing obvious direct or indirect benefits.

3.2 Mundane Tasks

Some articles suggest that long-term contributors are more inclined to take on community-related tasks. For instance, Shah (2006) finds that they take on mundane and unattractive tasks such as rewriting source code in order to keep code simple and extensible, while newcomers tend to work on attractive and creative tasks. Hence, these groups appear to form a symbiotic relationship: new need-driven participants tend to provide new directions and challenges, while an old core provides support, creates the requested features, and integrates them into the source code, while also taking on further maintenance duties. Similarly, some studies suggest that participants may be inclined to take on more challenging tasks the more experienced they are within a specific project and the more knowledge they have gained over time (Hann et al., 2013; Hertel et al., 2003). Certain project governance structures can help to make community-related and mundane work more manageable: O'Mahony & Ferraro (2007) argue that peer production systems structure themselves in ways that are beneficial for community-related tasks such as coordination of efforts and integration.

The necessity of performing mundane community-related tasks leaves space for the involvement of commercial actors providing extrinsic incentives, such as payment, or providing internalized extrinsic incentives. For instance, Alexy & Leitner (2012) conclude that using financial rewards could compensate for the absence of intrinsic motivation. *Bounty programs* (Krishnamurthy & Tripathi, 2006) where users pledge financial support for someone else performing certain tasks are another way of providing such extrinsic motivation. In some open source communities with defined roles, individuals accept certain mundane tasks by occupying respective (more or less formal) roles (Hann et al., 2013).

Some studies find internalized extrinsic motives to be drivers to perform mundane or tedious tasks. For instance, some scholars consider peer reputation and kinship as major drivers to motivate "Wikipedians" to perform community-related tasks (Gallus, 2016; Kittur, Pendleton, & Kraut, 2009; Xu & Li, 2015). Another study finds learning to be a major driver (Lakhani & von Hippel, 2003). Specifically, they examine the reasons why contributors in the Apache community are providing technical support to others. They find that answering questions on Usenet help forums was mainly driven by anticipated learning benefits. Accordingly, community members spent considerable time scanning and reading questions in forums that may also have been relevant for their own projects.

3.3 Summary of Findings

Altogether, we can see that peers' motivation partly determines the type of task they will self-allocate. In order to perform all required tasks (both creative and community-related) a mix of various incentives and motivations is required in a healthy commons-based peer production system, as it is a mix of a diverse set of actors. Specifically, extrinsic and internalized extrinsic motives seem to play a crucial role to impel individuals to perform mundane tasks in peer production systems.

4 Peer Production as a Social Practice

While self-determination theory provides some insights into the psychological processes that may adopt impulses external to the individual as inputs, leading to immediate observable behaviors, it neglects the wider context that may explain, for example, why individuals maintain their participation and contributions over time (von Krogh et al., 2012). Peer motivations are not detached from influences springing from an individual's environment, as shown by the dynamically evolving motivations of peers embedded in the social structure of a peer production system (Shah, 2006). Rather than identifying individual-level motivations in isolation, an alternative promising approach is to view peer production systems from a practice perspective (Feldman & Orlikowski, 2011).

Some theoretical approaches started to view peer production systems as social movements (McCarthy & Zald, 1977) or linked them to the collective action literature (Ostrom, 1990, 1999) as they share many traits. By becoming part and member of a social practice, contributions stop being seen as expensive investments and collaboration instead become part of the intrinsic incentive itself. In this regard, Jon Elster remarks:

> Cooperation reflects a transformation of individual psychology so as to include the feeling of solidarity, altruism, fairness, and the like. Collective action ceases to become a prisoner's dilemma because members cease to regard participation as costly: It becomes a benefit in itself, over and above the public good it is intended to produce. (1986, p. 132)

One comprehensive approach considers peer production as social practice, which explicitly accounts for the interrelation of collective activities with institutions (Ostrom, 1990; von Krogh et al., 2012). The approach builds upon MacIntyre's seminal work, which describes a practice as

> any coherent and complex form of socially established cooperative human activity through which goods internal to that form of activity are realized in the course of trying to achieve those standards of excellence which are appropriate to, and partly definitive of, that form of activity, with the result that human powers to achieve excellence, and human conceptions of the ends and goods involved, are systematically extended. (MacIntyre, 1981, p. 187)

Following this view, peer production involves the creation of internal goods with public goods characteristics, such as source code or encyclopedia articles, which are produced by members of the practice. The collective of contributors follows certain general principles (so-called *standards of excellence*) that are determined and shared by the whole collective (von Krogh et al., 2012). As individuals decide to contribute, they gradually adopt the general norms and principles of the social practice (Rullani & Haefliger, 2013) and over time collectively adapt them. Internal goods are defined by the social practice and do not only benefit individual contributors but also other members in the social practice as well as the wider community (von Krogh et al., 2012).

Institutions house these practices and provide external goods, such as status or capital that enable and extrinsically motivate contributors (MacIntyre, 1981; von Krogh et al., 2012). Institutions can be seen as "sustainable forms of human cooperation" (von Krogh et al., 2012, p. 660), including companies and foundations, that are governed by certain organizations (e.g., the community), rules (e.g., coordination), and routines (von Krogh et al., 2012).

By drawing attention to social practices, the focus shifts from short- and mid-term motivation – going back to direct rewards – towards the long-term motivation of participants, as the social practice becomes intertwined with their lives, creating the perception of a moral obligation associated with the pursuit of the *unity of life* (von Krogh et al., 2012). In particular, the theory postulates that individuals attempt to reach and maintain the uniformity of their actions, forming a consistent journey that emphasizes values such as personal development (cf. von Krogh et al., 2012). In this way, peer production as a social practice explains interview statements, such as "this is just how open source programmers are supposed to act" or "it is kind of a moral obligation to contribute." Social practices frame peer production as a school of virtue in which norms, attitudes, and standards are concurrently being created with the internal goods themselves. Peer production becomes a lifestyle.

4.1 Social Exposure

Research suggests that social exposure in a peer production community is crucial. Community participation and kinship are powerful drivers aligning individuals' activities with overall goals and social norms inherent in the community. Thus, the exposure to the community may positively affect the overall contribution level of participants (Rullani, 2007).

As such, literature considers the architecture of social practices as comprising a core and a periphery, where individuals start as passive lurkers and observers situated at the periphery, eventually starting to make small contributions while gradually tending to be dragged towards the core (Rullani, 2007; Rullani & Haefliger, 2013; Ye & Kishida, 2003). Reaching the core, they tend to focus more on maintaining the social practice by taking supportive tasks or helping to educate new members (Shah, 2006).

In the process of joining, an individual socializes with the community (Rullani & Haefliger, 2013) and acquires an identity related to the social practice that intertwines individual motivation with experiences related to the membership (von Krogh et al., 2012). While membership is typically granted by a consensus vote by the core group, studies found a tendency of lurkers to attempt to prove their abilities to that core (Midha & Bhattacherjee, 2012). Specifically, Midha & Bhattacherjee (2012) find that lurkers tried to complete assigned mundane maintenance tasks as quickly as possible to get attention and acceptance.

Further, on the one hand, the community core may exert social pressure by using collective sanctions to enforce certain social norms and standards of excellence (Sagers, 2004). As outlined before, norms within the group such as the refusal of payments can have a significant effect on certain motives (Alexy & Leitner, 2012). On the other hand, communities may reward members for their past performances, i.e., meritocratic self-organization. In this regard, Roberts et al. (2006) find that promotions within the Apache meritocracy led to higher intrinsic and extrinsic motivation and to higher participation levels as status increases.

Apart from that, some scholars illuminated the effects of two socialization outcomes, namely social identification and social integration, on motivation. Social identification refers to the extent to which an individual identifies with a certain community, whereas social integration denotes the perception of being accepted and trusted by the community. Several studies demonstrate that they can positively influence contributors' behaviors within the community towards kinship amity and lead to an increase in task performance (Carillo, Huff, & Chawner, 2017; Gallus, 2016; Spaeth, von Krogh, & He, 2015).

4.2 Institutional Frameworks

In a social practice, institutions and institutional frameworks impact on the peers' behavior, with impacts ranging from the enabling (by providing infrastructure and support or by remunerating them) to the corrupting (by introducing conflicts of interests or formal restrictions).

4.2.1 Governance structures

One such institutional framework is the governance structure of a peer production system. It plays a crucial role in influencing individuals' motivations. Governance can be defined as "the means of achieving the direction, control, and coordination of entirely or partially autonomous individuals and organizations on behalf of a [peer production] project to which they jointly contribute" (Markus, 2007, p. 152). Markus (2007) highlights the importance of governance as it may solve collective action problems as well as coordination problems and determines the climate for contributors. The relevant literature identified different governance structures and examined their relation with individual motivation (Di Tullio & Staples, 2013; Klapper & Reitzig, 2018; Shah, 2006).

In this regard, studies show that the formalization of conflict management and development processes can lead to a higher task performance and a better community climate, if overall goal-definition remains within the decision-making scope of the project members (Di Tullio & Staples, 2013; Klapper & Reitzig, 2018). A complementary study by Ho & Rai (2017) shows a further positive effect of input and output control on the continued participation intention of contributors. Specifically, they argue that a formal accreditation process of members and leaders based on their skills (input control) as well as a careful code review and acceptance (output control) can effectively signal high standards of excellence, attracting further volunteers to the social practice.

4.2.2 *Sponsorship*

Moreover, the organizational integration of the peer production system into a private firm appears crucial. In that regard, research distinguished two governance structures, namely "open" and "gated" communities. While open communities produce true public goods, gated communities constitute hybrid forms of collective development and private ownership and are controlled by a focal company (cf. Shah, 2006; Spaeth et al., 2015).

A study by Shah (2006) found that particularly in the long run, developers who were primarily motivated by use-value were more inclined to contribute to gated communities, whereas developers motivated by enjoyment (i.e., hobbyists) tended to contribute to open communities. Furthermore, the sponsor's specific characteristics as well as their perception by the contributors can play a crucial role in influencing their motivation by affecting social identification (Spaeth et al., 2015; Stewart et al., 2006). Contributors may assess institutions such as sponsoring companies or foundations and consider them as either supportive or restrictive depending on their adherence to the standards of excellence (von Krogh et al., 2012). For instance, attributes such as expertise and trustworthiness as well as an attitude towards mutual knowledge exchange were found to have a positive impact (Spaeth et al., 2015). Overall, scholars suggested that firm sponsors should avoid extracting too much benefit from the peer production community as this may depress community morale and work (Haruvy et al., 2003; von Krogh et al., 2012).

4.2.3 *Licensing*

Another institutional framework to consider are the licenses of the goods produced. Generally, research focuses on the restrictions embedded in the applied license. The famous GNU General Public License (GPL) imposes two restrictions: (1) modified versions of the source code need to be open as well (*copyleft*), and (2) the code can only be combined with programs distributed under the same license (*viral* provision). The intuition that more intrinsically motivated individuals tend to prefer more restrictive licenses, while less restrictive licenses attract more extrinsically motivated ones is reflected by empirical findings (Allyn, Misra, & Kozyreva, 2008).

Furthermore, contributions may differ across both types of licenses. For instance, Fershtman & Gandal (2007) find that output per contributor in open source projects is much higher when licenses are less restrictive and more commercially oriented, such as the Berkeley Software Distribution (BSD) type license, while the number of contributions is higher when licenses are more restrictive such as the GPL. It highlights the differences among individual motivations and the role of appropriate protection mechanisms for intellectual property in influencing contributors' behaviors.

It is noteworthy, however, that the effect of licenses may depend on the type of sponsor. In particular, Stewart et al. (2006) find that the presence of non-market sponsors such as non-profit foundations may reduce concerns about the project's future. Thus, development activity and user interest were the highest when the application was distributed under a non-restrictive license and the sponsor was a non-market organization.

4.3 Summary of Social Practice View

Altogether, we can say that a social practice perspective complements an isolated analysis of peers' individual motivations as it allows investigating the social practice and the influence it has on individuals. Thus, we showed that social pressure exerted by the community may either sanction or promote certain behaviors, influencing individual motives. Furthermore, social integration and identification seem to be crucial factors influencing the individuals' behavior as well as performance. The social practice view further allows to identify institutions (such as financial sponsors) and institutional arrangements, such as the project governance structure and the produced good's license as well as their impact on peers' motivation.

As the social practice view stresses, the overall social context surrounding an individual, one should observe that participation in peer production systems as reflected by the reviewed studies is a privilege of Global North nations.

5 Conclusion

Peers are motivated by a number of individual motivations, ranging from the intrinsic (ideology, altruism, kinship amity, enjoyment and fun), to the internalized extrinsic (reputation, reciprocity, learning, own-use value), to the purely extrinsic (pay, career benefits) although boundaries between these categories can be somewhat blurry at times. Studies have generally found a mix of motivations in various empirical peer production systems. Diversity is high though, and motivations have been shown to change over time (e.g., from own-use problem solving to pure enjoyment and kinship amity), which can be explained by considering peer production systems as a social practice (the self-given term "Wikipedians" nicely illustrates the strong community identification in such systems). Commercial involvement in these systems is not necessarily bad *per se*, as we have shown, and empirical evidence of crowding out of intrinsic motivation is scarce. However, domination of a gated community by a commercial entity and controlling core aspects of processes and goods produced can indeed prevent voluntary contributions in the first place.

Most individuals will be consumers and free-riders of commons-based peer production systems. Still, in many cases and under the right conditions enough peers can be motivated to achieve impressive output. Success seems to depend on the right mix of motivations and types of peers in peer production settings. Fortunately, as Elinor Ostrom (2000, p. 138) succinctly put it, "the world contains multiple types of individuals, some more willing than others to initiate reciprocity to achieve the benefits of collective action."

References

Alexy, O., & Leitner, M. (2012). A fistful of dollars: Are financial rewards a suitable management practice for distributed models of innovation? *European Management Review*, *8*(3), 165–185. https://doi.org/10.1111/j.1740-4762.2011.01017.x.

Allyn, M., Misra, R., & Kozyreva, A. (2008). The effect of license type and status hierarchy on developer motivation in open source communities. *Proceedings of the Northeast Business & Economics Association*, 76–81.

Benkler, Y. (2002). Coase's Penguin, or Linux and the nature of the firm. *Yale Law Journal*, *112*(3), 369–447.

Benkler, Y. (2017). Peer production, the commons, and the future of the firm. *Strategic Organization*, *15*(2), 264–274.

Bergquist, M., & Ljungberg, J. (2001). The power of gifts: Organising social relationships in open source communities. *Information Systems Journal*, *11*(4), 305–320. https://doi.org/10.1046/j.1365-2575.2001.00111.x.

Carillo, K., Huff, S., & Chawner, B. (2017). What makes a good contributor? Understanding contributor behavior within large free/open source software projects – a socialization perspective. *The Journal of Strategic Information Systems*, *26*(4), 322–359. https://doi.org/10.1016/j.jsis.2017.03.001.

Chandler, C. L., & Connell, J. P. (1987). Children's intrinsic, extrinsic and internalized motivation: A developmental study of children's reasons for liked and disliked behaviours. *British Journal of Developmental Psychology*, *5*(4), 357–365. https://doi.org/10.1111/j.2044-835X.1987.tb01072.x.

David, P. A., & Shapiro, J. S. (2008). Community-based production of open-source software: What do we know about the developers who participate? *Information Economics and Policy*, *20*(4), 364–398. http://dx.doi.org/10.1016/j.infoecopol.2008.10.001.

Deci, E. L., Koestner, R., & Ryan, R. M. (1999). A meta-analytic review of experiments examining the effects of extrinsic rewards on intrinsic motivation. *Psychological Bulletin*, *125*(6), 627–668.

Deci, E. L., & Ryan, R. M. (1985). *Intrinsic motivation and self-determination in human behavior*. New York, NY: Springer Science+Business Media.

Deci, E. L., & Ryan, R. M. (2000). The "what" and "why" of goal pursuits: Human needs and the self-determination of behavior. *Psychological Inquiry*, *11*(4), 227–268. https://doi.org/10.1207/S15327965PLI1104_01.

Di Tullio, D., & Staples, D. S. (2013). The governance and control of open source software projects. *Journal of Management Information Systems*, *30*(3), 49–80. https://doi.org/10.2753/MIS0742-1222300303.

Elster, J. (1986). *An introduction to Karl Marx*. Cambridge: Cambridge University Press.

Feldman, M. S., & Orlikowski, W. J. (2011). Theorizing practice and practicing theory. *Organization Science*, *22*(5), 1240–1253. https://doi.org/10.1287/orsc.1100.0612

Fershtman, C., & Gandal, N. (2007). Open source software: Motivation and restrictive licensing. *International Economics and Economic Policy*, *4*(2), 209–225. http://dx.doi.org/10.1007/s10368-007-0086-4.

Frey, B. S. (1997). *Not just for the money: An economic theory of personal motivation*. Cheltenham: Edward Elgar Publishing.

Frey, B. S., & Jegen, R. (2001). Motivation crowding theory. *Journal of Economic Surveys, 15*(5), 589–611. https://doi.org/10.1111/1467-6419.00150.

Gallus, J. (2016). Fostering public good contributions with symbolic awards: A large-scale natural field experiment at Wikipedia. *Management Science, 63*(12), 3999–4015. https://doi.org/10.1287/mnsc.2016.2540.

Ghosh, R. A., Glott, R., Krieger, B., & Robles, G. (2002). *Free/libre and open source software: Survey and study*. Retrieved from https://www.math.unipd.it/~bellio/FLOSS%20Final%20Report%20-%20Part%204%20-%20Survey%20of%20Developers.pdf.

Hann, I.-H., Roberts, J. A., & Slaughter, S. A. (2013). All are not equal: An examination of the economic returns to different forms of participation in open source software communities. *Information Systems Research, 24*(3), 520–538.

Hardin, G. (1968). The tragedy of the commons. *Science, 162*(3859), 1243–1248. https://doi.org/10.1126/science.162.3859.1243.

Hars, A., & Ou, S. (2002). Working for free? Motivations for participating in open-source projects. *International Journal of Electronic Commerce, 6*(3), 25–39.

Haruvy, E., Prasad, A., & Sethi, S. P. (2003). Harvesting altruism in open-source software development. *Journal of Optimization Theory and Applications, 118*(2), 381–416. http://dx.doi.org/10.1023/A:1025455523489.

Hausberg, J. P., & Spaeth, S. (2020). Why makers make what they make: Motivations to contribute to open source hardware development. *R&D Management, 50*(1), 75–95. https://doi.org/10.1111/radm.12348.

Hemetsberger, A. (2004). *When consumers produce on the Internet: The relationship between cognitive-affective, socially-based, and behavioral involvement of prosumers*. Retrieved from http://opensource.mit.edu/papers/hemetsberger1.pdf.

Hertel, G., Niedner, S., & Herrmann, S. (2003). Motivation of software developers in open source projects: An internet-based survey of contributors to the Linux kernel. *Research Policy, 32*(7), 1159–1177.

Ho, S. Y., & Rai, A. (2017). Continued voluntary participation intention in firm-participating open source software projects. *Information Systems Research, 28*(3), 603–625. https://doi.org/10.1287/isre.2016.0687.

Ke, W., & Zhang, P. (2008). Motivations for participating in open source software communities: Roles of psychological needs and altruism. *12th Pacific Asia Conference on Information Systems (PACIS 2008)* (pp. 136–148).

Kittur, A., Pendleton, B., & Kraut, R. E. (2009). Herding the cats: The influence of groups in coordinating peer production. *Proceedings of the 5th International Symposium on Wikis and Open Collaboration, 7*, 1–79. https://doi.org/10.1145/1641309.1641321

Klapper, H., & Reitzig, M. (2018). On the effects of authority on peer motivation: Learning from Wikipedia. *Strategic Management Journal, 39*(8), 2178–2203. https://doi.org/10.1002/smj.2909.

Krishnamurthy, S., Ou, S., & Tripathi, A. K. (2014). Acceptance of monetary rewards in open source software development. *Research Policy*. https://doi.org/10.1016/j.respol.2013.10.007.

Krishnamurthy, S., & Tripathi, A. K. (2006). Bounty programs in free/libre/open source software. *The economics of open source software development*. Amsterdam: Elsevier.

Lakhani, K. R., & von Hippel, E. (2003). How open source software works: "Free" user-to-user assistance. *Research Policy, 32*(6), 923–943.

Lakhani, K. R., & Wolf, R. G. (2005). Why hackers do what they do: Understanding motivation and effort in free/open source software projects. In J. Feller, B. Fitzgerald, S. Hissam, & K. R. Lakhani (Eds.), *Perspectives on free and open source software* (pp. 3–22). Cambridge, MA: MIT Press.

Lerner, J., & Tirole, J. (2000). The simple economics of open source. *NBER Working Paper No. 7600*. https://doi.org/10.3386/w7600.

Lerner, J., & Tirole, J. (2002). Some simple economics of open source. *Journal of Industrial Economics, 50*(2), 197–234.

MacIntyre, A. (1981). *After virtue – a study in moral theory* (1st ed.). Notre Dame, IN: University of Notre Dame Press.

Markus, L. M. (2007). The governance of free/open source software projects: monolithic, multidimensional, or configurational? *Journal of Management and Governance, 11*(2), 151–163. http://dx.doi.org/10.1007/s10997-007-9021-x.

McCarthy, J. D., & Zald, M. N. (1977). Resource mobilization and social movements: A partial theory. *American Journal of Sociology, 82*(6), 1212–1241. https://doi.org/10.1086/226464.

Midha, V., & Bhattacherjee, A. (2012). Governance practices and software maintenance: A study of open source projects. *Decision Support Systems, 54*(1), 23–32. https://doi.org/10.1016/j.dss.2012.03.002.

Moilanen, J. (2012). Emerging hackerspaces – peer-production generation. *Open source systems: Long-term sustainability* (pp. 94–11)1. https://doi.org/10.1007/978-3-642-33442-9_7.

Olson, M. (1965). *The logic of collective action: Public goods and the theory of groups.* Harvard University Press.

O'Mahony, S., & Ferraro, F. (2007). The emergence of governance in an open source community. *Academy of Management Journal, 50*(5), 1079–1106. https://doi.org/10.5465/amj.2007.27169153.

Osterloh, M., & Frey, B. S. (2000). Motivation, knowledge transfer, and organizational forms. *Organization Science, 11*(5), 538–550.

Osterloh, M., & Rota, S. G. (2007). Open source software development – just another case of collective invention? *Research Policy, 36*(2), 157–171.

Ostrom, E. (1990). *Governing the commons: The evolution of institutions for collective action.* Cambridge: Cambridge University Press.

Ostrom, E. (1999). Coping with tragedies of the commons. *Annual Review of Political Science, 2*(1), 493–535. https://doi.org/10.1146/annurev.polisci.2.1.493.

Ostrom, E. (2000). Collective action and the evolution of social norms. *Journal of Economic Perspectives, 14*(3), 137–158.

Raymond, E. S. (1999). The magic cauldron. *The cathedral and the bazaar.* Retrieved from http://www.catb.org/~esr/writings/magic-cauldron/.

Roberts, J. A., Hann, I.-H., & Slaughter, S. A. (2006). Understanding the motivations, participation, and performance of open source software developers: A longitudinal study of the Apache projects. *Management Science, 52*(7), 984–999.

Rullani, F. (2007). Dragging developers towards the core. How the free/libre/open source software community enhances developers' contribution. *LEM Working Paper Series 2006/22.* Retrieved from http://opensource.mit.edu/papers/rullani_socenv.pdf.

Rullani, F., & Haefliger, S. (2013). The periphery on stage: The intra-organizational dynamics in online communities of creation. *Research Policy, 42*(4), 941–953. https://doi.org/10.1016/j.respol.2012.10.008.

Sagers, G. (2004). The influence of network governance factors on success in open source software development projects. *ICIS 2004 Proceedings, 34,* 427–438.

Schroer, J., & Hertel, G. (2009). Voluntary engagement in an open web-based encyclopedia: Wikipedians and why they do it. *Media Psychology, 12*(1), 96–120. https://doi.org/10.1080/15213260802669466.

Shah, S. K. (2006). Motivation, governance, and the viability of hybrid forms in open source software development. *Management Science, 52*(7), 1000–1014.

Spaeth, S., Haeflinger, S., Von Krogh, G., & Renzl, B. (2008). *Communal resources in open source software development.* Retrieved from http://InformationR.net/ir/13-1/paper332.html.

Spaeth, S., von Krogh, G., & He, F. (2015). Perceived firm attributes and intrinsic motivation in sponsored open source software projects. *Information Systems Research, 26*(1), 224–237. https://doi.org/10.1287/isre.2014.0539.

Spence, M. (1976). Competition in salaries, credentials, and signaling prerequisites for jobs. *The Quarterly Journal of Economics, 90*(1), 51–74. https://doi.org/10.2307/1886086.

Stewart, D. (2005). Social status in an open-source community. *American Sociological Review, 70*(5), 823–842.

Stewart, K. J., Ammeter, A. P., & Maruping, L. M. (2006). Impacts of license choice and organizational sponsorship on user interest and development activity in open source software projects. *Information Systems Research, 17*(2), 126–144. https://doi.org/10.1287/isre.1060.0082.

Torvalds, L., & Ghosh, R. A. (1998). What motivates free software developers? (Interview). *First Monday,* 3(*3*).

von Krogh, G., Haefliger, S., Spaeth, S., & Wallin, M. W. (2012). Carrots and rainbows: Motivation and social practice in open source software development. *MIS Quarterly, 36*(2), 649–676.

von Krogh, G., & Spaeth, S. (2007). The open source software phenomenon: Characteristics that promote research. *Journal of Strategic Information Systems, 16*(3), 236–253. https://doi.org/10.1016/j.jsis.2007.06.001.

von Krogh, G., Spaeth, S., & Lakhani, K. R. (2003). Community, joining, and specialization in open source software innovation: a case study. *Open Source Software Development, 32*(7), 1217–1241. https://doi.org/10.1016/S0048-7333(03)00050-7.

von Krogh, G., & Von Hippel, E. (2006). The promise of research on open source software. *Management Science, 52*(7), 975–983. https://doi.org/10.1287/mnsc.1060.0560

Wasko, M. M., & Faraj, S. (2005). Why should I share? Examining social capital and knowledge contribution in electronic networks of practice. *MIS Quarterly, 29*(1), 35–37.

Wu, C.-G., Gerlach, J. H., & Young, C. E. (2007). An empirical analysis of open source software developers' motivations and continuance intentions. *Information & Management, 44*(3), 253–262. https://doi.org/10.1016/j.im.2006.12.006

Xu, B., Jones, D. R., & Shao, B. (2009). Volunteers' involvement in online community based software development. *Information & Management*, *46*(3), 151–158. http://dx.doi.org/10.1016/j.im.2008.12.005.
Xu, B., & Li, D. (2015). An empirical study of the motivations for content contribution and community participation in Wikipedia. *Information & Management*, *52*(3), 275–286. https://doi.org/10.1016/j.im.2014.12.003.
Ye, Y., & Kishida, K. (2003). Toward an understanding of the motivation of open source software developers. *Proceedings of 2003 International Conference on Software Engineering (ICSE2003)* (pp. 419–429).
Zeitlyn, D. (2003). Gift economies in the development of open source software: Anthropological reflections. *Research Policy*, *32*(7), 1287–1291. https://doi.org/10.1016/S0048-7333(03)00053-2.

11

Governing for Growth in Scope

Cultivating a Comparative Understanding of How Peer Production Collectives Evolve

Rebecca Karp, Amisha Miller, and Siobhán O'Mahony

1 Introduction

One of the early challenges for any peer production collective is how to govern the growth of new members or contributors. Since the goal of peer production communities is to foster the collaborative generation of knowledge, and peer production communities are largely open to self-directed volunteers, most are founded with few barriers to participation. When peer production collectives convene through a common platform (e.g., Parker & Van Alstyne, 2017), some form of founding governance mechanisms establish the rules that initially guide open and collective participation (e.g., Gawer, 2014; West & O'Mahony, 2008) as well as the scope of the platform's purpose (Gawer & Henderson, 2007). These early mechanisms help define governance as "the rights available to members of an organization, their modes of representation, and the structures and processes that conduct power and resources" (O'Mahony & Bechky, 2008, p. 16).

When peer collectives grow in size, new governance mechanisms tend to emerge over time, distributing rights that enable contributing participants to either vote or give voice to increasing administration or coordination responsibilities (e.g., Dahlander & O'Mahony, 2011; O'Mahony & Ferraro, 2007; Shaw & Hill, 2014). However, an influx of new members can strain both the infrastructure and the norms upon which any peer production collective is founded. For example, Ferraro and O'Mahony (2012) showed how open boundaries to the Debian free and open source operating system project were flooded to the point that the community stopped accepting new members and redesigned how members could participate. Thus, mechanisms to address the unregulated growth of participants and their contributions often emerge (Aaltonen & Lanzara, 2015) in ways that distinguish core vs. periphery participation (e.g., Brown & Duguid, 2001; Lakhani & Von Hippel, 2003; Lave & Wenger, 1991) and segment participant rights based on these differences (Dahlander & O'Mahony, 2011). Often governance roles are allocated to select participants and there is some evidence that as participants take on new coordinating or administrating roles, they devote more time to this type of work (Dahlander & O'Mahony, 2011; Shaw & Hill, 2014). While the relationship between governance mechanisms and the growth of new contributors has been examined (West, 2003), little research explores how peer production collectives govern their growth in scope.

Scholars define an organization's scope as the range of different activities (markets, services, product offerings) an organization is actively engaged in (Jones & Hill, 1988; Peng, Lee, & Wang, 2005). Building on Jones and Hill (1988), scope can be adapted in one of two ways: vertically or horizontally. When a collective grows its vertical scope, it takes on new activities in the innovation lifecycle or the range of activities supporting the emergence, development, or implementation of new innovations or knowledge – from generating concepts to implementing them (Garud, Tuertscher, & Van de Ven, 2013). Alternatively, collectives can also expand their horizontal scope when they take on new domains or address new types of audiences.

The Handbook of Peer Production, First Edition. Edited by Mathieu O'Neil, Christian Pentzold, and Sophie Toupin.

Scope growth was not a topic of concern when scholars were focused on understanding the emergence of peer production collectives as a phenomenon and on identifying the conditions enabling their growing prevalence. But, with the emergence of peer production communities addressing domains as diverse as problem solving at NASA (Lakhani, Lifshitz-Assaf, & Tushman, 2013; Lifshitz-Assaf, 2018); to engaging in other innovative and technical endeavors (Boudreau, Lacetera, & Lakhani, 2011; Mollick, 2014), the time has come to treat the topic of scope growth with the import it deserves. Scope growth is important to understanding the current and future challenges that maturing peer production collectives face for several reasons. First, as Aaltonen and Lanzara argued, a peer production collective "must learn to execute new and progressively complex tasks and steer production toward shifting objectives" (2015, p. 1653) in order to sustain its growth. Second, the range or scope of activities a collective can support influences the participants' motivations to contribute – further fueling or restricting collective growth.

While scope growth in peer collectives has not been explicitly studied, the triggers of scope growth have been examined by scholars interested in the expansion of firm boundaries (Helfat & Eisenhardt, 2004; Santos & Eisenhardt, 2009; Sorenson, 2000; Williamson, 1991). Unlike firms, in collectives, decision-making is less hierarchical (Adler, 2001) and most likely the product of a growing body of participants and their evolving interactions (e.g., Aaltonen & Lanzara, 2015). With new contributors, come new ideas that can expand or shift a community's founding scope in ways that are difficult to anticipate ex ante (e.g., Benkler, 2016). As peer production collectives often have open or permeable boundaries (West, 2003), the growth of new members can expand production scope in ways that stretch a collective's founding purpose (e.g., Grodal & O'Mahony, 2017).

If peer production collectives are to be taken seriously as either a rival or complement to firm based modes of production (Baldwin & von Hippel, 2011; Benkler, 2017; O'Mahony & Lakhani, 2011; Seidel, Langner, & Sims, 2016; von Hippel, 2005), then scholars need to examine what leads collectives to change their scope and how governance mechanisms evolve to support scope changes. In doing so, we can inform the next frontier in the maturity of peer production. We argue that by examining the shifting nature of innovation tasks performed (e.g., Barley & Kunda, 2001) in peer production collectives, we can learn how peers adapt to take on "increasingly complex activities" (Aaltonen & Lanzara, 2015). With an activity-based perspective, we identify five different roles that peer collectives fulfill in the innovation lifecycle, from idea generation to post-production review and testing. Building on Garud, Tuertscher, and Van de Ven (2013), we define the innovation lifecycle as activities that foster the emergence, development, and implementation of novel ideas. With published sources, historical and publicly available data, we conducted a multi-case comparative study of 12 mature peer collectives, in existence for at least a decade, to understand how their scope of innovative activities evolved over time.

With this cursory review, we found that some peer collectives expanded their scope vertically to take on new activities in the innovation lifecycle, while others expanded their scope horizontally to address new domains. Some experienced both types of scope growth while others maintained their existing scope. However, none narrowed their scope – possibly because we selected mature, high growth collectives. We identified three different modes of peer production (*Collaborative, Co-creative, or Aggregative*) and traced the evolution of these modes at a high level. Drawing on this analysis, we articulate a research agenda that treats peer production as a mature rather than an emergent phenomenon and encourages a shift from single case studies to a comparative case approach (Bechky & O'Mahony, 2015) to foster an appreciation for the varied ways in which peer collectives evolve over time.

2 Case Analysis

This comparative research draws from published sources and historical data to explore how the activities of peer production collectives have evolved. Even experts disagree on what constitutes a peer production collective – while some forms of peer production compensate contributors when they compete for prizes (such as Kaggle) or complete tasks (as in MTurk); other forms are purely voluntary (such as Wikipedia) (Benkler, 2016). To advance the research frontier, our aim was to examine a variety of collectives at a mature stage (ten or more years) and assess how the tasks they engaged in across the innovation lifecycle changed over time. This approach recognizes the variety of innovative activities that collectives engage in: from a single activity such as idea generation, to engaging in all activities across the innovation lifecycle – through post production review and idea refinement. With this activity-based approach, we guard against assuming that peer collectives are all the same.

We identified five different types of roles peer production collectives engaged in throughout the innovation process. Collectives involved in ***Idea Generation*** generated a wide range of ideas to support innovation challenges. Collectives engaged in ***Innovation Search and Reference*** generated content to create a public or private informational resource, whether that be a curated collection of images or an online encyclopedia. Collectives engaged in ***Funding and Sourcing*** accumulated investments to fund or support new innovations. Collectives engaged in ***Post Production*** conducted peer reviews, user testing, and provided feedback which often informed the next iteration of innovations (Seidel, Hannigan, & Phillips, 2020). ***End-to-End*** or full life cycle collectives performed all activities from idea generation to post production. We selected 12 mature peer production collectives across these five roles to compare, two per phase, and over sampled on the most prominent activity – ***Innovation Search and Reference***. Where possible, we sought to include collectives owned by a firm that managed the rules of participation (e.g., Shah, 2006; West & O'Mahony, 2008) as well as independent collectives that featured participant representation in governance and decentralized decision-making (e.g., O'Mahony, 2007). We chose mature collectives that offered a rich set of publicly available data. Most of the collectives in our sample were associated with a commercial business model – this may have been a function of selecting mature collectives. The majority produced publicly available information goods. Table 11.1 provides descriptive data on each collective at their founding, including their role in the innovation lifecycle and what each collective produced.

Table 11.1 Peer production collectives founding characteristics.

Role in innovation lifecycle	*Definition*	*Collective*	*Year founded*	*Ownership*	*What was produced Public or privately available*
Idea generation	Generate a range of ideas for focused or unfocused innovation challenges	Threadless	2000	Sponsored	T-shirt designs – ***public***
		InnoCentive	2001	Sponsored	Scientific/technological solutions – ***private***
Innovation search and reference	Generate or curate content to create a public or private informational resource	Wikipedia	2001	Community managed	Encyclopedic content – ***public***
		YouTube	2005	Sponsored	User-generated videos – ***public***
		Reddit	2005	Sponsored	Advice and aggregated content – ***public***
		Pinterest	2009	Sponsored	Curated collections of images / videos – ***public***
Funding and sourcing	Fund innovations, creative works, or small enterprises	Kiva	2005	Sponsored	Micro loans – ***public***
		Kickstarter	2009	Sponsored	Creative works and products – ***public***
Post production	Provide peer reviews, testing, and feedback for products and services	TripAdvisor	2000	Sponsored	Agglomerated travel information – ***public***
		Yelp	2004	Sponsored	Reviews of local businesses – ***public***
End to end	Manage the full innovation lifecycle from emergence to implementation	Eclipse	1998	Sponsored	Software development – ***public***
		Topcoder	2001	Sponsored	Software development – ***private***

To inform a provisionary, comparative analysis, we drew from publicly available historical and published sources to unpack observable changes in scope and governance for each of the 12 cases. We did not conduct a comprehensive history of each collective at a granular level. Rather, our intent was to compare the activities peer production collectives engaged in and how these evolved at a high level. We triangulated between scholarly texts and historical sources to identify critical changes over time, comparing the validity of each text with other available data (Golder, 2000; Kahl & Grodal, 2016). Following case-based analysis techniques (Eisenhardt & Graebner, 2007; Yin, 1994), we traced how the scope of innovation activities and governance rights changed over time. We first developed each case discretely and then compared across cases.

When collectives took on an expanded set of activities in the innovation lifecycle, we coded this as expansion in vertical scope. For example, when Pinterest transitioned from enabling contributors to share pinned and collated images to encouraging contributors to review and recommend recent purchases, we coded Pinterest as expanding their vertical scope. Pinterest shifted from creating an innovation resource to supporting post-production activities. When a collective expanded their scope to address new domains or audiences, we coded this as an expansion of horizontal scope. For example, when Yelp expanded from supporting restaurant bookings to also supporting bookings for hotels, we coded this as an expansion of horizontal scope. Yelp did not change the types of innovation activities conducted, but rather, expanded the domains to which those activities applied.

We also identified if each collective made a change in the allocation of governance rights over time. We examined how rights over technical control, decision-making, and ownership of intellectual property were distributed among peers contributing within the collective and the sponsor or hosting firm (if relevant) at founding and at present. We then assessed, to the best of our ability, if the collective gained or lost rights in these domains over time. Even if only some participants in the collective gained rights, we coded this as a gain in collective rights.

First, we coded technical control as the degree to which the collective gained some degree of control over the rules guiding peer participation. For example, the founders of Topcoder initially made all strategy decisions. However in 2018, participants could nominate "MVPs" who could participate in strategy meetings (D'Amato Ford, 2018).

Second, we examined if rights to make decisions about *what* the collective produced and *how* it was produced changed. For example, when Eclipse was managed by IBM, IBM decided who could become a member. However, after 2004, members gained the right to collectively determine the criteria that constituted membership (O'Mahony & Karp, 2020).

Third, we assessed if contributors to peer production gained any new intellectual property rights to what they produced. For example, in 2007, YouTube introduced a content management tool, Video ID, to help copyright owners determine whether to block their content or share in the revenue gained from the content (Kim, 2012).

We then compared which collectives experienced changes in vertical or horizontal scope and if any governance rights were distributed toward the collective and analyzed these data to identify patterns. We summed the changes in scope and governance rights for each collective and plotted them. This configurational approach helped cluster the 12 cases into three modes of production (collaborative, co-creative, or aggregative). *Collaborators* allow individuals to work together to support expanding roles in the innovation lifecycle. *Co-creators* build on each other's individual contributions to create a coherent body of work. *Aggregators* cooperate by contributing their own content to a collective resource without building upon the work of others. These modes of production help explain how collectives integrate or accumulate the work of those that participate to take on different roles in the innovation lifecycle. Based on this provisional framework, we explored plausible mechanisms and relationships to advance a research agenda, rather than to reach causal conclusions.

3 Scope of the Innovation LifeCycle

Most (9 of 12) of the collectives we sampled grew in either vertical scope (by taking on new types of innovative activities in the innovation lifecycle) or in horizontal scope (by addressing new domains or audiences) between the 10–20 years from their founding until 2018 (Table 11.2). Only three collectives

Table 11.2 Change in scope and governance rights.

	Initial scope				*Change in scope*					*2018 scope*				*Change in governance rights*		
	Founding role in the innovation lifecycle				***Vertical*** *Expansion into different activities across the innovation lifecycle*				***Horizontal*** *Expansion into new domains*	*Role in the innovation lifecycle*				**Technical control** *Technical rules governing participation*	**Decision rights** *What is being produced and how*	**IP rights** *Rights over intellectual property produced*
Collectives	*IG*	*ISR*	*FS*	*PP*	*IG*	*ISR*	*FS*	*PP*		*IG*	*ISR*	*FS*	*PP*			
Eclipse	Black	Black	Black	Black					Light Gray	Dark Gray	Dark Gray	Dark Gray	Dark Gray	Mid-Gray	Mid-Gray	Mid-Gray
Topcoder	Black	Black	Black	Black					Light Gray	Dark Gray	Dark Gray	Dark Gray	Dark Gray	Mid-Gray	Mid-Gray	Mid-Gray
Threadless	Black		Black						Light Gray	Dark Gray	Dark Gray	Dark Gray	Dark Gray	Mid-Gray	Mid-Gray	Mid-Gray
InnoCentive	Black		Black		Light Gray	Light Gray			Light Gray	Dark Gray	Dark Gray	Dark Gray		Mid-Gray	Mid-Gray	Mid-Gray
Wikipedia		Black						Light Gray	Light Gray	Dark Gray		Dark Gray		Mid-Gray	Mid-Gray	Mid-Gray
YouTube		Black									Dark Gray			Mid-Gray	Mid-Gray	
Reddit	Black	Black				Light Gray	Light Gray	Light Gray			Dark Gray	Dark Gray	Dark Gray			Mid-Gray
Pinterest		Black						Light Gray		Dark Gray	Dark Gray		Dark Gray	Mid-Gray	Mid-Gray	
Kiva			Black			Light Gray	Light Gray	Light Gray			Dark Gray	Dark Gray	Dark Gray		Mid-Gray	
Kickstarter			Black						Light Gray			Dark Gray				
TripAdvisor				Black								Dark Gray				
Yelp				Black					Light Gray				Dark Gray		Mid-Gray	

Legend
IG = Idea generation
ISR = Innovation search and reference
FS = Funding and sourcing
PP = Post-production
Note: End-to-end collectives are represented by illustrating all of their activities across the innovation lifecycle.
Black = Initial scope in the innovation lifecycle
Light Gray = Change in scope
Dark Gray = 2018 scope
Mid-Gray = Change in governance rights

(Wikipedia, Kickstarter, and TripAdvisor) experienced no significant change in scope, although they grew in other ways such as in the number of contributors or content produced (Lock, 2019; Szmigiera, 2019; Wikipedia, 2019).

3.1 Expanding Horizontal Scope

Half (6 of 12) of the collectives sampled expanded their horizontal scope (Eclipse, Topcoder, Threadless, InnoCentive, Kiva, and Yelp) – broadening their relevant domains of production. For example, in 2010, Kiva expanded from funding low income entrepreneurs to offering loans for higher education in low income countries. The Student Microloans program, initially launched in Bolivia, Paraguay, and Ecuador, was expanded further in 2014, to provide educational loans for poor and low-income students (Kiva, 2014). Yelp also expanded their horizontal scope from restaurant bookings to serve a broader variety of services (florists, lawyers, golf clubs, etc.). In making these changes, Kiva and Yelp expanded their relevance to address the needs of new audiences.

3.2 Expanding Vertical Scope

Approximately half (5 of 12) of the collectives (Threadless, InnoCentive, YouTube, Reddit and Pinterest) expanded their vertical scope by incorporating new activities in the innovation lifecycle. For example, Threadless expanded their vertical scope from idea generation to hosting individual artists' design collections. At founding, artists could generate new tee-shirt design ideas in online competitions and vote to determine a winner, but Threadless management ultimately chose what designs to produce (Lakhani & Kanji, 2008; Langner & Seidel, 2015). In 2016, Threadless began inviting artists to create their own "artist shops" and sell their own designs, manufactured and shipped by Threadless. This change addressed an unmet need for artists to produce their own work and was well received as indicated by the 100,000 shops created in the first year. One artist, James White, recounts: "I've been looking for something like this for over 7 years, an on-demand product company who could handle my wild colors professionally and swiftly. Artist Shops has genuinely altered the trajectory of my brand, allowing me to reach my audience with ease" (Threadless website, 2019). Threadless extended the innovation activities peers could engage in vertically: artists could now sell their collections directly without relying on Threadless deciding what to produce.

3.3 Maintaining Scope

Only three collectives maintained their scope during the period of our analysis (TripAdvisor, Kickstarter, and Wikipedia). While Kickstarter rapidly increased the number of individuals receiving funding on its website as well as the total funds amassed, its founders fought to retain focus on sourcing and funding creative projects. As reaffirmed by Kickstarter's leadership team (Strickler, Chen, & Adler, 2015): "From Kickstarter's inception, we've focused on serving artists, creators, and audiences to help bring creative projects to life." To maintain their horizontal scope, Kickstarter rejected projects that ventured into medical, health, food, or contraband territories. To retain their vertical scope on funding and sourcing, Kickstarter restricted engagement in peer review by insisting participants fund a project before providing evaluative feedback.

One could argue that a mission of empowering and engaging "people around the world to collect and develop educational content under a free license or in the public domain, and to disseminate it effectively and globally" (Wikimedia Foundation website, 2019) is so broad that it would be subsequently difficult to expand. Wikipedia has remained focused on its founding mission with efforts to improve the quality of encyclopedia entries produced (Aaltonen & Lanzara, 2015) rather than to produce more ways to apply that knowledge. As founder Jimmy Wales stated: "We're going from the era of growth to the era of quality" (Levine, 2006). Both Kickstarter and Wikipedia "stuck to their knitting," adopting new mechanisms to improve the quality of projects and entries while resisting expansion into new territories.

4 Governance Rights

Governance rights offer participants the opportunity to make decisions that affect the participation architecture, or the social, legal, and technical capabilities that guide interactions and exchange in a collective (West & O'Mahony, 2008, p. 146). These include technical control rights (the rules governing how participation can happen); decision rights (what is being produced and how) and intellectual property rights (rights over the intellectual property produced by both individuals and collectives). To encourage contributions, the founder or sponsor of a peer production collective may, as the collective matures, convey rights to contributing members to cultivate collective engagement (O'Mahony & Karp, 2020). As O'Mahony and Karp showed in their study of the Eclipse community, contributors can be reluctant to deepen their engagement to a collective without mechanisms that provide them voice to influence production activities.

All of the collectives sampled, except for two (TripAdvisor and Kickstarter), conveyed some governance rights (technical control, intellectual property rights, or decision rights) to enable contributing participants to vote or give voice to administration or coordination responsibilities over time. For example, the management team of Threadless were aware that if they wanted to generate the best "ideas," they needed to involve artist contributors in the strategic decision-making process: "if the business direction of the company ultimately does not meet the needs and interest of the community, then everything starts to break down" (Langner & Seidel, 2015, p. 534). Collectives that engaged early in the innovation life cycle (idea generation) or throughout the entire life cycle, were more likely to share governance rights with the collective than those supporting later-stage funding and sourcing or post-production activities.

4.1 Distribution of Governance Rights

Governance rights often became more unevenly distributed, as the most active or powerful participants gained rights over time while those on the periphery did not (Aaltonen & Lanzara, 2015; Piskorski & Gorbâtai, 2017). Both Wikipedia and Reddit transitioned new decision rights to core contributors to lead content co-creation. Reddit founders distributed decision rights to moderators to set participation rules for forums dedicated to specific topics and ban participants at their discretion. Some reddit users highlighted perceived power abuses at the hands of moderators in new forums such as r/subredditcancer. As one indignant contributor named Chernobyl (n.d.) argued: "I can't think of a single reason someone should be banned in one [subreddit] for posting in another." This raises the question of how the differentiation of rights for peripheral versus core contributors (Dahlander & O'Mahony, 2011) can create tiered governance (Gulati, Puranam, & Tushman, 2012) and affect the on-going vitality of the collective.

Gulati and colleagues suggest that tiered models reduce coordination complexity and create a motivational mechanism – which may be critical to setting aspirations for collectives reliant on volunteers (2012, p. 578). Although much research has showed that tiered governance emerges over time (Aaltonen & Lanzara, 2015; Piskorski & Gorbâtai, 2017) or how a tiered model emerges (Dahlander & O'Mahony, 2011; Ferraro & O'Mahony, 2012; O'Mahony & Ferraro, 2007), little research has explored how tiered governance models affect the generation or accumulation of ideas (e.g., Murray & O'Mahony, 2007). Do more mechanisms of control, such as filtering and editing, emerge as a core group gains more power? Shaw and Hill (2014) suggested that as people gain more wiki editorial rights, they are more likely to wield their power – and revert the entries of others over time. In contrast, Dahlander and O'Mahony (2011) showed that once assuming a leadership role, free and open source contributors significantly increased their coordination efforts, which invited others on the periphery to engage. When governance rights are distributed to a small segment of the collective, it is not clear if the segmentation produced fosters ossification and the creation of oligarchy (e.g., Shaw & Hill, 2014) or enhances the coordination of complexity (Gulati et al., 2012). Further research could investigate the relationship between the distribution of governance rights and the innovating capacity of the collective.

4.2 Types of Governance Rights

Many (10 of 12) of the collectives we examined allocated some governance rights to participants, most commonly, decision rights about what was produced (9 of 12). For example Topcoder and Eclipse distributed decision rights to allow participants to become involved in reviewing submissions. Technical control rights (6 of 12) were more rarely distributed. Changes in technical control typically allowed participants to revise the rules guiding participation (Boudreau, 2010; O'Mahony & Karp, 2020) or make changes in the participation architecture. Six of the collectives studied experienced changes in technical control over the platform – four of these engaged in most phases of the innovation lifecycle (Topcoder, Eclipse, InnoCentive, and Threadless). For example, initially only IBM could decide when to accept a new project to the Eclipse platform. After creating an independent, representative governing body, Eclipse members designed a review and approval process for anyone to propose a new project – which put decisions over technical scope in the hands of project members (O'Mahony & Karp, 2020). As explained by the foundation's Executive Director, the process by which requirements are formulated is "bottom up." This enabled contributors to "lobby for the things they need to see happen in Eclipse." Collectives engaged in all stages of the innovation lifecycle were more likely to distribute technical control to contributors than collectives with more limited scope.

Finally, at founding, most collectives did not develop rules around intellectual property rights. Five collectives distributed new intellectual property rights to contributors over time (Topcoder, Eclipse, InnoCentive, Threadless, and YouTube). For example, Eclipse transitioned to a new license that gave users more rights and both Threadless and YouTube gave artists more rights to the works they created over time. Solvers in the InnoCentive collective initially worked on idea generation challenges and assigned their intellectual property rights to seekers through a license (Allio, 2004; Lakhani, Jeppersen, Lohse, & Panetta, 2007). When a new electronic Request for Proposals was created, solvers were granted rights to negotiate their own terms directly with seekers. Only three collectives were proactive in developing policies around intellectual property rights, making the assignment of intellectual property rights an ex ante requirement of participation (Topcoder, Threadless, and InnoCentive). Collectives like YouTube and Pinterest were more reactive: they retrospectively monitored contributions to assess potential violations rather than establishing the provenance or ownership of content as a condition for participation. Collectives that generate more scientific, technical, or creative content may be more proactive rather than reactive in ascertaining intellectual property rights. Proactive vs. reactive intellectual property requirements reveal different assumptions about individuals' contribution behavior and may shape the norms of what is viewed as permissible in a collective, but little research has examined this difference.

5 Patterns of Growth

As shown in Figure 11.1, collectives that maintained their scope (like Kickstarter and TripAdvisor) were less likely to experience a change in governance rights. Collectives that made changes in either vertical or horizontal scope were more likely to have allocated some governance rights to the collective. Or, quite possibly, collectives that redistributed governance rights were more likely to experience a change in scope – we cannot determine causal primacy from these data. For example, Threadless and InnoCentive made both vertical and horizontal scope changes and distributed all three types of decision rights. More fine-grained, longitudinal data is needed to explain the relationship between scope change and the distribution of governance rights. Figure 11.1 provides a provisional framework, plotting changes in scope on the x axis and changes in governance on the y axis.

This analysis helped cluster the 12 peer production collectives into three modes of production (*Collaborators*, *Co-creators*, and *Aggregators*) to understand how peer collectives accumulate or integrate their activities to fulfill different roles in the innovation lifecycle. These three categories lie on a continuum – from the least integrative, *Aggregators*, to the most integrative, *Collaborators*. We elaborate

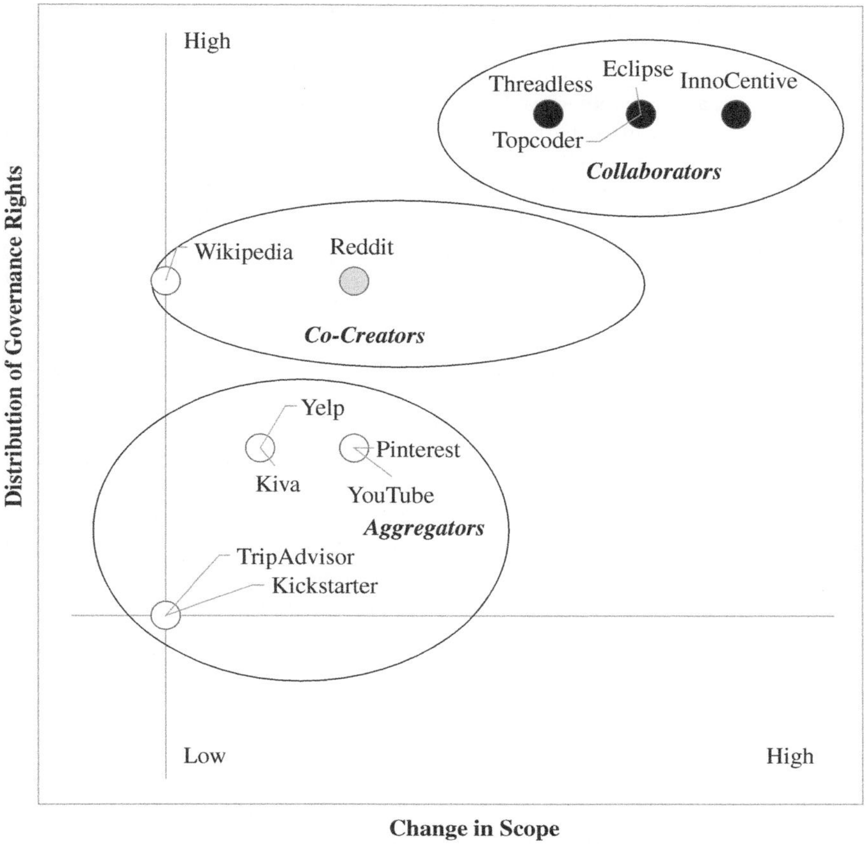

Note: Modes of production are circled and vary based on degree of integration required

Black circle – indicates collective transitioned to this mode of production over time
Gray circle – indicates collective transitioned to this mode of production very shortly after founding
White circle – indicates collective used this mode of production at founding and currently

Figure 11.1 Modes of collective production.

on these three modes to better explicate how collectives either sustain or expand their role in the innovation lifecycle and use this framework to help advance a research agenda that considers the dynamics of scope growth.

5.1 Collaborators

Peer production collectives are collaborative when individuals have the right to work together to support expanding activities in the innovation lifecycle. InnoCentive, Topcoder, Eclipse, and Threadless did not initially support collaborative modes of production at their founding, but in 2018, all leveraged a collaborative mode of peer production – where individuals can integrate their contributions directly with each other. At its founding, InnoCentive was concerned that collaborative work to solve innovation challenges would muddy the provenance of ideas (Allio, 2004; Lakhani, 2009). As the problem-solving community grew in scope to address a broader range of scientific and technical challenges, InnoCentive shifted towards a more collaborative mode of production, and in 2010, began to invite individuals to solve challenges in teams. As shared by InnoCentive on their website: "[s]ome challenges inherently call for insight from multiple [individuals] with diverse knowledge and skills . . . Creating solutions with other [individuals] will likely increase the quality of the solution." At Threadless, artists demanded new forums for feedback, iteration, and collaboration, fostering the creation of artist studios.

Similarly, as the scope of innovation challenges stretched the capabilities of founding contributors, Topcoder (Lakhani, Garvin, & Lonstein, 2010) and Eclipse found themselves breaking down barriers to individuals collaborating with each other. Eclipse encountered provisioning problems (e.g., Ostrom, 1990) where growing numbers of people contributed code in surprising and uncoordinated ways that threatened software quality (O'Mahony & Karp, 2020). To manage these issues and foster collaboration among diverse contributors, the foundation began requiring members from different firms to collaborate on new projects. The Executive Director of the foundation recognized that more collaborative modes of production could create friction, but also improve quality: "anything that any group is doing which is complex and strives for high quality requires some institutional friction. Collaboration creates that friction . . . people pushing back, that process is normal and healthy."

While these collectives started by sourcing new ideas through the broadcast search of diverse sources (Jeppesen & Lakhani, 2010), over time, they created new means to recombine diverse knowledge (Fleming, 2001, 2007; Murray & O'Mahony, 2007), largely to improve the integrative quality of solutions offered. Gulati et al. argue that designers of collaboration systems must consider who can collaborate with whom and what linkages are desired to shape knowledge sharing (2012, p. 581). More research on how the design of the participation architecture underlying peer production either inhibits or fosters collaboration among contributors would be helpful. For example, how did the creation of team rooms on the InnoCentive website influence the subsequent quality or complexity of submissions? Future research would do well to examine what prompts a transition to collaborative modes of production and what norms, standards, processes, and infrastructure change to support this shift.

5.2 Co-Creators

When peer production collectives co-create, individuals share their work in a cooperative manner and may build on each other's work by following design rules (e.g., Baldwin & Clark, 2000) without necessarily integrating individual content into a cohesive body of work. For example, Wikipedia relies on contributors to edit and build on each other's entries to update changing information – but these entries remain modular and, although they maintain a common format, do not have to integrate like software. Contributors need not collaborate to put forth new entries or add information, and some can maintain control of specific entries (Benkler, Shaw, & Hill, 2015; Forte, Larco, & Bruckman, 2009). Jimmy Wales explains that co-creative modes of production, where individuals build on one another's work enable deeper investigation than individual efforts could: "You can't put up a website called the *Denver Guardian* and say 'Pope Endorses Trump' and trick the Wikipedia community. They're information obsessives. They're very sophisticated. They spend a lot of their time evaluating the quality of sources." This was the case when Wikipedia captured more accurate and rapid information about the Grenfell fire in London, than the mainstream press (Thornhill, 2017).

While Wikipedia always supported co-creative modes of production, Reddit transitioned from an aggregative mode of production to a co-creative mode of production when they introduced the ability for contributors to comment back and forth. Both Reddit and Wikipedia have an interest in maintaining the quality of content produced and have distributed governance rights to empower the collective to participate not just in creating content, but also in monitoring the quality of that content (e.g., Ostrom, 2007; Simcoe, 2012). Larger samples are needed to assess how co-creating collectives balance quality concerns with maintaining open participation (e.g., Ferraro & O'Mahony, 2012) as they grow. Co-creative modes of peer production like Wikipedia and Reddit have experienced high growth in contributors and content, but still maintained a stable scope. Largely, this appears to be the product of deliberative decisions to maintain scope, which reflects governance work in and of itself. What is unexplored are the decision-making processes that allow growing communities like Wikipedia and Reddit to maintain a stable scope in the face of participant growth.

5.3 Aggregators

When peer production collectives aggregate production, individuals cooperate by contributing their own content to a collective resource without necessarily building upon each other's work. Aggregation constitutes the weakest form of integration, as peers need not coordinate standards, norms, formats, or efforts to support activities in the innovation lifecycle but just comply with modest submission design rules. Half (6 of 12) of the collectives we studied (TripAdvisor, Kickstarter, YouTube, Yelp, Pinterest, and Kiva) engaged in aggregation as the mode of production at founding and maintained this as their current mode of production, with minimal changes in scope. Aggregators differ from co-creators in fundamental ways as contributions need not follow a standard format or design rules.

For example, TripAdvisor has created a publicly accessible compendium of ranked travel reviews written by individuals all using very different standards, formats, styles, and judgment. As expressed by Orlikowski and Scott (2015, p. 885): "The evaluation categories that TripAdvisor provides its reviewers are open ended and variably interpreted." Unlike Reddit and Wikipedia, peers in Aggregator collectives cannot modify each other's contributions nor determine how content is sorted or prioritized. For example, TripAdvisor's algorithms determine how reviews are displayed. This separation between the provision of content and how content is prioritized is part of the value Aggregator sponsors believe they create (e.g., Orlikowski & Scott, 2015), but this is not co-creation. TripAdvisor, YouTube, Yelp, and Pinterest have an interest in maintaining perceived "objectivity" over how content is curated and propagated and manage this perception through algorithms inaccessible to those who contribute individual reviews and evaluations. This may help explain why aggregating collectives distribute fewer governance rights: the rules governing the display of content are not always clear and cannot be changed by participants. It would be interesting to examine what happens to content curation when *Aggregators* distribute more governance rights to the collective.

Peer collectives focused on funding and financing face a different set of regulatory concerns that can inhibit the free sharing that could foster co-creation and collaboration. For example, Kickstarter funders cannot affect each other's investments nor collaborate through the site to research innovators or make funding decisions. Kickstarter encourages potential funders to conduct due diligence independently off their site: "If you're not sure about something, you can look elsewhere on the web" (Kickstarter website, 2019). Kickstarter has been open about their desire to ensure that investment advice is not offered as a way to minimize fiduciary responsibility associated with giving advice. This interest in minimizing risk in the face of regulatory concerns might help explain Kiva and Kickstarter's sustenance of an aggregative mode of production that limits collaboration among peers. More research is needed to examine how these policies affect participants' on-going ability to share and accumulate knowledge. We do not know if contributors to aggregating collectives develop alternative means to find each other and connect. Clearly aggregating collectives play a vital role in fostering innovation by culminating raw materials for recombination and making them accessible (e.g., Murray & O'Mahony, 2007), but little research has examined how *Aggregators* foster innovation in larger markets or fields.

6 Advancing a Research Agenda

Our approach in this chapter focused on how peer production collectives changed the scope of innovation activities they supported over time. Studies of peer collectives often focus on one collective at a time, which has inhibited the development of more generalizable theoretical frameworks for understanding how peer production growth patterns vary. Building on Aaltonen and Lanzara's (2015) call for a more dynamic perspective, we started with the premise that changes in the nature of the work (Barley & Kunda, 2001) conducted by collectives could inform the evolution of their governance. We show that peer production collectives varied in whether and how they grew the scope of their innovative activities. Based on the transitions made in our sample, we identified three modes of production that differ in the participation rights given to contributors and the degree of integration possible among them. Table 11.3 compares the three modes and advances new research questions.

While *Collaborators* allow contributors to create and collaborate directly and build on each other's ideas, *Co-creators* primarily foster cooperation rather than collaboration. *Aggregators* do not require either but ask contributors to comply with modest submission rules. Future research could examine how and why these governing decisions were made. While imprinting theories would suggest that a collective's founding purpose shapes the rights extended to others, we observed many collectives making substantial changes to their scope and governance rights over time. Why do some collectives expand their scope while others retain theirs? What triggers a change in scope? Changes in scope can influence the type of creators or inventors that are attracted to contribute to a collective, but little research has examined this relationship. How do changes in scope affect the subsequent type and rate of incoming contributors to a collective? If new participants are likely to introduce new ideas, then we would expect participant growth to drive scope growth, but we also observed several high growth collectives with stable scope. Is this the product of strong leadership or collectives with a strong mission orientation?

The question of what determines an organization's scope is central to strategy scholars (Grossman & Hart, 1986; Hart, 1995; Hart & Moore, 1990; Kekre & Srinivasan, 1990; North, 1986; Rumelt, 1982; Sorenson, 2000; Sorenson, McEvily, Ren, & Roy, 2006; Santos & Eisenhardt, 2009; Williamson, 1991) but has not yet been empirically applied to peer production collectives. Perhaps what is needed is research collaborations that marry expertise from both worlds to create novel integrative research designs. For example, a research design could predict scope evolution for peer collectives that focus on a particular set of innovation activities, such as idea generation, to better capture competitive dynamics within a segment of the innovation lifecycle. From a research perspective, one of the great benefits of peer collectives is that their interactions are largely recorded in publicly available forums. Discourse analysis or machine learning techniques could be leveraged to analyze how the structure and narrative of participants' dialogue informs scope decisions and their associated consequences for participation, growth, and innovation capability.

When the distribution of governance rights evolves, the expectations of participants in how they can interact with others, create, collaborate, and produce work or support their own interests can be either upgraded or upended. We examined only three types of governance rights but there are many other domains of governance to be considered – some more difficult to observe. For example, we have not considered the role non-human actors play in the governance of peer production (Orlikowski, 1992, 2000). Yelp, TripAdvisor, and YouTube employ ranking algorithms to recommend the most helpful or most relevant content to each user. Future research could examine how these rankings contribute to differentiation among contributors and whether and how contributors game algorithms to achieve prominence.

Our approach takes to heart Benkler and colleagues' suggestion to advance the "science of the mechanisms, procedures, and techniques necessary for the effective production of networked informational resources" (2015, p. 9). One limitation of our study is that, like much of the work on peer production, we selected on success. We focused on changes in governance rights and scope in a small comparative sample of mature collectives, to the neglect of identifying what predicts the emergence or failure of new types of peer production collectives. There are many other domains such as culture and founding motivation that can also shape how governance evolves (e.g., Benkler, 2016) as peer production collectives mature.

Hopefully, what we contribute is a broadened conception of the roles peer production collectives engage in and advance a research agenda to explore scope changes in collective production. To us, the interesting future research questions to address are not static ones that identify categories such as: "What are all the different types of peer production collectives?" Rather, we urge scholars to embrace a dynamic approach to growth and explore: "What explains how peer production collectives vary in the ways they grow?" A comparative perspective that considers how changes in scope and governance affect the long-term innovative vitality of a collective will help further the next phase of peer production research, with implications for practice.

Table 11.3 Governing for growth: advancing a research agenda.

Mode of production	*Definition*	*Distribution of governance rights*	*Example*	*Advancing a research agenda*
Collaborating	Coordinating efforts to create content with other participants in the collective.	High	On InnoCentive, individuals have solved challenges in teams since 2010. For example, teams can collaborate in specialized team rooms to collaborate to solve complicated problems such as the eradication of polio.	• Is the move to a collaborative mode of production triggered by demands from the collective or by industry shifts? • How does a firm's strategy inform what types of collaboration are permitted by a peer collective's participation architecture? • Are contributors with high levels of technical or creative expertise more likely to demand opportunities for collaboration than less specialized contributors? Is the collaboration model best suited for high levels of expertise?
Co-creating	Producing content which builds directly upon the work of other contributors without coordinating individual efforts.	Medium	On Wikipedia, contributors edit and build on each other's entries. These entries are modular: contributors add work and can control specific entries as individuals, and they need not collaborate to put forth new entries or add information (Benkler et al., 2015; Forte et al., 2009). "Editing most Wikipedia pages is easy": *Wikipedia: FAQ/Editing, 2019.*	• How do co-creating collectives balance the need to maintain quality with open participation? • Are co-creative models more likely to have a broad mission? Or, is a broad mission more likely to produce a co-creative model? • How does increasing core versus peripheral differentiation affect the vitality of co-creation?
Aggregating	Contribute to a collective resource without coordinating content, standards, norms, formats, or efforts, but often complying with modest submission rules.	Low	On TripAdvisor, individuals contribute travel reviews using different standards, formats, styles, and judgment. Peers do not coordinate evaluation categories or standards and do not edit or rank each other's work (Orlikowski & Scott, 2014).	• How does the control of content creation affect contributors' ability to share and accumulate knowledge? • How does reactive versus proactive pursuit of intellectual property rights affect content aggregation? • How do Aggregators foster innovation in the markets or fields they target?

References

Aaltonen, A., & Lanzara, G. F. (2015). Building governance capability in online social production: Insights from Wikipedia. *Organization Studies, 36*(12), 1649–1673.

Adler, P. S. (2001). Market, hierarchy, and trust: The knowledge economy and the future of capitalism. *Organization Science, 12*(2), 215–234.

Allio, R. J. (2004). CEO interview: The InnoCentive model of open innovation. *Strategy & Leadership, 32*(4), 4–9.

Baldwin, C. Y., & Clark, K. B. (2000). *Design rules: The power of modularity* (Vol. *1*). Cambridge, MA: MIT Press.

Baldwin, C., & Von Hippel, E. (2011). Modeling a paradigm shift: From producer innovation to user and open collaborative innovation. *Organization Science, 22*(6), 1399–1417.

Barley, S. R., & Kunda, G. (2001). Bringing work back in. *Organization Science, 12*(1), 76–95.

Bechky, B. A., & O'Mahony, S. (2015). Leveraging comparative field data for theory generation. In K. D. Elsbach & R. M. Kramer (Eds.), *Handbook of qualitative organizational research: Innovative pathways and methods* (pp. 168–176). New York, NY: Routledge

Benkler, Y. (2016). Peer production and cooperation. In J. M. Bauer & M. Latzer (Eds.), *Handbook on the economics of the internet* (pp. 91–119). Cheltenham: Edward Elgar.

Benkler, Y. (2017). Peer production, the commons, and the future of the firm. *Strategic Organization, 15*(2), 264–274.

Benkler, Y., Shaw, A., & Hill, B. M. (2015). Peer production: A form of collective intelligence. In T. W. Malone & M. S. Bernstein (Eds.), *Handbook of collective intelligence* (pp. 175–208). Boston, MA: MIT Press.

Bond, J. M. (2018, September 2). Understanding Reddit: A beginner's guide to the front page of the internet. *The Daily Dot.* Retrieved from https://www.dailydot.com

Boudreau, K. (2010). Open platform strategies and innovation: Granting access vs. devolving control. *Management Science, 56*(10), 1849–1872.

Boudreau, K. J., Lacetera, N., & Lakhani, K. R. (2011). Incentives and problem uncertainty in innovation contests: An empirical analysis. *Management Science, 57*(5), 843–863.

Brown, J. S., & Duguid, P. (2001). Knowledge and organization: A social-practice perspective. *Organization Science, 12*(2), 198–213.

Chernobyl (n.d.). [Comment.] Retrieved from www.reddit.com/r/subredditcancer

Dahlander, L., & O'Mahony, S. (2011). Progressing to the center: Coordinating project work. *Organization Science, 22*(4), 961–979.

D'Amato Ford, J. (2018, January 4). Be a Topcoder MVP: An all new program. Retrieved from www.topcoder.com/topcoder-mvp-new-program/

Eisenhardt, K. M., & Graebner, M. E. (2007). Theory building from cases: Opportunities and challenges. *Academy of Management Journal, 50*(1), 25–32.

Ferraro, F., & O'Mahony, S. (2012). Managing the boundaries of an "open" project. In J. F. Padgett & W. W. Powell (Eds.), *The emergence of organizations and markets* (pp. 545–565). Princeton, NJ: Princeton University Press.

Fleming, L. (2001). Recombinant uncertainty in technological search. *Management Science, 47*(1), 117–132.

Fleming, L. (2007). Breakthroughs and the "long tail" of innovation. *MIT Sloan Management Review, 49*(1), 69.

Forte, A., Larco, V., & Bruckman, A. (2009). Decentralization in Wikipedia governance. *Journal of Management Information Systems, 26*(1), 49–72.

Garud, R., & Tuertscher, P., & Van de Ven. A. H. (2013). Perspectives on innovation process. *Academy of Management Annals, 7*(1), 775–819.

Gawer, A. (2014). Bridging differing perspectives on technological platforms: Toward an integrative framework. *Research Policy, 43*(7), 1239–1249.

Gawer, A., & Henderson, R. (2007). Platform owner entry and innovation in complementary markets: Evidence from Intel. *Journal of Economics & Management Strategy, 16*(1), 1–34.

Golder, P. N. (2000). Historical method in marketing research with new evidence on long-term market share stability. *Journal of Marketing Research, 37*(2), 156–172.

Grodal, S., & O'Mahony, S. (2017). How does a grand challenge become displaced? Explaining the duality of field mobilization. *Academy of Management Journal, 60*(5), 1801–1827.

Grossman, S. J., & Hart, O. D. (1986). The costs and benefits of ownership: A theory of vertical and lateral integration. *Journal of Political Economy, 94*(4), 691–719.

Gulati, R., Puranam, P., & Tushman, M. (2012). Meta-organization design: Rethinking design in interorganizational and community contexts. *Strategic Management Journal, 33*(6), 571–586.

Hart, O. (1995). *Firms, contracts, and financial structure.* Oxford: Clarendon Press.
Hart, O., & Moore, J. (1990). Property rights and the nature of the firm. *Journal of Political Economy, 98*(6), 1119–1158.
Helfat, C. E., & Eisenhardt, K. M. (2004). Inter-temporal economies of scope, organizational wiki modularity, and the dynamics of diversification. *Strategic Management Journal, 25*(13), 1217–1232.
Jeppesen, L. B., & Lakhani, K. R. (2010). Marginality and problem-solving effectiveness in broadcast search. *Organization Science, 21*(5), 1016–1033.
Jones, G. R., & Hill, C. W. (1988). Transaction cost analysis of strategy-structure choice. *Strategic Management Journal, 9*(2), 159–172.
Kahl, S. J., & Grodal, S. (2016). Discursive strategies and radical technological change: Multilevel discourse analysis of the early computer (1947–1958). *Strategic Management Journal, 37*(1), 149–166.
Kekre, S., & Srinivasan, K. (1990). Broader product line: A necessity to achieve success? *Management Science, 36*(10), 1216–1232.
Kickstarter website. (2019). Retrieved from www.kickstarter.com/trust
Kim, J. (2012). The institutionalization of YouTube: From user-generated content to professionally generated content. *Media, Culture & Society, 34*(1), 53–67.
Kiva and Vittana Join Forces to Expand Student Loans in the Developing World (2014) [Press release]. Retrieved from www.kiva.org/cms/pressrelease_20140225.pdf
Lakhani, K. R. (2009). *InnoCentive.com (A).* HBS No. 608170. Boston, MA: Harvard Business School.
Lakhani, K. R., Garvin, D. A., & Lonstein, E. (2010). *Topcoder (A): Developing software through crowdsourcing.* HBS No. 610032. Boston, MA: Harvard Business School.
Lakhani, K. R., Jeppersen, L. B., Lohse, P. A., & Panetta, J. A. (2007). *The value of openness in scientific problem solving.* HBS Working Paper No. 07-050. Boston, MA: Harvard Business School.
Lakhani, K. R., & Kanji, Z. (2008). *Threadless: The business of community.* HBS No. 608707. Boston, MA: Harvard Business School.
Lakhani, K. R., Lifshitz-Assaf, H., & Tushman, M. (2013). Open innovation and organizational boundaries: Task decomposition, knowledge distribution and the locus of innovation. In A. Grandori (Ed.), *Handbook of economic organization: Integrating economic and organizational theory* (pp. 355–382). Northampton MA: Edward Edgar.
Lakhani, K. R., & Von Hippel, E. (2003). How open source software works: "Free" user-to-user assistance. *Research Policy, 32*(6), 923–943.
Langner, B., & Seidel, V. P. (2015). Sustaining the flow of external ideas: The role of dual social identity across communities and organizations. *Journal of Product Innovation Management, 32*(4), 522–538.
Lave, J., & Wenger, E. (1991). *Situated learning: Legitimate peripheral participation.* Cambridge: Cambridge University Press.
Levine, R. (2006, August 7). The many voices of Wikipedia, heard in one place. *NYTimes.com.* Retrieved from www.nytimes.com
Lifshitz-Assaf, H. (2018). Dismantling knowledge boundaries at NASA: The critical role of professional identity in open innovation. *Administrative Science Quarterly, 63*(4), 746–782.
Lock, S. (2019, March 11). Number of user reviews and opinions on TripAdvisor worldwide from 2014 to 2018 (in millions). Retrieved from www.statista.com/statistics/684862/tripadvisor-number-of-reviews/
Mollick, E. (2014). The dynamics of crowdfunding: An exploratory study. *Journal of Business Venturing, 29*(1), 1–16.
Murray, F., & O'Mahony, S. (2007). Exploring the foundations of cumulative innovation: Implications for organization science. *Organization Science, 18*(6), 1006–1021.
North, D. C. (1986). The new institutional economics. *Journal of Institutional and Theoretical Economics (JITE)/ Zeitschrift für die gesamte Staatswissenschaft, 142*(1), 230–237.
O'Mahony, S. (2007). The governance of open source initiatives: What does it mean to be community managed? *Journal of Management & Governance, 11*(2), 139–150.
O'Mahony, S., & Bechky, B. A. (2008). Boundary organizations: Enabling collaboration among unexpected allies. *Administrative Science Quarterly, 53*(3), 422–459.
O'Mahony, S., & Ferraro, F. (2007). The emergence of governance in an open source community. *Academy of Management Journal, 50*(5), 1079–1106.
O'Mahony, S., & Karp, R. (2020). From proprietary to collective governance: How platform participant strategies adapt. *Strategic Management Journal.* https://doi.org/10.1002/smj.3150
O'Mahony, S., & Lakhani, K. R. (2011). Organizations in the shadow of communities. In *Communities and Organizations* (pp. 3–36). Bingley: Emerald Group.

Orlikowski, W. J. (1992). The duality of technology: Rethinking the concept of technology in organizations. *Organization Science, 3*(3), 398–427.
Orlikowski, W. J. (2000). Using technology and constituting structures: A practice lens for studying technology in organizations. *Organization Science, 11*(4), 404–428.
Orlikowski, W. J., & Scott, S. V. (2015). Exploring material-discursive practices. *Journal of Management Studies, 52*(5), 697–705.
Ostrom, E. (1990). *Governing the commons: The evolution of institutions for collective action.* Cambridge: Cambridge University Press.
Ostrom, E. (2007). Institutional rational choice: An assessment of the institutional analysis and development framework. In P. A. Sabatier (Ed.), *Theories of the policy process* (pp. 21–64). Boulder, CO: Westview.
Parker, G., & Van Alstyne, M. (2017). Innovation, openness, and platform control. *Management Science, 64*(7), 3015–3032.
Peng, M. W., Lee, S. H., & Wang, D. Y. (2005). What determines the scope of the firm over time? A focus on institutional relatedness. *Academy of Management Review, 30*(3), 622–633.
Piskorski, M. J., & Gorbâtai, A. (2017). Testing Coleman's social-norm enforcement mechanism: Evidence from Wikipedia. *American Journal of Sociology, 122*(4), 1183–1222.
Rumelt, R. P. (1982). Diversification strategy and profitability. *Strategic Management Journal, 3*(4), 359–369.
Santos, F. M., & Eisenhardt, K. M. (2009). Constructing markets and shaping boundaries: Entrepreneurial power in nascent fields. *Academy of Management Journal, 52*(4), 643–671.
Seidel, V. P., Hannigan, T., & Phillips, N. (2020). Rumor communities, social media, and forthcoming innovations: The shaping of technological frames in product market evolution. *Academy of Management Review, 45*(2), 304–324.
Seidel, V. P., Langner, B., & Sims, J. (2017). Dominant communities and dominant designs: Community-based innovation in the context of the technology life cycle. *Strategic Organization, 15*(2), 220–241.
Shah, S. K. (2006). Motivation, governance, and the viability of hybrid forms in open source software development. *Management Science, 52*(7), 1000–1014.
Shaw, A., & Hill, B. M. (2014). Laboratories of oligarchy? How the iron law extends to peer production. *Journal of Communication, 64*(2), 215–238.
Simcoe, T. (2012). Standard setting committees: Consensus governance for shared technology platforms. *American Economic Review, 102*(1), 305–336.
Sorenson, O. (2000). Letting the market work for you: An evolutionary perspective on product strategy. *Strategic Management Journal, 21*(5), 577–592.
Sorenson, O., McEvily, S., Ren, C. R., & Roy, R. (2006). Niche width revisited: Organizational scope, behavior and performance. *Strategic Management Journal, 27*(10), 915–936.
Szmigiera, M. (2019, October 2). Number of total and repeat Kickstarter project backers as of October 2019 (in millions). Retrieved from www.statista.com/statistics/288345/number-of-total-and-repeat-kickstarter-project-backers/
Strickler, Y., Chen, P. & Adler, C. (2015, September 21). Kickstarter is now a Benefit Corporation. [web log comment]. Retrieved from www.kickstarter.com/blog/kickstarter-is-now-a-benefit-corporation
The Easiest Way to Sell Your Art Online. (2019). Retrieved from www.threadless.com/artist-shops/signup/art
Thornhill, J. (2017, December 1). Can Wikipedia's Jimmy Wales fix the news? *Financial Times.* Retrieved from www.ft.com
Von Hippel, E. (2005). Democratizing innovation: The evolving phenomenon of user innovation. *Journal für Betriebswirtschaft, 55*(1), 63–78.
West, J. (2003). How open is open enough?: Melding proprietary and open source platform strategies. *Research Policy, 32*(7), 1259–1285.
West, J., & O'Mahony, S. (2008). The role of participation architecture in growing sponsored open source communities. *Industry and Innovation, 15*(2), 145–168.
Wikimedia Foundation Mission. (n.d.). Retrieved from https://wikimediafoundation.org/about/mission/
Wikipedia (2019). Modelling Wikipedia's growth. Retrieved from https://en.wikipedia.org/wiki/Wikipedia:Modelling_Wikipedia%27s_growth
Williamson, O. E. (1991). Comparative economic organization: The analysis of discrete structural alternatives. *Administrative Science Quarterly, 36*(2), 269–296.
Yin, R. K. (1994). *Case study research: Design and methods.* Thousand Oaks, CA: Sage Publications.

Part IV

Cases: Realizing Peer Production

12

Free and Open Source Software

Stéphane Couture

1 Introduction[1]

Free software, often used interchangeably with the term "open source software" or "free and open software" (FOSS), is software whose source code can be freely accessed, shared, modified, and improved. Because of these characteristics that enable the free access and sharing of knowledge, free and open source software has given rise to new models of software development based on peer-production practices, that may involve hundreds, if not thousands, of developers. For many authors, free and open source software is often seen as a prime example, if not the archetype, of the large-scale communal production of information, knowledge, and culture (Bauwens et al., 2012; Benkler, 2006, p. 5).

Today, free and open source software is used in multiple applications and forms much of the backbone of Internet infrastructures, like web or mail servers, or the core of operating systems like Android and Mac OSX. Beyond technologies, free and open source software may also be seen as a sharing culture, a community of actors collaborating online, and as a political movement for the freedom of knowledge on the Internet. Introduced in the 1980s in opposition to the proprietary software model, the success of free and open source software has garnered commercial interest to such an extent that many companies such as IBM, RedHat, Oracle and more recently Google and even Microsoft have focused their business strategy on supporting and taking advantage of the forms of collaboration at work in open source software communities. Public administrations around the world have also embraced free and open source software as a way to reduce costs or assert their sovereignty, and many social activists use it to develop autonomous infrastructures. Finally, these collaborative and moral dimensions of the free and open source movement have inspired many other "peer-production" projects, the most famous of them probably being Wikipedia, the "free encyclopedia."

This chapter first considers free and open source software from a historical perspective, by examining how these terms came about and are still sometimes opposed to each other. It then describes how free and open source software "works," that is, the collaborative and legal practices on which some of the most successful free and open source software projects rely. Then, it presents the mainstream adoption of free and open source software, and ends by addressing some of the challenges this peer-production model faces, in particular when related to tensions arising from its marketization as well as from participants' lack of diversity.

2 Genesis of a Movement

2.1 The Hacker Ethic at the Origin of Free Software

It is generally recognized that the free and open source software movement originated in the hacker culture of the early 1960s. While the meaning of the term "hacker" is today often associated with illegal practices, the term originally referred – and still refers within many free and open source communities – to what Coleman describes as an ethic and aesthetics of tinkering (Coleman, 2012),

The Handbook of Peer Production, First Edition. Edited by Mathieu O'Neil, Christian Pentzold, and Sophie Toupin.

similar to that described by the metaphor "art for art's sake," where technical activity is not a means or an instrument, but rather an end in itself (Riemens, 2006). Citing Levy (1985), Coleman writes that people who called themselves hackers in the 1970s "placed the desire to tinker, learn, and create technical beauty above all other goals" (Coleman, 2012, p. 17).[2]

Raymond (2000), in his brief *history of hackerdom*, situates the beginnings of hacker culture in the 1960s, more specifically at the Tech Model Railroad Club (TMRC), an MIT student association still active today in which people tinker with model railroads and other technical objects, including computers. Many members of the TMRC at the time were also members of the MIT Artificial Intelligence Laboratory, which facilitated the spread of "hacker culture" in the realm of computer programming. Several authors (Coleman, 2012; Himanen, 2001; Levy, 1985) have attempted to explore this hacker culture whose importance to cooperation and information sharing has played, according to Manuel Castells (2003), a decisive role in the development of the Internet and its potential for innovation.

Steven Levy (1985) identified several values associated with the hacker ethic, one being the idea that "All information should be free." Consistent with this value, hackers attached great importance to sharing the software and computer programs they created. However, with the reduction of hardware costs in the late 1970s, many computer corporations began to restrict this sharing of software, in order to increase their profits. Bill Gates, today at the head of one of the largest fortunes in the world, said: "The royalty paid to us, the manual, the tape and the overhead make it a break-even operation. One thing you do is prevent good software from being written. Who can afford to do professional work for nothing?" (Gates, 1976). This gave rise to the "proprietary" software model we know today, where the use and sharing of source code are restricted.

While many hackers decided to subscribe, often unwillingly, to this view of proprietary software development, others sought to develop technological alternatives based on sharing software and its source code. The most radical initiative proposed as an alternative to "proprietary" software was the creation of the concept of free software by Richard Stallman, whom Steven Levy (1985) described in his book as the "last of the hackers."

2.2 The GNU Project and the Free Software Foundation

In the late 1970s, hackers at MIT's Artificial Intelligence Laboratory had formed their community around the handling of a minicomputer and its co-developed operating system. After purchasing a new printer in 1982, the Laboratory's authorities decided to use a proprietary operating system, the license for which prohibited the modification of its source code. For Richard Stallman, a member of this hacker community, these new rules, being incompatible with the software-sharing ethic of the hacker community, were unacceptable and antisocial: "This meant that the first step in using a computer was to promise not to help your neighbor. A cooperating community was forbidden" (Stallman, Lessig, & Gay, 2002, p. 13). Rather than subscribing to this proprietary model, Stallman decided to create a new computer system, based only on free software, with the hope that a new sharing community would emerge around it. Three crucial milestones are attributed to Stallman, which significantly contributed to booting up the free software movement:

First, the project of a new completely free computer system was launched. The project, called *GNU* which stands for Gnu is Not Unix, a recursive acronym (popular in hacker humor) referring to the idea of a system that would be *like* a Unix system, but non-proprietary. The goal was to build an alternative to the then very popular Unix operating system, but that could be freely modified, used, studied, or redistributed. Stallman started to build the system with some software he was already developing, and made a call to contribute to the project with other pieces of software. The initial announcement of this project was on September 27, 1983 on a Usenet Group.

Second, the Free Software Foundation was created two years later, in 1985, foremost in order to finance and support the development of the GNU project and other "free software." This Foundation still exists today, but its activities have broadened beyond software development, to encompass advocacy activities involved in free software and related issues.

Third, Stallman, along with the Free Software Foundation, published in 1989 the first version of the GNU General Public License (GPL), a licensing agreement guaranteeing the freedom for users to use, change, and modify software, unlike many proprietary software licenses that aim to restrict the use of software (to a single machine for instance). This provided a strong legal framing for the creating and sharing of free software. Since then, three major versions of the GPL have been published, and these different versions are still in use in many free and open source software projects.

So what exactly is "free software?" While the formal definition has changed over time, its modern definition is often described using the ideas of four freedoms that users have in relation to software: using, studying, copying, improving (see Table 12.1). The free software definition is often characterized by the idea of *copyleft*, a play on words to signify an inversion of copyright: instead of using copyright to restrict the use and sharing of a work, it is used to guarantee this very possibility. One motto often stated by free software advocates is that free software is "free as in freedom, not as in free beer," to distinguish it from the idea of "zero price."

An important aspect in this definition of free software is the idea of "access to source code," as described in "freedom 1" and "freedom 3." Source code can generally be described as the human-readable instructions that specify the functioning of software.[4] Source code is what programmers make and interact with when they create or modify software (see Figure 12.1). The idea of free software is not that the "executable" source code should be free (as in free beer), but that everybody should be allowed to access and modify the source code. Indeed, source code is sometimes conceptualized

Table 12.1 Free software definition[3].

A program is free software if the program's users have these four essential freedoms:

- The freedom to run the program as you wish, for any purpose (freedom 0).
- The freedom to study how the program works, and change it so it does your computing as you wish (freedom 1). Access to the source code is a precondition for this.
- The freedom to redistribute copies so you can help others (freedom 2).
- The freedom to distribute copies of your modified versions to others (freedom 3). By doing this you can give the whole community a chance to benefit from your changes. Access to the source code is a precondition for this.

Source: What is free software?, GNU.org. © 2020, Free Software Foundation, Inc. Retrieved from: https://www.gnu.org/philosophy/free-sw.html (retrieved May 2nd, 2019). Licensed under CC BY-ND 4.0.

```
/**
 * Simple HelloButton() method.
 * @version 1.0
 * @author john doe <doe.j@example.com>
 */
HelloButton()
{
  JButton hello = new JButton( "Hello, wor
  hello.addActionListener( new HelloBtnList

  // use the JFrame type until support for t
  // new component is finished
  JFrame frame = new JFrame( "Hello Button"
  Container pane = frame.getContentPane();
  pane.add( hello );
  frame.pack();
  frame.show();              // display the fra
}
```

Figure 12.1 Source code. Source: Image by Dr. E.F. Tymac, used under license. CC BY-SA 2.5.

by free software advocates as a form of speech and as such, sharing source code is considered as freedom of speech (Byfield, 2006; Coleman, 2009).

It is important to emphasize that the idea of free software is foremost grounded in the moral principle of sharing. This quote from an interview given by Stallman captures the essence of this ethical principle:

> For me, free software is above all a question of freedom and community. We need free software so that computer users are free to cooperate. It is for this reason alone that I decided to reject the nonfree software. That free software also leads to efficient and powerful software has been a surprise to me, and I am delighted. But it's a bonus. I would have chosen free software, even if it had been less efficient and less powerful – because I do not sell my freedom for simple matters of convenience. (Gleizes & Papathéodou, 2000; free translation)

The development of free software continued throughout the 1980s and mobilized several hundreds of programmers who collaborated through various telecommunication networks of the early Internet. This gave rise to a new model of software development that later sparked the interest of commercial firms.

2.3 Disseminating to Companies: The Creation of "Open Source"

On August 25, 1991, Linus Torvalds, a Finnish student, launched a call for contributions for a new project he called *Linux*. In the months that followed, this call received growing interest from many free software developers, not least because Linux was the missing piece of the GNU project that started some years earlier and finally saw the establishment of a completely free computer system. Torvalds' leadership and charisma, combined with momentum in the development of the Internet catalyzed the free software community's efforts to develop a product that would soon reach a certain maturity.

Enthralled by the success of Linux, Eric Raymond wrote an essay in 1997, later published as a book (Raymond, 2001), titled *The Cathedral and the Bazaar*. It is considered by some to be the first sociological analysis of free and open source software collaboration and development. In his essay, Raymond describes the software development style of Linux as a "bazaar" where the goal is foremost to create a community of contributors to a bazaar of ideas, and to choose from this bazaar the best ideas and integrate them in the new software. The metaphor of the "bazaar" is contrasted with that of the "cathedral" where software is developed in a top-down matter by a grand architect.[5]

This perspective is of great ideological significance. Unlike the previous legitimization of free software, Raymond does not attach any ethical importance to software freedom or sharing. It is to the extent that free software allows the open participation of a large number of programmers and users in the bazaar model that software must remain free to access. What is then at stake is simply the technological and managerial success of a new form of software development described as "the bazaar." This discourse, devoid of moral concerns, was much more appealing to private firms, compared to Stallman's moral variant. Inspired by Raymond's perspective, Netscape announced in 1998 its intention to publish the source code of its web browser to obtain the collaboration of the free software movement, so they could destabilize their competitor Microsoft (Williams, 2002, p. 165). This would later become today's Firefox browser. Following this decision, several business leaders and advocates decided to overcome the ambiguity, in the English language, of the terms "free" and "free software," that are too often associated with lack of cost. This is why Christine Peterson, the chair of a Silicon Valley think tank, advanced the term "open source" to replace "free software" in order to make it more friendly to commercial business (Williams, 2002).

The creation of open source has, however, created some controversies. For Stallman and many of his followers, "The Free Software movement and the Open Source movement are like two political camps within the free software community" (Stallman, Lessig, & Gay, 2002 p. 43). Many proponents still insist today on using the term "free software," implying that open source, while proving technically successful, is erasing the moral dimension of their project. The acronyms FLOSS (free/libre open source software), or FOSS (free and open source software) are sometimes used, especially by researchers, to grasp the phenomenon as a whole while avoiding taking a position in the controversy.

3 How it Works: The Economy and Sociology of Free and Open Source Software

In an article addressing "bottom-up innovation," French sociologist Dominique Cardon (2005) characterized free and open source software as having two axes of "coherence": a normative axis, centered around hacker ethics and the value of sharing knowledge, and an organizational axis, first described by the "bazaar" metaphor, put forward by Raymond. In this section, I will go through these two "axes of coherence" and then present a survey of some of the research to date on free and open source software.

3.1 Sharing Code: The Legal and Normative Model of Free and Open Source Software

As I have mentioned before, free software was born foremost within the hacker value of freely sharing information. While this ethos might seem less prominent today, it is still an important part of the engagement within free and open source software. This ethos is actualized legally through different software licenses that frame the possibilities to use and share source code, each of them being more or less permissive concerning the integration of free or open source software and proprietary software.

The most radical license innovation is the notion of "copyleft," which refers to granting the right to use, modify, or improve some works and distribute them, ***under the same conditions as the original terms.*** The most notable expression of the idea of copyleft is the GNU General Public License, created by Richard Stallman, that I have described earlier, which allows the sharing of source code but forbids its integration in proprietary software.

Other permissive licenses have been created, in particular to favor collaboration with private firms. They are sometimes considered as "non-copyleft" licenses. For instance, in contrast to the GPL, the *MIT license* permits the modification of source code, but also its integration into proprietary software. This means that it is also possible to integrate MIT-licensed source code under the GPL, but not the other way around. The MIT license is currently the most used license on GitHub, a popular open source hosting platform (see Table 12.2).

Other innovative license models have appeared, inspired by the models of free and open source. For instance, the idea of adding a license granting access to source code, except for military use (a so-called "non-military clause"), has been discussed in the past.[6] However, proponents of free and open source software have been reluctant to embrace this approach as it goes against the right to access and modify source code "for any use." Similarly, the Creative Commons "non-commercial" license which permits redistribution of the work for non-commercial purpose only, is considered a non-free software license as it restricts the rights of some users (i.e., commercial ones).

Table 12.2 Most used licenses on GitHub (April 2, 2019).

Rank	*License*	*% of projects*
1	MIT	44.69%
2	Other	15.68%
3	GPLv2	12.96%
4	Apache	11.19%
5	GPLv3	8.88%

Source: From Balter, B., Open source license usage. The GitHub Blog, March 9, 2015, 2015, GitHub, Inc. Retrieved from: https://github.blog/2015-03-09-open-source-license-usage-on-github-com/. Licensed under CC 1.0.

3.2 The Bazaar: The Sociology of Free and Open Source Software

Several research projects, sometimes inspired by *The Cathedral and the Bazaar*, have explored the forms of network collaboration within free and open source software and sought to highlight its innovative character. Some works try to capture the motivations that lead people to become involved in its development (Lakhani & Wolf, 2003) or try to grasp the conditions allowing for the development of a network innovation. Eric von Hippel, a researcher at MIT, has extensively studied the development of free and open source software, which he characterizes as a case of "innovation by use" or "innovation by user networks" (von Hippel, 2005, 2007). Cardon (2005) has reused this model, which he calls "bottom-up innovation" (or "innovation ascendante," in French), to distinguish it from the "top-down" innovation that would come from laboratories or large software firms. Contrary to the usual business model of proprietary software in which developers are paid for their work through the sale of software, free and open source software is instead based on massive collaboration on the Internet, proceeding from contributions from users or volunteers.

In the most mature software, development is based on the sometimes minimal contributions (a comment, a "patch") from what may be several thousands of developers. Cardon distinguishes three levels of contributors in bottom-up innovation networks. First, on the inside, the "core of innovators" who are often the ones who created the project or are leading it. Second, on the outside, the "nebula of contributors," each of whom make small contributions. And, third, in between, the "circle of reformers," who act as mediators between the previous categories by selecting the best contributions and rearranging them so they fit with the vision of the innovators. This is what Cardon calls the "organizational coherence" of free and open source software (see Figure 12.2).

Some authors and analysts also characterize free software as *a commons* (Aigrain, 2005; Benkler, 2003; Birkinbine, 2014; Boyle, 2008; Schweik, 2007). This characterization sometimes refers to Elinor Ostrom's work on the economy and governance of commons, which studied institutional forms that rely on common resources rather than exclusive use, which typify property relations.[7] Indeed, Ostrom, along with Charlotte Hess, published an edited book titled *Understanding Knowledge as a Commons,* which includes a chapter that argues that free and open source software could serve as a model for collaboration built around intellectual property (Schweik, 2007). Philippe Aigrain, for his part, considers the free and open source software movement as a key moment in the development of the paradigm of "informational commons," as opposed to that of "information as property" (Aigrain, 2005). Benkler (2006), in his book *The Wealth of Networks,* coined the term *commons-based peer production* when analyzing emerging collaborative forms such as free software as well as Wikipedia. These commons practices, for Benkler, constitute a new economic paradigm that emerges at the heart of the capitalist economy.

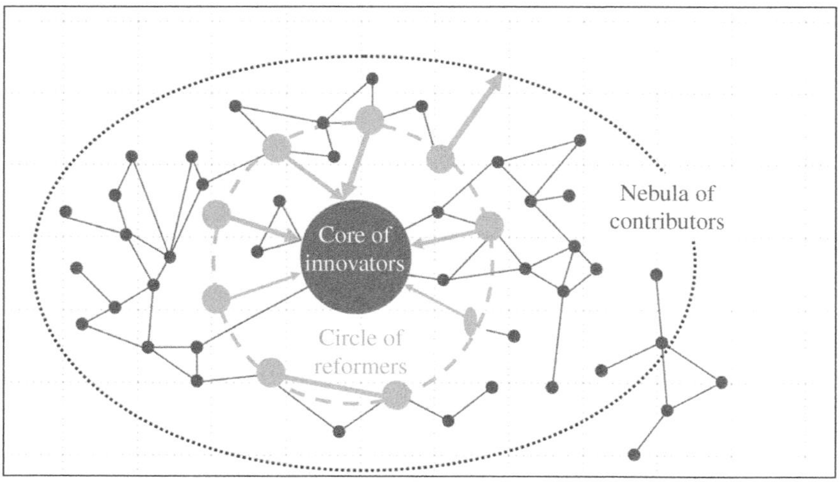

Figure 12.2 Bottom-up innovation in free software development (Cardon, 2005). Source: Cardon, D., Innovation by Usage. In A. Ambrosi, V. Peugeot, & D. Pimienta (Eds.), Word Matters. Multicultural perspectives on information societies. © 2005, C&F Editions.

It should be noted here that although free and open source software can be characterized as commons-based, it does not mean that all contributors work for free or solely on a voluntary basis. In many cases, companies such as IBM, Sun, or more recently Google or even Microsoft, play a leading role in securing the salaries of some contributors. In 2006, a study estimated that more than 500,000 *employees* worldwide were working on open source projects (Ghosh, 2006). However, in all cases, the software subsequently created remains "free" and is not subject – in theory – to any restrictions of use. The idea is then that sharing and commons-based production should serve the mutual interests of concerned actors, whether they are individuals, companies, or non-profit organizations. At least, this is the ideal which, in reality, can create tensions, as we will see in Section 5.1.

4 Beyond Experimentation: Mainstream Adoption and Social Significance

In this section, I will briefly discuss the current place of free software, which now extends to different social groups and fields of activity.

4.1 Free and Open Source as the "Glue" of the Internet

Contemporary free and open source software has reached a level of maturity that makes it a good alternative to many popular end-user software programs: computer systems like Ubuntu or Red Hat (called "distributions") instead of Windows or MacOS; LibreOffice instead of MS-Office, Chromium instead of Chrome. But most importantly, free and open source software is also an important part of the Internet infrastructure. This might be difficult to grasp, as infrastructure is by definition most of the time invisible to common users (Star, 1999). However, free and open source software is used in servers for sending emails, running websites, or creating web applications.

For example, Table 12.3 shows that two free and open source web servers – Apache and Nginx – are used in more than 80% of all web servers. The same goes for "Server-side Programming Languages" for which PHP, licensed as free software, accounts for almost 80% of the market. Finally, we should also note that Android is, by far, the most popular Mobile Operating System. While we tend to associate Android with Google – and indeed Google strongly controls it, as we will see below – it is important to remember that Android is actually a piece of software that is, in part, licensed under the GNU General Public License (GPL). In short, one can argue that free and open source software is the "glue" that holds the Internet together.

In the next sections, I will describe in more detail some of the aspects of free and open source software.

4.2 Public Policies and Governmental Adoption

The success of the development of free and open source software has prompted many firms and public bodies to consider adopting them from an end-user perspective. From the beginning of the millennium, studies have found free and open source software to be economically and technologically viable, and several governments have established policies favoring free and open source software or compelling

Table 12.3 Free and open source software as the Internet infrastructure.

Server-side programming languages		*Most popular web servers*		*Most popular mobile OS*	
PHP	*79.0%*	*Apache*	*43.8%*	*Android*	*86,8%*
ASP.NET	11.4%	Nginx	41.6%	iOS	13,2%
Java	4.0%	Microsoft-IIS	8.7%		
Ruby	2.5%	LiteSpeed	4.0%		

Sources: W3Techs.com. 1 April 2019 and (for Android): https://www.idc.com/promo/smartphone-market-share/os.

Table 12.4 Some experiences of migration to free and open source software in Europe.

Year	*Organization*	*Details*
2003	Finnish Union of Practical Nurses	300 migrations to GNU/Linux and OpenOffice
2003	French Ministry of Interior	15,000 migrations to OpenOffice
1997 –2005	The Spanish region of Extremadura	100,000 computers using *gnuLinEx*, a version of Linux
2005	The Dutch city of Haarlem	2,000 migrations to OpenOffice
2005	French customs	16,000 migrations to OpenOffice
2005	Bristol City Council in the United Kingdom	5,500 migrations to StarOffice
2003 – 2010	Ministry of Justice, Finland	10,000 migrations to OpenOffice
2005 – 2013	French Gendarmerie Nationale	70,000 computers partly using OpenOffice, Firefox, and Thunderbird

Source: Modified from Couture, S. Logiciels libres : réduction des coûts et souveraineté numérique. Institut de recherche et d'informations socio-économiques. https://iris-recherche.qc.ca/publications/logicielslibres.

its adoption outright. In the 2000s, the Center for Strategic and International Studies routinely updated a report on these policies around the world. The last version (Lewis, 2010) listed 354 policy initiatives that had been proposed or approved worldwide since 2001 regarding free open source software, divided into four categories: research (looking at the possibilities of using open source), mandatory (requiring the use of open source software), preferences (preferring open source software), and advisory (permitting its use). The significant savings offered by free and open source software are often the main political rationale for adopting these policies, but other arguments are put forward that emphasize public responsibility, such as the adaptability of free and open source software to meet the particular needs of public administration, the increased security of free and open source software that allows the possibility of analyzing its source code, and technological independence with regards to a particular technology or provider (e.g., Microsoft or Oracle) (Oram, 2011).

This last concept, *technological independence*, sometimes framed as "digital sovereignty," is also emphasized in several studies insisting on the capacity of states to ensure their sovereignty. In 2013, the French Minister for the Digital Economy also noted that free and open source software was a real guarantee of the "digital sovereignty" of a country (Pellerin, 2013). This term is additionally being used to support the need to implement a digital industrial policy aimed at counteracting the hegemony of the United States of America (USA) in the sector.

Several so-called "mandatory" policies, sometimes involving massive migration to free and open source software, have been adopted. Table 12.4 summarizes several European experiences of free and open source software adoption. Note that several countries in Latin America have also adopted free and open source software, especially Brazil, which was for a few years recognized as a leader in the field, especially at the beginning of Lula da Silva's left-wing government (Benson, 2005; Bollier, 2011).

4.3 Social Movements, "Tech Activists," and Free and Open Source Software

Social movements and so-called "tech activists" have also long been interested in the use of free and open source software. Since its inception, there has always been a discourse that associated – rightly or wrongly – free and open source software (especially the "free software" perspective) with a left-wing or progressive ideology. In 1999, Eben Moglen – who later provided significant support and legal advice to the Free Software Foundation – characterized the emergence of free software as "Anarchism triumphant" (Moglen, 1999). In the early 2000s, Indymedia, a global network of alternative media created as part of the anti-globalization movement, was technically built with free and open source software and influenced by its culture (Coleman, 2004). Free software activists were also always present within the

World Social Forum, an annual meeting of progressive and environmentalist social movements around the world (Juris, Caruso, Couture, & Mosca, 2013), though not always in a straightforward way.

Many "tech activists" are still engaged alongside or in support of social movements around the world. Many "radical servers" or alternative infrastructures have been built, like riseup in North America, Koumbit in Canada, or "dégooglisons Internet" (de-google Internet) in France, all of which are based on free software and aimed at social movements and non-profit organizations. While infrastructures built by these groups are not as extensively developed as Google's commercial products, they serve as cultural resources or a "concrete utopia" (Broca, 2013) to participate in global discussions about technical choices. Activist-researcher Alex Haché (2017) has proposed to use the term "technological sovereignty" to describe efforts by civil society to develop autonomous infrastructures and protect itself from commercial and governmental surveillance, notably by using free and open source software (Couture & Toupin, 2019).

4.4 Studying Free and Open Source Software: The Proliferation of Academic Work

Numerous academic articles and research projects have been produced about free and open source software. For instance, a search on ProQuest Central for the term "open source" between 2000 and 2018 yielded 73,956 peer-reviewed articles, while a search for the term "free software" for the same period yielded 9,093 peer-reviewed articles, and "free and open source software" yielded 1,121 articles. While many of these articles have not been produced from a social science perspective, these numbers show the significant academic interest that this phenomenon evokes.

Early research on free and open source software has, for instance, studied the economics of free and open source software, and analyzed why developers and businesses sometimes contribute for free (Lerner & Tirole, 2000). Other research has focused on the cultural significance of free and open source software by looking at the hacker ethics and values that permeate free and open source software communities (Auray, 2000; Coleman, 2012; Broca, 2013). Others have looked at the phenomenon from the perspective of knowledge construction, for instance by analyzing conventions and forms of regulation that ensure their cohesion (Demazière, Horn, & Zune, 2007), or at the diversity of contexts in which free and open source software is deployed (Lin, 2005). In the mid-2000s, Tuomi (2004) and Lin (2005) argued – and criticized – that much of the research into free and open source software was based on a conception of a homogeneous free and open source software community with relatively shared values and motivations. Other authors have analyzed, from a critical perspective, the link between the production of free and open source software and contemporary capitalism (Birkinbine, 2014; Coris, 2006; Dafermos & Söderberg, 2009; Moulier-Boutang, 2007). Similarly, other researchers have sought to capture the conditions that allow the appropriation of free and open source software in non-technical environments (Jullien & Zimmerman, 2007).

5 Challenges

The development of free and open source software faces several challenges. I will present two of them: the tension between capitalist appropriation and community sustainability; the weak participation of women and broader lack of diversity.

5.1 Capitalist Appropriation vs. Community Sustainability

An important challenge that faces free and open software is the opposition between capitalist appropriation of technology and community sustainability. It is important to look at this as a tension, since the problem is not so much that the free and open source software model is not sustainable, but rather that the "community form" of free and open source software is greatly struggling, while its "commercial form" is increasingly articulated in what can be characterized as informational capitalism.

Various analyses have critically studied the role of private companies in free and open source software development. For example, several analyses have looked at how companies succeed in channeling the free work (as in gratis work) of volunteers in order to extract some commercial profit (Broca, 2013; Dafermos & Söderberg, 2009; Terranova, 2000). For Moulier-Boutang, free and open source software can be apprehended as a "signifier" of cognitive and post-industrial capitalism where entrepreneurship consists in "converting the wealth already there in digital space, into economic value" (Moulier-Boutang, 2007, p. 167, as cited by Broca, 2013, p. 244).

Coris (2006, 2009) for her part sees free and open source software as an example of "absorption by the market sphere." Coris focuses more specifically on the case, in the French context, of what she calls "Free Software Services Companies" (FSSCs) created in opposition to traditional information technology service companies (ITSC),[8] that did not include free and open source software in their product offerings at the time of their creation. Coris (2009) notes that, on the one hand, "pure" free open software service companies have been unable to strive and change the software industry. At the same time, free and open source software has been further integrated into pre-existing and more traditional IT service offerings, because of its benefits such as reduced costs and flexibility. In short, the emergence of an industrial context favorable to free and open source software has contributed to quell the early community-oriented approach rather than perpetuate it. In addition, free and open source software that is embedded in "traditional" enterprises is generally limited to that whose maturity and recognition is comparable to similar proprietary software. Essentially, Coris finds that free and open source software has managed to become dominant at the cost of a form of "industrialization" that challenged the initial "community" model: "Born from opposition to the capitalist system – exemplified by the "proprietary" model protection of intellectual property – yet free and open source software seems to succeed: the software companies end up being forced to integrate free software into their offerings. At the same time, free and open source software software services companies are, in turn, forced to industrialize" (Coris, 2006, p. 21; free translation).

5.2 Low Participation of Women and Lack of Diversity

Lack of diversity is another important challenge in free and open source communities. The low participation of women, in particular, has been noted several times. In 2001, a study funded by the European Union found that only 1,1% of free and open source software developers were women (Ghosh, Glott, Krieger, & Robles, 2002, p. 8).[9] Since then, many other studies – whether informal or formal – have echoed this finding. For instance, a study of the Debian community showed that only 1,8% of its developers were women (Reina, Langasek, & Ruiz, 2013). Following the European Union study, another follow-up study was conducted to better understand the issue, this time from a qualitative and ethnographic point of view (Nafus, Leach, & Krieger, 2006a, 2006b). The findings of this study emphasized behaviors within free and open source software development that "actively discouraged" women to participate. More recently, in 2017, a survey was created by GitHub, a software collaboration platform very popular among free and open source practitioners, to better understand the practices and demography of these actors (GitHub, 2017). Again, this survey found out that only 5% of respondents were women, while 22.6% of professional computer programmers in the US are women (Finley, 2017).

Despite these numbers, the situation may have improved in different ways. First, the very fact that research was conducted to study the phenomenon is in itself a positive signal. Second, many initiatives were developed to raise awareness on the low participation of women, and potentially discriminatory or sexist behaviors. Indeed, many female-oriented or explicitly feminist groups have emerged throughout the years for women to mutually support each other, and to encourage others to participate in these projects. LinuxChix,[10] for instance, is a group founded in 1999, to bring together and support women interested in developing or using Linux. Another group, Debian-Women was founded in 2004 to actively engage women in the Debian Project.[11] Feminist hackerspaces have also raised interest in recent years (Savic & Wuschitz, 2018; Toupin, 2014) as sites engaging women in hacker culture and free and open source software development. Many project leaders have also since taken a step back following accusations or recognition of sexist behaviors. The most famous of them is probably Linus Torvalds, the creator of Linux who, in 2017, decided to take a break from his leadership of the project to reflect on his aggressive behavior (Cohen, 2018). Another famous case is Jacob Applebaum who retired from his leadership in the TOR project after allegations of sexual abuse (Greenberg, 2016). More recently, Richard Stallman himself has resigned from his position as president of the Free Software Foundation and from its board of directors, following a much-criticized message he sent to a mailing list concerning underage victims of sex trafficking (Evangelho, 2019).[12] These actions could be interpreted as proof of enduring sexism and machismo, but they could also be signs of growing intolerance towards such behavior and thus point to an improvement in terms of openness for women.

In terms of racial, ethnic, or linguistic equity, however, much less work has been done to study or tackle the subject. First, there seems to be a hegemonic understanding that computer programming is done in English. Apart from some less popular software projects written in other languages (Couture, 2017), the vast majority of free and open source projects are written in English. Concerning racial and ethnic diversity, the previously cited survey conducted by GitHub found that "16 percent of respondents said they belonged to ethnic or national groups that are in the minority in the country they live in [. . .] as compared with 34 percent of programmers in the US" (Finley, 2017; GitHub, 2017). Dunbar-Hester, while praising initiatives advocating for diversity in free and open source software communities, also noted that these initiatives are often restricted to gender and miss the opportunity to tackle diversity more broadly. As she writes, this "serves to perpetuate the marginalized status of poor white women and women of color in technical cultures" (Dunbar-Hester, 2019, p. 93).

6 Conclusion

Most observers today recognize the undeniable ability of the free and open source software model to produce mature and stable technologies, many of which presently form a critical part of Internet infrastructures, like mail and web servers, or significant parts of widely adopted devices, such as Android phones. Contrary to the so-called "proprietary" software, free and open source software can be characterized as a commons-based peer-production model, founded on normative and organizational pillars. The normative pillar refers to the ethics of sharing and its surrounding legal arrangements, more specifically "copyleft" licenses that guarantee the right to share source code rather than limiting it. The organizational pillar refers to the "bazaar" form of making software, centered around the assemblage of a multitude of distributed contributions into a whole by the project leaders.

While free and open source software is today strongly backed by commercial interests, it must be recalled that its principles were first developed for ethical reasons, based on the desire to preserve a culture of sharing and self-help in computer programming, as opposed to restrictions of use imposed by commercial software editors. Although motivations to become involved in the development of free and open source software are now quite diverse, an ethical dimension continues to drive many participants who often take part in peripheral causes, such as providing access to government data, the protection of privacy on the Internet, or the right to share cultural works online. The cultures, philosophies, and models of free and open source software development have also inspired similar models of collaborative production or the legal guarantee of the preservation of digital works in the public domain, such as the Creative Commons licenses or the well-known Wikipedia online encyclopedia. However, this model – and especially its ethical component – is in strong tension with commercial appropriation. Free and open source software development has also been criticized for its lack of diversity and its hostility towards women, which can in turn have implications in terms of software quality. In short, while free and open source software has succeeded in being adopted by dominant players in the field and is today used for much of the infrastructure of the Internet, struggles are still ongoing to address the lack of diversity within its contributors and to keep alive the ethical ideas at the origin of the movement.

Notes

1 This chapter is based on previous work of the author on the subject (Couture, 2007, 2013, 2015). Thanks to Geneviève Szczepanik for the grammatical revisions and to the editors for their comments.

2 Beyond this basic characterization of hacking, Coleman insists however on the diversity of hacking genres (Coleman, 2012, p. 18; Coleman & Golub, 2008), the main one being the difference between FOSS hackers, who are oriented towards transparency and collaboration, and the "hacker underground," that is more oriented towards secrecy and spectacle.

3 www.gnu.org/philosophy/free-sw.html (retrieved May 2, 2019).
4 For a more complex and situated analysis of this notion, see Couture (2019).
5 *The Cathedral and the Bazaar* is often understood as a comparison between free and open source software, and proprietary software. However, it is rather Stallman's style of software development that is associated with the cathedral style in this essay.
6 www.linux.com/news/open-source-project-adds-no-military-use-clause-gpl
7 Although Ostrom has most studied "natural commons" (such as water or land), one of her last books with Charlotte Hess addresses knowledge commons, including free and open source software (Hess & Ostrom, 2007).
8 In French : "Sociétés de services en ingénierie informatique" or SSII.
9 It has to be noted that this survey was aimed at "developers" and authors of free and open source software. As I showed elsewhere (Couture, 2019), the way in which we define software development and contributions to source code can have implications for how we value and give more visibility to one type of work over others. In other words, if the survey had been aimed more broadly at "participants" rather than "developers," it could have yielded a higher proportion of female participants.
10 www.linuxchix.org/
11 www.debian.org/women/
12 Stallman's message was sent to the MIT Computer Science and Artificial Intelligence Laboratory (CSAIL) mailing list, and concerned someone who was accused of assaulting an underage girl connected to the Epstein case. In his message, Stallman stated that "*the most plausible scenario is that she [the underage victim] presented herself to him as entirely willing*" which sparked much criticism on the mailing list and elsewhere. While Stallman did resign from his several roles following many requests to do so, he still considers the situation to be a "series of misunderstandings and mischaracterizations of what I have said," and does not recognize any wrongdoing on his part. It has to be noted however that Stallman's statement is actually one of his many documented and questionable acts perceived as sexist, ableist, or legitimatizing child sexual abuse. See for instance https://geekfeminism.wikia.org/wiki/Richard_Stallman (accessed on October 24, 2019).

References

Aigrain, P. (2005). *Cause commune: l'information entre bien commun et propriété*. Paris: Fayard.

Auray, N. (2000). Le savoir en réseaux et l'empreinte inventive. *Revue Alice*, *3*, 78–97. Retrieved from www.multitudes.net/le-savoir-en-reseaux-et-l/

Balter, B. (2015, March 10). Open source license usage. Retrieved from https://github.blog/2015-03-09-open-source-license-usage-on-github-com/

Bauwens, M., Lacomella, F., Mendoza, N., Burke, J., Pinchen, C., Léonard, A., & Mootoosamy, E. (2012). Synthetic overview of the collaborative economy, P2P foundation. Retrieved from http://p2p.coop/files/reports/collaborative-economy-2012.pdf

Benkler, Y. (2003). Freedom in the commons: Towards a political economy of information. *Duke Law Journal*, *52*, 1245–1276.

Benkler, Y. (2006). *The wealth of networks: How social production transforms markets and freedom*. New Haven, CT: Yale University Press.

Benson, T. (2005, March 29). Brazil: Free software's biggest and best Friend. *The New York Times*. Retrieved from www.nytimes.com/2005/03/29/technology/brazil-free-softwares-biggest-and-best-friend.html

Birkinbine, B. (2014). *Incorporating the commons: A political economic analysis of corporate involvement in free and open source software* (PhD, University of Oregon). Retrieved from https://scholarsbank.uoregon.edu/xmlui/handle/1794/18513

Bollier, D. (2011). *Brazil: No longer the "free culture" nation*. David Bollier news and perspectives on the commons. Retrieved from www.bollier.org/brazil-no-longer-free-culture-nation

Boyle, J. (2008). *The public domain: enclosing the commons of the mind*. New Haven, CT: Yale University Press.

Broca, S. (2013). *Utopie du logiciel libre: du bricolage informatique à la réinvention sociale*. Lyon: Le passager clandestin.

Byfield, B. (2006, May 20). FSF reaches out to social activists. *Linux.com*. Retrieved from www.linux.com/news/fsf-reaches-out-social-activists/

Cardon, D. (2005). Innovation by usage. In A. Ambrosi, V. Peugeot, & D. Pimienta (Eds.), *Word matters. Multicultural perspectives on information societies*. Caen: C&F Editions.

Castells, M. (2003). *The Internet galaxy: Reflections on the Internet, business, and society* (1st ed.). Oxford: Oxford University Press.

Cohen, N. (2018, September 19). After years of abusive e-mails, the creator of linux steps aside. *The New Yorker*. Retrieved from www.newyorker.com/science/elements/after-years-of-abusive-e-mails-the-creator-of-linux-steps-aside

Coleman, G. (2004). Indymedia's independence: From activist media to free software. *Planetwork Journal*, *1*(1). Retrieved from https://mediastudiesfa12.files.wordpress.com/2012/12/coleman_indymedias-independence.pdf

Coleman, G. (2009). CODE IS SPEECH: Legal tinkering, expertise, and protest among free and open source software developers. *Cultural Anthropology*, *24*(3), 420–454. https://doi.org/10.1111/j.1548-1360.2009.01036.x

Coleman, G. (2012). *Coding freedom: The ethics and aesthetics of hacking*. Princeton, NJ: Princeton University Press.

Coleman, G., & Golub, A. (2008). Hacker practice: Moral genres and the cultural articulation of liberalism. *Anthropological Theory*, *8*(3), 255–277. https://doi.org/10.1177/1463499608093814

Coris, M. (2006). Chronique d'une absorption par la sphère marchande: les sociétés de services en logiciels libres. *Gérer & Comprendre*, *84*, 12–24.

Coris, M. (2009). Free software on the market-side: The failure-story of free software services companies in France. *Economics of Innovation and New Technology*, *18*(6), 547–564. https://doi.org/10.1080/10438590802231556

Couture, S. (2007). *Logiciel libre, activité technique et engagement politique: la construction du projet GNU en Argentine* (Master's dissertation, Université du Québec à Montréal). Retrieved from https://archipel.uqam.ca/3301/1/M9669.pdf

Couture, S. (2013). *Logiciels libres: réduction des coûts et souveraineté numérique*. Institut de recherche et d'informations socio-économiques. Retrieved from https://iris-recherche.qc.ca/publications/logicielslibres

Couture, S. (2015). Le contrôle des communs numériques à des fins commerciales: le cas des logiciels libres. *Éthique Publique*, *17*(2). doi:10.4000/ethiquepublique.2275

Couture, S. (2017). Values and configuration of users in the design of software source code. *International Journal of Communication*, *11*, 1112–1132. Retrieved from http://ijoc.org/index.php/ijoc/article/view/4179

Couture, S. (2019). The ambiguous boundaries of computer source code and some of its political consequences. In J. Vertesi & D. Ribes (Eds.), *digitalSTS: A field guide for science and technology studies* (pp. 136–156). Princeton, NJ: Princeton University Press.

Couture, S., & Toupin, S. (2019). What does the notion of "sovereignty" mean when referring to the digital? *New Media & Society*, *21*(10), 2305–2322.

Dafermos, G., & Söderberg, J. (2009). The hacker movement as a continuation of labour struggle. *Capital & Class*, *33*(1), 53–73. https://doi.org/10.1177/030981680909700104

Demazière, D., Horn, F., & Zune, M. (2007). The functioning of a free software community: Entanglement of three regulation modes – control, autonomous and distributed. *Science Studies*, *20*(2), 34–54.

Dunbar-Hester, C. (2019). If "diversity" is the answer, what is the question? Understanding diversity advocacy in voluntaristic technology projects. In J. Vertesi & D. Ribes (Eds.), *digitalSTS: A field guide for science and technology studies* (pp. 81–98). Princeton, NJ: Princeton University Press.

Evangelho, J. (2019, September 17). Richard Stallman has resigned as president of the Free Software Foundation. *Forbes*. Retrieved from www.forbes.com/sites/jasonevangelho/2019/09/17/richard-stallman-has-resigned-as-president-of-the-free-software-foundation/

Finley, K. (2017, June 2). Diversity in open source is even worse than in tech overall. *Wired*. Retrieved from www.wired.com/2017/06/diversity-open-source-even-worse-tech-overall/

Gates, B. (1976). Bill Gates – an open letter to hobbyists. Retrieved from www.adp-gmbh.ch/personal/histoire/lettergates.html

Ghosh, R. A. (2006). *Economic impact of floss on innovation and competitiveness of the EU ICT sector*. Maastricht: UNU-MERIT. Retrieved from https://la-rem.com/wp-content/uploads/2013/11/2006-11-20-flossimpact_en.pdf

Ghosh, R. A., Glott, R., Krieger, B., & Robles, G. (2002). *Free/libre and open source software: Survey and study*. Retrieved from www.math.unipd.it/~bellio/FLOSS%20Final%20Report%20-%20Part%204%20-%20Survey%20of%20Developers.pdf

GitHub. (2017). *Open source survey*. Retrieved from https://opensourcesurvey.org/2017/#about

Gleizes J., & Papathéodou, A. (2000). La passion du libre entretien avec Richard Stallman (interview by J. Gleizes et A. Papathéodou). *Multitudes*, *1*(1), 377–391.

Greenberg, A. (2016, June 6). Tor developer Jacob Appelbaum resigns amid sex abuse claims. *Wired*. Retrieved from www.wired.com/2016/06/tor-developer-jacob-appelbaum-resigns-amid-sex-abuse-claims/

Haché, A. (2017). Technological sovereignty. *Mouvements*, *79*(3), 38–48. www.cairn-int.info/revue-mouvements-2014-3-p-38.htm

Hess, C., & Ostrom, E. (2007). *Understanding knowledge as a commons: From theory to Practice*. Cambridge, MA: MIT Press.

Himanen, P. (2001). *The hacker ethic and the spirit of the information age*. New York, NY: Random House.
Jullien, N., & Zimmerman, J.-B. (2007). Peut-on envisager une écologie du logiciel libre favorable aux nuls? *Cahier de recherche, môle armoricain de recherche sur la société de l'information et les usages d'internet (MARSOIN)*, (9). https://bit.ly/31Faadi
Juris, J., Caruso, G., Couture, S., & Mosca, L. (2013). The cultural politics of free software and technology within the social forum process. In J. Juris & A. Khasnabish (Eds.), *Insurgent encounters: Transnational activism, ethnography, and the political* (pp. 342–366). Durham, NC: Duke University Press.
Karjalainen, M. (2010). *Large-scale migration to an open source office suite: An innovation adoption study in Finland* (PhD, Tampere University). Retrieved from http://tampub.uta.fi/handle/10024/66646
Lakhani, K., & Wolf, R. G. (2003). Why hackers do what they do: Understanding motivation and effort in free/open source software projects. *SSRN Electronic Journal*. https://doi.org/10.2139/ssrn.443040
Lerner, J., & Tirole, J. (2000). The simple economics of open source. *Journal of Industrial Economics, 52*, 197–234. Retrieved from www.nber.org/papers/w7600
Levy, S. (1985). *Hackers: Heroes of the computer revolution*. New York, NY: Dell.
Lewis, J. A. (2010). Government open source policies. *Center for Strategic and International Studies*. Retrieved from http://csisdev.forumone.com/files/publication/100416_Open_Source_Policies.pdf
Lin, Y. (2005). The future of sociology of FLOSS. *First Monday*. https://doi.org/10.5210/fm.v0i0.1467
Moglen, E. (1999). Anarchism triumphant: Free software and the death of copyright. *First Monday, 4*(8).
Moulier-Boutang, Y. M. (2007). *Le capitalisme cognitif: la nouvelle grande transformation*. Paris: Editions Amsterdam.
Nafus, D., Leach, J., & Krieger, B. (2006a). Gender: Integrated report of findings. *Free/libre and open source software: Policy support*. Retrieved from https://bit.ly/2odFVMK
Nafus, D., Leach, J., & Krieger, B. (2006b). Gender: Policy recommendations. *Free/libre and open source software: Policy support*. Retrieved from https://bit.ly/2pX2G87
Oram, A. (2011). Promoting open source software in government: The challenges of motivation and follow-through. *Journal of Information Technology & Politics, 8*(3), 240–252. https://doi.org/10.1080/19331681.2011.592059
Pellerin, F. (2013, June 15). Le discours de Fleur Pellerin dans les nouveaux locaux Mozilla à Paris. *Framablog*. Retrieved from https://framablog.org/2013/06/15/fleur-pellerin-logiciel-libre-mozilla/
Raymond, E. S. (2000). A brief history of hackerdom. Retrieved from www.catb.org/~esr/writings/cathedral-bazaar/hacker-history/
Raymond, E. S. (2001). *The cathedral and the bazaar*. Sebastopol, CA: O'Reilly Media.
Reina, L. A., Langasek, P., & Ruiz, M. (2013). Women in Debian 2013. DebConf13, Vaumarcus, Switzerland. Retrieved from http://penta.debconf.org/dc13_schedule/attachments/243_women_in_debian_2013.pdf
Riemens, P. (2006). Some thoughts on the idea of "hacker culture." *Complément Numérique de La Revue Multitude, 8*. Retrieved from www.multitudes.net/some-thoughts-on-the-idea-of/
Savic, S., & Wuschitz, S. (2018). Feminist hackerspace as a place of infrastructure production. *Ada, 13*. Retrieved from https://adanewmedia.org/2018/05/issue13-savic-wuschitz/
Schweik, C. M. (2007). Free/open-source software as a framework for establishing commons in science. In C. Hess & E. Ostrom (Eds.), *Understanding knowledge as a commons: From theory to practice* (pp. 277–309). Cambridge, MA: MIT Press.
Stallman, R., Lessig, L., & Gay, J. (2002). *Free software, free Society: Selected essays of Richard M. Stallman*. Boston, MA: Gnu Press Free Software Foundation.
Star, S. L. (1999). The ethnography of infrastructure. *American Behavioral Scientist, 43*(3), 377–391. https://doi.org/10.1177/00027649921955326
Terranova, T. (2000). Free labor: Producing culture for the digital economy. *Social Text, 18*(2), 33–58.
Toupin, S. (2014). Feminist hackerspaces: The synthesis of feminist and hacker cultures. *Journal of Peer Production, 10*. Retrieved from http://peerproduction.net/editsuite/issues/issue-5-shared-machine-shops/peer-reviewed-articles/feminist-hackerspaces-the-synthesis-of-feminist-and-hacker-cultures/
Tuomi, I. (2004). Evolution of the Linux credits file: Methodological challenges and reference data for open source research. *First Monday, 9*(6). Retrieved from http://firstmonday.org/issues/issue9_6/tuomi/index.html
von Hippel, E. (2005). *Democratizing innovation*. Cambridge, MA: MIT Press.
von Hippel, E. (2007). Horizontal innovation networks – by and for users. *Industrial and Corporate Change, 16*(2), 293–315. Retrieved from http://icc.oxfordjournals.org/content/16/2/293.short
Williams, S. (2002). *Free as in freedom: Richard Stallman's crusade for free software* (1st ed.). Sebastopol, CA: O'Reilly Media.

13

Wikipedia and Wikis

Jutta Haider and Olof Sundin

1 Introduction

WikiWikiWebs or wikis constitute the core platform of peer production. Wikis are software programs allowing for flexible collaboration without necessarily having a defined content owner or leader. A wiki is a user-editable website or content management system. Wikis might use different programming languages and licenses, but they apply the same model for cooperation, which means that collaborators can modify content, insert hyperlinks, and change the structure of a document directly in any web browser. Edits are usually archived and open to revision for all collaborators. The most popular and successful wiki-based project by far is Wikipedia. Established in 2001, today this encyclopedia is one of the most popular sites on the Web; what is more, with its data supporting other applications and commercial platforms like Google, Wikipedia has taken on an infrastructural role in the contemporary Internet. This chapter lays the groundwork for a basic technical understanding of wikis and Wikipedia, sometimes with reference to Wikipedia itself, but also to reviews of research from a number of different disciplines, including (but not limited to) computer science, media and communication studies, and information studies. It delineates their most central design principles, history, and affordances.

Wikis are described as content management systems that allow for flexible collaboration without a defined content owner or leader. Users can modify the content and structure of documents directly in their web browser. Edits are usually archived and open to revision. This chapter pays particular attention to the most successful wiki-based system, the non-profit, online encyclopedia Wikipedia. As the chapter explains, Wikipedia actually contains a broad range of more or less individual wiki projects and has inspired a plethora of other endeavors, both open source and proprietary. The particular peer-production model employed in Wikipedia is elucidated, and in the course also complicated and critically assessed. The chapter concludes by highlighting a number of tensions emerging from a wiki-based peer-production model, between amateurs and experts, human editors and bots, lay knowledge and academic knowledge and the shaping of trust through external actors.

There is a substantial amount of literature on Wikipedia (for research overviews, see Jullien, 2012; Mesgari et al., 2015; Okoli et al., 2012, 2014) and therefore this chapter will only review selected highlights. Furthermore, the focus is on wikis and Wikipedia in the context of peer production; accordingly, it leaves out many other aspects of Wikipedia research (such as studies on the quality of content *per se* or on Wikipedia usage in different settings) except for when those aspects can be related to peer production. It should also be noted that the academic interest in wikis and Wikipedia peaked around 2015, after about a decade of intense research activity, and has since then has decreased, as Figure 13.1 illustrates.

This diminishment of scholarly interest might result from the fact that wikis, and especially Wikipedia, are nowadays taken for granted and their mode of peer production is no longer considered noteworthy. Wikipedia itself has an article for research on Wikipedia with links to ethical guidelines (Research, 2019, March 24), and some of its administrators also engage in research on and development of Wikipedia or other wikis. The foundation behind Wikipedia, Wikimedia, has developed a specific wiki system, the MediaWiki, that operates a number of other wikis based on this software and also employs a research team that works to improve their various wikis (Wikimedia Research, n. d.).

The Handbook of Peer Production, First Edition. Edited by Mathieu O'Neil, Christian Pentzold, and Sophie Toupin.

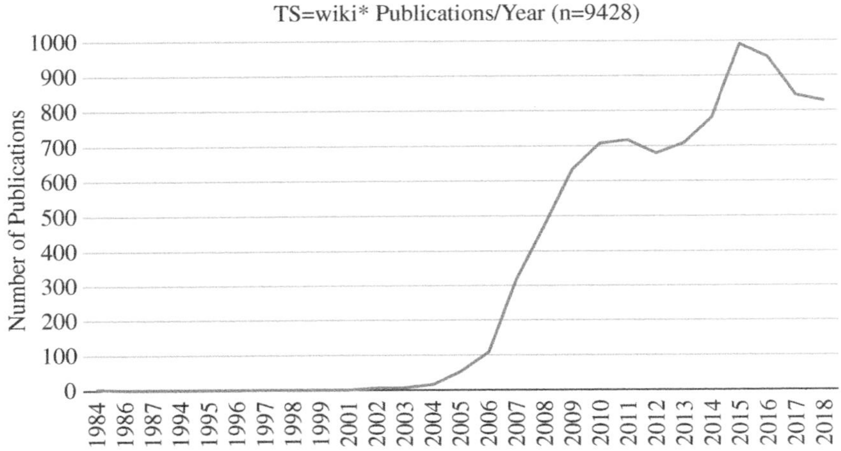

Figure 13.1 Articles in World of Science database 1994–2018 with wiki* in the article title.

2 The Wiki Platform

The initial development of wikis can be traced back to 1994 when American programmer Ward Cunningham started developing a software intended to simplify collaboration during programming. This software came to be the first wiki (WikiWikiWeb). The name *wiki*, coined by Cunningham in the 1990s, comes from Hawaiian and means "quick." Since its inception, countless wikis have been set up for many different purposes. Many of these are open for anybody to contribute to, although the subjects addressed in many cases are very specialized, which limits public interest. Other wikis, such as those that have been set up to support cooperation within an organization, require registration and logging in. Different wikis have different designs and different rules for engagement, but the overall principles, guided by the wiki software, are similar in all of them.

But what exactly is a wiki? A wiki is a software (sometimes referred to as an application), a website, or a platform that allows users in different locations to collaborate on certain specific content (that is, to discuss, write, edit, link, delete, search, create new pages, and reorganize) from within an ordinary web browser. A wiki can also be described as a content management system in which users of each wiki can define their own structures and organization, and invent and enforce their own rules for user participation. Users find information they want to access through browsing and following hyperlinks or, in some wikis, through a dedicated search engine (Wiki, 2019, January 19) and, increasingly (as in the case of Wikipedia), through Google.

Typically, a wiki page has four access points: (1) the *page* itself, which contains the published information on the topic in question, (2) a *talk page* where editors can discuss the content of the page, (3) an *edit page* on which editors can make their contributions, and (4) a *history page* on which it is possible to see all changes or edits applied to a specific page. Any change to a page is made with an easy-to-use tool (such as mark-up language or a WYSIWYG interface), and the change automatically creates a new version in the history of the page. As a result, it is always possible to re-create any earlier version – to roll back time, metaphorically speaking. A page that has 100 edits can be thought of as existing simultaneously in 100 layers, where each layer is always accessible.

In addition to the particular content that a wiki is dedicated to, there are community pages that usually include other information and instructions about the wiki, such as rules and guidelines for user contributions, and other meta-information. These pages can usually also be changed in exactly the same way as the content pages. Registered, signed-in editors can create user pages where they can present themselves for other editors or readers. History pages show which users have contributed to a page, what their contributions were, and when those contributions were made. Taken together, history pages and talk pages can thus provide glimpses into disagreements and discussions between editors, and can show how a topic solidified into a certain factual representation. Usually these pages are rather dry and bureaucratic, but occasionally they record a history of veritable editing wars. Either way, as Pentzold (2009) suggests, when discussing Wikipedia as a "global memory place," these pages constitute residues of cultural memory in their own right, specific to the particular peer-production model employed (see also Gustafsson, 2019).

In addition to Wikipedia, there are numerous other projects built on the wiki software. Most public ones are also encyclopedic in nature and they are usually about very particular subjects, ranging from gardening, traveling, and programming to news, Internet memes, sharia law, lyrics, and pornography. Unsurprisingly, Wikipedia has a comprehensive list of popular public wikis, including projects such as Heroes Wiki, LyricWiki, Citizendium, WikiTribune, and numerous others (List of Wikis, 2019, March 31).

Wikis are grounded in the open source community and this origin is reflected in a wiki's information architecture and in its technological affordances for collaboration. Its technical features convey certain ways of how users can (and cannot) work together and these are derived from hidden assumptions and values stemming from the open source community. Accordingly, most public wikis use creative commons (CC) licenses for their content. Some, however, are more restrictive and even have their content protected by copyright, such as for instance the Chinese encyclopedia Baidu Baike, which is owned by the Chinese search engine giant Baidu. Assumptions of openness are built into wiki software, which can limit the usefulness of wikis for knowledge work in other contexts. For instance, in a study on adoption of wikis in the corporate setting, Yeo and Arazi (2012) demonstrated the lack of directly visible attribution in the edited text (which is otherwise a clear advantage in an encyclopedia or a software manual) conflicted with certain corporate norms, for instance with the norm of personal career advancement. Lack of attribution could also hinder the adoption of wikis in certain educational situations where grading, guidance, and formative feedback from authority figures are valued and need to be considered.

3 Wikipedia: Background and History

The best-known wiki by far is Wikipedia. According to the Web traffic analysis company Alexa Internet (The top 500 sites on the Web, n. d.), the English-language version of Wikipedia is, at the time of writing, the fifth most popular site on the Web. It is safe to say that the printed encyclopedias of the past, although well-known and much used, never came close to this level of general usage. Wikipedia's front page greets the user with the following words – "Welcome to Wikipedia, the free encyclopedia that anyone can edit" – in the upper left corner. Immediately after the greeting, the number of articles that Wikipedia currently contains is mentioned. At the time of writing, in September 2020, that number for the English-language Wikipedia is 6,161,546 (ed. note: number updated). The front page thus highlights some of Wikipedia's key characteristics: it is based on a wiki; it is free to use; it is an encyclopedia that anyone can edit, and its size is enormous.

Wikipedia stands in a tradition of thought grounded in the Enlightenment ideal, emphasizing rationality and the benefits of universal access to knowledge (Haider & Sundin, 2010). In line with these ideals, Jimmy Wales, (co-)founder of Wikipedia, once summarized the utopian goal of Wikipedia: "Imagine a world in which every single human being can freely share in the sum of all knowledge" (Roblimo, 2004, July 28). As such, Wikipedia is only one project out of many attempts to summarize, organize, and make accessible all possible knowledge, akin to historical projects such as Denis Diderot and Jean le Rond d'Alembert's *Encyclopédie*, Paul Otlet's *Encyclopedia Universal Mundial*, Gottfried Wilhelm von Leibniz's *Characteristica Universalis*, and Otto Neurath's *International Encyclopedia of the Unified Sciences* (Haider & Sundin, 2010). Historically, it was assumed that this overarching aim had to rely on experts. Wikipedia's peer-production model with its digital platform has certainly revolutionized both the production and distribution of encyclopedic content. At the same time, Wikipedia also continues the long tradition of encyclopedism and as Loveland and Reagle emphasize, "it is important not to exaggerate the discontinuities between Wikipedia and its predecessors" (2013, p. 1305).

Wikipedia started in 2001 as a side project to another encyclopedic project, Nupedia (1999–2003), which was funded by the company Bomis. The two encyclopedic projects Wikipedia and Nupedia were founded by Jimmy Wales and Larry Sanger, and both were built on voluntary contributors. Nupedia leaned on traditional peer review, while Wikipedia used a wiki-based open editor system from the start (History of Wikipedia, 2019, March 19). Wales and Sanger ended their collaboration because they had differing views on the requirements for contribution; Wales, who had argued for an open platform, continued with Wikipedia. In 2007, Sanger started a wiki called Citizendium, which, like Nupedia, also uses a more traditional peer-review model to vet information. Wikipedia grew quickly and in its first year got substantial media coverage by high-profile mainstream outlets like *The New York Times* as well as in more specialized online communities, including for instance the social news website *Slashdot* (History of Wikipedia, 2019, March 19.). A number of additional language versions of Wikipedia were set up almost immediately, including German, French, Catalan, Japanese, Chinese, Dutch, Hebrew, Italian, Portuguese, Russian, Spanish, Swedish, Polish, Afrikaans, Norwegian, and Esperanto in the same year that Wikipedia was founded (History of Wikipedia, 2019, March 19.).

The presence of volunteer editors makes Wikipedia possible, but Wikipedia also requires physical infrastructure, which in turn mandates direct financial backing. Since 2003, these needs have been met by donations to the non-profit organization Wikimedia Foundation. According to Wikimedia, "[t]he mission of the Wikimedia Foundation is to empower and engage people around the world to collect and develop educational content under a free license or in the public domain, and to disseminate it effectively and globally" (Mission Statement, 2007, September 1). In other words, Wikimedia provides the infrastructure (including servers and software development) for a number of wiki projects, such as Wikiquote, Wikinews, Wiktionary, and Wikipedia (Wikimedia Foundation, 2019, March 28). The different projects are developed using the MediaWiki software, which is available under the GNU General Public License.

4 Wikipedia and its Contributors

Wikipedia is "peer production *par excellence*" (Pentzold, 2018, p. 455). It is difficult to conceive of another decentralized peer production effort of this magnitude in terms of number of collaborators and amount of output. In early 2019, the English-language Wikipedia alone had 35,488,240 registered users (Statistics, 2019, March 27). Obviously, not all registered users are equally active in Wikipedia's production; rather, Wikipedia is dependent on a considerably smaller number of very prolific editors. For instance, in February 2019, just 3,445 editors had performed more than 100 edits per month. Of these very active editors, 312 were actually bots (Statistics, 2019, March 27) (for more about bots and Wikipedia, see Section 5). There appears to be a friction between, on the one hand, the ideal of "the free encyclopedia that anyone can edit" and, on the other hand, a reality where the project depends on a rather limited number of so-called elite editors. However, as Kittur and colleagues (2007a) showed in their study on the development of the early dominance of elite editors, there has been an increase in "low-edit users" (p. 7). Two years later, Suh and colleagues (2009) also identified a decline in the numbers of editors. Halfaker and colleagues (2013, p. 683) discussed the reasons for this decline, concluding:

> Wikipedia has changed from the encyclopedia that anyone can edit to the encyclopedia that anyone who understands the norms, socializes himself or herself, dodges the impersonal wall of semi-automated rejection, and still wants to voluntarily contribute his or her time and energy can edit.

This somewhat polemic statement chimes well with the work of Shaw and Hill (2014) in which they show that often when wikis (not just Wikipedia) grow in size and become more complex, an "organizational oligarchy" forms whereby original participants tend to control a wiki's further development at the expense of a strict democratic model.

Some have found a correlation between the number of edits and distinct editors and the quality of articles. Wilkinson and Huberman (2007) assert that Wikipedia entries with more contributing editors and more edits are of higher quality than those with fewer editors. Yet other research suggests that the different ways that editors coordinate (or do not coordinate) collaborations among themselves need to be considered to fully understand this connection. Kittur and Kraut (2008) distinguished two different ways that editors coordinate writing, namely implicit and explicit coordination. ***Implicit coordination*** occurs when a limited number of editors carry out the majority of the work, and most other editors only participate in a limited way. ***Explicit coordination*** describes situations with visible divisions of labor, where negotiations over content and planning of work occur on the talk page of the respective article. Kittur and Kraut (2008, no page numbers) conclude that "implicit coordination was especially valuable for articles and time periods with many contributions, while explicit coordination was especially valuable for articles and time periods with few contributors." The way that talk pages function as an arena for planning and coordinating the writing of an article has been elucidated by Viégas and colleagues (2007), who visualize the talk pages' functions for socializing users by acclimating them to the policies and guidelines of Wikipedia. They argue that meta-activities, such as writing on talk pages, increase in significance as Wikipedia becomes larger and more structurally complex. In other words, while the number of distinct editors and edits might be indicative of quality to a degree, it is important to also allow for the role of coordination.

The incentives for editors to contribute to Wikipedia have been likened to those thought to be at work in science, that is, an interest in contributing to the creation of knowledge – of "true facts" as for instance Forte and Bruckman (2005, no page number) argue. Clearly, the reward for most voluntary, human Wikipedia editors is recognition rather than financial rewards (van Dijck 2013, p. 148), in contrast to commercial platforms built on user contributions. Different editors have different levels of authorization regarding their interaction with Wikipedia. In the English-language Wikipedia, users are divided into numerous categories, authorization types, and user access levels, such as administrators or bureaucrats, unregistered, registered, confirmed or extended confirmed users, founders or researchers, rollbackers, various types of patrollers or mass message senders, to name just a few (User access levels, 2019, July 30). Clearly, the types are not strictly hierarchical and a single user often fits into several categories. Of course, a large number of non-registered users (who nevertheless can easily be identified by their IP numbers) contribute to Wikipedia. Their editorial permissions differ slightly in different language versions. Additionally, there is yet one more user category that often remains unmentioned: the blocked user. This user, as for instance Niederer and van Dijck (2010) discuss, has the least authorization to contribute to Wikipedia. Moreover, different pages have different rules for editing: usually anyone (except blocked users) can contribute, but on what terms, in what way exactly, and with what authority differs considerably.

The number of contributor categories and the rather hierarchical permissions system that comes with those categories make clear that Wikipedia is a socially complex, tightly controlled system thoroughly infused with bureaucracy, quite different from the straightforward "anyone can edit" rhetoric often associated with it (van Dijck, 2013, pp.145–147; see also Tkacz, 2014). As Ford and Wajcman (2017, p. 522) show, "being a Wikipedian . . . involves acquiring particular forms of sociotechnical expertise and authority that constitute the knowledge or epistemological infrastructure of Wikipedia. Beneath the rhetoric of amateurism," they write "a new form of expertise, and hence power, is being constituted, but that it is once again gender-coded as male." This male gender-coding also has to be seen in relation to the community of human Wikipedia editors being largely male, something which is also reflected in the choice and representation of topics in Wikipedia (Ford & Wajcman, 2017). This resonates with Hargittai and Shaw's (2015) work on the Wikipedia gender gap and biased access. They moreover show that, besides gender, a variety of other aspects, including age, educational background, Internet skills, employment status, income, or race/ethnic identity, affect who participates in the Wikipedia project as content producers, but also as readers (Shaw & Hargittai, 2018). The various inequalities in Wikipedia have also been described as a "politics of exclusion" (Graham, 2012). It contributes to entrenching the further marginalization of indigenous epistemologies and knowledge systems (e.g., van der Elden, 2012) as well as of content and authors from non-Western countries (e.g., Ford, 2012; John, 2012; Luyt, 2018).

As mentioned previously, working with the content of articles is not the only work that editors engage in (Bryant, Forte, & Bruckman, 2005; Littlejohn & Hood, 2018). A core feature of Wikipedia is that all articles (even internal Wikipedia policies) and user pages have talk pages. That is, each article about a topic has a talk page on which the contributing editors discuss the content of that specific article, and how to collaborate and improve it. It has been suggested that this talk page function is one of the keys for getting such a large encyclopedic project to work despite the seemingly leaderless structure and the high number of number of editors (Viégas et al., 2007). There is also a community portal that includes something called a *Village Pump*, where editors can discuss all kinds of questions, projects of various kinds, dispute resolutions regarding article content, and so on (Community portal, 2019, March 27). For the purpose of assessing the credibility of specific articles, the back-end activities and discussions are in many ways just as interesting as the front-end encyclopedic articles, answering questions such as who has contributed, has there been there a discussion about the content in the talk page, has the article received "featured article" status, and so forth (Sundin & Francke, 2009). Kittur and colleagues (2007b) noted that almost 40% of all edits concern contributions other than to front-end encyclopedia entries, and they distinguish between direct work and indirect work, where the former relates to working with the content of articles and the latter relates to all the other kinds of meta-activities in Wikipedia. Kittur and colleagues argued that the direct work on articles decreased due to what they called "increasing coordination costs" (no page number), which results from the sheer size of the Wikipedia project.

In a similar line of thought, O'Neil (2010) identifies a range of further costs involved in crowdsourcing knowledge production by ridding it of editorial oversight and specifically in crowdsourcing Wikipedia. He distinguishes between costs, "directly affecting the quality of the product" on the one hand, more precisely uncertainty, lack of perspective, as well as irresponsibility, and indirect costs on the other hand. Most notably the latter derive from editorial conflicts and edit wars, "which divert resources from the task of building an encyclopaedia" (O'Neil, 2010, p. 3).

5 Wikipedia and Its Content

When Wikipedia content is discussed, the focus is most often on issues of quality and the various facets of quality, including comprehensiveness, timeliness, readability and style, and reliability (Mesgari et al., 2015). Of these, reliability, understood in terms of the accuracy of the content presented as factual, has probably attracted most attention, both in terms of scholarly consideration and attention in the public debate. When Wikipedia first became popular, it did not just pose a challenge to traditional encyclopedias by threatening their financial models based on subscriptions or sales – it also challenged their very production model. Wikipedia is often said to have profoundly challenged and changed the role of experts, specifically the role of named authors, who adorned with certain credentials, are entitled to express themselves as experts (e.g., Weinberger, 2007). We will nuance this picture in the upcoming sections, but it is safe to say that Wikipedia has contributed considerably to the success of a novel way of producing and reviewing the content of codified knowledge.

The success of this new model quickly led to discussions of Wikipedia's reliability and accuracy, and content comparisons between Wikipedia and traditional encyclopedias started to proliferate. In a much-discussed article (Giles, 2005), the quality of Wikipedia was assessed by letting experts compare 42 articles with the corresponding articles in the Encyclopaedia Britannica. The author found the accuracy of the two encyclopedias to be more or less equivalent (see also Chesney, 2006). In addition, within a short time Wikipedia became considerably more comprehensive than traditional encyclopedias (at least the English-language version), continuously expanding and adding new articles. Moreover, Wikipedia had a clear advantage regarding how quickly errors could be corrected. The question of quality is transparent in the sense that it is possible to follow discussions about quality while those discussions are happening among the editors (Stvilia et al., 2008).

Another way of assessing quality in Wikipedia has been by investigating references. The need to verify knowledge claims by backing them up with references to external sources, *verifiability*, is one of the key policies (Huvila, 2010; Nielsen, 2007; Sundin, 2011). By referring to sources outside Wikipedia, the credibility of a certain article comes to depend on other, often scholarly, publications rather than peer production as such. As in academic knowledge production, the higher the status of the cited sources, the stronger the knowledge claim becomes. Besides, roughly one article per thousand in the English-language Wikipedia is labeled as a *featured article*, that is, it is an article of particularly high quality (Featured articles, 2019, April 1). The decision concerning which articles should be included amongst featured articles is achieved by a community vote. However, there are a number of criteria that must be met before an article is considered for feature article status, one of which is whether the article is properly researched and refers to external sources. Voting is also the practice for deciding whether an article should be deleted and other important decisions, like policy decisions. Citing practices and verifiability of content is discussed further in section 7, which concerns Wikipedia and the role of experts

Wikipedia content is not only produced by humans, but also by bots and other non-human actors, such as policies, templates, menus or other navigational items, and information architecture (Geiger & Ribes, 2010; Niederer & van Dijck, 2010). A bot is a computer program that functions as "an automated tool that carries out repetitive and mundane tasks to maintain the 46,939,344 pages of the English Wikipedia" (Bots, 2018, December 29). Niederer and van Dijck (2010, p. 1369) consider Wikipedia a socio-technical system in which humans and technology work together: "It is the intricate collaboration between large numbers of human users and sophisticated automated systems that defines Wikipedia's ultimate success as a knowledge instrument." Some of the automated tasks are blocking spam, detecting vandalism, spell-checking, data-mining for detecting copyright violation (Niederer and van Dijck, 2010, p. 1378; see also Geiger & Ribes, 2010). An example of a bot is the MediaWiki extension WikiOpener which automatically queries external databases (Brohée et al., 2010). In fact, several thousand bots are active on Wikipedia, and these bots have user pages just like human editors do. In total, in early 2019, bots had created more than 16 million articles (across all language versions). So far, the most productive bot has been one called Lsjbot. Lsjbot was particularly active on the Swedish, the Cebuano Wikipedia, and the Waray Wikipedia where it created millions of articles and made tens of millions of edits (Bot article only, 2019, January 31). Lsjbot's impressive productivity is also the reason that the Cebuano Wikipedia ranks number two and the Swedish Wikipedia ranks as number three in total number of articles among all languages, ranked higher than for instance Spanish-language, Japanese, Chinese, or Arabic Wikipedias, all of which are much more widely spoken languages. Lsjbot has started out by writing short articles about animals and plants that provide references in accordance with Wikipedia's citation rules (Guldbrandsen, 2013), but has since also contributed articles in other areas.

At the core of Wikipedia is the question of how the quality of content can be vetted and guaranteed. Above we discussed the various policies and guidelines providing the frame for producing articles of good quality. Another important aspect is how the peer-review process, or more accurately post-publication-review process, is carried out. Notably, these processes differ somewhat in Wikipedia's different language versions. The original principle for contribution to wikis and thus also to Wikipedia was that anyone could edit without being a registered user, and that all edits are visible to all users as soon as they are saved. When an already existing article is edited, those changes are immediately visible for other users. Furthermore, bots as non-human contributors constantly examine changes to find and root out instances of vandalism.

Originally, Wikipedia's reviewing principle builds entirely on a review process carried out after publication. However, over the years exceptions to this have emerged, most notably in the German-language Wikipedia. In order to come to grips with vandalism, that is purposeful destruction of Wikipedia content, and other types of misuse, such as for political or commercial purposes (Osman, 2014), and to raise the level of quality, an additional layer of peer-review or editorial control has been tested, first in the German-language Wikipedia and thereafter also in the English-language one. In the German instance, edits are marked for editors to review through the MediaWiki extension "FlaggedRevs." After a test period accompanied by intense debate in the English-language Wikipedia community, flagged protection for certain articles (for instance, articles about living persons) has been introduced when they are written by very new and very inactive users. The issue has been discussed frequently within the community (Pending changes, 2019, January 8). In the German-language Wikipedia, a similar peer-review system (Gesichtete Versionen, 2019, March 19) has been in use since 2008, but for all edits. The system's primary purpose is to guard the encyclopedia against vandalism. Editors with the right to review a change must have passive or active reviewer status. This status is assigned automatically when a user fulfills certain sets of criteria, mostly expressed in numbers of edits during certain periods. There is also the possibility of manually assigning and reversing reviewer status. In addition to different types of reviewing rights the German-language Wikipedia has also introduced a voting system. And while even bots can have the status of reviewers, at least passive ones, they are not entitled to vote. In the German-language Wikipedia, having the right to vote requires that you have been a registered user for two months, that you have made 200 edits in one namespace, and that you are a human editor. Even here a hierarchy is in place. There are two different types of voting power, a general right to vote, and the more restricted right to vote on arbitration committees. For the latter, stricter criteria are applied.

6 The Wikipedia Community

Wikipedia has been studied not only as an encyclopedia with advantages and shortcomings, but also as a culture in its own right (e.g., Konieczny, 2009; Reagle, 2011; Sundin, 2011). Notably, Reagle (2011, p. 3) includes "both the textual artifact and the community producing it" into his conception of what constitutes Wikipedia, and he talks about its particular collaborative approach as an expression of "good faith culture." Similarly, Pentzold (2011, p. 458), in an attempt to capture the core idea of collaboration in Wikipedia, describes its working methods as "a form of collectivism." The Wikipedia community, or more broadly the Wikimedia community, builds on notions of transparency and egalitarianism as advanced by the open source philosophy, and the Wikimedia community's existence is dependent on the work of collaborating volunteers acting as editors.

Wikipedia is intimately connected to a notion of *open* specifically as developed within the open source movement and ideas connected to free software development. Knowledge, in this context, is considered "open if anyone is free to access, use, modify, and share it – subject, at most, to measures that preserve provenance and openness" (Open Knowledge Foundation, n.d.). The project needs to be understood within the context of a proliferation of different notions of "open" and thus as advancing a "politics of openness" as Tkacz (2014) elucidates in much detail, which is conditioned and curtailed by the circumstances the current economic system creates. That is, despite its conceptual alliances with "transparency, collaboration, competition and participation, and its close ties with various enactments of liberalism" (Tkacz, 2012, p. 399), the notion of openness gives way to closures that remain largely invisible. This way openness itself transmutes into a near-irrefutable, catch-all concept obstructing the view towards the diverseness of issues it is applied to and the various power relations shaping those issues.

Connected to this, Wikipedia has been described as promoting an "uneven openness" (Graham & Hogan, 2014), which applies to the topics represented in Wikipedia, to the ways in which certain topics are framed and represented (Luyt, 2019), and also to who has access to the encyclopedia. Between 2012 and 2018, the Wikimedia Foundation ran a project called Wikipedia Zero, which through subsidized mobile plans enabled users in developing economies to get free access to Wikipedia (Gebhart, 2016). Being in breach of the tenet of net neutrality, this practice was described as colonial (Toupin, 2019). However, it was only discontinued after users in the Global South creatively re-appropriated the infrastructure provided by Wikipedia Zero to share copyright protected files (Toupin, 2019).

Wikipedia is more than an encyclopedia, i.e., a textual artifact or database. It is also a site for people to meet, collaborate, discuss, and debate. These meetings are not constrained to meetings on the platform – active editors sometimes also meet in person in local or global events, such as most notably the yearly Wikimedia Conference, and also at smaller national conventions. The writing of Wikipedia articles is occasionally made public and performed collectively in the same space and at the same time. *Editathons* (edit-a-thons) have been organized the world over to increase contributions to Wikipedia in certain topics and by certain people who are underrepresented in the regular community of editors. Most notably, the community as it currently stands has been shown to be largely male (Hill & Shaw, 2013), which has led to Wikipedia being called out as representing "the sum of all male knowledge" rather than *all* knowledge as it otherwise claims (Ford & Wajcman, 2017, p. 522). Editathons are often organized by cultural heritage institutions, such as museums or libraries, and occasionally also in tandem with an organization hosting a Wikipedian in residence. A Wikipedian in residence is a further instantiation of Wikipedia intersecting with traditional institutions and physical space; the British Library in 2010 was the first institution to host a person with the dedicated aim of strengthening cooperation between that institution and the Wikipedia community (Wikipedian in residence, 2019, March 27).

Wikipedia is a place for disputes, but also a place that values the idea that disputes can be solved through discussions and democratic processes. Assume Good Faith (AGF) is one of the leading guidelines for Wikipedia, the essence of which is captured in the following quote:

> When disagreement occurs, try to the best of your ability to explain and resolve the problem, not cause more conflict, and so give others the opportunity to reply in kind. Consider whether a dispute stems from different perspectives, and look for ways to reach consensus. (Assume good faith, 2019, March 24)

Despite this ideal, not all disagreements can be easily solved. Sometimes disagreements occur at very high levels between and within communities, which points to profound discrepancies about how openness and transparency are framed and how hierarchies are organized. For instance, a controversy between the community of editors and the community of managers at the Wikimedia Foundation became evident when plans for creating a Wikimedia search engine were revealed in 2016, and the Executive Director of Wikimedia left her position as a result of the conflict (Sentence, 2016, March 3). The conflict concerned the perceived secretive nature of the project, which was regarded as being incompatible with the guiding principle of transparency. That is, there was no open process involving the community. It was furthermore considered "a move away from human-curated and authored knowledge towards automatically generated content," which was seen as undesirable (Sentence, 2016, March 3).

Wikipedia is constantly in motion and none of its rules is written in stone. However, there are policies and guidelines steering peer production, and these are specifically listed on Wikipedia's About page (italics added by the authors):

> *Wikipedia is not*, which summarizes what belongs in Wikipedia and what does not;
> *Neutral point of view*, which describes Wikipedia's mandatory core approach to neutral, unbiased article-writing;
> *No original research*, which prohibits the use of Wikipedia to publish the personal views and original research of editors and defines Wikipedia's role as an encyclopedia of existing recognized knowledge;
> *Verifiability*, which explains that it must be possible for readers to verify all content against credible external sources . . .;
> *Reliable sources*, which explains what factors determine whether a source is acceptable;
> *Citing sources*, which describes the manner of citing sources so that readers can verify content for themselves;
> *Manual of Style*, which offers a style guide – in general editors tend to acquire knowledge of appropriate writing styles and detailed formatting over time.
> (About, 2019, March 6 [italics added by the authors])

Together, these rules constitute the cornerstones of Wikipedia. Different scholars have emphasized different policies and guidelines when they have discussed the uniqueness and success of the collaborative Wikipedia project. Reagle (2011) stressed the collaborative culture of Wikipedia as leaning on the policy of Neutral Point of View (NPOV) and the above discussed guideline Assume Good Faith (AGF).

When transferring the notion of collaboration as it is used in software development, to collaboration when writing about subjects with many perspectives and even divergent knowledge claims, another take on how to reach consensus is necessary. In computing and engineering, consensus can often be reached by testing *what works*, which is basically a pragmatic epistemology in the common-sense understanding of pragmatic, rather than *what is true*. In an encyclopedia, the potential for reaching consensus is often not even the goal; rather, the issue is which aspects of a topic should be represented, to which extent, and in what form. Wikipedia tries to approach this by combining the AGF and NPOV policies. A neutral point of view is defined as "representing fairly, proportionately, and, as far as possible, without editorial bias, all of the significant views that have been published by reliable sources on a topic" (Neutral point of view, 2019, March 18). The definition does not mean that all views should be given equal space, but that differing views should be given proportionate space according to the general understanding of the proportion of "reliable sources" for a particular view. The engineering notion of *what works* has here been translated to *what can survive the editing process.* At the same time, conflicts about content (often referred to as "edit wars") do happen, and although the talk page is not supposed to be used for solving conflicts, they are often full of debates that are at times longer than the article in itself. Occasionally conflicts or repeated vandalism lead to different levels of protection of an article, which in fact are restrictions to the "anyone can edit" ideal (Protection policy, 2019, March 31). Also, editors can, through the "Maintained" template, claim territoriality of a certain section of an article in attempts to strengthen that section's quality (Thom-Santelli et al., 2009).

7 Wikipedia and the Role of Experts

Wikipedia is based on the principle of "the wisdom of crowds" rather than the expertise of a few. That is, the guiding principle is that due to the many contributors will the quality of Wikipedia be upheld. The concept of "the wisdom of crowds" was coined by James Surowiecki in the 2004 book *The Wisdom of Crowds: Why the Many Are Smarter Than the Few and How Collective Wisdom Shapes Business, Economies, Societies and Nations*, and Wikipedia is often regarded as a premium example of this principle.

The Wikipedia project has repeatedly been described as challenging experts in favor of amateurs (e.g., Keen, 2007). Yet, these depictions often overly simplify the complex ways in which Wikipedia relates to and validates different knowledge claims and thus require nuance. Tkacz (2014, p. 110) discusses how the policy of NPOV goes hand in hand with "No original research" and "Verifiability" as "the three core content policies which are designed to work in unison." The policy of *verifiability* is one way that is used to overcome the fact that many editors write about subjects they are not always particularly familiar with. The policy stresses that any knowledge claim must be supported by a credible source outside Wikipedia itself. Together with the policy of *No Original Research*, "verifiability means that other people using the encyclopedia can check that the information comes from a reliable source" (Verifiability, 2019, March 24). In fact, citing sources is seen as a core policy in Wikipedia, which means that Wikipedia is actually dependent on traditional publishing (Forte & Bruckman, 2005; König, 2013; Sundin, 2011; Tkacz, 2014). Likewise, Forte and colleagues (2018) describe Wikipedia citation practices as the means to argue for certain content as *information fortification*. In a similar line of thought, König (2013, p. 172) asserts that "[a]lthough laypeople apparently play a significant part in the text production, this does not mean that they favor lay knowledge."

An important characteristic of the Wikipedia community is a strong belief in the power of knowledge, a belief that it shares with earlier utopian knowledge projects (Reagle, 2011; Sundin, 2011). Specifically, the idea that anyone can contribute should be considered against the background established by the call for verifiability and more precisely of how to achieve verifiability by means of citing reliable sources rather than personal views or experiences, or even original research. The reliability and trustworthiness of these sources in turn is established in relation to other vetting systems, most notably those recognized in science and in the academy more broadly. It could even be argued, as for instance Sanger (2009, p. 62) does, that "if anything, Wikipedia's own policies actually reinforce the epistemic prerogatives of experts." This process has also been described as a type of "source positivism" (Haider & Sundin, 2010). In this way, the Wikipedia community is prone to a far more traditional view of knowledge than is often posited (see also O'Neil, 2010). How Wikipedia is produced might challenge the role of experts, but the manner by which knowledge is sourced and stabilized does not (Sundin, 2011).

The Wikipedia community has identified citing sources as one of the main ways of strengthening the encyclopedia's trustworthiness. There is a template for keeping track of missing references that is used for collaboration during article writing. On encounter of an unsupported knowledge claim, editors can either add a reference themselves or use the template for others to add a reference. There is even a bot called CitationHunt specifically for identifying missing citations, another example of the significant role non-human contributors have in Wikipedia. Sundin (2011) showed how citing external sources is a tool for instituting knowledge claims very much akin to the way that Latour (1987, p. 41) described the making of facts in science: "Fact construction is so much a collective process that an isolated person builds only dreams, claims and feelings, no facts." More specifically, Latour identifies three strategies for constructing facts: (1) "Bringing in friends," (2) "Referring to former texts," and (3) "Being referred to by later texts" (Latour, 1987, pp. 31–44). In the context of Wikipedia, the latter two points are most relevant. Referring to former texts is exactly what citing valid sources in Wikipedia implies, and to a certain degree, its peer-production model constitutes a structured form of collaborative referencing. Inversely, while Wikipedia articles are not generally referred to as sources in research and science (Lindgren, 2014), they are referred to from the outside by the various platforms such as Google, YouTube, and Facebook, to an extent and in a form that has contributed to instituting Wikipedia as one of the most fundamental parts of today's networked media ecology and one that has come to signal trustworthiness and factuality, which we discuss in the next section.

8 Wikipedia and the Media Ecology

The mutual interdependence between Wikipedia and search engines also needs to be understood within the broader context of the contemporary digital media ecology, which is dominated by a handful of giant commercial platforms. One particularly powerful actor in this media ecology is Google. Part of Wikipedia's popularity results from the fact that search engines, and specifically Google, prioritize links to Wikipedia (Höchstötter & Lewandowski, 2009). At the same time, Google and other platform enterprises owned by the same holding company use Wikipedia's open content in their efforts to not only convey links to other resources, but to develop into fact providers themselves. A prominent example of this is the Google Knowledge Graph, which compiles open data sources, such as Wikipedia, and provides users with factual information in response to a query directly on the search results page.

According to Alexa Internet, about 55% of Wikipedia's visitors arrive there through a search engine; no other website within the 20 most popular websites in the world has so many visitors from searches (Wikipedia.org: Competitive Analysis, Marketing Mix and Traffic, n.d.). The trust placed in Wikipedia depends to a large extent on how visible Wikipedia is in search engines, which Shirky (2009) describes in terms of the "algorithmic authority" attributed to Google rather than to Wikipedia itself. Most often, ordinary users in everyday life accept the selection that the search engine provides without carrying out any further assessment of the credibility of the sources presented (Haider & Sundin, 2019). Increasingly, for the typical Google user, a quick look at the right upper side of the screen of the search results often provides a sufficient answer for many queries. Here, Google presents factual information compiled from different open data sources, but mostly it draws on Wikipedia. In many cases, this listing on the Google search results eliminates the need to go to the actual Wikipedia article. As a result, the organic traffic from Google to Wikipedia has decreased (Hamit, n.d.). Since the Wikipedia project is not financed by advertising, this lack of traffic might not have immediate repercussion, but it is likely to have implications for the number of contributors, potentially weakening its peer production model. The success of Wikipedia as a freely available, CC licensed (CC-BY-SA), collaboratively-constructed encyclopedia also makes it vulnerable. Wikipedia content can be used for any purposes, including commercial ones, as long as the content is not legally or technically locked in. Google is clearly aware that Wikipedia and various other Wikimedia projects and peer-production models are crucially important, and Google is on the list of major donors to the Wikimedia Foundation alongside Apple and Microsoft (Benefactors, 2020, January 10).

Collaboratively-created Wikipedia content is circulated also beyond Google through its algorithms to other sites. Langlois and Elmer (2009) showed that Wikipedia content is collected, recycled, and published together with advertisements on other sites as a way to attract traffic from search engines. The authors call these commercial sites "Wikipedia leeches" and argue that "automated content reproduction is not so much about the large-scale dissemination of meaning as it is about the redistribution of flows of users and flows of traffic through new commercial channels" (p. 780). Wikipedia is an open source project both in terms of software and content, but Wikipedia is also firmly embedded in a capitalist media system (Lund, 2017; Tkacz, 2014). Indicative of this ambiguous position are of course the donations by Google, Apple, or Microsoft, but increasingly, Wikipedia is also being held up by commercial platforms in its role as the bearer of vetted, trustworthy knowledge. In their attempts to address problems with misinformation and the spread of what is in the public debate often called "fake news," social media sites YouTube and Facebook use Wikipedia to combat misinformation; for example, anti-vaccination videos on YouTube now include information panels with links to Wikipedia articles on the subject (O'Donovan, 2019, February 22). Not only do these sites use Wikipedia's data, they promote Wikipedia's role as a source of truthfulness. The non-commercial peer production model, which at the turn of the century was seen to be the encyclopedia's weakness, has (discursively) turned into a stable and accurate knowledge base drawn upon to correct the work of algorithmic recommender systems and to provide the data for artificial intelligence systems. Yet this also puts Wikipedia in an exposed position of power with clear (geo-) political implications, as for instance China's growing interest in controlling narratives and processes of fact-creation concerning Chinese interests shows (Miller, 2019, October 5).

9 Conclusion

Wikis and Wikipedia can be characterized by a number of tensions emerging from its peer-production model. Some of these concerns pertain to wikis in general, while others pertain to Wikipedia more specifically. The first tension is the one between amateurs and experts. Wikis and Wikipedia came about when a number of Internet applications were built on user-generated content, and legacy media and the traditional role of experts have been challenged constantly ever since. The second important tension is that between openly licensed content and proprietary content that is subject to copyright.

The Wikipedia community consists of various groups with different levels of authority, where a small group of elite editors are doing most of the content work while the majority of editors are considerably less active. There is another tension between human editors and bots, although Wikipedia is an excellent example of how technology can enhance knowledge production where humans and bots are peers working side by side. Wikis are in general built on a kind of post-publication-review, where the content is tested only after publication, but, particularly in the German-language Wikipedia, a kind of pre-publication peer review has been introduced to avoid vandalism. There is also a tension between lay knowledge and academic knowledge, where the core policies of Wikipedia give priority to academic publication, thus showing the dependence of Wikipedia on peer production and traditional academic writing (peer-reviewed or not). Finally, trust in Wikipedia is not only shaped by Wikipedia itself, but it is also promoted by other online actors that refer to Wikipedia either implicitly though search results or explicitly through links.

References

About. (2019, March 6). In *Wikipedia*. Retrieved from https://en.wikipedia.org/w/index.php?title=Wikipedia:About&oldid=886394833

Assume good faith. (2019, March 24). In *Wikipedia*. Retrieved from https://en.wikipedia.org/w/index.php?title=Wikipedia:Assume_good_faith&oldid=889253693

Benefactors. (2020, January 10). In *Wikipedia*. Retrieved from https://wikimediafoundation.org/support/benefactors/

Bot article creations only. (2019, January 31). In *Wikipedia*. Retrieved from https://stats.wikimedia.org/EN/BotActivityMatrixCreates.htm

Bots. (2018, December 29). In *Wikipedia*. Retrieved from https://en.wikipedia.org/w/index.php?title=Wikipedia:Bots&oldid=875795955

Brohée, S., Barriot, R., & Moreau, Y. (2010). Biological knowledge bases using Wikis: Combining the flexibility of Wikis with the structure of databases. *Bioinformatics, 26*(17), 2210–2211.

Bryant, S. L., Forte, A., & Bruckman, A. (2005). Becoming Wikipedian: Transformation of participation in a collaborative online encyclopedia. In *Proceedings of the 2005 International ACM SIGGROUP Conference on Supporting Group Work* (pp. 1–10).

Chesney, T. (2006). An empirical examination of Wikipedia's credibility. *First Monday*, 11(*11*). Retrieved from http://firstmonday.org/issues/issue11_11/chesney/index.html

Community portal. (2019, March 27). In *Wikipedia*. Retrieved from https://en.wikipedia.org/w/index.php?title=Wikipedia:Community_portal&oldid=889673439

Featured articles. (2019, April 1). In *Wikipedia*. Retrieved from https://en.wikipedia.org/w/index.php?title=Wikipedia:Featured_articles&oldid=890378998

Ford, H. (2012). The missing Wikipedians. In G. Lovink & N. Tkacz (Eds.), *Critical point of view: A Wikipedia reader* (pp. 258–268). Amsterdam: Institute of Network Cultures.

Ford, H., & Wajcman, J. (2017). "Anyone can edit," not everyone does: Wikipedia's infrastructure and the gender gap. *Social Studies of Science, 47*(4), 511–527.

Forte, A., Andalibi, N., Gorichanaz, T., Kim, M. C., Park, T., & Halfaker, A. (2018). Information fortification: An online citation behavior. In *Proceedings of the 2018 ACM Conference on Supporting Group Work* (pp. 83–92).

Forte, A., & Bruckman, A. (2005). Why do people write for Wikipedia? Incentives to contribute to open-content publishing. *GROUP'05*, Sanibel Island, FL.

Gebhart, G. (2016). Zero-rating in emerging mobile markets: Free Basics and Wikipedia Zero in Ghana. In *Proceedings of the Eighth International Conference on Information and Communication Technologies and Development* (p. 16).

Geiger, R. S., & Ribes, D. (2010). The work of sustaining order in Wikipedia: The banning of a vandal. In *Proceedings of the 2010 ACM Conference on Computer Supported Cooperative Work* (pp. 117–126).

Gesichtete Versionen. (2019, March 19). In *Wikipedia*. Retrieved from https://de.wikipedia.org/w/index.php?title=Wikipedia:Gesichtete_Versionen&oldid=186748273

Giles, J. (2005). Internet encyclopaedias go head to head. *Nature, 438*, 900–901.

Graham, M. (2012). Wiki space: Palimpsests and the politics of exclusion. In G. Lovink & N. Tkacz (Eds.), *Critical point of view: A Wikipedia reader* (pp. 269–280). Amsterdam: Institute of Network Cultures.

Graham, M., & Hogan, B. (2014). *Uneven openness: Barriers to MENA representation on Wikipedia*. Oxford: Oxford Internet Institute Report. Retrieved from https://papers.ssrn.com/sol3/papers.cfm?abstract_id=2430912

Guldbrandsen, L. (2013). Swedish Wikipedia surpasses 1 million articles with aid of article creation bot. Retrieved from https://blog.wikimedia.org/2013/06/17/swedish-wikipedia-1-million-articles/

Gustafsson, K. (2019). Chinese collective memory on the Internet: Remembering the Great Famine in online encyclopaedias. *Memory Studies, 12*(2), 184–197.

Haider, J., & Sundin, O. (2010). Beyond the legacy of the Enlightenment? Online encyclopaedias as digital heterotopias. *First Monday, 15*(1). Retrieved from https://firstmonday.org/ojs/index.php/fm/article/view/2744/2428

Haider, J., & Sundin, O. (2019). *Invisible search and online search engines: The ubiquity of search in everyday life*. Abingdon: Routledge.

Halfaker, A., Geiger, R. S., Morgan, J. T., & Riedl, J. (2013). The rise and decline of an open collaboration system: How Wikipedia's reaction to popularity is causing its decline. *American Behavioral Scientist, 57*(5), 664–688.

Hamit, J. (n.d.). Wikipedia's organic traffic is on the decline because of Google. *Improve my search* ranking.com. Retrieved from www.improvemysearchranking.com/wikipedias-organic-traffic-is-on-the-decline-because-of-google/

Hargittai, E., & Shaw, A. (2015). Mind the skills gap: the role of Internet know-how and gender in differentiated contributions to Wikipedia. *Information, Communication & Society, 18*(4), 424–442.

Hill, B. M., & Shaw, A. (2013). The Wikipedia gender gap revisited: Characterizing survey response bias with propensity score estimation. *PLoS ONE 8*(6): e65782.

History of Wikipedia. (2019, March 29). In *Wikipedia*. Retrieved from https://en.wikipedia.org/w/index.php?title=History_of_Wikipedia&oldid=890010240

Höchstötter, N., & Lewandowski, D. (2009). What users see: Structures in search engine results pages. *Information Sciences, 179*(12), 1796–1812.

Huvila, I. (2010). Where does the information come from? Information source use patterns in Wikipedia. *Information Research, 15*(3) paper 433. Retrieved from http://InformationR.net/ir/15-3/paper433.html

John, G. (2012). Wikipedia in India: Past, present, future. In G. Lovink & N. Tkacz (Eds.), *Critical point of view: A Wikipedia reader* (pp. 283–287). Amsterdam: Institute of Network Cultures.

Jullien, N. (2012). What we know about Wikipedia: A review of the literature analyzing the project(s). Retrieved from https://hal.archives-ouvertes.fr/hal-00857208/document

Keen, A. (2007). *The cult of the amateur: How the democratization of the digital world is assaulting our economy, our culture, and our values.* New York, NY: Doubleday Currency.

Kittur, A., Chi, E. H., Pendleton, B. A., Suh, B., & Mytkowicz, T. (2007a). Power of the few vs. wisdom of the crowd: Wikipedia and the rise of the bourgeoisie. In *Proceedings of the 25th Annual Conference on Human Factors in Computing Systems (CHI 2007)*, San Jose, CA.

Kittur, A., & Kraut, R. E. (2008). Harnessing the wisdom of crowds in Wikipedia: Quality through coordination. In *Proceedings of the 2008 ACM Conference on Computer Supported Cooperative Work* (pp. 37–46).

Kittur, A., Suh, B., Pendleton, B. A., & Chi, E. H. (2007b). He says, she says: Conflict and coordination in Wikipedia. In *Proceedings of the SIGCHI Conference on Human Factors in Computing Systems* (pp. 453–462).

Konieczny, P. (2009). Governance, organization, and democracy on the Internet: The iron law and the evolution of Wikipedia. *Sociological Forum, 24*(1), 162–192.

König, R. (2013) Wikipedia: Between lay participation and elite knowledge representation. *Information, Communication & Society, 16*(2), 160–177.

Langlois, G., & Elmer, G. (2009). Wikipedia leeches? The promotion of traffic through a collaborative web format. *New Media & Society, 11*(5), 773–794.

Latour, B. (1987). *Science in action: How to follow scientists and engineers through society.* Milton Keynes: Open University Press.

Lindgren, S. (2014). Crowdsourcing knowledge: Interdiscursive flows from Wikipedia into scholarly research. *Culture Unbound: Journal of Current Cultural Research, 6*, 609–627.

List of wikis. (2019, March 31). In *Wikipedia.* Retrieved from https://en.wikipedia.org/w/index.php?title=List_of_wikis&oldid=890323321

Littlejohn, A., & Hood, N. (2018). Becoming an online editor: Perceived roles and responsibilities of Wikipedia editors. *Information Research, 23*(1). Retrieved from http://www.informationr.net/ir/23-1/paper784.html

Loveland, J., & Reagle, J. (2013). Wikipedia and encyclopedic production. *New Media & Society, 15*(8), 1294–1311.

Lund, A. (2017). *Wikipedia, work and capitalism: A realm of freedom?* Cham, Switzerland: Palgrave Macmillan.

Luyt, B. (2018). Wikipedia's gaps in coverage: Are Wikiprojects a solution? A study of the Cambodian Wikiproject. *Online Information Review, 42*(2), 238–249.

Luyt, B. (2019). Representing African cities in Wikipedia: The case of Lagos and Kinshasa. *Information Development, 35*(3), 388–396.

Mesgari, M., Okoli, C., Mehdi, M., Nielsen, F. Å., & Lanamäki, A. (2015). The sum of all human knowledge: A systematic review of scholarly research on the content of Wikipedia. *Journal of the Association for Information Science and Technology, 66*(2), 219–245.

Miller, C. (2019, October 5). China and Taiwan clash over Wikipedia edits. *BBC Click.* Retrieved from www.bbc.com/news/technology-49921173

Mission statement. (2007, September 1). In *Wikipedia.* Retrieved from https://foundation.wikimedia.org/w/index.php?title=Mission_statement&oldid=21859

Niederer, S., & van Dijck, J. (2010). Wisdom of the crowd or technicity of content? Wikipedia as a sociotechnical system. *New Media & Society, 12*(8), 1368–1387

Nielsen, F. (2007). Scientific citations in Wikipedia. *First Monday, 12*(8). Retrieved from www.firstmonday.org/ojs/index.php/fm/article/view/1997/1872

Neutral point of view. (2019, March 18). In *Wikipedia.* Retrieved from https://en.wikipedia.org/w/index.php?title=Wikipedia:Neutral_point_of_view&oldid=888312035

O'Donovan, C. (2019, February 22). YouTube just demonetized anti-vax channels. *Buzzfeed.* Retrieved from www.buzzfeednews.com/article/carolineodonovan/youtube-just-demonetized-anti-vax-channels

Okoli, C., Mehdi, M., Mesgari, M., Nielsen, F., & Lanamäki, A. (2012). *The people's encyclopedia under the gaze of the sages: A systematic review of scholarly research on Wikipedia.*

Okoli, C., Mehdi, M., Mesgari, M., Nielsen, F. Å., & Lanamäki, A. (2014). Wikipedia in the eyes of its beholders: A systematic review of scholarly research on Wikipedia readers and readership. *Journal of the Association for Information Science and Technology, 65*(12), 2381–2403.

O'Neil, M. (2010). Shirky and Sanger, or the costs of crowdsourcing. *Journal of Science Communication, 9*(1), 1–6.

Open Knowledge Foundation. (n.d.). Retrieved from https://opendefinition.org/od/2.1/en/

Osman, K. (2014). The free encyclopaedia that anyone can edit: The shifting values of Wikipedia editors. *Culture Unbound: Journal of Current Cultural Research, 6*, 593–607.

Pending changes. (2019, January 8). In *Wikipedia*. Retrieved from https://en.wikipedia.org/w/index.php?title=Wikipedia:Pending_changes&oldid=877375019

Pentzold, C. (2009). Fixing the floating gap: The online encyclopaedia Wikipedia as a global memory place. *Memory Studies, 2*(2), 255–272.

Pentzold, C. (2011). Imagining the Wikipedia community: What do Wikipedia authors mean when they write about their "community"? *New Media & Society, 13*(5), 704–721.

Pentzold, C. (2018). Grounding peer production in practice: Editorial routines and everyday engagement in the "Free Encyclopedia Anyone Can Edit." *Communication, Culture and Critique, 11*, 455–474.

Protection policy (2019, March 31). In *Wikipedia*. Retrieved from https://en.wikipedia.org/w/index.php?title=Wikipedia:Protection_policy&oldid=890354581

Reagle, J. M. (2011). *Good faith collaboration: The culture of Wikipedia*. Cambridge, MA: MIT Press.

Research. (2019, March 24). In *Wikipedia*. Retrieved from https://en.wikipedia.org/w/index.php?title=Wikipedia:Research&oldid=889254557

Roblimo (2004, July 28). Wikipedia founder Jimmy Wales responds. *Slashdot*. Retrieved from https://slashdot.org/story/04/07/28/1351230/wikipedia-founder-jimmy-wales-responds

Sanger, L. M. (2009). The fate of expertise after Wikipedia. *Episteme, 6*(1), 52–73.

Sentence, R. (2016, March 3). Everything you need to know about Wikimedia's "Knowledge Engine" so far. *Search Engine Land*. Retrieved from https://searchenginewatch.com/2016/03/03/everything-you-need-to-know-about-wikimedias-knowledge-engine-so-far/

Shaw, A., & Hargittai, E. (2018). The pipeline of online participation inequalities: The case of Wikipedia editing. *Journal of Communication, 68*(1), 143–168.

Shaw, A., & Hill, B. M. (2014). Laboratories of oligarchy? How the iron law extends to peer production. *Journal of Communication, 64*(2), 215–238.

Shirky, C. (2009). A speculative post on the idea of algorithmic authority. Retrieved from www.shirky.com/weblog/2009/11/a-speculative-post-on-the-idea-of-algorithmic-authority/

Statistics. (2019, March 27). In Wikipedia. Retrieved from https://en.wikipedia.org/w/index.php?title=Wikipedia:Statistics&oldid=889757200

Stvilia, B., Twidale, M. B., Smith, L. C., & Gasser, L. (2008). Information quality work organization in Wikipedia. *Journal of the American Society for Information Science and Technology, 59*(6), 983–1001.

Suh, B., Convertino, G., Chi, E. H., & Pirolli, P. (2009). The singularity is not near: Slowing growth of Wikipedia. In *Proceedings of the 5th International Symposium on Wikis and Open Collaboration*. ACM., Article No 8, 1–10.

Sundin, O. (2011). Janitors of knowledge: Constructing knowledge in the everyday life of Wikipedia editors. *Journal of Documentation, 67*(5), 840–862.

Sundin, O., & Francke, H. (2009). In search of credibility: Pupils' information practices in learning environments. *Information Research, 14*(4). Retrieved from http://InformationR.net/ir/14-4/paper418.html

Surowiecki, J. (2004). *The wisdom of crowds: Why the many are smarter than the few and how collective wisdom shapes business, economies, societies and nations*. London: Little, Brown.

Thom-Santelli, J., Cosley, D. R., & Gay, G. (2009). What's mine is mine: Territoriality in collaborative authoring. In *Proceedings of the SIGCHI Conference on Human Factors in Computing Systems* (pp. 1481–1484).

Tkacz, N. (2012). From open source to open government: Critique of open politics. *Ephemera: Theory & Politics in Organization, 12*(4), 386–405.

Tkacz, N. (2014). *Wikipedia and the politics of openness*. Chicago: University of Chicago Press.

The top 500 sites on the web. (n.d.). *Alexa Internet*. Retrieved from www.alexa.com/topsites.

Toupin, S. (2019). "Hacking in" Wikipedia zero in Angola. *Canadian Journal of Communication, 44*, 351–359.

User access levels. (2019, July 30). In *Wikipedia*. Retrieved from https://en.wikipedia.org/w/index.php?title=Wikipedia:User_access_levels&oldid=908480259

van der Elden, M. (2012). When knowledges meet: Wikipedia and other stories from the contact zone. In G. Lovink & N. Tkacz (Eds.), *Critical point of view: A Wikipedia reader* (pp. 236–255). Amsterdam: Institute of Network Cultures.

van Dijck, J. (2013). *The culture of connectivity: A critical history of social media*. Oxford: Oxford University Press.

Verifiability. (2019, March 24). In *Wikipedia*. Retrieved from https://en.wikipedia.org/w/index.php?title=Wikipedia:Verifiability&oldid=889194614

Viégas, F. B., Wattenberg, M., Kriss, J., & Van Ham, F. (2007). Talk before you type: Coordination in Wikipedia. In *Proceedings of the 40th Hawaii International Conference on System Sciences*. Washington, DC: IEEE Computer Society.

Weinberger, D. (2007). *Everything is miscellaneous: The power of the new digital disorder*. New York, NY: Times Books.

Wilkinson, D. M., & Huberman, B. A. (2007). Cooperation and quality in Wikipedia. In *Proceedings of the 2007 International Symposium on Wikis* (pp. 157–164).

Wiki. (2019, January 19). In *Wikipedia*. Retrieved from https://sv.wikipedia.org/w/index.php?title=Wiki&oldid=44283812

Wikimedia Foundation (2019, March 28). In *Wikipedia*. Retrieved from https://en.wikipedia.org/w/index.php?title=Wikimedia_Foundation&oldid=889873909

Wikimedia Research. (n.d.). Retrieved from https://research.wikimedia.org/

Wikipedian in Residence. (2019, May 27). In *Wikipedia*. Retrieved from https://en.wikipedia.org/w/index.php?title=Wikipedian_in_residence&oldid=889765635

Wikipedia.org: Competitive Analysis, Marketing Mix and Traffic. (n.d.). *Alexa Internet*. Retrieved from www.alexa.com/siteinfo/wikipedia.org

Yeo, M. L., & Arazy, O. (2012). What makes corporate Wikis work? Wiki affordances and their suitability for corporate knowledge work. In K. Peffers, M. Rothenberger, & B. Kuechler (Eds.), *Design science research in information systems: Advances in theory and practice* (pp. 174–190). Berlin: Springer.

14

Participatory Cartography

Drones, Countermapping, and Technological Power

Adam Fish

1 Introduction: New Materialism and the Technologies of Hacker Cartography

> Researching mapping as an infrastructure needs to foreground the materiality of production. . . (From Dodge et al., 2009)

Geographer Alan McConchie (2015, p. 878) coined the term "hacker cartography" to describe the "geoweb-based practices of collaboratively creating and curating crowdsourced geographic data and representations, using a mixture of open software and repurposed tools and data." By privileging "hacker" McConchie emphasizes volunteer and collaborative socio-technical experimentation in participatory mapping. Many different "hackers" exist (Coleman & Golub, 2008) but most express the liberal values of sharing, openness, decentralization, access to computers, and world improvement through technology. With these ethics plus a "spirit of playful cleverness" (Stallman, n.d.), cartographic hackers evolved from a material foundation consisting of computers, networks, software, as well as the Global Positioning System (GPS) and atmospheric platforms such as satellites and drones. Hacker cartography is an expansive field of cultural production inclusive of OpenStreetMap contributors, indigenous countermappers, entrepreneurs of GPS apps, geographical hardware hackers like DIY drone communities, and spatial artists in the geohumanities. Discourses of freedom, the valuing of peer production, and the open sharing of data influence hacker cartographic culture; while material technologies provide the means of production and the reason for association. By focusing on the material technologies, I arrive at the dependencies, as opposed to the autonomy, of peer production within hacker cartography.

By "material" I am not referencing historical materialism, a Marxist theory of historical change, capitalist exploitation, and proletariat revolution (Eagleton, 2016) but rather *new materialism*, a theory which emphasizes matter over consciousness, physicality over idealism, and realism over social constructivism (Coole & Frost, 2010; Fox & Alldred, 2017). New materialism is anti-anthropocentric (Braidotti, 2013). Several different new materialisms exist including vitalism which is inspired by Gilles Deleuze (DeLanda, 2013), agential realism that sees humans, technologies, and non-humans "intra-acting" in complex "entanglements" (Barad, 2007), and media archeology that finds forgotten genealogies hardcoded into media infrastructures (Kittler, 2009). Other materialisms see humans, as well as objects, networks, animals, and plants as having agency (Latour, 2005). The new materialism is useful for understanding the dependencies of hacker cartography begin with the technologies and physical realities that make these assemblages possible. From here, a theory is developed that examines the implications of this technological dependency and capture of power coded in technological infrastructures.

Benkler and Nissenbaum (2006, p. 405; also in this volume) famously argued that "[n]o matter what other demands constrain their lives, participation in peer production constitutes an arena of

The Handbook of Peer Production, First Edition. Edited by Mathieu O'Neil, Christian Pentzold, and Sophie Toupin.

autonomy, an arena where they are free to act according to self-articulated goals and principles." The four examples in this chapter – GPS, 3DR, OpenStreetMap, and indigenous countermapping – tell a different story, not one of hacker freedom but hacker dependency upon costly and opaque infrastructures: GPS-enabled systems, manufactured atmospheric platforms like satellites and/or drones, and proprietary and capitalized software. GPS satellites, for instance, are "high-tech, high-capital and high-maintenance devices that are seemingly beyond the purview of the popular and the personal" (Parks, 2001, pp. 210–211). The free labor of participatory mapmakers adds proprietary value to firms and reinforces power asymmetries around access to technology. While hacker cartography can be a manifestation of peer production, the suite of materials it requires demands substantial professional invention, development, and manufacturing.

These four cases are more than convenient examples of technological power from the history of cartography. Rather, the new materialistic approach taken in this chapter emphasizes the entanglements of technology, culture, and politics and the movement of power across these different domains. However, that peer production is dependent on extant forms of technological power does not extinguish its political potential. On the contrary, the materialism approach invites proponents of peer production to better respond to the inequities associated with elite domination of technological power.

2 Technological Power

> Power is a transient, fluctuating phenomenon – a momentary exercise by one relation over another; only if replicated in multiple events over time and space, does it acquire a more regular patterning. (Fox & Alldred, 2017, p. 27)

New material studies of power foreground how it flows across distributed systems, not how it is held indefinitely by certain actors over others. As new materialist scholars Fox and Alldred (2017, p. 27) explain above, power is one key line-of-flight within "participatory assemblages" (Grove & Pugh, 2015) constituted by humans, technologies, non-humans, and larger ecologies. Thus, while human power is downplayed and considered diffused within new materialism, it exists in dynamic trajectories and geometries within the assemblage (Massey, 1999).

A working definition of technological power in participatory assemblages can be constructed by mobilizing the work of digital sociologist Tim Jordan (1999, pp. 2–19) who builds his theory of cyberpower through a convergence of concepts from Max Weber, Barry Barnes (1988), and Michel Foucault (1980). Weber's definition of power is possessed by elites and executed through imposition. Barnes' power is pervasive; it can be mobilized by either those in dominant or subordinate positions. Similarly, for Foucault power can also be held and performed by several actors but it is ultimately unequally distributed in the assemblage. Jordan argues that all three forms of power exist in the digital economy and throughout information infrastructure. However, technological power arrays differently in different assemblages. While a new material theory of power foregrounds its flow, Weber's power is defined by control through static possession. An example would be professional cartographic systems which exercise control over hardware and software and platform developers, investors, and owners. Power in the social order is hardcoded into hardware and dictated by control of information flows between individuals. In this manner, "cyberpower" occurs through possession, social order, and domination (Jordan, 1999, p. 210). While a Weberian theory of power that foregrounds its possession by a few over the many is not new materialistically, Barnes' and Foucault's theories of power align with new materialism's highlighting of how power is diffused across a field of production within and outside human agency.

Foucault's notion is that power is nested within a system. While Foucault's theories of the body are useful for new materialism of embodiment and affect, his theories of power are less often applied to new materialism that focuses on the relation of things, technologies, and other species. Barad (2007) for instance, sees Foucault as an anthropocentric anti-materialist, a scholar who finds society, humans, and discourse as the only elements with agency and the ability to wield power. However, Lemke (2015) disagrees and identifies in Foucault's concept of the "governance of things" – a reference to how power is enacted throughout biopolitics – as latent new materialism. He quotes Foucault's references to "environmentalism," "rivers, marshes, hills," and "climate, dryness, fertility" as a "sort of complex of men and things" (all Foucault quotes in Lemke, 2015, p. 9). He sees in the later Foucault a "move beyond a concept of biopolitics as limited to the physical and biological existence to a government of things that takes into account the interrelatedness and entanglements of men and things, the natural and the artificial, the physical and the moral" (Lemke, 2015, p. 13). This conceptualization allows Foucault's theories of power to unfold within an assemblage of relations. Thus, for Foucault power is diffuse but it can also be controlled by organized actors. As an example of this concept, consider GPS as a complex global assemblage of multiple technologies, practices, and people that makes possible a range of practices from tracking endangered species with radio-collars to self-driving vehicles. The technological power in this state-funded GPS assemblage is distributed à la Barnes, possessed as in Weber's model, and both distributed and possessed as in Foucault's model. Resources and relations within this GPS assemblage flow from the US state military to innumerable organized publics and for-profit firms needing geospatial tracking features, as well as social justice actors and humanitarian actions.

Jordan applies his theory of cyberpower to describe the co-constituting relationship between individuals and elites. Writing in 1999, his example is how Sun Microsystems – much like Apple today – promised to empower individuals by providing computers and software less complex than that offered by Microsoft and Intel at that time. In this manner, individual consumers and elites at Sun both become empowered – but in vastly different ways. Interpreting this Jordan writes, "[i]ndividuals exist in cyberspace by using the tools that depend on a technological infrastructure controlled by an elite, but that technological infrastructure exists to provide tools for individuals to use for empowerment" (Jordan, 1999, p. 211). Thus, cyberpower is pervasive, and thus culturally-embedded across individuals, elites, and technological infrastructure. It is also irreducible to any of these components. The control of the infrastructure emboldens modes of domination; while other abilities and potentials are made possible by that very infrastructure. In this manner, the flow of cyberpower may pool in particular relations at certain times but never indefinitely.

I am informed by Jordan's concept of cyberpower but prefer to use a different term to describe a similar phenomenon. I deploy "technological power" to describe the flow of power within socio-technical assemblages, in this instance hacker cartographies. A new materialism of technological power cannot be Weberian or exclusively dominating, but is more akin to being Barnsian, and accessible to numerous actors in the assemblage. "Technology," in this case, is preferable to "cyber," because technology emphasizes the materiality of hardware over the virtuality of software.

Technological power is a force that develops from the synergistic interaction of actants. Technology power moves through the connections between actants to subjugate others in the assemblage. It originates in human constructions and may impact humans but the theory of technological power is not anthropocentric; it is not centered around human action but affects other technologies, species, and environments. The technological power of hacker cartography may begin with state-funded GPS, for instance, but it impacts human users, mapped geographies, species sensed by citizens, and geospatially marked ecologies. As I will illustrate in the four case studies that follow, new materialism is particularly well-suited as an instrument for understanding technological power and participatory mapping. Each of the cases that follow exhibit different facets of how technological power manifests within participatory assemblages. For example, technological power manifests in the form of the state in the GPS case, corporations in the examples of OpenStreetMap and DIY Drones, and through traditional forms of domination in Indonesian countermapping, each described below.

3 OpenStreetMap

> despite the open ethos, most projects using OSM data do not differ greatly from conventionally produced mapping. . . . The political economy of the wider geoweb limits counter-hegemonic applications. (Perkins, 2011, p. 6)

One mode of cultural production infused with technological power is mapmaking. Mapmaking is a political activity that has traditionally reinforced power relations. Maps reveal who rules what, hide forms of oppression through omission, and enshrine in legible forms the control over space by certain people and classes. As such, any effort to produce alternative maps using unorthodox methods and containing unique cultural information from the periphery of the hegemony is an act of resistance (Harley, 1989). Alternative maps have various names including indigenous, citizen, participatory cartography or peer-to-peer mapmaking. Before portable tangible or digital maps indigenous people had manifold ways of communicating geographical information – non-literate peoples used stories and various pictorial representations to memorize and transmit geospatial knowledge (Chapin et al., 2005). The emergence of contemporary participatory mapping – that is, maps made collectively by communities and within the dominant nomenclature and sign system – is the result of not only the spreading of mapmaking skills outside of elite communities but also of the diffusion of digitization and allied technologies which "democratize" mapmaking.

A surprising example of mapmaking and technological power is OpenStreetMap (OSM). Fueled by "volunteered geographic information" (VGI) (Goodchild, 2007), OSM is the largest aggregation of geographical information marked-up by citizens. Celebratory readings of OSM see it as the democratization of mapmaking (Wood, 2015), a "technoliberal" view that the inequities of capitalism can be ameliorated by new technologies (Fish, 2017). As a form of hacker cartography, OSM draws to it advocates for open information, free and open source software, and radical transparency. Others see less potential and more problems in OSM. Leszczynski (2012), for example, situates OSM within the neoliberal roll-back of the state paired with the corporate roll-out of the market. Where once the state was actively producing maps – often for its colonial projects – now corporations and their allies are in the for-profit business of mapmaking.

Leszczynski (2012) argues that OSM is an example of the "coming together of information technology and markets, a historically specific articulation of technoscience and capitalist exchange" (p. 73). She takes the counter-intuitive position that instead of being a site of non-profit peer production, important facets of peer-produced OSM dovetail with the capitalist economy. As the state retreats from its cartographic project – linked as it is with legibility for taxation – a market gap was created. Into this potential came technology companies with their ability to collect, harvest, and monetize VGI. To make this point, Leszczynski (2012) focuses on the material infrastructure (how the data is stored and handled) of OSM and how it generates value for an allied corporation, Bytemark – and as many as 31 other companies and universities which host the OSM servers. In exchange these "partners" are thanked on the home page and a link is provided to their Internet sites.[1]

Other important evidence for the capitalization of OSM's VGI comes from OSM founder Steve Coast who started with collaborators CloudMade, a for-profit company that makes software development kits (SDKs) that enable people to develop applications for intelligent car companies using OSM. The more data OSM has the better CloudMade can be and the more profitable it may become. Microsoft's map service, Bing Maps, includes the OSM data as a layer, adding value to their offering (Leszczynski, 2012, p. 85). Other companies such as Alphabet, for instance, followed Bing's model and allow for the crowdsourcing of waypoints and tags on Google Maps. In this manner, peer production inspires corporate production and "offer(s) an alternative economic model to professional cartography" (McConchie, 2015, p. 885).

Furthermore, the demographic of peer producers on OSM is not diverse. It consists of a population that is more wealthy, educated, middle class, Western, and male. This results in more peer-produced maps being made of the regions where this demographic lives, works, and travels: cities in the Western northern hemisphere (Bittner, 2017). Also, volunteering data to OSM requires some advanced technological skills often associated with this demographic. In the early days use of portable GPS units was necessary. Today, some use voice recorders and collect video and still images which they subsequently tag and upload. Others remix geographical data with sophisticated coding and tagging techniques (Perkins, 2011, p. 6). While OSM is not a for-profit entity it exhibits how peer-productive activities are enabled by demographic-specific technological literacy, contained by technological power, and provides collateral for firms.

A new materialistic approach focuses on the tangle of forces that constitute participatory mapmaking – user affordances, storage infrastructure, different actors at work on the system, and the manners in which Bing and Google import OSM into their corporate sites. This investigation requires answers to questions of ownership that run contradictory to the ethos of peer production. This results in insights into how technological power is harnessed by certain actors within the OSM assemblage. In this example, Microsoft's Bing, CloudMade, Bytemark are each elite actants who acquired technological power within the participatory assemblage that includes the user-generated data of OSM.

4 Global Positioning System

> For what is GPS . . . if not the very surface of the planet itself as a total chess board, every inch rendered discrete, finite, and bounded. (Wark, 2012, p. 94)

Traditionally cartography required sophisticated tools and an ability to work these tools into accurate depictions of topography. Cartography is an ancient and elite job. The Romans enshrined the profession and the ancient Greek Ptolemy wrote a major text on the art in 150 CE. William the Conqueror funded the drafting of an early English survey, the Domesday Book in 1086. These surveys were made by hand and guided by the unaided eye until new technologies made producing more accurate maps possible. In 1640, the English astronomer William Gascoigne developed a telescope with cross-hairs that enabled him to survey distances between points. By the late 1700s, the triangulation technique made it possible to accurately map entire countries. This tactic was put to use in colonial projects such as the 1801 Great Trigonometric Project in India (Keay, 2000). In the United States, the Land Ordinance of 1785 instigated a massive survey of the territories taken from the indigenous Native Americans. Similar projects occurred in Australia and colonial Africa (Blakemore & Harley, 1980). By the mid-twentieth century, rangefinding devices using electronic pulses of light to check distances were being used, replacing the older chain and rope technique. By the late 1970s, the GPS system had been developed using time and space-positioned satellites which could be linked to total stations, mobile phones, and other devices. Thus, maps became tools for elites: kings, colonists, and the US military which funded and deployed GPS satellites. From prehistory to today, cartography has been an exceptional practice requiring access to sophisticated technologies of space and time, and professional knowledge of their application. Efforts to democratize mapmaking have to contend with this history.

GPS, probably the most important innovation for the development of hacker cartography, is a large and expensive operation. Most clearly, it illustrates the necessity of state investment for peer production. In 1974, the US military launched the first of 24 satellites with the ability to transmit their time and space coordinates and therefore be able to triangulate a precise location of a receiver on the Earth. This was NAVSTAR and it evolved into GPS. Between 1978 and 1985, the military launched more sophisticated satellites with atomic clocks on-board, capable of even more precise locating abilities. Then the US government invested in making the receivers on Earth smaller and more mobile. By 1989, the first commercial handheld GPS receiver was on the market, the Magellan NAV 1000. In 1994, 24 satellites were rotating around the Earth twice a day, with three satellites in orbit for backup. Around this time, the Clinton administration and the Federal Aviation Authority gave the commercial airline industry the ability to use the system to locate airplanes. In 1999, the first GPS phone, the Benefon Esc!, was made available and the US Defense Department stopped intentionally degrading GPS (a precaution designed to not give adversaries an advantage) so capitalist industries could develop (Sullivan, 2012). Since this time, the science and technology has remained relatively similar with satellite updates taking place every decade.

The GPS satellite is the quintessential black box technology as it "occupies a remote orbital position beyond users' reach and outside of the field of vision; . . . [and has] historically been controlled by states and used in myriad ways without citizens' knowledge, involvement or consent" (Parks, 2001, pp. 210–211). Despite this, media geographer Lisa Parks implores us to see "the satellite as an interactive technology" (Parks, 2001, p. 210). Advancements in science, agriculture, aviation, marine work, public safety, and transportation have developed that rely upon GPS-provided location, navigation, tracking, mapping, and timing. Today, GPS is applied in preventing attacks by GPS-tagged sharks, tracking stolen prescription drugs, finding lost dogs, geocaching parties, guiding driverless cars, and dating applications. Citizen cartographic projects are reliant upon GPS and the immense investment of humans, capital, resources, and technologies – provided mainly by the United States and other nations.

GPS has become a necessary means of relation within an assemblage of participatory and countermapping. It is a material system with its foundation in 24 orbiting satellites which send and receive atomic clock time signatures from microwaves to moving terrestrial and atmospheric humans and non-humans. In friction-free near space, these solar-powered satellites will indefinitely exist beyond human lifespans. GPS-tagged orca whales, self-driving cars, and other non-humans report their geospatial locations via these satellites. GPS is necessary to monitor species populations, avoid airplane and automobile collisions, and to track the global flow of goods from producers to consumers. In this manner, the materialism of GPS brings us in contact with the technological power that is distributed throughout the assemblage. Participatory cartography is but one ancillary beneficiary of this state-supported technological power. All subsequent peer-produced innovations developed from GPS are reliant upon that most centralized and dominant force, the state.

5 DIY Drones and 3D Robotics

> Just as the PC emerged from the Homebrew Computer Club hobbyists and eventually overturned mainframe-based corporate computing in the 1980s, I suddenly saw how the same sort of movement would bring robots to the skies. (Anderson, 2017)

Not only is participatory mapmaking dependent upon state-financed technologies like GPS. It can also become indebted to other technologies such as unpersonned aerial vehicles (UAVs or drones), themselves dependent upon GPS as well as corporate development and a business culture designed to capitalize on peer production. In this fashion, the practice of so-called "DIY drones" and the culture of participatory and activist mapmaking it facilitates is reliant upon the state and corporations. The productivity of peer production also fuels capital accumulation through the capitalist appropriation of user-creativity. Where once the state was the reason for the advances, corporate America and Chinese manufacturing have become the driving force for peer-to-peer mapmaking from the atmosphere. The empirics for this argument are how the American drone company 3D Robotics developed from the peer community DIY Drones.

Former *Wired* magazine editor Chris Anderson created the community platform DIY Drones in 2007, a site (diydrones.com) where drone tinkerers, makers, hackers, and hobbyists could trade concepts, code, prototypes, hacks, and other work-arounds. From DIY Drones, Anderson and colleagues started 3DR, a for-profit robotics firm focused on the drone industry. 3DR was motivated to fill an investment niche and an atmospheric volume between satellites and the Earth. Anderson was inspired by what he saw as the "democratization of earth observation" (Anderson, 2017). Like the trend described by Leszczynski (2012) in reference to OSM and the roll-back of the state and the roll-out of the market, the state was the financier of GPS satellites and mapping in the latter half of the 20th century whilst in the early 21st century corporations assumed this role, particularly in terms of air and space science, innovation, and development. This follows the rise of private air and space firms – SpaceX, Blue Origin, and Virgin Space for instance – and the decline in state funding for space programs such as NASA. But before Anderson could arrive at that lofty position in suborbital space much hardware and software peer production was necessary.

Initially, DIY Drones featured a chaotic yet productive peer production. This consisted of sharing design files as a form of "nerd braggadocio": a liberal advocacy for information freedom, radical transparency, a mercurial anti-establishmentarianism, and the joy of tinkering (Coleman, 2013). Anderson wanted a system with which to capitalize on this collective creativity. He formalized the structure for code authoring and joked "behind every great open source or open innovation project, there is a malevolent dictator" (Stuart & Anderson, 2017). After this confession, Anderson shared his dictatorship with CEO Jordi Muñoz, a 20-year-old Mexican citizen who contributed significantly to DIY Drones including prototypes of an innovative hack of Arduino, a plan for a drone controlled by Nintendo Wii, and successful drone stabilizing code. By 2014, DIY Drones was the largest robotics community on the planet, numbering 50,000 active members and millions of monthly page views. With this base of peer producers, Anderson and Muñoz understood that the moment was right to start a company. Soon 3DR had a $360 million valuation – a value largely created by peer producers. The manner in which DIY Drones morphed into 3DR is indicative of the relationship between peering publics or "organized publics" that become formal enterprises (Fish et al., 2011, Fish & Srinivasan, 2012). This peer-to-peer process developed value for enterprises including Alphabet and Microsoft through their mapping subsidiaries as well as Bytemark, and CloudMade.

By April 2015, 3DR were ready to manufacture the Solo, their first drone. The Solo had a wealth of fascinating features – it was packed with scripted flight plans and it was open source so that developers could create software for the hardware. But the roll-out did not go smoothly. The GPS on the Solo failed to connect. The gimbal, the key component for the making of smooth aerial videos – the favored application for most drone enthusiasts – was not ready and the first Solos were shipped without the component. Worse still, the drone would dangerously fall from the sky. By mid-2015 they had sold only 22,000 – half the estimate – and tens of thousands remained in the inventory. Soon 3DR's drone manufacturing division was broke and bankrupt (Mac, 2016) – and with it the idea that drone peer production could result in profitable participatory mapmaking technologies.

But 3DR has not entirely left the drone business; it just abandoned manufacturing to Chinese companies like DJI, which control 80% of the civilian drone market. 3DR now primarily develop Site Scan, a mapmaking software suite that captures and analyzes aerial data for industrial and mapping applications. As the only account that has access to the entire data set – as a private company they are bound to protect the privacy of their clients – Site Scan is far from its DIY Drones ethos of camaraderie, peer production, transparency, and code-sharing. Accessing Site Scan's tools for advanced orthomosaic and three-dimensional mapping is costly, at $7,500 for an annual subscription. This is no longer hacker cartography.

DIY Drones transformed into 3DR, a company that was modeled on a peer community, transitioned into and failed as a manufacturing company, and become a closed, professional, and expensive mapping platform. This story constitutes a reverse mirror-image of the trajectory explored in the earlier section on GPS. Instead of professional and state investment creating the platform for subsequent peer production, peer producers initiated the drone industry in the United States. Instead of the example of the state, a formal social enterprise seeding the organized publics that developed around GPS, DIY Drones emerged organically as a public only to be capitalized on later by the enterprise that became 3DR. Eventually, the global drone market was dominated by a company with no connection to peer production: DJI, the Shenzen (China) company which dominates the drone manufacturing market. Frank Wang, DJI's founder, did not need a peer producing public – just a good design, a small grant from his university, and his parent's garage to make the first DJI drones, and eventually become the youngest tech billionaire in Asia.

As a classic example of peer-to-peer historiography, Anderson's drone does not begin with the military industrial complex but rather in advances associated with commonly available technologies, namely, mobile phones, remote sensors, lens systems, microprocessors, and improved battery efficiency. Anderson sees this as "an appropriation from below in order to 'demilitarize and democratize'" drones (Jablonowski, 2015, p. 5). This appropriation does not remain democratized, however, but is effectively transformed into private wealth for DJI and, perhaps eventually, for 3DR. As 3DR shows even the software that stitches together the orthomosaic maps used by participatory mapmakers can be prohibitively expensive.

This new materialism of drones points to the social effects of hardware and its failures in forming and eroding technological power and the economic power that comes with it. This is a story of economic churn, market brutalism, and military inscription. Drone hacker-cartographers, while once the beneficial recipients of the democratization of GPS, the peer-to-peer DIY drone network, and open hardware designs were omitted from the commercialization of drone hacker cartography. Technological power in drone assemblages, with its controversial origins and social asymmetries, diffused from DIY cultures to professional manufacturers first in California and on to China. This is a far cry from the ideals of peer production as autonomous from capitalism, and as virtuous because of its collectively created methodology.

I now turn to my final example, to illustrate how a new materialist approach to hacker cartography shows how participatory mapping projects evolve with networked forms of technological power.

6 Countermapping in Indonesia

> Many of us who study new media still proceed too often from the assumptions that peer production is radically participatory, egalitarian, efficient, and psychologically fulfilling. (Kreiss et al., 2011, p. 255)

As the section above on OSM illustrates, cartography is a discipline that has been recently transformed by peer networks consisting of citizens and activists coming together around open platforms in order to challenge professionally-made and standardized maps and mapmaking. In this section, I investigate one adroitly political manifestation of hacker cartography: countermapping – the collective building of alternative geographies by local communities and indigenous people. The purpose of a counter-map is to challenge how existing economic and political power is concretized in actual space by the virtual means of a map. I will discuss how despite its activist intentions countermapping reinforced technological power inequities.

In Indonesia, participatory mapping began in 1992, borne out of an emphasis placed on community-based forest management and the rise of the environmental movement and efforts to reclaim customary rights to lands (Peluso, 1995). Countermapping took root in this context and in light of the Indonesian government's usurping of customary land rights for industrial timber exploitation. In 1996, the Community Mapping Network (Jaringan Kerja Pemataan Partispatif, JKPP) in league with the Alliance of Indigenous Peoples of the Archipelago (Alinasi Masyarakat Adat Nusantara, AMAN) was formed and in a decade 510 villages and 2.5 million hectares of customary lands had been mapped. The center of countermapping was with the Dayak people of West Kalimantan where hundreds of villages or *kampungs* had been mapped (Radjawali et al., 2017, pp. 821–822). These efforts required indigenous people, allied NGOs and their funding backers, to work closely with GPS technologies, in community consortiums, to create maps of customary lands.

The contemporary movement in countermapping was facilitated by the Indonesian government in May 2013 when they rejected the 1999 Forest Law. The new law asserted that instead of the state government it was the indigenous people who had access and land rights to customary forests (Tsing, 2004). With this change to the law AMAN and JKPP used the method of countermapping to register 4.8 million hectares of customary lands. But this represents not even 20% of 74,000 villages and surrounding lands that need to be mapped. Conflict continued because the remaining lands were informally identified, in a manner with little to no legal value. Adding to this problem was that the Geospatial Information Body (Badan Informasi Geospasial, BIG), the only official mapmaker in Indonesia, refused to recognize participatory maps. It is into this moment of opportunity and confusion that the contemporary drone and participatory countermapping movement in Indonesia began.

As an action-research project with seven "Participatory Hydro-Political Appraisals" (PHPAs), between 2013 and 2014, a series of participatory drone workshops engaged communities in the Kapuas River area of Kalimantan, Indonesia, in the use of drones for the production of countermaps. The PHPAs consisted of citizen research groups which collected data on place stories, as well as river and customary forest locations. Drones were not only tools for the collective production of high-resolution spatial information; they also gave villagers visual access to areas otherwise restricted by timber industry police and security forces. Most importantly, the participatory labor with drones instigated a collaboration with NGOs with the aim of developing a more transparent mapping process (Radjawali et al., 2017, p. 823).

The most substantial impact of drone-based participatory mapping occurred in the Tayan area of Kalimantan, Indonesia, in the sub-district of Tayan Hilir. As a result of immense concessions given by the Indonesian government to a mining corporation, bauxite was being extracted for shipment to China where it was to be refined into aluminum. The project at Tayan involved 30 individuals from Sejotang and Subah villages of which ten – consisting of both women and men, Dayak and Malay, young and old – joined the Citizen Research Group. Together they used drones to document the activities of the bauxite mining company, map customary fruit trees, and produce maps that disputed the state's designation of village lands and farms as "forest."

A new materialist account of drone countermapping in Tayan Hilir would array the following forces: bauxite, Chinese refineries, forest ecologies, fruit trees, drones, communities using and communities not using drones, and so on. This assemblage was fraught with power asymmetries. The bauxite mine had security personnel that intimidated indigenous activists wanting to document ecological destruction. The drone provided a vertical vantage-point from which to overcome the hidden topographies; their semi-inclusive participatory methodology brought others into the web of interactive activism. However, a closer look at social inequities instructs our inquiry into the relation of technological power and peer-to-peer and drone countermapping.

Seasoned practitioners of drone countermapping and drone workshops with communities in Indonesia, Radjawali et al. (2017) critique their own work in countermapping via epistemological, socio-economic, and political arguments.

The *epistemological argument* is that indigenous knowledge is inherently different from Western knowledge and non-translatable in static two-dimensional forms. It could be argued that "hypertext cartographies" – with the ability to polyvocally represent images, video, sound, and text – may be a more appropriate model for indigenous or even collectively-authored maps (Fish, 2011, pp. 100–101). However, epistemologically, maps, particularly when they are designed by communities and contain cultural information, are reductive of traditional knowledge, omitting important information and geospatial concepts in the process. The concern is that if communities "pigeonhole" their knowledge into Western-style maps they will concretize their process-based values, perceptions, and orientation towards landscapes and land management. Also, countermaps may be used by the state to liquidate customary lands. This solidification will make it easier for exploitative forces to identify and legally work-around these static categories.

The *socio-economic critique* of countermapping is concerned with how the practice requires technological literacy and is therefore an exclusive practice within indigenous communities. GPS-enabled mobile devices, WiFi reception, drones, and mapping software remain expensive for much of the developing world. The costs associated with access may be exacerbated by the competition for socio-technical resources which the countermapping project may require. Access can mean access to not only technologies to produce countermaps but also access to the products of countermaps – the maps themselves – as the process can result in the transfer of the maps themselves from villagers to NGOs. Linked to the socio-economic inequalities of technological access is the political problem of inclusion.

In small communities a difference of opinion prevails; gender, class, prestige, and age inequities exist and can be reinforced through mapping exercises. Radjwali et al. write, "[p]articipatory and countermapping may change internal social dynamics, may open new avenues for co-optation and corruption, and can lead to new hierarchies" (2017, p. 820). Mapping can be an exclusive exercise and reinforce pre-existing inequalities. Women are commonly excluded, for instance. Epistemological concerns are also present in these political issues as motivations and definitions once collectively devised can be transformed by this exclusive group so as to shift the original motivation towards something more legible for the state.

Finally, the *political critique* of countermapping is that it confirms rather than rejects pre-existing forms of social inequalities. This is a persuasive argument. The drone school failed to be inclusive having to work through the Dayak Customary Council, a group of male elders, and the majority of the peer-participants in the drone school were young males. The involvement of women, young and old, remains elusive, and thus the drone school reinforced the asymmetry of gender in traditional Dayak communities.

Villagers hoping to protect their customary lands must map, and they must do this as a peer-to-peer community. These maps cannot overturn the wrongs of colonial, corporate, and state violence on lands and people. Maps can only be a part of the solution. But they can be substantial. In late 2014, the countermaps accompanied testimony about environmental degradation caused by the mining company in the Indonesian Constitutional Court. The court ruled against the mining company. The drone images proved that the mining operation was in violation of the law, and the precedent was established that drone-produced peer-to-peer maps were valid in Indonesian courtrooms.

This new materialist investigation into countermaps acknowledges how technological power adheres to pre-existing forms of power through possession, social embeddedness, and domination and thus can contribute to the perpetuation of pre-existing forms of epistemic, socio-economic, and political power. But countermaps can also form acute interventions wherein technological power shifts to other actants within the assemblage. A new materialism of indigenous countermapping sees technological power as diffuse within relations. In this example, indigenous countermapping mobilizes the technological power of drones to challenge state control of customary lands.

7 Conclusion

> The first requirement is the existence of a technological infrastructure that operates on peer to peer processes and enables distributed access to "fixed" capital. (Bauwens, 2005)

This new materialism of peer production has sought to link together the technologies, historical trajectories, and human and non-human actors that come to constitute several important instantiations of participatory mapping. In the process, the role of technological power as opportunity and problem for the more radical decentralizing potentials of peer production has been illustrated. OpenStreetMap and its links with Bytemark and CloudMade, the early days of the US drone hacker network succumbing to 3DR and DJI, and Dayak countermapping and problems around control, access, and visibility/legibility offer examples of how the autonomy of peer production can be limited by elite technological power. Finally, GPS is an example of counter-technological power emerging from a centralized state-funded service.

The mapmaking methods I investigated are generated through peering practices – participatory, indigenous, and counter-methods – yet the underlying infrastructures are constructed through the imposition, social order, and domination of asymmetrical and elite-produced technological power. This critical understanding of technological power in hacker cartography focuses on peer production as a form of surplus value creation that is generated thanks to the materiality of cartographic infrastructure: GPS, atmospheric platforms like satellites, software, and in some instances, drones. Two, OSM and 3DR, exist because of code freely given by volunteers. GPS and indigenous countermapping are platforms made possible by professional, corporate, state, or traditional cultural organizations. GPS is funded by the US military, and some argue that the technologies behind small civil drones also emerged from the US military and are coopted by elite firms. Countermapping projects reinforce economic, class, gender, technology access, and other forms of community-based inequalities.

In this manner, participatory mapping may reify elite hegemony while also providing instances of citizen empowerment within these assemblages. As the countermapping example proves, technological power is not exclusively situated within Indonesian state elites. For instance, the technological power of GPS is not exclusively retained by the American state which originally funded it. It is not only exerted from drone firms like DJI and 3DR, neither is it total and conclusive. A new materialist approach to peer production exhibits how technological power has the potential to be differentially arrayed across technologies, cultures, and practices.

Let us return to Jordan's triptych of technological power. Barnes' theory of power is open-ended and an available resource for many actants in the assemblage. An example is indigenous countermapping in which drones are the tool for the production or attraction of technological power which is executed throughout an assemblage of relations: a gender imbalanced community of producers in Indonesia using drones to make orthomosaic maps of environmental degradation for court proceedings about the re-licensing of a bauxite mine. In this example of the making of indigenous maps participation is channeled by formal enterprises networked within the assemblage. Thus, technology power in this example of indigenous countermapping is mobile and fluid, located between the relations of a bauxite mine, environmental degradation, security forces, drones, photos, maps, and courts.

According to Jordan (1999), Weber's theory of power is centralized onto specific forces and is actively deployed across the assemblage by elites; examples include OSM and DIY Drones. OSM contributors can donate waypoints, polygons, and tags on self-collected geospatial data. DIY Drones participants share best practices, encouragements, code, and hacks. But these volunteered contributions to a centralized site are ultimately commodified. The participation stops with important spin-off firms that result directly from these peer communities. Here, the organized public provides the data for the formal social enterprise to come. Bytemark, CloudMade, Alphabet's and Microsoft's map subsidiaries profit off of this free labor. 3DR emerged explicitly to capitalize on DIY Drones, quickly became worth 100s of millions of dollars, before going bankrupt and re-emerging as an expensive boutique spatial data analytics company. Throughout the lifespans of these examples, technological power within the assemblages were controlled by actants and strategically deployed through relations.

A new material approach to peer production emphasizes technological power but also signals its mobility and how it fuels modes of resistance, community development, and the emergence of counter-intuitive formations. Those who advocate for peer production as a more autonomous, egalitarian, community-driven, or counter-hegemonic practice need not be disheartened by this conclusion. Rather, by highlighting technological power and how it variously manifests and is diffused across technologies, cultures of technological production, and political motivations, I seek to expose the materialistic breaking points of hegemonic power while realistically illuminating where dependencies on elites reside. In this manner, this scholarship is designed to enhance peer production studies by showing how bottom-up and top-down cultures are never static in their power geometries, but rather entangled in matter and through time.

Note

1 At the time of Leszczynski (2012) writing on OSM, the Bytemark logo was displayed on the OSM user-interface. As of late 2018, this is no longer the case.

References

Anderson, C. (2012). *Makers: The new industrial revolution*. New York, NY: Crown Business.

Anderson, C. (2017). Drones go to work. *Harvard Business Review*. Retrieved from https://hbr.org/cover-story/2017/05/drones-go-to-work

Barad, K. (2007). *Meeting the universe halfway: Quantum physics and the entanglement of matter and meaning*. Durham, NC: Duke University Press.

Barnes, B. (1988). *The nature of power*. Champaign, IL: University of Illinois Press.

Bauwens, M. (2005). The political economy of peer production. *Ctheory*. Retrieved from https://journals.uvic.ca/index.php/ctheory/article/view/14464/5306

Benkler, Y., & Nissenbaum, H. F. (2006). Commons-based peer production and virtue. *Journal of Political Philosophy, 14*(4), 394–419.

Bittner, C. (2017). OpenStreetMap in Israel and Palestine – "Game Changer" or reproducer of contested cartographies? *Political Geography, 57*, 34–48.

Blakemore, M., & Harley, J. B. (1980). Concepts in the history of cartography: A review and perspective, monograph 26. *Cartographica, 17*(4), 17–23.

Braidotti, R. (2013). *The posthuman*. Cambridge: Polity.

Chapin, M., Lamb, Z., & Threlkeld, B. (2005). Mapping indigenous lands. *Annual Review of Anthropology, 34*, 619–638.

Coleman, E. G. (2013). *Coding freedom: The ethics and aesthetics of hacking*. Princeton, NJ: Princeton University Press.

Coleman, E. G., & Golub, A. (2008). Hacker practice: Moral genres and the cultural articulation of liberalism. *Anthropological Theory, 8*(3), 255–277.

Coole, D., & Frost, A. (2010). *New materialisms: Ontology, agency, and politics*. Durham, NC: Duke University Press.

DeLanda, M. (2013). *Intensive science and virtual philosophy*. New York, NY: Bloomsbury.

Dodge, M., Perkins, C., & Kitchin, R. (2009). Mapping modes, methods and moments: A manifesto for map studies. In M. Dodge, R. Kitchin, & C. Perkins (Eds.), *Rethinking maps* (pp. 220–244). New York, NY: Routledge.

Eagleton, T. (2016). *Materialism*. Princeton, NJ: Yale University Press.

Fish, A. (2011). Indigenous digital media and the history of the internet on the Columbia Plateau. *Journal of Northwest Anthropology, 45*(1), 89–110.

Fish, A. (2017). *Technoliberalism and the end of participatory culture in the United States*. London: Palgrave Macmillan.

Fish, A., Murillo, L., Nguyen, Panofsky, A., & Kelty, C. (2011). Birds of the internet: A field guide to understanding action, organization, and the governance of participation. *Journal of Cultural Economy, 4*(2), 157–187.

Fish, A., & Srinivasan, R. (2012). Digital labor is the new killer app. *New Media and Society, 14*(1), 137–152. doi:10.1177/1461444811412159.

Foucault, M. (1980). Truth and power. In C. Gordon (Ed.), *Power/knowledge* (pp. 109–134). New York, NY: Pantheon Books.

Fox, N., & Alldred, P. (2017). *Sociology and the new materialism*. Los Angeles, CA: Sage.

Goodchild, M. F. (2007). Citizens as sensors: The world of volunteered geography. *GeoJournal, 69*(4), 211–221.

Grover, K., & Pugh, J. (2015). Assemblage thinking and participatory development: Potentiality, ethics, and biopolitics. *Geography Compass, 9*(1), 1–13.

Harley J. B. (1989). Deconstructing the map. *Cartographica, 26*, 1–20.

Jablonowski, M. (2015). Drone it yourself! On the decentring of "Drone Stories." *Culture Machine, 16*. Retrieved from http://culturemachine.net/index.php/cm/article/viewFile/589/594

Jordan, T. (1999). *Cyberpower: The Culture and Politics of Cyberspace*. New York, NY: Routledge.

Keay, J. (2000). *The great arc: The dramatic tale of how India was mapped and Everest was named*. New York, NY: HarperCollins.

Kittler, F. (2009). Towards an ontology of media. *Theory, Culture & Society, 26*(2–3), 23–31.

Kreiss, D., Finn, M., & Turner, F. (2011). The limits of peer production: Some reminders from Max Weber for the networked society. *New Media and Society, 13*(2), 243–259.

Latour, B. (2005). *Reassembling the social: An introduction to actor-network-theory*. Oxford: Oxford University Press.

Lemke, T. (2015). New materialisms: Foucault and the "government of things." *Theory, Culture & Society, 32*(4), 3–25.

Leszczynski, A. (2012). Situating the Geoweb in political economy. *Progress in Human Geography, 36*, 72–89.

Levy, S. (1984). *Hackers: Heroes of the computer revolution*. Garden City, NY: Doubleday.

Mac, R. (2016). Behind the crash of 3D Robotics, North America's most promising drone company. *Forbes*. Retrieved from www.forbes.com/sites/ryanmac/2016/10/05/3d-robotics-solo-crash-chris-anderson/#1808dc883ff5

Massey, D. (1999). Imagining globalization: Power-geometries of time-space. In A. Brah, M. J. Hickman, & M. M. Ghaill (Eds.), *Global futures. Explorations in sociology*. (pp. 27–44). London: Palgrave Macmillan.

McConchie, A. L. (2015). Hacker cartography: Crowdsourced geography, OpenStreetMap, and the hacker political imaginary. *ACME: An International E-Journal for Critical Geographies, 14*(3), 874–898.

Parks, L. (2001). Cultural geographies in practice: Plotting the personal: Global positioning satellites and interactive media. *Cultural Geographies, 8*(2), 209–222.

Peluso, N. (1995). Whose woods are these? Counter-mapping forest territories in Kalimantan, Indonesia. *Antipode, 27*(4), 383–406.

Perkins, C. (2011). Researching mapping: Methods, modes, and moments in the (im)mutability of OpenStreetMap. *Global Media Journal, 5*(2).

Radjawali, I., Pye, O., & Flitner, M. (2017). Recognition through reconnaissance? Using drones for counter-mapping in Indonesia. *The Journal of Peasant Studies, 44*(4), 1–17.

Stallman, R. (n.d.) On hacking. Retrieved from https://stallman.org/articles/on-hacking.html

Stuart, T., & Anderson, C. (2017). 3D robotics: Disrupting the drone market. *Berkeley Haas Case Series*, Retrieved from http://sk.sagepub.com/cases/3d-robotics-disrupting-the-drone-market

Sullivan, M. (2012). A Brief history of GPS. *PCWorld*. Retrieved from www.pcworld.com/article/2000276/a-brief-history-of-gps.html

Tsing, A. L. (2004). *Friction: An ethnography of global connection*. Princeton, NJ: Princeton University Press.

Wark, M. (2012). *Telesthesia*. Cambridge: Polity.

Wood, M. (2015). The society of university cartographers – its growth and international status. *The Cartographic Journal, 52*(3), 224–227.

15

P2P Learning

Panayotis Antoniadis and Alekos Pantazis

1 Introduction

There are numerous terms referring to horizontal and collaborative learning processes and methodologies, including peer learning, p2p learning, cooperative learning, collaborative learning, connected learning, networked learning, experiential learning, self-organized learning, project/problem-based learning, DIY learning, informal learning, social learning, situated learning, connected learning, critical pedagogy, radical pedagogy, hacker pedagogy, hip-hop pedagogy, horizontal pedagogy, post-pedagogy, andragogy, peeragogy, democratic education, progressive education, open education, direct education, popular education, free education, freedom schools, free schools, deschooling, and more. While all of these terms refer in one way or another to the same ideal, the granting of more independence and freedom to learners, they can differ significantly when it comes to the learning situation for which they were devised and thus are most relevant for. Perhaps the most important difference concerns the age of the learners, which allows for different assumptions regarding agency and self-determination (e.g., between children vs. adults). Other important differentiating assumptions concern the type of knowledge (e.g., explicit vs. tacit) and the overall sociocultural context (e.g., Global North vs. Global South). Finally, technology has given rise to new possibilities extending existing terms and concepts or even introducing new ones.

We use the term peer-to-peer (p2p) to describe such processes, thus making an explicit reference to technological p2p systems. The reason is that technology is being portrayed today as a key actor for the democratization of education (Benkler, 2006; Cronin, 2019; Knox, 2016; Williamson, 2015) and we believe that it is critical to better understand how technological tools can support p2p learning processes and the more or less hidden power asymmetries that lie behind the design and management of digital platforms themselves.

The term p2p was invented in the early 2000s (Oram, 2001) with the appearance of Napster, a p2p file sharing application that revolutionized online content sharing. It instigated a big shift toward the decentralization of the Internet at many levels, but also toward stricter policies on copyright infringement. Computers at the edge of the networks were (re-)imagined as "peers" performing identical functions, instead of "clients" connecting to powerful "servers." Napster itself was not a truly p2p application in the sense that it depended heavily on a central server for indexing and searching content. It was only the actual transfer of content that was happening directly from edge computer to edge computer, without the mediation of a central server. What is interesting for our comparative work on p2p learning, is that the engineering analogy with p2p systems helps to see clearly how centralization and power asymmetries are often more or less "hidden" and can appear in different dimensions of a system.

When applied to learning, the concept of peer-to-peer can also evoke many different interpretations and systemic challenges. For example, there are many cases where p2p learning and knowledge production methodologies could be established by a powerful entity in regards to a subordinate group, such as the employees of a company. More subtle influences and power structures could also exist in

The Handbook of Peer Production, First Edition. Edited by Mathieu O'Neil, Christian Pentzold, and Sophie Toupin.

deliberately horizontal groups for various reasons, like a strong personality influencing a learning group in a certain direction, or a digital platform promoting certain types of activities rather than others.

At a higher level, there are numerous cultural, political, social, and other power structures that significantly influence learning. From the very early days of formalized, and conservative, education systems there have been liberating forces toward more "progressive" education, like those described in Rousseau's famous *Émile*; and later by famous progressive educators such as Pestalozzi, Montessori, and Steiner who celebrated the individuality of children and their capacity to learn, stressed the role of family life and the wider society, but also of nature and impulses, and in general the importance of knowledge "beyond words." As noted by education historian John Howlett (2013) "Rousseau's Émile grows up in an environment designed, manipulated, and controlled entirely by the tutor." This pattern acknowledging the liberating features of p2p learning approaches toward one dimension of learning but stressing power imbalances in others, will frequently be observed in this chapter.

It is important to stress that this long thread of innovation around progressive education concerns mostly child-centered education. For adult learners, much less has been written and formalized. Andragogy (Knowles, 1980) is perhaps the closest concept to the idea of a p2p learning group, and Plato's "Symposium" one of the first documented peer-learning processes between adults. To structure the discussion, we have chosen three important dimensions of learning to refer to while analyzing various examples of methodologies, tools, and practices.

1. *Curriculum selection:* the choice of the learning objectives of an intentional or unintentional learning group.
2. *Learning process:* the different roles and interactions between different actors involved in the learning project, and their evolution over time.
3. *Knowledge abstraction:* the production of knowledge in the form of encyclopedic entries, tutorials, guidelines, tools, methodologies, and patterns.

Note that there is already a lot of existing work aiming to summarize, classify, or analyze learning processes, which often focus on one or two of these dimensions (Topping, 2005; Wegner, 1991). Some of these high-level analyses are contextualized, like the education system in the United States (Giroux, 2011; Monchinski, 2008) or Sweden (Laginder et al., 2013). Others address the technological dimension (Deimann & Peters, 2016; Williamson, 2015), while others focus more on political aspects (Haworth, 2012; Means, 2014).

There is also a body of literature on peer production, peer-to-peer and the commons. Benkler (2006) has coined the term "commons-based peer production" and discussed different areas of peer production. Bauwens, Kostakis, and Pazaitis (2019) have explicitly linked the term peer-to-peer with the commons, envisioning a societal transformation based on commons-based peer production and producing thorough research and concrete proposals.

While our analysis will be based on work carried out in different educational settings and contexts, our main focus and contribution will be on intentional adult learning groups. We highlight four specific projects, which share an important characteristic – they are examples of p2p learning processes that include face-to-face interactions:

- Two digital platforms, the P2P University (P2PU) and Openki.net, for creating and supporting self-organized learning groups in localities.
- Two physical spaces, the Tzoumakers rural makerspace in Greece and L200, a central self-organized space in Zurich, Switzerland.

In the following, we introduce the three learning dimensions we previously identified (Sections 2, 3, and 4) analyzing critically a selection of existing approaches in terms of visible or hidden power asymmetries. In Section 5 we analyze the underlying infrastructure, digital and physical, needed to support p2p learning projects. Section 6 concludes the chapter with a discussion on the need for a continuous struggle for democratic governance and empowerment, for which one of the most important tools is "reflection in action."

2 Curriculum Selection

There are different actors, such as the state, the family, the market, the community, that affect in different ways what people choose to learn, while their powerful influence is not always visible. For example, Ivan Illich (1971) warns us that "Everywhere the hidden curriculum of schooling initiates the citizen to the myth that bureaucracies guided by scientific knowledge are efficient and benevolent."

Making, therefore, such choices more consciously and more independently should be part of the essence of p2p learning. However, entering the terms "p2p learning" in an online search engine brings up among the top results a White Paper by the Versal Group (2016) which sees p2p learning as a means to "help your organization transition into a 'learning organization'; become more agile and competitive; heighten collaboration and productivity" (p. 2), focusing on narrow potential benefits for a specific company which of course is the one to determine the topic of learning for its employees. This is in contrast with, for example, P2P University (P2PU)'s wider value-driven vision aiming to create "a rich learning environment in which everyone simultaneously teaches and learns, acts and observes, speaks and listens," empowering "individuals to achieve something greater than they could have on their own" (P2PU, n.d., p. 5).

In his seminal work, the *Pedagogy of the Oppressed*, educator and philosopher Paulo Freire (2014 [1970]) has clearly set out as his mission toward a more participatory curriculum selection process: "How can the oppressed, as divided, unauthentic beings, participate in developing the pedagogy of their liberation?" (p. 48). However, Firth and Robinson (2017) challenge such "pedagogies of moulding," and criticize patronizing educational processes with "goals" or "missions" that aim to transform learners "in a direction desired by the knowing subject [. . .] conceived as necessary to meet social goals, to produce a particular kind of ethical subject, or to help the learner 'succeed' relative to social criteria" (p. 57). In his last book *Pedagogy of Freedom,* Paulo Freire (1998) defends the role of the educator when protecting the freedom of the students. In his own words: "Freedom is not the absence of limits. What I have always sought is to live the tension, the contradiction, between authority and freedom so as to maintain respect for both. To separate them is to provoke the infraction of one or the other" (Freire, 1998, p. 99). The fields of critical pedagogy and radical pedagogy, among others, address in depth such questions of power in terms of the object of learning.

Especially for adult education, technology can play a key role for empowering learners in their curriculum selection. For example, there are today online tools to implement the vision of the "learning webs" by Ivan Illich (1971), long time before the Internet, which included "Peer-Matching – a communications network which permits persons to describe the learning activity in which they wish to engage, in the hope of finding a partner for the inquiry" (p. 56). Digital platforms like P2PU and Openki.net are precisely designed to play the role of a modern learning web by facilitating, among others, the creation of ad hoc learning groups around different topics of interest.

The basic frame developed by P2PU through the overall narrative and design details of the platform is that there is a huge amount of knowledge available on the Internet, on various online learning platforms like Coursera, Khan Academy, and the like. Then the so-called "learning circles" are presented as a tool that can help groups to decide together what to learn from this wealth of knowledge and to help each other achieve their common learning objectives using p2p learning methodologies (Figure 15.1).

Learning circles are imagined as groups of people meeting once a week for 6–8 weeks; the facilitator, who does not have to be a domain expert, is central for bringing the group together. In terms of selecting the actual content, special guidance is offered through the *P2PU Facilitator Handbook*. P2PU (n.d.) directly addresses a candidate facilitator of a new course: "What knowledge gaps exist in your community? Who do you expect to attend? What do you like to learn?" (p. 8). To facilitate the collective process of creating learning circles, P2PU maintains a database of online learning resources. As P2PU's Chief of Stuff Grif Peterson (personal communication, November 2019) explains: "The goal is not to be a search engine of every online course, but a community-curated tool of high-quality free and open access materials," which is not static but always open for the addition of existing courses or the creation of new ones.

Openki.net is a project based in Zurich with similar objectives, which has developed over more than eight years a custom free and open source platform. The Openki platform looks more like a search engine than a forum, as in the case of P2PU. The goal is to create an attractive online place where

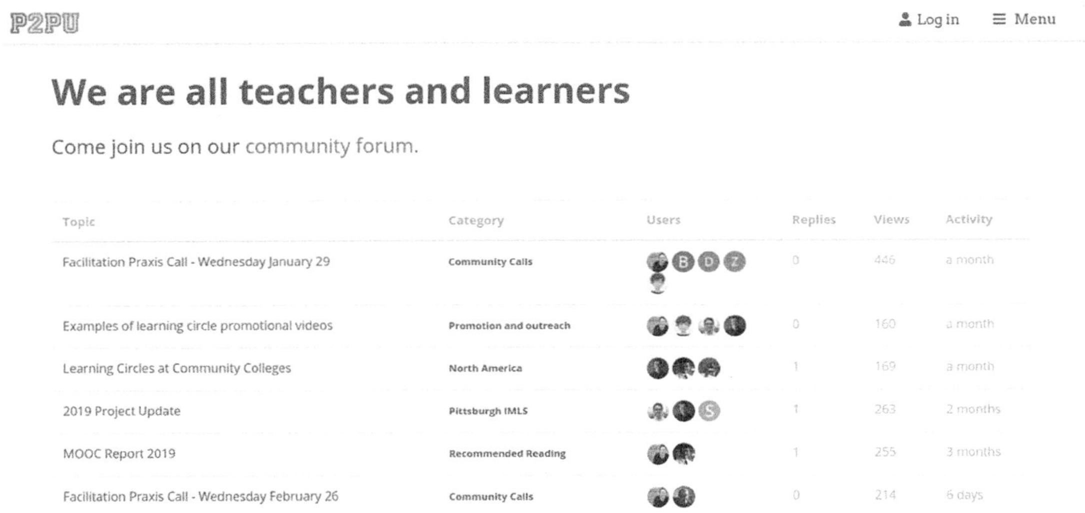

Figure 15.1 P2P University community forum <p2pu.org> stresses the duality between teaching and learning, "we are all teachers and learners" (accessed: 3 March, 2020). Source: Peer 2 Peer University.

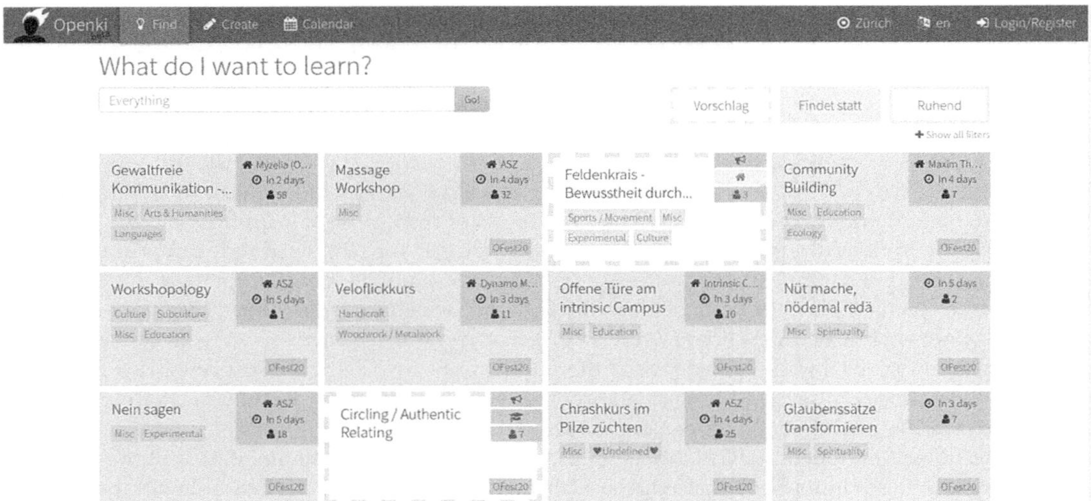

Figure 15.2 The Openki.net home page in the Zurich region, encouraging the visitors to think "what do I [really] want to learn?" (accessed, 3 March, 2020). Source: Openki.net.

people build their learning groups and learn things that are not necessarily already available as online courses. One of its novelties is that the platform is designed in a way to strongly encourage the creation of courses in the platform by those that just want to learn rather than those that have the capacity to teach or would like to play the role of the facilitator. Literally, it is a bottom-up learning group creation process. When such a project grows significantly, however, the need for filtering will become more and more apparent. Then an important source of power becomes indeed the prioritization of certain learning topics over others, or even excluding those judged as inappropriate (Figure 15.2).

What is interesting about both P2PU and Openki is that they put in place technological tools for facilitating face-to-face interactions instead of strictly online learning processes like MOOCs (Knox, 2016). This type of "blended learning" is very important to guarantee that p2p learning processes are really democratic and can form defenses against powerful actors like the mass media, which can create "hidden curricula" and influence significantly seemingly free choices. As Dewey

(1927) stated "Democracy must begin at home, and its home is the neighborly community" (p. 213).

Study circles in Sweden is another project that shares this local perspective of creating p2p learning groups (Laginder et al., 2013). However, no special care is taken to facilitate the creation and organization of such groups, which were traditionally formed in the workplace. Today group formation is left to the participants, and special institutions provide support to already formed groups. Funding is an ambivalent dimension, which on the one hand significantly stimulates such processes but also adds an important power element to the overall structure.

For learning projects that narrow down the learning scope, as participatory design/making projects do, there is also a lot of room for selecting the actual object of learning, or better learning by doing, as in the case of the *"Tzoumakers" project*. Tzoumakers is a rural makerspace initiated by the *P2P Lab* research collective. More specifically, Tzoumakers connects small-scale farmers with similar communities around the globe with the aim of collaboratively designing and manufacturing open source tools for their everyday life and work. The choice of tools is up to them, inspired by like-minded people (e.g., practitioners from *L'Atelier Paysan* community in France) working together with the locals to translate their needs into solutions and their solutions into tools. In the case of Tzoumakers, it is interesting that the facilitation team has no expertise concerning the technical aspects of the making process, but their main task is to develop methodologies regarding the process and to codify the knowledge produced to make it easily applicable and replicable.

3 Learning Process

Assume now that a group has been formed and has selected, democratically or not, a topic to be learned in collaboration. How is the learning process to be organized, in what way, and to what extent could it be described as "peer-to-peer"?

3.1 Roles

There are different roles in a learning process, those of learner/student/participant and teacher/mentor/expert/facilitator being the most prominent in terms of function, but most importantly, when one explores the p2pness of a learning process, in terms of the interactions and power relationships between them. On the learning side, an interesting analysis and comparison of the words "learner" vs. "student" by Biesta (2010) considers the word "student" as more empowering since students gain themselves, through study, the necessary knowledge and the teacher only needs to encourage them to use their own skills; in the extreme case an "ignorant schoolmaster" (Rancière, 1991). Biesta proposes the term "speaker" to signify an active agent in the learning process who not only listens but also speaks. We chose the word "learner," as "student" is connected to traditional education and to the "classroom." The term "speaker" is a qualitative term and it is actually our objective to explore settings, methodologies, and technologies that enable learners to become "speakers."

On the teaching side, the key question is to what extent an expert, a teacher, or a mentor is actually needed in a p2p learning process. Illich (1971) observes that skilled experts are not necessarily the most suitable actors for transferring their knowledge. This knowledge and/or the observed real life activities of experts could be considered instead as useful "sources" of information, which a p2p learning group can process independently with the optional support of a facilitator. The facilitator might have no knowledge of the subject itself but can help the group to perform the learning task in collaboration, as suggested in the "facilitator handbook" by P2PU (n.d.), among many other similar toolkits. In the context of p2p learning, a mentor could be seen as a knowledgeable facilitator who aims more to inspire and empower than to "teach" in the traditional sense, close to the image of Paulo Freire as described above.

In the vocabulary of the P2P University, "content" is an additional core element expressing already developed online learning material which could act as an "expert source" of learning for a p2p group. Openki.net explicitly names more "operational" roles for an Openki course, like the host or organizer, and ongoing development of the platform aims to introduce even more predefined or customized roles like the note-taker, researcher, communicator, and more. The *Peeragogy Handbook* (Corneli et al., 2016) provides a long list of such potential roles in peer-learning projects: "Team Member,

Manager, Leader, Reviewer, Editor, Author, Content Creator, Presenter, Designer, Graphic Artist, Technologist, Participant, Coordinator, Planner, Mediator, Moderator, Facilitator, Proponent, Advocate, Representative, Contributor, Activist" (p. 17).

Of course, many roles are possible depending on the specific topic and the context, and do not have to stay fixed. A learning group can start with an expert taking the key role of mentor, but along the way the learning process could be gradually more empowering for the rest of the group so as to feel ready to take the mentor role in the same or different groups. The learning methodology could also explicitly give the teacher role to the learners as an effective way to learn.

In the Tzoumakers makerspace (Figure 15.3), the participatory making of designed global and manufactured local tools entails a variety of roles from "external experts" invited to a hackathon and local experienced craftsmen to "facilitators," local designers, programmers and coordinators of an event. Moreover, the local community often draws ideas from the designs of tools that other communities worldwide have shared freely via the Internet. Codified knowledge in the form of freely distributed mechanical design or a YouTube video inspires and teaches a distant community in various p2p ways to build actual tools, improving their everyday life.

There are often important roles that are not particularly visible in the actual learning process, like that of a funding body which can significantly influence the formation of groups and the topic of learning. The case of study circles in Sweden is very characteristic in this respect, "an activity whose charter is "free and voluntary" but which, for its very existence, depends on state subsidy" (Laginder et al. 2013, p.19). In Tzoumakers, funding plays a delicate role since it usually places the initiators and facilitators of the process in a powerful position. For example, participatory action research entails both orienting and observing the whole process at the same time, but the opinion of the group that injected most of the funding for the realization of the project, indubitably weighs more. The equilibrium in such circumstances should and can be found most often through developing mutual trust.

Figure 15.3 A moment at the Tzoumakers.gr makerspace depicting the friendly relationship between local experts and community members toward building common experiences and exchanging knowledge. Photo: Nicolas Garnier.

Figure 15.4 The entrance of L200 at Langstrasse 200, Zurich, which stresses the multiple potential uses of the space. Photo: Panayotis Antoniadis.

P2p learning processes can also take place between "groups" rather than individuals, as in interdisciplinary or transdisciplinary research, and mutual learning or knowledge exchange programs (Boehm et al., 2017; Healey & Upton, 2010; INURA, 1999). Such roles are always representational, in the sense that the groups that engage in the interaction represent a bigger group, a discipline, a traditional practice in a certain country, and so on. NetHood Zurich, for example, has acted as a facilitator for knowledge exchange and transdisciplinary research on various areas of commoning through a series of EU projects, namely, COMPARE (2015), MAZI (2016–2019), and netCommons (2016–2019), which resulted in the design of a new collective space in Zurich, *L200*, co-founded by NetHood and now run without any external support (Figure 15.4). The ability to engage in an action research project without external dependencies is a particularly luxurious situation both for the researcher and activist roles, which allows experimentation with a structured laissez-faire methodology of participatory design (Apostol & Antoniadis, 2020), which does not pose any constraints on the use of the space, except for a strict rule of non-domination of the identity of the space.

L200 offers today a hybrid digital and physical platform for citizens to express their needs and ideas, and to learn together, not through answering an online questionnaire or by raising their hands in a public meeting, but through direct actions of their choice. The main role of the researcher(s) in action is to be reflective and to analyze the process informed by different fields and disciplines, toward the generation of knowledge for the co-creation of the space itself. As the space coordinator, the main role of the activist is to make sure that the members of the association experience the space as their own, as free to use as they wish, through carefully designed tools, rules, and processes, both physical and digital, that are supporting the commoning activities that take place on "top" of the common infrastructure that the space provides (Apostol & Antoniadis, 2020).

3.2 Interactions

Moving from the different roles that peers can assume in a learning process to the actual enacted relationships and interactions that take place during the process itself, there is a wide variety of learning methodologies to consider. These methodologies follow in principle the "learning as participation" instead of the "learning as acquisition" metaphor formulated by Sfard (1998), which assumes that "learning is more akin to apprenticeship" instead of considering the mind as "a container to be filled through knowledge and learning."

There are many variations depending on the quality of the relationship both for "links" between actors assuming some sort of hierarchy, like between learners and experts. For example, Paulo Freire was in his classroom an uncontestable authority, but was guiding his students toward autonomy. And even when the students are children, there are ways to transform the learning process into a more empowering experience (Howlett, 2013).

Similarly, even in the most horizontal models there are often inherent asymmetries of knowledge, skills, and position, and it might be more harmful than beneficial if they are not acknowledged from the start (Freeman, 2013). Often in groups, it is more difficult to deal with informal hierarchies than with clearly stated ones. But "dealing" does not mean eliminating special roles, since they are often necessary, but making sure that such roles are not exclusive, concealed, or authoritarian.

The most "traditional" p2p learning methodologies are clearly hierarchical in that a strong figure (teacher, instructor, mentor) is leading the process and the focus is on the benefits of encouraging peer interactions among the learners. In short, the main idea is that students can learn better the object of learning (the content) when they help each other in groups or when they are brought in the position of the teacher, being asked to explain the freshly learned content to others as if they were the teacher. Peer instruction, Peer Tutoring, Cooperative Learning, Collaborative Learning, Learning by Teaching, all fall into this category and when one reads the corresponding definitions it is hard to perceive what differentiates them.

When there is no expert, nor a powerful authority to dictate the process, a facilitator could help peers to reach a certain learning goal and make sure that they do not only "learn" but that they also "speak" (Biesta, 2010). Going back to our technical p2p file sharing systems, they do not only "download" content, but also "upload."

Experiential learning (Dewey, 1997; Kohonen, 2012; Kolb, 2015) is an example of a methodology that promotes p2p interactions and a participatory approach to learning. Experiential learning focuses on a specific collective activity or experience, whose outcome is not a priori defined. Learning by doing, active participation, and experimentation are the main pillars of this approach. Kolb's circle proposes a cycle of multiple brakes for reflective observation, a rotation between abstract conceptualization, active experimentation, concrete experience, and reflective observation (Kolb, 2015, p. 4). Thus, Kolb's circle drags up the experienced learning from the subconscious to the conscious level, making learning accessible and taking experiential learning one step further. The role of facilitators is then to design the common experience, the moments of reflective observation, and the conceptualization process. One of their key tasks is to balance the excitement of the collective action, and the need for self-reflection and deeper understanding.

In other patterns of p2p learning interactions, such as the simple sharing of personal experiences in a group, facilitators need to make sure that all voices are heard and respected. For example, "Consciousness-raising (CR) groups," an example of a "pedagogy without moulding" (see Section 2), are defined by Firth and Robinson (2017) as "voluntary, usually women-only, regular discussion groups". In this case the collective learning process includes the "recounting and interpreting the experiences of participants, generally by presenting members' experiences around a defined topic and then drawing out similarities and structural relations to the oppression of women" (p. 66).

Another pattern of p2p learning is the existence of an intermediate object, like the so-called *boundary object*, a term coined by Star and Griesemer (1989) to mean a shared space, a common object "both plastic enough to adapt to the local needs and constraints of the several parties employing them, yet robust enough to maintain a common identity across sites" (p. 393). These different groups are often referred to as "social worlds" and the basic assumption is that "consensus is not necessary for cooperation nor for the successful conduct of work" (p. 388). Facilitation in this case is more demanding and requires the balance of power asymmetries that are often hidden, the lack of a common (disciplinary) language, and time constraints, among others (Apostol & Antoniadis, 2020).

All the above examples of methodologies and patterns of p2p interactions make sense independently of technology, which significantly affects the content of learning and identities of the learners and their "way of thinking" (Williamson, 2015). But it also affects the process itself, giving rise to new patterns of interactions such as the so-called connected or networked learning, which take place in dedicated online educational environments, including Massive Open Online Courses (MOOCs) and many other learning digital platforms, or "in the wild," through participation in social media platforms and a wide variety of online groups and communities. The Internet has become a global educational platform, which on the one hand empowers people, giving them access to information and knowledge, but on the other leaves learners alone: self-learners find themselves decontextualized and in effect powerless.

4 Knowledge Abstraction

It is not difficult to argue that how knowledge is abstracted, documented, and represented plays a key role in how it will be consumed, integrated and further produced. There are different forms of knowledge, others more explicit (knowing that) and others more tacit (knowing how). Texts, songs, drawings, images, graphics, films, formulas, recipes, social norms, institutions, laws, tools, patterns, blueprints, objects, statutes, machines, language itself, are all abstractions of knowledge produced and developed over the history of human civilization. It is very often the case that learning groups base their common seeking of knowledge on existing books, or online information. But learning also produces knowledge along the way, even for very technical topics. It might not develop further the actual theory, for example, but could reveal new ways to learn about it, interesting associations, interpretations, and more.

Focusing on such groups, or "communities of practice," Wegner (1991) uses an extended interpretation of the term "reification" to describe knowledge as a simplification of practice like "abstractions, tools, stories, terms, and concepts that reify something of that practice in a congealed form" (p. 61), stressing that, another key concept, participation "is essential to repairing the potential misalignments inherent in reification" (p. 64). His learning theory puts a high emphasis on the duality between participation and reification, without going into very much depth on how reification actually takes place, and who is responsible for producing the abstractions of knowledge developed over time in a community of practice.

Traditionally, it was a key actor, the author, the scientist, the educator, the designer, the politician, the priest, the artist, the planner, the architect, who was responsible for collecting information, observing, and reifying the collective social learning processes. It was dependent on the special qualities of this actor to what extent the final result was representative of the collective, whose only involvement in the abstraction process was to judge the way "their representative" has analyzed, interpreted and eventually reified, their individual contributions to the common pool of knowledge, through their participation in everyday life. The easier it is for someone to contribute to this common pool of knowledge the more critical it becomes to have democratic evaluation mechanisms in place to filter the available subject to different constraints, like cost (e.g., printing), space (e.g., the size of a library), or attention (e.g., the time required to process content). To this end, the peer review system in science is an example of a sophisticated peer evaluation mechanism between "authors" whose numerous variations showcase both the importance and difficulty of the task (Tennant et al., 2017).

Digital technology has made such peer evaluation processes of all types more and more expressive and accessible. Now derivative works, adaptations, or memes get created at a very fast pace, a process that is largely facilitated by the creative commons license. Multi-layered evaluation systems, where reviewers are also evaluated, are easy to implement, like the innovative online reputation-based review system of the Slashdot platform. Moreover, all types of knowledge abstractions can gain reputation and visibility (but unfortunately not credibility) by being "viewed," "liked," "voted," or "shared." Perhaps the most dangerous misconception regarding the digital platforms that mediate such p2p learning and evaluation processes is that they are neutral in their design and people are free to "like" and "share" what they want. The reality is that such platforms hide many biases that are far from "random" but very carefully engineered toward manipulating their users according to the commercial and political objectives of the corporations that run them (Antoniadis, 2018).

Advanced digital tools allow also for modes of peer production of knowledge which are unique in history, like the co-creation of a single learning object, through collaborative editing, designing, making, coding. Clearly, Wikipedia is the flagship p2p project in this context, an online encyclopedia edited freely by thousands of Internet users who produced millions of articles (Haider & Sundin, this volume). Tkacz (2015) provides a very comprehensive analysis of how Wikipedia's principles generate certain "frames" and "statements," which promote some views over others about which types of knowledge are legitimate, acting as a source of power in the conflict resolution process "from which the authority of commands can be established ('This is an encyclopedia, therefore Wikipedia Art must be deleted')" (Tkacz, 2015, p. 85). Casemajor et al. (2019) provide another example through which they argue convincingly that the very design of Wikipedia, assuming its content is unconditionally "open," "categorized," and unidimensional, promotes a Western mindset that is not compatible with indigenous culture.

The question then arises: why should someone insist on using Wikipedia and not simply create another platform based on the wiki technology with appropriate design and rules? The term "forking" has been established in the software programming world to describe this possibility, which often plays the role of a "liberogenic" device (Tkacz, 2015, p.136). The truth is that projects like Wikipedia cannot be easily forked in their entirety but it is perfectly possible to do so for similar smaller scale projects, and there are many examples of successful "forks" in the digital world like LibreOffice, NextCloud, and many Linux variants. But notice that the choice to fork Wikipedia is not merely a technical one, as it is in the case of "stand-alone" systems like LibreOffice or Linux. Wikipedia is a socio-political system whose value goes beyond the personal consumption of knowledge. Indeed, the desire for one's perspective to be represented in a system like Wikipedia, could be attributed to the visibility and credibility of such a project. If it is on Wikipedia, there is a high chance that someone has read and potentially reviewed an entry. However, such globalized and unidimensional knowledge is by design limited and de-contextualized, and is more relevant to explicit than tacit knowledge.

One of the motivations behind our selected case studies was to bring back the focus to localities, and to help create learning processes that not only consume globalized knowledge but also produce local knowledge. For this it is important to add the "knowledge abstraction" part explicitly in the activities of p2p learning groups such as a recent open call the Tzoumakers space circulated to its members and supporters to build methodologies for documenting the design of the tools designed and constructed in the lab. This action is part of the "design global, manufacture local" (Kostakis et al., 2015, 2016) mode of commons-based peer production where groups or individuals freely collaborate to design tools, share the knowledge as a global commons while building them locally with local conditions in mind.

It is only when the right balance between consuming global knowledge and co-creating local knowledge is maintained that many of the hidden power asymmetries can be reduced and learners become speakers. For this, platforms like P2PU and Openki should consider integrating in their design functionalities and processes that encourage collective knowledge abstraction, not only of the content itself but also of the p2p learning processes required for its appropriation and further development.

To this end, design blueprints, DIY toolkits, FOSS software and patterns are very important knowledge abstraction devices that can help to create a balance between the global and the local. They could be seen as flexible seeds, developed as configurable prototypes or patterns at the global level through international networks of p2p learning, which then need to be carefully customized and planted in the local environment. As Christopher Alexander (1979) introduces his "pattern languages" in architecture,

> It is the process which creates the organism – and it must be so. No thing which lives can possibly be made in any other way. If you want to make a living flower, you don't build it physically, with tweezers, cell by cell. You grow it from the seed.

What makes things complicated in the case of learning is that it is not only the knowledge seeds that need to be designed in a way to be easily appropriated by local communities but also the "ground" where such seeds can grow, the underlying digital and physical infrastructure.

5 Learning Infrastructures

5.1 Digital Infrastructure

Digital technologies have an impact on different aspects of human life, including of course education. All three dimensions of the learning process discussed above can be facilitated with the use of digital platforms of different types. The more powerful these digital platforms become in storing, processing, and filtering information, the less critical appears to become the role of the teacher, the educator, the facilitator.

But as more power is moving from traditional actors to digital platforms, the more important it becomes to question how these platforms work and who owns, who designs and who controls them. There is an abundant literature on the serious threats caused by the increasing dependence on digital technology, especially when it is big corporations that shape its functionality and exploit the immense amount of behavioral data produced toward commercial and political objectives. For example, Cronin (2019) warns us that "the suppression of privacy lies at the heart of the business models of most digital and social media platforms – which rely directly on the appropriation of data for profit." Most importantly, "many of the tools and platforms we use to engage in social connection and open educational practices have bias and inequality built into them – they are designed to allow and encourage forms of participation, and prevent others."

The e-book edited by Williamson (2015) again provides a comprehensive analysis of the critical role that software code plays in shaping educational processes, among others, and the hidden and dangerous biases that an algorithmic treatment of education, as manifested through "learning analytics," "automated learning," and other "smart" educational platforms, can impose on "users." The big question is, what can we do? How can we engage deeply with the reality of the Internet and the digital platforms running on it?

At the software level, free and open source software (FOSS) provide a credible alternative to corporate platforms and offers significant flexibility for democratic appropriation and the adaptation of existing tools to serve local needs; what is called "infrastructuring" in the field of participatory design (Ehn, 2008). There are also specialized platforms that can facilitate asynchronous collaboration for digital creations (e.g., GitHub, GilLab). Such platforms are based on the experience of p2p digital collaboration, integrating features and practices that enhance the capability of communities to carry collaborative projects.

Openki.net is a platform built with exactly these issues in mind. Similarly, as noted by Grif Peterson (personal communication, November, 2019), P2PU initially considered the use of MeetUp but this option was rejected because there were doubts that the project could scale on its own terms if depending on such a platform. So, the P2PU team built their own tool for event management and a complimentary API. This offered to the learning circle communities a ready-to-go solution for running learning circles, without giving up the flexibility for people who are looking for a self-hosted solution.

Notice that software is only a part of the infrastructure required to run a digital platform. Community Networks (Dulong de Rosnay et al., 2019; see also Shaffer, this volume) are examples of networking infrastructure, including routers, antennas, and servers, which are owned and managed by local communities as a commons. Antoniadis (2018) proposes the concept of the "organic Internet" to describe such infrastructure, stressing its capacity to be installed and deployed where the community using the corresponding software services is actually located. What is crucial in this scenario is that the co-creation of the digital platforms designed to support p2p learning processes is itself a very emancipatory learning process about how the Internet works and why digital sovereignty is linked to fundamental human rights.

5.2 Physical Infrastructure

Formal education includes in general a very well-defined space where learning activities take place, the classroom. Peer-learning processes of the type described in this chapter also need spaces where people can meet to create learning groups, organize their learning processes, interact, and so on. The design and governance of those spaces also play a key role in the process, and they can also be more or less "peer-to-peer," or in this context commons-based.

The P2PU has developed a strong collaboration with public libraries, which are natural locations for learning activities, although mostly imagined as solitary places for study. More dedicated spaces for p2p learning around making and innovation are makerspaces, hackerspaces, or fablabs (i.e., fabrication laboratories). There is a wide range of people that you can meet in these places: architects, designers, engineers, programmers but also woodworkers, machinists, hobbyists, or just curious individuals. All of them engage in the process of learning by doing, often helping and collaborating to develop an idea into an actual physical or digital artifact (Sheridan et al., 2014; see also Boeva & Troxler, this volume).

While the most prominent aspect in makerspaces is the final artifact, the social connections and interactions that frame the making and learning experiences are of equal importance (Telhan et al., 2016), but in general the corresponding processes that lead to these experiences are neither predefined or regulated (Halverson & Sheridan, 2014). As Blikstein and Worsley (2016) mention, "novices coming into a maker lab need a considerable amount of onboarding and facilitation before they can start 'hacking' and learning by themselves" (p. 71). However, in reality preconceptions on social groups that are not considered "technology people" (Blikstein & Worsley, 2016) often lead to self-exclusion and indeed well-educated, affluent, white males are by far the main actors (Halverson & Sheridan, 2014).

In general, governance rules can play a key role in sustaining the participatory spirit. In the case of the Tzoumakers project, space rules are formed from both the local rural community, the P2P Lab research collective that initiated this project, the local municipality that offers the building and, for the first years, from the Interreg EU project that funds a number of its activities. Does this reflect on the learning processes that take place? Of course it does, as p2pness consists of a mixture of typologies involving learners, facilitators, and experts in a space that incorporates functional characteristics from all the formal or informal bodies that constitute this special makerspace.

Notice that in the case of makerspaces, the p2p learning process is always anchored in the space itself which serves as the focal point for the corresponding community of practice. The same holds in the case of learning circles that in most cases are anchored in libraries. In the case of Openki.net the matchmaking platform is the focal point and physical space is seen as a flexible infrastructure which can be offered by one of the participants. The idea that p2p learning activities can take place in any urban space, indoors or outdoors, as part of everyday life, significantly increases the target audience and the overall impact. Seen from this perspective, L200 is a collective space in Zurich that tries to combine these two perspectives. On the one hand, L200 is a very central space that acts as a focal point, between a wide community of citizens that participate in its governance as a commons and in its sustainability through cost sharing. On the other hand, it explicitly avoids creating a specific identity for the activities through a sustained effort to make the space as inclusive as possible, open to always new and unexpected uses. Placing the question "What is this?" prominently in the space's facade (Figure 15.4) reminds everyone that failing to keep the space neutral and open to all types of usages would be against its main goal. The L200 project is also interesting because it is perceived as a hybrid (digital and physical) space designed as a prototype whose success depends on its ability to replicate; its core identity, even its name and URL is defined by its location, Langstrasse200.ch. Through such design choices the message is clear that "forking" is not considered as an "exit strategy" that helps to build "consensus," (Tkacz, 2015, p. 136) but as a necessary step for transforming p2p learning into an everyday activity.

6 Conclusion

In times when digital technology has been promoted by industry and governments as a game changer for enabling p2p learning beyond borders and constraints (Benkler, 2006; Cronin, 2019; Deimann & Peters, 2016; Knox, 2016; Williamson, 2015), we highlight the irreplaceable role of face-to-face interactions and the conditions under which technologies can empower participants. Although technology still has an important role to play, it brings many threats and more layers of power imbalances, in addition to race, gender, culture, class, location, sexuality and other structural inequalities. When all these are analyzed in detail, the concept of peer-to-peer looks more like a utopian than a realistic goal. But instead of becoming paralyzed by the impossibility of a truly egalitarian and p2p way to learn together, we advocate for a continuous effort to engage with p2p learning in everyday life, in the "neighborly community," even when this seems extremely difficult. The aim is to use technology as a means and as a complement, not as an end. For example, local communities can co-produce global knowledge, share it openly, and thus empower other communities to adequately adapt and address their needs without imposing external values on them. Our selected case studies are examples of the type of tools and institutions that we believe can help in this direction.

We would add to this toolbox of p2p learning "reflection in action" which acknowledges our "capacity for reflection on our intuitive knowing in the midst of action and, sometimes uses this capacity to cope with the unique, uncertain, and conflicted situations of practice" (Schön, 1983, pp. viii–ix). Pettit (2006) argues similarly that "understanding and addressing power, calls for more innovative learning processes, [. . .] the conceptual and rational re-evaluation of one's assumed perspective [. . .] making sense of one's experiences of power, and of realizing one's capacity to shift power." Ending then with a self-reflection exercise, we would like to highlight the fact that works like the *Handbook of Peer Production*, full of arguments and analyses on the benefits of openness and collaboration, are distributed behind a paywall. Of course, there are no easy solutions. But perhaps we should take a moment to think about why people like us, criticizing the current publishing industry, do not find a more open and p2p way to publish our work and still be respected in our professional circles.

Acknowledgments

Panayotis Antoniadis has been partially supported by the project Heteropolitics that has received funding from the European Research Council (ERC) under the European Union's Horizon 2020 research and innovation program (grant agreement 724692). Alekos Pantazis acknowledges financial support from the European Research Council (ERC) under the European Union's Horizon 2020 research and innovation program (grant agreement No. 802512).

References

Alexander, C. (1979). *The timeless way of building*. Oxford: Oxford University Press

Antoniadis, P. (2018). The organic Internet: Building communications networks from the grassroots, co-designing economies in transit. New York, NY: Palgrave Macmillan.

Apostol, I., & Antoniadis, P. (2020). Central urban space as a hybrid common infrastructure. *Journal of Peer Production, 14*, 62–86.

Bauwens, M., Kostakis, V., & Pazaitis, A. (2019). Peer to peer: The commons manifesto. London: University of Westminster Press.

Benkler, Y. (2006). *The wealth of networks*. New Haven, CT: Yale University Press.

Biesta, G. (2010). Learner, student, speaker. *Educational Philosophy and Theory, 42* (5–6), doi:10.1111/j.1469-5812.2010.00684.x

Blikstein, P., & Worsley, M. (2016). Children are not hackers, building a culture of powerful ideas, deep learning, and equity in the maker movement. In K. Peppler, E. R. Halverson, & Y. Kafai (Eds.), *Makeology: Makerspaces as learning environments (Volume 1)* (pp. 64–79). New York, NY: Routledge.

Boehm, K., James, T., & Petrescu, D. (2017). Learn to act: Introducing the Eco Nomadic School. Retrieved from www.rhyzom.net

Casemajor, N., Coocoo, C., & Gentelet, K. (2019). Openness, inclusion and self-affirmation: Indigenous knowledge in open knowledge projects. *Journal of Peer Production, 13*. Retrieved from http://peerproduction.net/issues/issue-13-open/peer-reviewed-papers/openness-inclusion-and-self-affirmation/.

Corneli, J., Danoff, C. J., Pierce, C., Ricaurte, P., & Snow MacDonald, L. (Eds). (2016). *The peeragogy handbook* (3rd ed.). Chicago, IL: PubDomEd/. Retrieved from http://peeragogy.org

Cronin. C. (2019). Open education: Design and policy considerations. In H. Beetham & R. Sharpe (Eds.), *Rethinking pedagogy for a digital age: Principles and practices of design*. London: Routledge. Retrieved from http://eprints.teachingandlearning.ie/4343/1/Cronin%20chapter%202019.pdf

Deimann, M., & Peters, M. A. (2016). *The philosophy of open learning: Peer learning and the intellectual commons*. New York, NY: Peter Lang.

Dewey, J. (1927). *The public and its problems*. New York, NY: Holt.

Dewey, J. (1997). *Experience and education*. New York, NY: Free Press.

Dulong de Rosnay, M., Tréguer, F., Antoniadis, P., Apostol, I., Aubrée, V., Burns, A., Michalis, M., Neumann, J., & Roca i Tió, R. (2019). *Telecommunications reclaimed: A hands-on guide to networking communities*. Reston, VA: Internet Society.

Ehn, P. (2008). Participation in design things. In *PDC '08: Proceedings of the Tenth Conference on Participatory Design, Bloomington, Indiana* (pp. 92–101). New York, NY: ACM Press.

Halverson, E. R., & Sheridan, K. (2014). The maker movement in education. *Harvard Educational Review, 84*(4), 495–504.

Freeman, Jo. (2013). The tyranny of structurelessness. *WSQ: Women's Studies Quarterly, 41*(3–4): 231–246. https://doi.org/10.1353/wsq.2013.0072

Firth, R., & Robinson, A. (2017). From the unlearned un-man to a pedagogy without moulding: Stirner, consciousness-raising, and the production of difference. In R. H. Haworth & J. M. Elmore (Eds.), *Out of the ruins: The emergence of radical informal learning spaces.* Oakland, CA; PM Press.

Freire, P. (2014) [1970]. *Pedagogy of the oppressed* (30th anniversary ed.). New York, NY: Bloomsbury.

Freire, P. (1998). *Pedagogy of freedom.* Lanham, MD: Rowman & Littlefield.

Giroux, H. A. (2011). *On critical pedagogy.* London: Bloomsbury.

Haworth, R. H. (2012). *Anarchist pedagogies: Collective actions, theories, and critical reflections on education.* Oakland, CA: PM Press.

Healey, P., & Upton, R. (Eds.). (2010). *Crossing borders.* London: Routledge.

Howlett, J. (2013). *Progressive education: A critical introduction.* London: Bloomsbury.

Illich, I. (1971). *Deschooling society.* New York, NY: Harper & Row.

INURA (Eds.). (1999). *Possible urban worlds. Urban strategies at the end of the 20th century.* Basel: Birkhäuser.

Knowles, M. S. (1980). *The modern practice of adult education: From pedagogy to andragogy.* Chicago, IL: Follett.

Knox, J. (2016). *Posthumanism and the massive open online course: Contaminating the subject of global education.* New York, NY: Routledge.

Kohonen, V. (2012*). Learning to learn through reflection: An experiential learning perspective.* Tampere: University of Tampere.

Kolb, A. D. (2015). *Experiential learning: Experience as the source of learning and development* (2nd ed.). Upper Saddle River, NJ: Pearson.

Kostakis, V., Niaros, V., Dafermos, G., & Bauwens, M. (2015). Design global, manufacture local: Exploring the contours of an emerging productive model. *Futures, 73,* 126–135.

Kostakis, V., Roos, A., & Bauwens, M. (2016). Towards a political ecology of the digital economy: Socio-environmental implications of two value models. *Environmental Innovation and Societal Transitions, 18,* 82–100.

Laginder, A., Nordvall, H., & Crowther, J. (Eds.). (2013). *Popular education, power and democracy: Swedish experiences and contributions.* Cardiff: Niace.

Means, J. A. (2014). Educational commons and the new radical democratic imaginary. *Critical Studies in Education.* https://doi.org/10.1080/17508487.2014.903502

Monchinski, T. (2008). *Critical pedagogy and the everyday classroom.* New York, NY: Springer.

Oram, A. (2001). *Peer-to-peer: Harnessing the power of disruptive technologies.* Sebastopol, CA: O'Reilly.

P2PU. (n.d.). *Learning circles: Facilitator handbook.* Retrieved from www.p2pu.org/en/handbook/

Pettit, J. (2006). Power and pedagogy: Learning for reflective development practice. *IDS Bulletin, 37*(6), 69–78.

Rancière, J. (1991). *The ignorant schoolmaster.* Stanford, CA: Stanford University Press.

Schön, D. A. (1983). *The reflective practioner: How professionals think in action.* New York, NY: Basic Books.

Sfard, A. (1998). On two metaphors for learning and the danger of choosing just one. *Educational Researcher, 27*(2), 4–13.

Sheridan, K., Halverson, E. R., Litts, B., Brahms L., Jacobs-Priebe L., & Owens, T. (2014). Learning in the making: A comparative case study of three makerspaces. *Harvard Educational Review, 84*(4): 505–531. https://www.hepgjournals.org/doi/abs/10.17763/haer.84.4.brr34733723j648u

Star, S. L., & Griesemer, J. R. (1989). Institutional ecology, "translations" and boundary objects: Amateurs and professionals in Berkeley's Museum of Vertebrate Zoology, 1907–39. *Social Studies of Science, 19,* 387–420.

Telhan, O., Kafai B. Y., & Litts, K., B. (2016). Designing for connected making, supports for collaboration and community building in crafting activities. In K. Peppler, E. R. Halverson, & Y. Kafai (Eds.), *Makeology: Makerspaces as learning environments (Volume 1)* (pp. 223–236). New York, NY: Routledge.

Tkacz N. (2015). Wikipedia and the politics of openness. Chicago, IL: University of Chicago Press.

Tennant, J. P. et al. (2017). A multi-disciplinary perspective on emergent and future innovations in peer review. *F1000Research, 6,* 1151. https://dx.doi.org/10.12688%2Ff1000research.12037.3

Topping, K. (2005). Trends in peer learning. *Education Psychology, 25*(6), 631–645.

Versal Group. (2016). 15 ways peer-to-peer learning makes organizations agile and Collaborative. *White Paper.* Retrieved from https://versal.com/assets/docs/versal-p2p-whitepaper.pdf

Wegner, E. (1991). *Communities of practice: Learning, meaning, and identity.* Cambridge: Cambridge University Press.

Williamson, B. (Ed.) (2015). *Coding/learning: Software and digital data in education.* Stirling: University of Stirling. Retrieved from https://codeactsineducation.wordpress.com/codinglearning-e-book/

16

Biohacking

Morgan Meyer

1 Introduction

Over the past few years, a plethora of terms have emerged to describe scientific activities in the life sciences that happen outside traditional institutions: biohacking, garage biology, DIY biology, DIY genetics, DIY medicine, DIY science, and so on and so forth. These movements have attracted a growing number of enthusiasts, from young students to professional scientists, from artists to aspiring entrepreneurs, from people with no technological background to computer hackers.

The diversity of people interested in DIY science has also generated a large diversity of activities: extracting DNA for genetic testing, producing bioreactors, creating fermentation kits (i.e., for home-brewing), doing bio-art projects, developing biosensors (i.e., to detect the presence of contaminants in the environment or in food), giving lectures and organizing workshops, and fabricating cheaper alternatives to scientific equipment. The DIYbio movement has become an object of interest for journalists, for museums and, occasionally, for decision makers. The DIYbio movement can also be of interest for people who analyze peer production since similar dynamics can be observed (i.e., the emphasis put on flat hierarchies, openness, and decentralization). However, while peer production has become the subject of a large body of academic literature – analyzing for instance practices in the fields of computer software and hardware – the field of biology is usually not examined in this literature.

In this chapter, I first provide an overview of the history, and the various practices of DIY biology. Thereafter I offer a few examples of DIY medicine. I then discuss and analyze some of the key issues of DIY biology: the openness of the movement (Section 5), what the "yourself" in do-it-yourself stands for (Section 6), and concerns with ethics and governance (Section 7). The penultimate section of the chapter looks at the economic aspects and the valuations of DIY biology. In the conclusion, I will argue that further academic work could look into the geographies and fragilities/instabilities of DIY biology, and its relation with the public.

2 History

The history of DIY biology is both a recent and a long one. In terms of its recent history, 2008 is usually mentioned as being its year of birth. The first DIYbio meeting took place in a pub in Cambridge, Massachusetts, on May 1 of that year. Several months later, in February 2009, the Hackteria network was created, a network linking people interested in open source bio-art from countries such as the Czech Republic, Switzerland, and several Asian countries. From 2009 onwards, stories such as those of Kay Aull, a student who managed to carry out a homemade test to detect hemochromatosis (a genetically transmitted disease that affects the metabolism of iron), became widely reported in the media. And in 2010, the *Biopunk Manifesto* was released by Meredith Patterson, one of the most visible members of the biohacking movement. But even before 2008/2009, terms such as "garage biology" and "DIY biology" began to circulate. In 2005, for instance, biologist Rob Carlson (2005:

The Handbook of Peer Production, First Edition. Edited by Mathieu O'Neil, Christian Pentzold, and Sophie Toupin.

unpaged) predicted that "The advent of garage biology is at hand. Skills and technology are proliferating, and the synthesis and manipulation of genomes are no longer confined to ivory towers." In 2006, biologist and computer scientist Attila Chsordash (2006, p. n.d.) wrote about "the coming world of personal biotech," signifying that the "beautiful retro idea of tinkering with digital devices in a garage [. . .] can be extended to biotech too"; he also used the term "bioDIY."

Among the first laboratories dedicated to DIY biology, we can mention BiologiGaragen in Copenhagen (2010), Genspace in New York (2010), Biocurious in Santa Clara (2010), and La Paillasse in Paris (2011). At the time of writing (June 2020), the website diybio.org lists 109 groups across the world. Most of these groups are located in the Global North (40 in Europe, 42 in the United States, 8 in Canada) with a smaller number in Asia (8) and in South America (6). Roughly half of these groups have created their own laboratories.

Despite this recent history, the genealogy of DIY biology is more complex, and needs to be traced back in connection to various other developments within – but also outside – of science. First, as the term biohacking suggests, there is a close connection between biohacking and hacking. The connection is not only semantic but also ethical and philosophical: ideals such as openness, access, sharing, and decentralization are key principles in both fields. There is also a spatial proximity between these areas of practice. Numerous activities and groups did start off within already existing hacker labs. For example, in 2011 several DIY biology projects were launched at the MadLab in Manchester (created in 2008 – see Bell, Fletcher, Greenhill, Griffiths, & McLean, 2014) and there are numerous accounts of people, such as Berlin resident Lisa Thalheim, who moved from computer hacking to biohacking (Charisius, Friebe, & Karberg, 2013). This does not mean that there was an easy co-habitation. At the Noisebridge hackerspace in San Francisco, for instance, computer hackers complained about smells and were allegedly disgusted by the wet-lab work carried out by DIY biologists (Charisius et al., 2013, p. 53).

DIY biology is embedded within the larger peer production and open source movements that have developed since the 1990s. DIY biology is not the only field that has roots in the hacking and open source movements: open source agriculture (Chance & Meyer, 2017), open source ecology (Thomson & Jakubowski, 2012), and open source architecture (see Parvin, 2013) are other examples. The rise of DIY biology is also closely connected to the maker movement and the magazine MAKE which have provided a "hospitable" forum (Tocchetti, 2014, p. 136). The maker movement, in turn, can be linked to the do-it-yourself tradition, that developed from the 1950s and 1960s onwards (including dedicated magazines, books, shops and TV shows). Some DIY biologists have also been involved in setting up the *Gathering for Open Science Hardware* in 2016 and, thereafter, the launch of the *Journal of Open Hardware* in 2017.

Apart from these links to hacking, making, and open source, there is also a connection between DIY biology and synthetic biology. The vision promulgated by synthetic biology – that engineering principles can be applied to biology in order to create new substances or organisms – has been influential. Numerous founders of DIY biology laboratories have met at the international Genetically Engineered Machine (iGEM) competition for students in synthetic biology. Some of the key features of the iGEM competition – a mix between entrepreneurial spirit, fun, team-work, reflections on ethical and social issues, and a positive attitude towards science – can also be found in DIY biology laboratories. However, until 2013, DIY biology laboratories were excluded from the competition – issues of security, responsibility, and governance were put forward as the reasons. It was not until 2014 that a "community labs" track was opened at the competition in order to welcome projects from community laboratories.

Finally, the rise of DIY biology can be placed within the longer history of the contribution of amateurs and citizens to science. Four examples can be mentioned here. A first example is what has been called popular epidemiology, "a process by which laypersons gather scientific data and other information, and direct and marshal the knowledge and resources of experts to understand the epidemiology of disease" (Brown, 1992). Second, AIDS treatment activists are also involved in the production and evaluation of biomedical knowledge (Epstein, 1995). In this case also, laypeople engaged in scientific practice and thus challenged the idea that only experts can engage in research practices. A third example is the French Association of Muscular Dystrophy (Callon, Lascoumes, & Barthe, 2001), an association mainly composed of patients and their families that is sometimes actively involved in scientific research and collaborates with professionals. And, fourth, in natural history there is a long tradition of amateurs doing fieldwork and producing knowledge (Alberti, 2001). These are just a few sites of what we could call the spaces of amateur science. The spaces of amateur science are, on the one hand, related to specific disciplinary fields: natural history (including botany, zoology, entomology, ornithology), astronomy, epidemiology, and so on. On the other hand, there are specific physical locations in which we can observe amateurs producing and sharing knowledge: the field (to make observations, see & Kohler & Kuklick, 1996), the museum (to work with specimens, see Ellis & Waterton, 2005; Star & Griesemer, 1989), the pub or the coffee house (as a meeting place for learned societies, see Secord, 1994), or even the home.

Several authors have reflected about the links between DIY biology and amateurs in the natural sciences. Roosth (2010, p. 119) believes that the difference lies in "observation" as opposed to "making new things, building, tinkering, modifying." Historian of science Curry (2014), on the other hand, stresses historical continuity while arguing that both "share characteristics" and that "parallels can be found."

3 Practices and Identities

Within DIY biology, very diverse activities and projects are carried out, concerned with agriculture, the environment, health, scientific equipment and protocols, food, education, museums, art, and so on. The projects that have been presented at the iGEM competition (see Table 16.1) provide a sense of this heterogeneity.

It has been argued (Meyer & Wilbanks, 2020) that four main families of activities can be distinguished within the DIY biology movement:

- projects dedicated to the development of low-cost equipment
- projects concerned with the environment and health issues
- projects that fall in the category of bio-art
- educational activities such as workshops, introductory courses, conferences, and classes

In addition to these four categories, an evolution towards professionalization and entrepreneurialization has also been observed in the movement. Bagnolini (2018, p. 101, my translation) writes that: "we must avoid [. . .] characterizing the movement of biohackers as a set of homogeneous practices." Instead of a reductive and simplistic definition of biohacking, we must thus leave the definition of biohacking open and be attentive to the various trajectories, interests, practices, values, affiliations, identities, but also tensions and paradoxes.

According to a survey carried out in 2013 (Woodrow Wilson International Center for Scholars, 2013), DIY biologists work on average seven hours per week on their projects; they are well-educated, more than half of them are fully employed (besides their DIY biology activities) and 25% are students. About two thirds are between 25 and 45 years old and three quarters are male. The number of DIY biologists is relatively difficult to estimate: the survey by the Woodrow Wilson International Center for Scholars (2013) had 359 respondents and on the DIYbio Google Group, in a discussion titled *How big is the DIYbio community?* (2017), several figures are given: "globally [. . .] something like 2000," "5,000 people in Europe." Trojok (2016, p. 155) states that within the movement we find natural scientists, engineers, artists, philosophers, most of whom have university degrees. In a similar vein, Charisius et al. (2013, p. 23) declare that the movement comprises nerds, entrepreneurs, hackers, professional scientists, and so on. As many people active in DIY biology are not amateurs or laypersons, we can talk about a "promised" citizen science, or a citizen science "in the making" (Meyer, 2013).

To the question that is often raised "are they really doing *science*?", we need to respond that DIY biology is an assemblage of various activities and practices – technical, environmental, entertaining, medical, artistic, educational – that do not necessarily translate into academic publications. DIY

Table 16.1 Projects by community laboratories that participated to the iGEM competitions between 2014 and 2019.

Year	*Name of the team*	*Name of the project*	*Description of the project*
2014	BUGSS Baltimore	Polymerase to the people!	Development of a new Biobrick for the Pfu polymerase by amplifying and cloning the gene of the polymerase, and by developing a kit for purification in order to reduce costs
	The Tech Museum	e.Mosaic	Creation of an activity for the visitors of the Tech Museum to become part of a museum team and participate in the experience of engineering bacteria
	Genspace	Open Lab	Creation of the complete set of tools, knowledge and resources needed to develop a community laboratory (an *Open Lab Blueprint*, open source equipment, fluorescent protein genes, etc.)
	LA Biohackers	Boot up a Genome	Use of Bacillus subtilus to incorporate the whole genome of Streptococcus thermophilus to demonstrate the use of Bacillus subtilus as a chassis to boot up an artificial genome
	London BioHackspace	JuicyPrint, a 3D printer using bacteria to print cellulose forms on demand	Creation of a 3D printer fed with fruit juice to print 3D structures of bacterial cellulose (for tissue engineering, textile design, experimental art)
	SF Bay Area DIYbio	Real Vegan Cheese	Creation of vegan cheese by using baker's yeast (S. cerevisiae) to express cheese proteins and thereby create a substitute to milk
2015	Genspace	SuperFUNd Gowanus!	Development of a biosensor to measure the pollution of the Gowanus canal and thereby give to the community real-time access to data on the state of the canal
	London BioHackspace	DIY Brew Kit – Synbio Brewery	Development of an accessible kit containing a variety of brewing yeast strains for the purpose of *homebrewing*
	SF Bay Area DIYbio	BioSunBlock – Evolved Sunscreen for Bacteria	Studying cyanobacteria that have developed microbial sunscreens to be able to survive in environments with high radiation (applications: alternatives to toxic synthetic sunscreens, protection of terraforming bacteria, markers for genetic engineering)
	Wellesley TheTech	BacPack for New Frontiers: Designing Interactive Museum Exhibits for Synthetic Biology	Development of an interactive exhibition that shows the basic principles of synthetic biology to the public, with digital and wet-lab components
2016	Denver Biolabs	An oxytocin diagnostic toolkit and other biotools for use in low-resource environments	Using yeast to detect oxytocin, a natural hormone that prevents postpartum hemorrhage during childbirth (and thereby reducing mortality)
	EMW Street Bio	Low-Cost Labs: Machines That Grow	Development of a minimal set of tools necessary to carry out biotechnological processes
	Genspace	Tardigrades as a model animal for stress-resistance and developmental biology	Study of tardigrades (water bears) for their capacity to survive in extreme dry and cold environments

Table 16.1 (Continued)

Year	*Name of the team*	*Name of the project*	*Description of the project*
	IngenuityLab Canada	DNA assisted assembly of modular nanowires	Fabrication of modular nanowires from ADN with the advantage, when compared to traditional nanomaterials, to be cheap, biocompatible and with a flexible structure
2017	Cadets2Vets	Affordable, Paper-based Assay For Detection Of Arsenic Contamination	Development of an affordable, portable and accessible means for testing arsenic contamination (by developing a plasmid that expresses the Green Fluorescent Protein)
	iTesla-SoundBio	Eliminating PCB pollution in the Puget Sound by genetically modifying E. coli	Decomposition of polychlorinated biphenyls – toxic substances that only degrade very slowly in the environment – by the bacteria Dehalococcoides mccartyi
	Moscow RW	Phytases piggy bank	Resolving the problem of the thermal destruction of the phytase of Citrobacter braakii during the production if granulated compounds for farm animals
2018	Bio Without Borders	Blueblood	Development of a simple device to create horseshoe crab blood, in order to do a test for endotoxins
2019	Bioriidl_Somaiya	Steriport – Making sterilization and dispensing system portable	Development of an easy to use and portable system for sterilization
	Tacoma_RAINmakers	Improving the Agricultural Potential of Rhizobia	Development of the microorganism rhizobia as a potential alternative to chemical fertilizers

biology does not neatly fit into the category "citizen science" and has even been the subject of museum exhibitions (see Vignette 16.1) and documentaries (Lassale, 2019; Schlichter & Karberg, 2012).

4 Examples of DIY Medicine

While DIY biology now enjoys a certain visibility, for DIY medicine and healthcare the situation is different. There are only very few laboratories specifically dedicated to it. It is rather around specific projects and issues, which are labeled "do-it-yourself" and/or "open source" medicine that practices are structured and collectives are formed. The *Nightscout Project*, for example, is dedicated to the development of mobile devices to measure type 1 diabetes (Greene, 2016). At the origin of this project is a father who, in 2013, developed a computer code to be able to remotely access the data of the glucose levels of his son. Since then, the project has attracted many users who have created a Facebook group with more than 23,000 members, developed sharing platforms, and made the source code open source.

Another project, developed at the Denver Biolabs in 2016 – and presented at the iGEM competition – is the *Oxytocin Diagnostic Toolkit* (see Table 16.1). This kit uses yeast to detect the presence of oxytocin, a natural hormone that prevents postpartum hemorrhage during childbirth and thus reduces maternal mortality, especially in poor countries. We can also mention the *DIY transcranial direct-current stimulation device* to stimulate the brain (Jwa, 2015; Lee, Hirschfeld, & Wedding, 2016); the development of an open source ultrasound probe; the Amplino project, a portable device for detecting

Vignette 16.1 Exhibitions featuring DIY biology

Exhibiting DIY biology in museums

The exhibition *Beyond the Lab – The DIY Science Revolution* was developed – and first shown – at the London Science Museum (from July 2016 to September 2016). The exhibition thereafter traveled to various other countries. The exhibition focused on three related themes: citizen science, health hacking and DIY biology. The curatorial team chose to represent *individual* persons and their projects. "Equipped with low-cost sensors, smartphone apps and the ability to share information with communities online, these DIY science pioneers are challenging our ideas of who a scientist is and what science and our societies will look like in the future" the exhibition summary reads. In some countries, the exhibition was adapted to local contingencies by including additional examples (i.e., in Luxembourg, the exhibition featured a portrait of a local student and his invention: a helmet capable of measuring the force of an impact).

The exhibition *Biohacking: Do-it-yourself!* took place at the Medical Museion in Copenhagen from January 2013 until summer 2014. For the exhibition, the museum built a laboratory with the help of BiologiGaragen, a DIY biology laboratory in Copenhagen (see Davies, Tybjerg, Whiteley, & Söderqvist, 2015). The exhibition was built in a "garage" spirit: alternatives were sought to build the laboratory with a minimal budget. But there were many differences between a "real" DIY biology lab and the museum exhibit: a DIY biology lab provides a certain "aesthetic feel" and produces a "vibrant atmosphere," while the museum is organized in a more orderly way, and the challenge was to "maintain the impression of a real and vibrant hackerspace" (Sørensen, 2012). To display a laboratory in a museum, it is necessary to transform, adapt, and rethink the laboratory. One may also wonder if, beyond the (temporary) transformations of lab into a "lab-in-the-museum," the museum itself can be transformed? According to the Director of the Medical Museion, "biohacking may have consequences for a museum. [...] I think we have much to learn about the culture of hacking [...] to help us rethink what a museum could be" (Söderqvist, 2013).

At the *Science Gallery* in Dublin, DIY biology was featured as part of an exhibition titled *Grow Your Own – Life after nature* (2013–2014). In order to do so, the Science Gallery installed a Community BioLab that was curated by Genspace. In this lab, various projects were presented and workshops and discussions organized. Members from Hackteria, (Art)ScienceBLR, La Paillasse, and MadLab were invited to organize workshops. La Paillasse, for instance, organized an event to produce biological ink by using bioproduction from soil bacteria. The production of biological ink was enacted in a specific form: it was not only displayed and celebrated, but workshop participants were taught how to use it. In other words, beyond the argumentation that ink can be made "yourself," the workshops delivered instructions for how to do so, with all the needed gestures, skills, and material practices.

Biohacking has also been featured in various exhibitions at *Ars Electronica* in Linz (Austria). The following projects have been displayed: the "DIY mobile gene lab" (developed by the Hackteria network), a collection of equipment to enable people to do biology at home; the "gene gun" (developed by Rüdiger Trojok), a device to inject gold particles coated with DNA into cells; and "Future Flora," featuring a DIY kit for vaginal flora (developed by Guilia Tomasello). The exhibition *Beyond the Lab – The DIY Science Revolution* has also been shown at *Ars Electronica*.

malaria; the Open insulin project, which develops an open source protocol for insulin production; Epidemium, a collaboration between La Paillasse and the Roche company to study large data sets on cancer; and EpiPencil (an epinephrine autoinjector) and the Apothecary MicroLab, both developed by the Four Thieves Vinegar Collective. Finally, let us also mention the *Open BioMedical Initiative*, a nonprofit initiative supporting the design and distribution of low-cost and open source devices such as mechanical hands and feet, as well as incubators. While these projects are praised for making care and medicine more affordable, more transportable, and more accessible, concerns are raised about risks and regulations. For instance, "troubling is the overlap with patients who self-diagnose and self-medicate while rejecting evidence-based expert advice" (Ball, 2016, p. 555). And despite the promise of autonomy, new dependencies can be created: "The liberatory promise of the new DIY may be real, but in offering freedom from one dependence (on doctors or the physical location of the clinic), it creates a new dependence on the app, the peripheral, and the speculatively financed startup firms that produce them" (Greene, 2016).

While these new dependencies can contradict the ideals of autonomy and empowerment, some collectives seek, nevertheless, to build attachments and links around certain health problems. This is the case of patients' associations, studied by sociologists Michel Callon and Vololona Rabeharisoa since the 1990s. These associations have often had to face the problem that rare diseases (also called "orphan" diseases) are little known by researchers and doctors, and the pharmaceutical industry is usually not interested in them. So while these associations have themselves produced scientific and medical knowledge about certain diseases – showing that "do-it-yourself" medicine is not such a recent phenomenon – they also aim to partner with health professionals, in order to better understand and cure diseases. The aim is therefore both epistemic – producing knowledge – and political: turning a problem into a priority for institutions and making it more visible.

5 Openness and Democratization

Openness, accessibility, transparency: these are some of the terms commonly used when describing DIY biology and, in particular, its ambition to transform science. The key argument put forward is that the life sciences (including molecular biology, genetics, but also medicine) should not be activities only carried out by professional researchers in their "ivory towers," but, instead, that other actors should also be involved (be they called "people," "citizens," "amateurs," etc.). The idea, simply put, is to *democratize* science. Democracy and democratization are, however, terms that need to be further problematized and unpacked to make them analytically operational. In order to "democratize" science, DIY biologists aim to open up biology spatially, technically, socially, and economically.

First, there have been efforts to create spaces for experimentation – or "autonomous laboratories" as Esquivel-Sada (2017) calls them – outside the confines of traditional scientific institutions. Fifty-six community laboratories have been created over the past ten years, and there are numerous laboratories in people's private homes.[1] In a very spatial sense, then, biology has been "communalized" and "domesticated." Second, there have been efforts to make biology more accessible, via the scientific and technical equipment needed in laboratories. Among the growing list of alternative equipment, we can mention: the use of webcams instead of microscopes, of the *OpenPCR* machine instead of a conventional PCR machine, the *DremelFuge* instead of a centrifuge, and the construction of electrophoresis gels, magnetic stirrers, autoclaves, and so on. In 2016, the *Essential Biohacker's Guide* was released. The guide, developed by a network of biohackers from Latin America, *Syntechbio*, lists alternatives for 25 techniques and laboratory tools (there are English, French, Spanish, and Portuguese versions of the guide). The availability and mutability of these objects are an important factor in the development of DIY biology. According to Trojok (2016, p. 115), it is now possible through this alternative equipment to set up a functional laboratory in molecular biology for about €2,000.

An important practice of DIY biologists is to work on "creative workarounds." That is, they try to find workarounds around objects in order to transform and combine them in novel ways, as well as around institutions, to circumvent established economic links between universities and industry. Such equipment is tinkered with and redesigned to serve as "convivial tools" (Illich, 1973), tools that can be more easily, more broadly and more flexibly used. Even more so, their "conviviality" is not only fostered by making them "open," but also by allowing them to be modified and improved. "DIYbio combines a sort of individual craftiness and self-determination *to do* things with a praxis in which things are left open, waiting for the next realization," writes Delgado (2013, p. 67).

The notion of "amaterialization" can be used to account for these constructions and transformations of laboratory equipment. This amaterialization is a process that is social (making equipment available to amateurs) and technical (transforming and recombining equipment). And this amaterialization is furthermore based on the co-construction of material equipment and "amaterial" versions of this equipment, that is, all the texts, pictures, videos, and schemata that are used to ease their diffusion and reproduction (Meyer, 2015a). The relationships between amateurs and professionals are thus not only located *in* disciplinary fields or specific places (the usual sites to locate the amateur/professional boundary), they are also made possible *through* objects. The affordances of objects – as much as their mobility and their malleability – thus need to be taken into account in analyses of DIY biology.

Third, DIY biologists have developed their laboratories and equipment with a key social imperative in mind: that persons from all kinds of social and professional backgrounds can join laboratories and online discussions (such as the DIYbio discussion forum on Google Groups). Neither diplomas nor specific expertise is needed in order to become a member. Organizational structures are based on a "do-ocracy," not a meritocracy. While making this social openness explicit, DIY biologists frequently voice criticisms towards institutionalized science for its "countless rules, obligations, hierarchies, dependencies and pressures to succeed" (Trojok, 2016, p. 167, my translation). Delfanti (2013, p. 125) reports on their "distrust for bureaucracies." This is also why accepting institutional funding is seen as problematic: "A part of the DIYbio community strongly upholds the philosophy of independence, freedom and decentralization, while fearing that accepting governmental grants may hinder their autonomy to speak freely or to pursue their work in their own choice and direction" (Nascimento, Pereira, & Ghezzi, 2014, p. 41).

Finally, DIY biologists also aim to democratize biology in economic terms. Community laboratories usually demand no fees, or at least very small ones, and one of the motives for building creative workarounds is to diminish the cost of laboratory equipment. DIY biologists can also save money by buying used equipment, and via donations (i.e., university laboratories or pharmaceutical companies donating their old equipment).

Despite presenting themselves as open and democratic, DIY biology laboratories might nevertheless face various difficulties and tensions. There have been reports of autocratic leadership, of visions not being shared among members, and of internal tensions regarding business models (Rivoire, 2017). Jen (2015) has noted that the dominant representation and promises of DIY biology are gendered: it is presented as a masculine universe of tinkerers and hackers, at the expense of feminine and feminist figures. Bagnolini (2018) has observed three problems within the biohacking movement: an overly strong focus on individuals and "self-made-men," auto-experimentations,[2] and the reduction of life to a machine. Concerning this last point, DIY biologists have in general a positive – and positivist – attitude towards GMOs and biotechnologies (Esquivel-Sada, 2017).

6 A Sociology of the "yourself"

When setting up a laboratory in a garage, kitchen, or basement – and thereby domesticating biology spatially, technically, socially, and economically – people depend on others interested in DIY biology, on scientific institutions, on the sharing of information, on the circulation of objects, on Internet platforms, on donations, and so on. In other words, the "yourself" has to be connected. Because of this need for connectivity and collectivity, the term "yourself" might appear as a misnomer and terms such as "do-it-together" or "do-it-with-others" to be more appropriate. Yet, "yourself" symbolizes more than just a passive, individual person. "Yourself" stands for someone that engages with biology and literally *does* things, a self that is active and that is juxtaposed to other sites and scales of science production: the university, the institution, the enterprise, and so on. The key idea is to make biology "personal" (Tocchetti, 2014). DIY biology thus aims to constitute a distinct and political form of self by providing people access, by enabling them to transform themselves into active producers of science, by making their bodies and environments more knowable, and by demonstrating that one can do it oneself.

Various scholars have noted that DIY biology can be analyzed in terms of *demonstrations*. Meyer (2013, p. 126) has argued that a DIY experiment is not only a demonstration of science per se, but also a socio-political demonstration in that it "attributes a social value, shows the feasibility of a process, and tries to convince and enroll actors." While looking at festivals, maker faires and blog documentations, Tocchetti (2014, p. 96) states that demonstrations are not only about biological problems but that the spectator and/or reader becomes "the witness in an experiment aimed at demonstrating that everyone can do biology." In a similar vein, Delgado and Callén (2017) speak of "technical and political demonstrations."

7 Ethics and Governance

The politics of openness of DIY biology is at once celebrated and dreaded. On the one hand, it is praised for empowering citizens, for fostering open innovation, for providing new means for education, and for being inclusive. Optimist accounts often compare DIY biology to the Home Brew Computer Club and to Steve Jobs – a promissory comparison that brings side by side a familiar success story and a story-in-the-making, and offers a narrative of innovation and revolution. On the other hand, DIY biology has also raised concerns about security, safety, and regulation. When the movement emerged, a number of media articles portrayed it in negative terms, and pondered on the potential threat of bioterrorism. DIY biology has been examined in several national reports (in the US, the UK, the Netherlands, Germany, France, etc.) and international reports (i.e., by the United Nations Interregional Crime and Justice Research Institute), usually in relation to synthetic biology. While these reports deal with very similar concerns – ethics, governance, risks, safety, regulation – they nonetheless reveal differences between countries in terms of regulatory environments, visibility/invisibility of surveillance by secret services, and policy and funding recommendations. In recent years, the use of new gene editing techniques, such as CRISPR/cas9, has raised additional concerns about the control of the fabrication of synthetic organisms.

There have been different kinds of responses from the DIY biology community to such safety and security concerns. The first response is discursive: responses by practitioners highlight that DIY biologists do not work with dangerous material and that potential terrorists do not need DIY biology to meet their aims. The second response has been ethical, by collectively establishing a code of ethics. The European code of ethics, first drafted in 2011, states that practitioners should "emphasize transparency," "adopt safe practices," "promote citizen science and decentralized access to biotechnology," "respect humans and all living systems," and so on (see Figure 16.1). The American code of ethics is similar, but shorter: it does not include principles such as modesty, respect, and responsibility, and it can thus be considered as being weaker than the European one (Eggleson, 2014, p. 191). The ethical principles of the DIY biology movement are similar to those of the *Gathering for Open Science Hardware* (GOSH) movement: GOSH "is accessible, makes science better, is ethical, changes the culture of science, democratizes science, has no high priests, empowers people, has no black boxes, is impactful tools, allows multiple futures for science."

The third response has been practical: on DIYbio.org, a portal has been set up through which people can ask questions about safety to a panel of biosafety experts and members of a professional association of biosafety; DIY Bio Europe (a network that was launched in 2012) has developed "Community Biolab Guidelines," and several laboratories have developed internal guidelines or courses dealing with safety.

The fact that some DIY biologists have been in touch with the FBI has been discussed in numerous texts (Charisius et al., 2013; Ledford, 2010; Meyer, 2015b; Tocchetti & Aguiton, 2015; Wolinski, 2016). In recent years, the FBI has developed a proactive attitude, presenting itself as the "new FBI." For example, the FBI organized a DIYbio outreach workshop from June 12 to 14, 2012 near San Francisco. Around 40 people attended, including FBI members and DIY biologists from around the world. Although the event was organized by an American security service, many European DIY biologists were invited and agreed to participate. Representatives from Paris, Copenhagen, Berlin, Helsinki, London, Manchester, Amsterdam, and Prague were present (but numerous persons also declined the invitation). During the workshop, the FBI stated that it wanted to "work with" DIY biologists, viewing them as "partners" in a "positive relationship." While security and responsibility remain its main concerns, the FBI aims to distinguish between "white hats" and "black hats," so that the DIY

DIYbio Code of Ethics
Draft from the European Delegation
09/07/2011

Transparency
Emphasize transparency and the sharing of ideas, knowledge, data and results.

Safety
Adopt safe practices.

Open Access
Promote citizen science and decentralized access to biotechnology.

Education
Help educate the public about biotechnology, its benefits and implications.

Modesty
Know you don't know everything.

Community
Carefully listen to any concerns and questions and respond honestly.

Peaceful Purposes
Biotechnology must only be used for peaceful purposes.

Respect
Respect humans and all living systems.

Responsibility
Recognize the complexity and dynamics of living systems and out responsibility towards them.

Accountability
Remain accountable for your actions and for upholding this code.

Figure 16.1 The European DIYbio code of ethics, drafted in 2011 in London. Source: Draft DIYbio Code of Ethics from European Delegation, 2011. diybio.org. © 2011, diybio.

biology community is "protected against harmful actors." The FBI also distributed a set of didactic cards called "biosecurity outreach cards" with information on various bacteria and viruses, such as Ebola, adenovirus, and smallpox. The stance of the FBI was that DIY biologists should themselves carefully monitor and report suspicious activities. The reaction of the DIY biology community to the FBI's strategy can be summed up as follows: while for US practitioners, it is seen as obvious and normal to collaborate with the FBI, the European delegates were suspicious, skeptical, and irritated (Charisius et al., 2013, pp. 196–202). What the relationship between DIY biologists and the FBI also reveals is that the boundary between experts and suspects is not clear-cut, and that biohacking can be differently interpreted (Tocchetti & Aguiton, 2015).

Other nuances and differences should be highlighted. For instance, regulation is less strict in the United States (where it easy to produce GMOs in a DIY biology laboratory) than in Europe (where it is very difficult to get an authorization to produce GMOs outside scientific institutions).[3] Further, in Europe there are stronger collaborations between DIY biologists, artists, and designers than there are in the United States (Seyfried, Pei, & Schmidt, 2014). It is also important to note the differences between the Global North and the Global South, as made clear by the following statement by the cofounder of Africa Open Science and Hardware, Thomas Hervé Mboa Nkoudou: "People engage in Western countries, because it's like a hobby. But in Africa, it's different. People are engaged in this movement to survive, to find a job, to solve a particular problem. [. . .] There are many hospitals in rural zones [. . .] in Africa, which don't have even the small equipment to run small diagnoses for patients. So biohacking can be helpful" (cited in *The Commoners Scientist* (Lassale, 2019)).

8 Economics and Valuation

Discussions do not only concern the ethics and (geo)politics of DIY biology, but economics and funding as well. We see both non-market and market rationales at work. In its purest form, there is a logic of "do-it-without": DIY biology is often portrayed as a reaction against pharmaceutical companies, standard/expensive equipment, patents, and, in general, against everything that keeps knowledge from being freely shared. Yet, DIY biology relies – at least partially – on markets and industries: by buying used equipment from commercial websites, tools and products from supermarkets, or seeking and accepting donations from companies. Also, various sources for funding DIY biology projects have been sought, be they via membership fees, crowdfunding, grant funding, or from local municipalities. There are also several entrepreneurial projects that have developed out of DIY biology (see *OpenPCR*, *Pearl Biotech*, *LavaAmp*, *Ginkgo Bioworks*, *Bento Lab*, *PILI*).[4] Scholars have therefore talked about a "cooptation" (Delfanti, 2014) and "recuperation" (Delfanti & Söderberg, 2018) of the movement. The potential relationships and co-existence – or, conversely, tensions and ruptures – between non-market and market rationales will be interesting to follow as the movement further develops.

Different forms of ***valuation*** can already be observed when looking at the moments and places where DIY projects are valued (such as during a competition or in front of potential investors) (Meyer & Wilbanks, 2020). On the one hand, there are non-market valuations of the movement, with the promotion of social, ethical, and cultural values (such as citizenship, democracy, openness, sharing, education, empowerment, etc.). On the other hand, there are also market valuations: projects that turn into start-ups; projects that, during a pitch in front of investors, require several hundred thousand euros; partnerships with private companies. These valuations are distributed and heterogeneous: products, practices, principles, and places are valued at the same time (Meyer & Wilbanks, 2020). They aim to generate an "interest" in a triple sense of the word: a general interest (i.e., a common and public good), an interest for a given public (its curiosity and awareness), and a monetary interest (by asking public or private actors to contribute financially to projects).

9 Conclusion

DIY biology and medicine are based on various practices and logics: amateur and DIY practices, the ethics of hacking and open source, the drive to domesticate molecular biology and genetics, the ideal of participation and citizen science. This democratization is a process that is at once spatial (construction of new spaces), technical (creative workarounds around equipment), social (establishment of accessible networks/laboratories) and political. It is therefore through their practices, gestures, and questions – tinkering, experimenting, demarcating, working around, amaterializing, ethicizing, comparing, valuating, and so on – that we need to grasp DIY science.

In the academic literature, several facets of DIY biology have already been quite thoroughly addressed: its history (Curry, 2014; Kelty, 2010; Tocchetti, 2014), materiality (Delgado, 2013; Meyer, 2013; Meyer, 2015a), identity (Esquivel-Sada, 2017; Meyer, 2016), ethics (Bagnolini, 2018; Eggleson, 2014; Keulartz & van den Belt, 2016), politics (Delfanti, 2013), and economics (Delfanti & Söderberg, 2018; Meyer & Wilbanks, 2020). The issues of the democratization of biology, as well as the links between hacking, biohacking, and synthetic biology, have been addressed in nearly all the texts cited in this chapter.

Several issues have, however, not been researched in great depth: While most work has looked at the various laboratories, devices, projects, and discourses that have been created, much less is known about the failures, dead-ends, and abandonments within the movement. An exception is the book by "experimental" journalists Charisius et al. (2013) that provides some insights into the difficulties and failures that can be encountered. But an analysis and theorization of these difficulties is still lacking. Research on the fragilities, instabilities, and ambiguities within the DIY biology movement would be a key contribution to the literature.

The geographical scope of academic work has been limited to either the United States or the European context. Kera (2012, 2015) is among the few scholars to have looked into Asia and the

Global South as other sites of DIY biology. Further research could thus look at the ethics, politics, and practices of biohacking in non-Western contexts in order to provide more comparative and situated accounts of DIY biology.

The fact that many DIY biology laboratories are engaged in educational and outreach activities has not been a research topic thus far. Apart from the examples from the museum world mentioned above (see Vignette 16.1), only one paper explicitly addresses the topic of education (Kallergi & Zwijnenberg, 2019). The authors of this paper examine a module that they teach on the ethics of biotechnology – a module in which they use hands-on participation as a pedagogical means to reflect upon responsible innovation. However, the many workshops, visits of school classes, conferences, and introductory courses organized by DIY biology laboratories still need to be analyzed more systematically and sociologically.

Academic work on DIY biology has so far mainly focused on four related themes: its social and material architecture, its genealogy and history, its ethical and political characteristics, and its entrepreneurial dynamics. The wider geographies, fragilities/instabilities, and relations to the public are topics that deserve further inquiry.

Notes

1 It is, however, difficult to provide exact figures of the latter. In the above-mentioned 2013 survey, 83 respondents declared working in their "home lab." But given that 82% of survey respondents were from the United States and only 10% from Europe, and given that between 2013 and 2019, the number of community labs has almost tripled, there are today most likely several hundred (if not more) laboratories in private homes across the world.

2 There have been recent controversies concerning people who have injected themselves with various substances.

3 Among the exceptions we can count laboratories in Graz (Austria), London (UK), and Cork (Ireland).

4 The *DIYbiosphere* initiative currently lists 32 start-ups linked to the DIY biology movement.

References

Alberti, S. J. (2001). Amateurs and professionals in one county: Biology and natural history in late Victorian Yorkshire. *Journal of the History of Biology*, *34*(1), 115–147.

Bagnolini, G. (2018). *À la marge des sciences institutionnelles, philosophie et anthropologie de l'éthique du mouvement de biohacking en France* (Doctoral dissertation, Université de Tours).

Ball, P. (2016). Medical hackers. *The Lancet*, *388*(10044), 554–555.

Bell, F., Fletcher, G., Greenhill, A., Griffiths, M., & McLean, R. (2014). Making MadLab: A creative space for innovation and creating prototypes. *Technological Forecasting and Social Change*, *84*, 43–53.

Brown, P. (1992). Popular epidemiology and toxic waste contamination: Lay and professional ways of knowing. *Journal of Health and Social Behavior*, *33*(3), 267–281.

Callon, M., Lascoumes, P., & Barthe, Y. (2001). *Agir dans un monde incertain: essai sur la démocratie technique.* Paris: Seuil.

Carlson R. (2005) Splice it yourself: Who needs a geneticist? Build your own DNA lab. *Wired Magazine*, *13*(5). Retrieved from www.wired.com/2005/05/splice-it-yourself/

Chance, Q., & Meyer, M. (2017). L'agriculture libre–les outils agricoles à l'épreuve de l'open source. *Techniques et culture*, *67*, 1–23.

Charisius, H., Friebe, R., & Karberg, S. (2013). *Biohacking: Gentechnik aus der Garage.* Munich: Carl Hanser.

Chsordash, A. (2006). What is bioDIY? Retrieved from https://pimm.wordpress.com/2007/01/24/what-is-biodiy/

Curry, H. A. (2014). From garden biotech to garage biotech: Amateur experimental biology in historical perspective. *The British Journal for the History of Science*, *47*(3), 539–565.

Davies, S. R., Tybjerg, K., Whiteley, L., & Söderqvist, T. (2015). Co-Curation as hacking: Biohackers in Copenhagen's medical museion. *Curator: The Museum Journal*, *58*(1), 117–131.

Delfanti, A. (2013). *Biohackers. The politics of open science.* Cambridge: Polity.

Delfanti, A. (2014). Is do-it-yourself biology being co-opted by institutions? In A. Bureaud, R. F. Malina, & L. Whiteley (Eds.), *Meta-life. Biotechnologies, synthetic biology, life and the arts.* Cambridge, MA: Leonardo and MIT Press.

Delfanti, A., & Söderberg, J. (2018). Repurposing the hacker. Three cycles of recuperation in the evolution of hacking and capitalism. *Ephemera, 18*(3), 457–476.

Delgado, A. (2013). DIYbio: Making things and making futures. *Futures, 48*, 65–73.

Delgado, A., & Callén, B. (2017). Do-it-yourself biology and electronic waste hacking: A politics of demonstration in precarious times. *Public Understanding of Science, 26*(2), 179–194.

Eggleson, K. (2014). Transatlantic divergences in citizen science ethics – comparative analysis of the DIYbio Code of Ethics drafts of 2011. *Nanoethics, 8*(2), 187–192.

Ellis, R., & Waterton, C. (2005). Caught between the cartographic and the ethnographic imagination: The whereabouts of amateurs, professionals, and nature in knowing biodiversity. *Environment and Planning D: Society and Space, 23*(5), 673–693.

Epstein, S. (1995). The construction of lay expertise: AIDS activism and the forging of credibility in the reform of clinical trials. *Science, Technology, & Human Values, 20*(4), 408–437.

Esquivel-Sada, D. (2017). *Un labo à soi: l'idéologie DIYbio de démocratie des biotechnologies et la conjonction entre facultés manuelles et autonomie.* (Doctoral dissertation, University of Montréal).

Greene, J. A. (2016). Do-it-yourself medical devices – technology and empowerment in American health care. *New England Journal of Medicine, 374*(4), 305–308.

Illich, I. (1973). *Tools for conviviality.* New York, NY: Harper & Row.

Jen, C. (2015). Do-it-yourself biology, garage biology, and kitchen science. In M. Wienroth & E. Rodrigues (Eds.), *Knowing new biotechnologies: Social aspects of technological convergence* (pp. 125–141). New York, NY: Routledge.

Jwa, A. (2015). Early adopters of the magical thinking cap: A study on do-it-yourself (DIY) transcranial direct current stimulation (tDCS) user community. *Journal of Law and the Biosciences, 2*(2), 292–335.

Kallergi, A., & Zwijnenberg, R. (2019). Educating responsible innovators-to-be: Hands-on participation with biotechnology. *EAI International Conference on Technology, Innovation, Entrepreneurship and Education* (pp. 79–94). Cham: Springer.

Kelty, C. (2010). Outlaw, hackers, Victorian amateurs: Diagnosing public participation in the life sciences today. *Jcom, 9*(1). doi.org/10.22323/2.09010303

Kera, D. (2012). Hackerspaces and DIYbio in Asia: Connecting science and community with open data, kits and protocols. *Journal of Peer Production, 2*, 1–8.

Kera, D. (2015). Open source hardware (OSHW) for open science in the global south: Geek diplomacy? *Open Science*, 133–156.

Keulartz, J., & van den Belt, H. (2016). DIY-Bio–economic, epistemological and ethical implications and ambivalences. *Life Sciences, Society and Policy, 12*(1), 7.

Kohler, R. E., & Kuklick, H. (Eds.). (1996). *Science in the field.* Chicago, IL: University of Chicago Press.

Lassale, P. (2019). *The commoner scientists.* Documentary Film, Philéas Films.

Ledford, H. (2010). Garage biotech: Life hackers. *Nature News, 467*(7316), 650–652.

Lee, J. M., Hirschfeld, E., & Wedding J. (2016). A patient-designed do-it-yourself mobile technology system for diabetes: Promise and challenges for a new era in medicine. *Jama, 315*(14), 1447–48.

Meyer, M. (2013). Domesticating and democratizing science: A geography of do-it-yourself biology. *Journal of Material Culture, 18*(2), 117–134.

Meyer, M. (2015b). Bricoler le vivant dans des garages. Le virus, le génie et le ministère. *Terrain. Anthropologie & Sciences Humaines, 64*, 68–83.

Meyer, M. (2015a). Amateurization and re-materialization in biology. In M. Wienroth, & E. Rodrigues (Eds.), *Knowing new biotechnologies. Social aspects of technological convergence* (pp. 142–157). London: Routledge.

Meyer, M. (2016). Steve Jobs, terrorists, gentlemen, and punks: Tracing strange comparisons of biohackers. In J. Deville, M. Guggenheim, & Z. Hrdlickova (Eds.), *Practising comparison: Logics, relations, collaborations* (pp. 281–305). London: Mattering Press.

Meyer, M., & Wilbanks, R. (2020). Valuating practices, principles and products in DIY biology: The case of biological ink and vegan cheese. *Valuation Studies, 7*(1), 101–122.

Nascimento, S., Pereira, A. G., & Ghezzi, A. (2014). From citizen science to do it yourself Science. *Joint Research Centre, European Commission*, Ispra, Italy.

Parvin, A. (2013). Architecture (and the other 99%): Open-source architecture and design commons. *Architectural design, 83*(6), 90–95.

Rivoire, A. (2017). Thomas Landrain n'est plus président de la Paillasse. Makery, 27 June 2017. Retrieved from www.makery.info/2017/06/27/thomas-landrain-nest-plus-president-de-la-paillasse/

Roosth, H. S. (2010). *Crafting life: A sensory ethnography of fabricated biologies* (Doctoral dissertation, Massachusetts Institute of Technology).

Schlichter, A., & Karberg, S. (2012). *Die Gen-Köche.* Documentary Film, BR Fernsehen.

Secord, A. (1994). Science in the pub: Artisan botanists in early nineteenth-century Lancashire. *History of science*, *32*(3), 269–315.

Seyfried, G., Pei, L., & Schmidt, M. (2014). European do-it-yourself (DIY) biology: Beyond the hope, hype and horror. *Bioessays*, *36*(6), 548–551.

Star, S. L., & Griesemer, J. R. (1989). Institutional ecology, translations' and boundary objects: Amateurs and professionals in Berkeley's Museum of Vertebrate Zoology, 1907–39. *Social Studies of Science*, *19*(3), 387–420.

Söderqvist, T. (2013). Opening the biohacking lab at Medical Museion. Blogpost. Retrieved from www.museion.ku.dk

Sørensen, A. P. (2012). Hacking the museum – thoughts on building a biohacker space. Blogpost. Retrieved from www.museion.ku.dk

Thomson, C. C., & Jakubowski, M. (2012). Toward an open source civilization (Innovations case narrative: Open source ecology). *Innovations: Technology, Governance, Globalization*, *7*(3), 53–70.

Tocchetti, S. (2014). *How did DNA become hackable and biology personal? Tracing the self-fashioning of the DIYBio network* (Doctoral dissertation, The London School of Economics and Political Science).

Tocchetti, S., & Aguiton, S. A. (2015). Is an FBI agent a DIY biologist like any other? A cultural analysis of a biosecurity risk. *Science, Technology, & Human Values*, *40*(5), 825–853.

Trojok, R. (2016). *Biohacking: Gentechnologie für alle*. Munich: Franzis.

Wolinsky, H. (2016). The FBI and biohackers: an unusual relationship. *EMBO reports*, *17*(6), 793–796.

Woodrow Wilson International Center for Scholars. (2013). *Seven myths and realities about do-it- yourself biology*. Washington, DC.

17

Makers

Yana Boeva and Peter Troxler

1 Introduction

In a value system based around creating artifacts, *makers* and *making* have been the focus of many bold promises for the last two decades. Back in the early 2000s, several prominent figures proposed that a "maker movement" would bring about a tangential development in technology design and production where "regular people [could] harness big factories at will to make what they want" (Anderson, 2012, p. 69). This "coming revolution" formed by personal fabrication (Gershenfeld, 2005) and peer collaboration foresaw the transformation of users into makers, inventors, and experts in total control over technology (Banzi, Cangiano, & Fornari, 2016). Across the world makers, equipped with microprocessors and given access to fabrication tools including 3D printers and laser cutters, exchange their knowledge on- and offline signaling a bottom-up process of technology design. Through their engagement with the different practices for peer production, "people not only make a plentitude of objects, they also make a specific social-technical-cultural context" (Balsamo, 2015, p. 84). In other words, maker activities and processes require little initial skill, knowledge, and expertise; rather, they enable their acquisition, opening them up to "everyone."

Given the multiple definitions of maker identities and what making could achieve, we tentatively define makers as individuals creating artifacts – most often in the form of practical hard- and software fixes to actual, real-world problems and challenges. They thereby alleviate an everyday need, even if only from the idiosyncratic view of an individual maker, in a practical, ingenious way with a concern for appropriateness, in reference to Nigel Cross's theory of designerly ways of knowing (1982). In this chapter on makers, we provide a critical overview of the different cultures of making, their motivations as well as the socio-technical infrastructures that encourage these forms of peer production. The chapter draws on a growing body of scholarship on making and hacking in the social sciences, in human-computer interaction (HCI), as well as in arts and design research.

2 A Short and Incomplete History of Making

The idea of makers and making in a peer-production context has surely developed a distinct profile in the 21st century. Making, however, is not such a new phenomenon per se. Florian Cramer cites "19th century romanticism as a counter-reaction to the industrial revolution" (2019, p. 63) as its epistemological origin, and Paul Atkinson situates its origins in the 19th century "as a leisure pursuit, a pastime, it emerged from a perceived need to 'keep idle hands busy'" (2011, p. 26). He argues that this pastime served to bring the Victorian work ethic from the factory into the home. In tracing making's lineage, popular accounts further appoint the Arts and Crafts movement as one of its many precursors. New approaches contingent upon craft and hands-on practices situate themselves partially in a historical discourse of handiwork and in the effects of industrial production modes on that handiwork. Such idealistic descriptions of making, however, present a somewhat mistaken impression upon a closer look

The Handbook of Peer Production, First Edition. Edited by Mathieu O'Neil, Christian Pentzold, and Sophie Toupin.

at its social structures characterized by a lack of discussion of institutions and political change (Morozov, 2014) and commercial ties that subtly put makers in the position of unpaid value creators in platform economies (coons, 2018).

While the 19th century's forms of making were mostly in the realm of crafts, the early twentieth century saw the emergence of several instruction-rich publications for the hobbyist. Magazines such as *Popular Mechanics* in the United States featured automotive, electronics, science, technology, home, outdoor, and do-it-yourself (DIY) topics for the inquisitive amateur. During and after war-times, however, self-sufficiency and repairing tools and machinery was often a sheer necessity. DIY practices proliferated owing to the scarcity of products, economic pressure, and a significant shortage of professionals and craftspeople to perform these activities (Franz, 2005; Kline & Pinch, 1996). Within only a few years of financial stabilization in many Western countries, the understanding of DIY and tinkering shifted from providing means to sustain oneself (financially) to giving purpose to meaningless employment – in other words, a shift from manual labor to manual leisure (Gelber, 1997; Maines, 2009). In the postwar boom, DIY became anew a pastime, promoted for example in Germany through the magazine *Selbst ist der Mann* or in the United States through the hobby magazine *CQ*, directed mainly at adult males and showcasing home decoration, furniture-making, and all sorts of technical subjects.

In the Northern-American counterculture of the 1960s, the *Whole Earth Catalogue* brought the topics of self-sufficiency, ecology, alternative pedagogy, and DIY, alongside reviews and recommendations of everyday tools to a new level. With its opening sentence, "[w]e are as gods and might as well get good at it," it set the tone for a topos that was repurposed decades later. The *Whole Earth Catalogue* can be seen as a seminal publication for the Californian ideology, "this unexpected collision of right-wing neo-liberalism, counterculture radicalism and technological determinism" (Barbrook & Cameron, 1996), serving as a source of inspiration for the likes of Steve Jobs and Kevin Kelly (founding executive editor of *Wired* magazine) and the harbinger of a proto-maker-culture (Cramer, 2019). The collective exchange about tinkering and DIY projects occurred in centralized and institutional initiatives as well. In the 1940s in Turkey, a number of agricultural schools, the village institutes, were established, teaching basic literacy and numeracy as well as agriculture, construction, and arts and crafts. In communist Czechoslovakia, the Svazarm were publicly accessible spaces equipped for the mechanical repair of agricultural and military machinery and radios. In East Germany, so-called stations for young naturalists and technicians (Station Junger Naturforscher und Techniker) existed as part of the Vladimir Lenin All-Union Pioneer Organization (Jeanrenaud, 1951). In the mid-1980s, the Greater London Council supported community-based prototyping workshops called "Technology Networks." By providing physical spaces, access to shared machine tools, and assistance from technical staff to local communities, workers, and cooperative enterprises, the Council hoped people could develop prototypes for new industrial products such as energy-saving devices for refrigerators, toys for schools, or electric bicycles. Some of those projects did indeed find developers and investors in the UK and other countries. Another stream of activities, though, was more oriented towards social mobilization. Eventually, funding for the program was withdrawn under Thatcher (Smith, 2014).

The 1980s and 1990s were marked by the appearance of the personal computer and computer networks, most notably the Internet. These technologies became the new backdrop for technology-based activism (aka "hacking") and technical pastime. The word "hacker" came to be in places such as MIT, purportedly as tech slang, meaning one who works like a hack at writing and experimenting with software, who enjoys computer programming for its own sake. The hacker subculture, spearheaded by the Chaos Computer Club (Germany) and the hacklabs and hacker spaces built on its model, introduced a DIY approach to information technology (IT) and IT security research. It was only in later developments that hackerspaces started gravitating towards the physical (maxigas, 2012).

2.1 The Year Zero of Making

In the early 21st century, making received a boost through the increasing availability of entry-level digital manufacturing equipment – from laser cutters and CNC mills to 3D printers and digital tinkering kits such as the Arduino and Raspberry Pi. Indeed, 2005 turned out to be a sort of "year zero" for self-production. Mark Frauenfelder and Dale Dougherty of O'Reilly Media launched *Make:*

magazine. Neil Gershenfeld promised personal fabrication as "the coming revolution on your desktop." Massimo Banzi and David Cuartielles presented their first Arduino boards that make microcontrollers easily accessible to artists and tinkerers. Eric Wilhelm launched the online platform *Instructables* where people could share instructions on how to make things such as tree houses or 3D printed bionic hands. Of course, these events were not planned to coincide, and some of them were just turning points in a longer development. For instance, Gershenfeld's fab labs evolved from an initiative by Shrinath Kalbag, the founder of an Ashram-type institute to promote non-formal education for rural youths in Pabal, India, who adopted some of Gershenfeld's tools in 2002 (2005, pp. 164–168). Arduino started as a fork of Hernando Barragán's master thesis project "Wiring" – Barragán graduated from the Interaction Design Institute Ivrea (IDII) in Italy in late 2004 under the supervision of Massimo Banzi.

Extending "year zero" into 2006, more making milestones would follow. Maker Media, publisher of *Make:* magazine, organized their first maker faire, a gathering, festival and market where makers would congregate, give talks, follow workshops, and sell the goods they made; and Jim Newton and Ridge McGhee opened the first TechShop location in Menlo Park, California. Through media, propaganda, and active colonization, this dream, and its related, mixed ideologies of individual empowerment through technology combined with a subtle but firm subjection to platform capitalism were exported to Europe, Asia, and Africa, promising a new economic blossoming after the end of the dot. com era and the 9/11 attacks in 2001, and the crisis following the collapse of Lehman Brothers in 2007–9 (Zuboff, 2019).

2.2 Making in the 21st Century

Making when producing hardware is bound to manufacturing infrastructure in the form of private or public workshops. These workshops typically are equipped with a range of prototyping and manufacturing equipment ranging from simple manual utensils and power tools to sophisticated machinery: machines for woodworking, metalworking, forges, ceramics workshops, sewing and embroidery machines, and looms for textile work, and digitally controlled laser cutters, CNC routers, 3D printers, and so on. Many workshops also have electronic workbenches with signal generators, oscilloscopes, and soldering equipment. Apart from manufacturing infrastructure, some places include shared office space, meeting rooms, even cafés or event areas. The public workshops are either accessible for free or after paying a monthly membership fee of a few hundred dollars or euros. Most sites also provide – or even require – training on the equipment.

In terms of their historic and ideological positioning, spaces differ considerably. There are at least five different flavors of self-production factories: first, there are traditional hobbyist workshops, stemming from pre-year-zero DIY initiatives – for example, the "offene Werkstätten" (open workshops) in Germany. Second, there are spaces that stand in the tradition of hacklabs and hacker spaces that have their origin in computer hacking, be it political (hacklabs) or leisurely (hacker spaces). Third, as of 2002 fab labs appeared, which followed the paradigm established by Neil Gershenfeld and Sherry Lassiter at MIT with the aim of "empowering inventors through sharing of capacities, people and projects" (CBA, 2012). Fourth, there are maker spaces which pretty much adhere to the Californian ideology of *Make:* magazine, the idea that hedonistic self-production leads to freedom and (commercial) success. Fifth, there were TechShops, a 2006–2017 saga of super-powered maker spaces following the pattern of bigger is better and with the ultimate goal of tapping into people's free time and money to create multiple businesses from making.

These workshops promise to be places to "make (almost) anything." This slogan suggests a certain depth of manufacturing, however, in practice mostly standard components and sheet and filament materials are used, and going back to produce raw materials completely eschews the realm of self-production – as Thomas Thwaites (2011) has shown. Even when it comes to designing products – a necessary predecessor to actual manufacturing – a large proportion of makers are using pre-existing designs that are propagated on YouTube and through platforms such as *Thingiverse.com* and *Instructables.com* (digitally) or *Make:* magazine (analogue). Makers reproduce these designs or adapt them to their taste, needs, and affordances. Making requires a certain skill-set – designing, manufacturing, electronics, coding – which is not that easy to acquire. Online instructions, at best, provide step-by-step instructions which might be incomplete, faulty, or only valid under some specific circumstances – e.g., having access to a US rather than a European or even an African supply of parts and materials (see Coban, 2018). They rarely furnish the conceptual understanding of what "proper" craftsmanship training delivers. And they mostly cover-up the things that might and will go wrong when manufacturing. So, taken with a pinch of salt, these workshops provide an infrastructure for the enthusiast and dedicated makers to actually produce many goods which often would not even be available on the (consumer) market. Yet in terms of production volume, in terms of availability of parts and raw materials, and in terms of the design and manufacturing skills needed, making is far from providing a straightforward alternative to the prevailing mass-manufacturing paradigm.

According to Neil Gershenfeld, fab labs in the Global North (in his words: the "developed world") enable the "market of one" – for goods that are not available (and probably never economically will be) under the mass-manufacturing paradigm. In the Global South (the "developing world") the market is one of "providing solutions to local problems" (Gershenfeld, 2006). It is relevant to notice that indeed in the Global South there thrives a different kind of self-production economy, e.g., jugaad in India or the Suame Magazine in Ghana. They create self-production solutions to local needs in ways that have become almost unthinkable in 21st-century Europe and North America. However, we are not going to discuss this latter aspect of peer production, as makers and making are concepts of the Global North, loaded with the ideology of the strong individualism of Randian descent (Žižek, 1998) that dominates the current US discourse of self-production and fuels the Californian ideology.

The typical markets for making are collective endeavors, the most prominent being the maker faires. Established in San Matteo, SF, in 2006, they were a franchise of O'Reilly's Maker Media. Part festival, part conference, part market, those events serve as a podium to promote Maker Media's self-production ideology, as a platform for makers to showcase their projects, and as a marketplace for anyone from lobbyists to purveyors of software, machines, tools, supplies, and kits to flog their merchandise to aspiring makers. Maker faires are modeled on the traditional country fairs – agricultural shows that "provided local people with an opportunity to celebrate achievements and enjoy a break away from day-to-day routine" with a combination of serious competition and light entertainment. These annual shows acknowledged and rewarded the hard work and skill of primary producers – similarly, maker faires acknowledge and reward the "work" of self-producing makers. Country fairs provided a venue for rural families to socialize – and so do maker faires for makers to meet. The faire image was of course chosen on purpose. It blended in with the revival of the self-sufficient, self-provisioning pre- and early industrial society, the romantic vision of a great America unfettered by globalization and dependence on (foreign) fuel reserves, raw materials, industrial goods, and commodities. The maker faire narrative corresponded to the restoring narrative of the First Build co-creation community – they produce high-end kitchen appliances locally and in small batches – and the revival of the image of the "American Maker" in Dale Dougherty's TED talk (2012).

As part of Maker Media's international franchising and colonization, maker faires have spilled all over the globe – and sometimes integrated into different national narratives. In Germany, technology hobby publisher Heise has engaged in an overall franchising agreement with O'Reilly's Maker Media to promote making, publish *Make:*, and run two maker faires in the country. However, Heise worked hard on relating *Make:* to German culture (Hepp, 2018) in which engineering is seen as one of the defining elements of national identity. This included publishing in-depth articles on technical issues (which have found an eager audience in the United States, much to the surprise of Maker Media) and aligning maker projects with the legal and regulatory environment in Germany, which is in certain aspects radically different from the situation in the United States, for instance when it comes to issues such as allowed and prohibited use of patent information or electrical safety regulations. Other obvious markets for makers are online platforms for promoting and selling products such as *Etsy* and similar platforms like the (defunct) *DaWanda* and *selekkt* in Germany, *folksy* in the UK, or *Artesanum* in Spain. On other platforms such as *xyz.com* and *3Dhub.com*, makers market their skills and capacities. The latter platform has become one of the world's largest 3D-print services – decentralized and local.

In making, not only products, tools and capacities are exchanged, but also "knowledge" – recipes, instructions, and tutorials on how to self-produce. The medium of choice for knowledge markets in self-production is the Internet, and various platforms to exchange manufacturing designs and instructions have been developed. Some of these platforms are (or have initially been) related to a specific technology. For instance, Arduino used to host a wiki with numerous Arduino-related projects; 3D printer manufacturer MakerBot Industries set up the platform *Thingiverse* for their customer to share 3D designs; *Instructables*, founded independently in 2005 and owned by Autodesk since 2011, is a website that allows its (registered) users to create and upload do-it-yourself projects; *wikiHow*, founded the same year, serves a similar purpose, though its contents sometimes tend to veer off the pure self-production realm.

The purpose of making is still up for debate, and four potential directions have been suggested: making as a mainly bourgeois pastime and token act of rebellion that at heart is just a form of entertainment and consumption; making as an innovation in technology education and a response to industry's fear of a declining technically skilled workforce; making as a new renaissance reuniting the liberal arts with science and engineering in a playful way; or making as a new industrial revolution replacing globalized mass manufacturing with local, boutique-style manufacturing and aiming at closing loops of material flows (Troxler, 2017).

3 Defining Making: Words Matter

Making is a catch-all idiom without obligations. Tracing the origins of the term might appear as needless – after all, it denotes a common verb for creating and forming something. But put into perspective with other peer-production practices, making gained a specific meaning in the past few years. In particular, the launch of *Make:* magazine by O'Reilly in 2005 and the promotional work performed by its co-founder Dale Dougherty popularized the terms making and maker worldwide. A particular reason for choosing the term "making" over "hacking," as Dougherty explains, was to demarcate a set of practices and actors, considered friendly and contributive to one's personal and social development, from hacking's more subversive and criminal associations (Dougherty, 2012; Stangler & Maxwell, 2012). The spectacle-character of maker faires and the communal meaning underpinning makerspaces, fab labs, and other types of "shared machine shops"[1] (Troxler & maxigas, 2014) emphasized the friendly, virtuous culture of makers over the potentially unlawful one of hackers.

In 2014, shortly before the financial collapse of TechShop, its CEO Mark Hatch would re-iterate the divine metaphor: "God is a maker, and he made us in his image. [. . .] we may not know much about God or humans, but we do know one thing for sure: we were made to make" (2014, p. 12). The rhetoric of "man is made to make," "all of us are makers," and "making engages us fully and deeply as human beings, and it satisfies our creative souls," repeated by Dougherty and Hatch as well as presidents Obama and Trump alike (Linneman, 2017) characterize the "maker movement" as something deeply US and as an almost magical source of creativity and innovation for rebuilding America's economy and a revival of the American Dream. Recalling David Noble's investigation of the Western relationship between religion and technology (1999), Fred Turner frames these prominent figures of making's "year zero" as story brokers building an "ideological archipelago" (2018). For Tom Jenkins and Ian Bogost, "[i]t's a deft rhetorical strategy: by construing anything whatsoever as 'making,' its proponents gain substantial momentum" (2015, p. 30). Based on a broad media ethnography of the core organizational figures behind Maker Media, several maker events and makerspaces, Andreas Hepp holds that the entire maker franchise follows a curated franchise strategy with the term "maker" conveying that anyone can belong to this diverse community (Hepp, 2018). The distinctive positioning aside from hacking ensured, therefore, at least in the United States, a closer alignment with economic goals, venture capital, and government interventions. Notable national activities such as the US presidential initiative *Nations of Makers* launched during Obama's presidency[2] and China's governmental backing of makerspaces as incubators and innovation hubs[3] convey this idea.

Despite the commercially-motivated origins of Maker Media and its franchise that helped popularize making, academic research, as well as popular accounts, have often portrayed making as emanating from hacker practices. From the point of view of computer-supported practices, maker and hacker cultures seem intertwined.[4] Both reside in the marginal areas of technology production, have a reputation of "nerdiness," and build upon highly social cultures (Coleman, 2013; Turkle, 1984).

In making's enfolding as a primarily commercial undertaking – a discernible development with, for example, the growing DIY electronic components economy including companies such as Arduino, adafruit, SparkFun, or littleBits – its potential origin in the hacker ethos and computer counterculture express a contradiction (Carelli, Bianchini, & Arquilla, 2014; Lindtner, Hertz, & Dourish, 2014). What could be observed in the product design of information and communications technologies is a "broader transition from hacking into making" (Lindtner et al., 2014).

Craft as a production practice is also paramount in maker discourses (e.g., see Buechley & Perner-Wilson, 2012; Cheatle & Jackson, 2015). Rarely does any idealized description of making go without mentioning *The Craftsman* (2008) by sociologist Richard Sennett. As part of the popular rhetoric of making as reviving (urban) manufacturing through the production for "a market of one," it tends to be explored in relation to different modes of production: from pre-industrial handiwork and craftsmanship through mass manufacturing to contemporary forms such as industry 4.0. Making, like craft, presupposes a complete workflow from design to production executed by the same person. By framing practices of making as craft (and craftwork), making also became positioned along established communities of practice respected for their proficiency in particular areas. Craft as a parameter, therefore, works to bestow agency upon makers. In fact, until the 18th century, craft with its guilds structure invested in power and secret knowledge described a form of politics and not a specific way of fabricating things. Possibly unsuspecting of this early definition, contemporary maker cultures instead position craft in opposition to mass manufacturing processes and products. In that sense, as Florian Cramer (2019) argues, making is a continuation and commodification of Thoreau's concept of civil disobedience (1854/1971).

Becoming mainstream and striving for significance, "year zero"-making further attempts to distinguish itself from labels such as DIY and tinkering, as both traditionally relate to leisure and hobbyism. Although examples such as ham radio hobbyists illustrate how the practice paved the way for individuals into professional realms (Haring, 2007), DIY and tinkering, in general, are concepts of practice that stand outside professional areas. The generic character of making thus denotes a different material activity from the one practiced by professional technology designers or by amateurs only. Making, in the words of Evan Barba, is "a form of design rhetoric" (2015, p. 639) which has different understandings and goals – the maker identities can be simultaneously multiple, diverse, and even contradictory. Rather than someone being simultaneously a maker and an amateur, a maker and a hacker, or a maker and a designer, making as a "palatable" practice defines the common ground between them.

4 Empowerment and Emancipation: Users Becoming Makers

Peer production overall and the communities, practices, and technologies for making, hacking, DIY, and craft, in particular, have been of growing interest in the areas of HCI, education, design research, as well as the social sciences. Under the premise of one of making's central commitments, the democratization of technology production, ongoing interest rests to some extent on the possibility of empowerment and emancipation of a passive consumer society: "[M]aking challenges whether there is (or needs to be) a gap between designers and engineers on the one hand, and between designers and end users on the other" (Lindtner, Bardzell, & Bardzell, 2016, p. 1398).

Research and practice in HCI, for instance, find value in making and maker cultures as these trouble the traditional concepts of design, production, and use through the structural component of user participation. Analyses of making (and digital fabrication) reference a range of subject matter in their efforts to situate them within established cultures and systems of technology design and production: the different motivations for making; the characteristics of communities and their cultures; the recognition of lay practice; the emancipatory promises of maker cultures for user empowerment and participatory citizenship; a critical assessment of the rhetoric of technosolutionism[5] and the expanding hegemony of globalizing franchises for making; down to the influence of craft and the actual material practice of making. Structured in this particular order, these analyses suggest an imaginary wayfinding from what attracts one to making in the first place to what problems arise as one becomes immersed in these cultures.

In these initial years after making's "year zero," several ethnographic and quantitative studies have investigated the motivations behind making (e.g., Davies, 2017; Kuznetsov & Paulos, 2010; Phillips, Silve, & Baurley, 2013). Earlier studies, often acknowledging the off-beat origins and the DIY character of maker practices, suggest that making appeals to people for its possibility of self-actualization and non-commercial orientation. Cultivated physical environments for making, regardless of their physical location and strategic orientation, also attract people with their programs for gaining literacy in digital technologies and processes and lessening the crisis of meaning for the individual (e.g., Di Roma, Minenna, & Scarcelli, 2017; Foster, 2017). Both issues are deterministically attributed to widespread digital transformations.

Understanding the motivations behind individuals' participation in maker cultures, however, requires an exposition of their geographical and economic context (Tanenbaum, Williams, Desjardins, & Tanenbaum, 2013; Lindtner et al., 2016). Makers' geographies and economic situation necessitate a differentiation between a pleasure-oriented and a utility-driven DIY practice if we want to understand ideas of democratization, innovation, and the difference between professionals and lay makers (Hertz, 2011). These examples help situate the emancipatory objective of making, but often they tend to leave out a differentiation between maker technologies as well-suited to prototyping and actual manufacturing technologies, and how that relates to the question of user emancipation. This to a degree resides in most studies' research foci on individual makers, smaller communities, and shared machine shops rather than on a connection to existing structures for (technology) design and production.

In recognition of the differences in the motivation and the practical experience of makers, a few studies in design scholarship studies on non-professional design and its relationship to professional practice call for the ennoblement of lay and amateur practice (e.g., Atkinson, 2010; Beegan & Atkinson, 2008). These design scholars argue that recognition is not about equating, but putting alongside and in dialogue. DIY practices and "digital tinkering" demonstrate that clear-cut boundaries around professional and non-professional design cannot be retained (Beegan & Atkinson, 2008). Making and digital fabrication move design towards a more inclusive "post-professional era" which gives agency to non-professionals through participation without making the other uses obsolete (Atkinson, 2010, p. 138). Empowerment, in this case, requires acknowledging the ruptures of social structures and labor hierarchies.

Technology appropriation and open source hacking provide a role for participatory citizenship and the potential emergence of user innovation. The emancipatory prospect of making recurrently tends to emphasize its contribution to marginalized individuals only without giving a description of their background or accounting for the stipulations for participation. In *FabLab: Of Machines, Makers and Inventors* (2013), for example, the function of fab labs is described as doubtless political: "FabLabs have a clear mission to empower people to participate and engage in technological development and to educate underprivileged people" (Carstensen, 2013, p. 61). As scholarship on making and hacking has emphasized, however, such political endeavors also often imply a level of expertise and knowledge about access to these activities and communities that is likely concealed by popular rhetoric (Powell, 2012; Roedl, Bardzell, & Bardzell, 2015). In her investigation of 3D printing's "political imaginaries," Jesse Adams Stein (2017) notes that it is essential to understand how such rhetoric around emancipation, empowerment, or economic revival become deployed and unquestioned social norms.

Elsewhere, information and social scientists had put up one of the earliest panels at the CHI conference on the contradictions between the emancipatory narrative and the actual state of maker cultures (see Ames et al., 2014). Concurrently, the first academic journal issue dedicated to shared machine shops appeared in the *Journal of Peer Production* 5 (Troxler & maxigas, 2014) questioning the myths of the "maker movement": "Shared Machine Shops are not new," "Sharing is not happening," or "Hackerspaces are not open," to name a few. The same journal extended these debates by looking at the questions of gender in an issue on feminism, making, and hacking (Bardzell, Nguyen, & Toupin, 2016), and how the increasing institutionalization of shared machine shops within universities and corporations redefines making's origins and prospects (Braybrooke & Smith, 2018). Looking back at Neil Gershenfeld's first "How to Make Almost Anything" MIT-course in 2003, it becomes evident that strategies for professionalization such as the integration into hardware startups and incubators (Lindtner et al., 2014) and the popularization of hackathons as means for job acquisition in tech and design industries (Irani, 2015) have always formed a major endeavor in establishing a maker culture.

5 Maker Communities and the "Movement"

Studies on characteristics of communities for making and their cultures draw upon ethnographic research of mostly single shared machine shops or multi-sited examinations within one geographical location (e.g., Hielscher, 2017; Kohtala & Bosqué, 2014). Their insights therefore often overlap. Other examples, however, highlight a critical broadening: for example, a study on the UK-based Men's Sheds[6] under the lens of making illustrates the role of open workshops in public life beyond innovation and peer education (Taylor, Hurley, & Connolly, 2016); the emergence of feminist hackerspaces challenges the technosolutionists norms of maker cultures (Fox, Ulgado, & Rosner, 2015; Toupin, 2014); or, Chinese maker communities challenge the "conceptual binaries of design as a creative process versus manufacturing as its numb execution" (Lindtner, Greenspan, & Li, 2014). These perspectives allow the evaluation of dissimilar projects focused on social contributions rather than innovation. Making, maker communities, and maker culture, as already illustrated, designate distinct instances not always sharing the same value system (Bardzell, Bardzell, & Ng, 2017). Frequently, commercial proponents like to portray makers as a global social movement opposing proprietary production schemes and entrepreneurial models, while the maker ecology avoids reflecting on its own neoliberal foundations depicted in individuals working on individual solutions, as well as ongoing corporate co-optation and mainstreaming.[7]

6 Discussion and Conclusion

As uncritical pieces of literature keep filling the shelves of bookstores, populate blogs and online video channels, and help reinforce the position of the Maker Media franchise and the fab lab revolution that promises a new way of "designing reality" (Gershenfeld et al., 2017), a critical scholarly assessment of the rhetoric of technosolutionism and the hegemony of these globalizing franchises also takes issue with those characterizations (e.g., Carelli, Bianchini, & Arquilla, 2014; Hunsinger, 2017). An early account of the technological utopianism inscribed in *Make:* magazine problematizes the absence of attention to political responsibility (Sivek, 2011). *Make:*'s assertive emphasis on self-actualization through technology aligns it with corporate interests rather than citizen empowerment (Sivek, 2011). Garnet Hertz (2012) has critiqued *Make:* for covering Natalie Jeremijenko's "Feral Robot Dogs," a political artwork and pedagogical project, in its inaugural issue, while later maintaining a completely apolitical focus.

Exposing maker culture's ambivalences requires mapping out the narratives, actors, relationships, developments, and histories as the characterization as a perpetual "grass-roots" movement rather than a mediated "pioneer community," as Hepp (2018) points out. Making is also bereft of an inward reflection – "research into" making is growing disproportionately to "research from within" making. Rarely do observations and proposals from empirical socio-technical research find their way into maker communities to address and contribute to actual inner concerns. By the same token, media content analysis and the prevalent emphasis on technical maker projects convey that there is little discourse within the culture of making about implicit assumptions on the society-technology divide. While the neutrality of technology is barely being disputed, the understanding of technology as part of a social assemblage, in Latourian terms technology as "society made durable" (1990), falls systematically through in the dominant maker narrative. Participation and peer production in making hardly work out of the box but depend on local culture, diversity, and the role of elites (Mansuri & Rao, 2004).

The homogeneity of maker typologies and projects substantially supported the expanse of the Maker franchise. The first nine years of *Make:* magazine portrayed a narrow definition of making – mostly electronics, vehicles, robots, and rockets – and of makers – 85 % men and boys, with no people of color (Buechley, 2013).[8] Having an identical demographic user group with similar technical interests and an alignment with innovation simplifies the search for venture capital from corporate partners. This kind of elite high-tech culture works to marginalize women, people of color, and the poor, which have always been at the core of diverse DIY, maker and repair activities with and without the use of technology (Rosner, Shorey, Craft, & Remick, 2018). Once the short-lived *Craft* magazine for the "non-technical," dubbed feminine, maker activities became incorporated as a few pages in *Make:*, Maker Media sent out a particular message about what counts as a valuable practice of making. Yet women are more likely to identify with makers than with technologists, hackers, or geek cultures.

Identification with this particular culture of making also depends on cultural backgrounds and access to funds and information (Foster & Boeva, 2018). In the Global South, but also in indigenous or poverty-stricken communities across the Global North, making or repairing things happens for many reasons such as the conservation of local craft traditions, due to limited commodities, or out of sheer necessity for survival. The diversity of these forms and practices of making and those who participate in them become neglected with the dominant focus on the presumed prospering maker technologists. However, as Maker Media ceased all operations after its bankruptcy in June 2019 – all coming unsurprisingly following the slow takeover of maker enterprises by large corporations, military research, government programs, and educational institutions in recent years – it is high time for making to account its long neglected communities, localities, and practices formed by notions of care.

Finally, for making to move beyond the circles of consenting nerds, it needs to contribute to the bigger challenges of society: becoming economically, socially, and ecologically sustainable, developing a convivial network, achieving equality, recognizing diversity, defying technocracy, and elaborating on the notion of lateral governance (Troxler, 2015, 2017). Eventually, if making wants to meaningfully contribute to a more equitable society within the remaining means of the planet (Raworth, 2017), it needs to abandon the market/state duopoly of the first and second industrial revolution, the market economy based on the assumption of unlimited growth, and the fair functioning of the free market. The principle of the commons has been proposed as a generative paradigm to step outside of the dominant discourse of the market economy and to build a social system for the long-term stewardship of resources that preserves shared values and community identity (Bollier, 2014, p. 175; also see Nissenbaum & Benkler, this volume). Making as social fabrication requires developing such stewardship for people and planet.

Notes

1 The term "shared machine shop" was first used by Karl Hess in his book *Community Technology* (1979) to describe communal workshops for tool sharing. It became applied to maker cultures in an issue of the *Journal of Peer Production*, 5 (2014) devoted to this topic.

2 See https://obamawhitehouse.archives.gov/nation-of-makers (accessed July 5, 2019).

3 See T. Saunders, T. and J. Kingsley (2016) *Made in China. Makerspaces and the Search for Mass Innovation. NESTA Report*, London.

4 The terms *makerspace* and *hacker space* have recently been included in the Oxford English Dictionary (2018). Interestingly, the only difference in the definition to hackerspace is the focus in computing and technology.

5 Technosolutionism describes the idea that the application and use of technology can solve social problems.

6 Men's Sheds are communal workspaces for older men to mitigate mental health concerns (see Taylor et al., 2016).

7 The persistent incorporation of making into corporate ventures has taken different forms from sponsorship by IT companies and Disney of the maker faires, through corporate partnerships of fab labs and makerspaces, e.g., Chevron, to the establishment of corporate fab labs, e.g., at Airbus or BMW, to boost employees' creativity and innovation (Hagel, Brown, & Kulasooriya, 2013). Simultaneously, the US military has been actively "infiltrating" making through DARPA sponsoring Maker Media's educational program or veterans' organizations investing in TechShop.

8 These numbers are uncritically reconfirmed with Maker Media's own study of attendance (2014).

References

Ames, M. G., Bardzell, J., Bardzell, S., Lindtner, S., Mellis, D. A., & Rosner, D. (2014). Making cultures: Empowerment, participation, and democracy – or not? In *CHI '14 Extended Abstracts on Human Factors in Computing Systems* (CHI EA '14) (pp. 1087–1092). doi:10.1145/2559206.2579405

Anderson, C. (2012). *Makers: The new industrial revolution.* London: Random House.

Atkinson, P. (2010). Boundaries? What boundaries? The crisis of design in a post-professional era. *The Design Journal, 13*(2), 137–155. doi:0.2752/175470710X12735884220817.

Atkinson, P. (2011). Orchestral manoeuvres in design. In B. van Abel, L. Evers, R. Klaasseen, & P. Troxler (Eds.), *Open design now: Why design cannot remain exclusive* (pp. 24–31). Amsterdam: Bis Publishers.

Balsamo, A. (2015). The cultural work of citizen R&D. In B. Timm Knudsen, D. R. Christensen, & P. Blemker (Eds.), *Enterprising initiatives in the experience economy: Transforming social worlds* (pp. 83–100). Abingdon: Routledge.

Banzi, M., Cangiano, S., & Fornari, D. (2016). Open technologies: The reasons for an issue. *Progretto Grafico, 30*, 8–9.

Barba, E. (2015). Three reasons why the future is in the making. *Science, Technology & Human Values, 40*(4), 638–650. doi:10.1177/0162243915572263.

Barbrook, R., & Cameron, A. (1996). The Californian ideology. *Science as Culture, 6*(1), 44–72.

Bardzell, S., Bardzell, J., & Ng, S. (2017). Supporting cultures of making: Technology, policy, visions, and myths. In *Proceedings of the 2017 CHI Conference on Human Factors in Computing Systems* (CHI '17) (pp. 6523–6535). doi:10.1145/3025453.3025975.

Bardzell, S., Nguyen, L., & Toupin, S. (Eds.) (2016). Feminism and (un)hacking. *Journal of Peer Production, 8.* Retrieved from http://peerproduction.net/issues/issue-8-feminism-and-unhacking/

Beegan, G., & Atkinson, P. (2008). Professionalism, amateurism and the boundaries of design. *Journal of Design History, 21*(4), 305–313. doi:10.1093/jdh/epn037.

Bollier, D. (2014). *Think like a commoner.* Gabriola Island: New Society Publishers.

Braybrooke, K., & Smith, A. (Eds.). (2018). Makerspaces and institutions. *Journal of Peer Production, 12.* Retrieved from http://peerproduction.net/issues/issue-12-makerspaces-and-institutions/

Buechley, L. (2013). Thinking about making. *FabLearn.* Keynote presented at the third FabLearn Conference, Stanford, CA. Retrieved from http://edstream.stanford.edu/Video/Play/883b61dd951d4d3f90abeec65eead2911d

Buechley, L., & Perner-Wilson, H. (2012). Crafting technology: Reimagining the processes, materials, *and cultures of electronics. ACM Transactions on Computer-Human Interaction, 19*(3), Article 21. doi:10.1145/2362364.2362369.

Carelli, A., Bianchini, M., & Arquilla, V. (2015). The "makers contradiction." The shift from a counterculture-driven DIY production to a new form of DIY consumption. STS Italia Conference "A Matter of Design: Making Society through Science and Technology." Paper presented at STS Italia Conference, Milan, Italy. Retrieved from http://www.stsitalia.org/?p=1548.

Carstensen, T. (2013). Gendered fablabs? In J. Walter-Hermann & C. Büching (Eds.), *FabLab: Of machines, makers and inventors* (pp. 53–64). Bielefeld: transcript Verlag.

CBA, The Center for Bits and Atoms. (2012). *The fab charter.* Retrieved from http://fab.cba.mit.edu/about/charter/

Cheatle, A., & Jackson, S. J. (2015). Digital entanglements: Craft, computation and collaboration in fine art furniture production. In *Proceedings of the 18th ACM Conference on Computer Supported Cooperative Work & Social Computing* (CSCW '15) (pp. 958–968). doi:10.1145/2675133.2675291

Coban, A. (2018). Making hardware in Nairobi: Between revolutionary practices and restricting imaginations. *Journal of Peer Production, 12.* Retrieved from http://peerproduction.net/issues/issue-12-makerspaces-and-institutions/peer-reviewed-papers/making-hardware-in-nairobi/

Coleman, G. (2013). *Coding freedom.* Princeton, NJ: Princeton University Press.

coons, g. (2018). Can one size fit one? A prospect for humane custom production. *Journal of Peer Production, 12.* Retrieved from http://peerproduction.net/issues/issue-12-makerspaces-and-institutions/peer-reviewed-papers/can-one-size-fit-one/

Cramer, F. (2019). Does DIY mean anything? A DIY attempt (=essay). *Anrikningsverket Journal, 1*, 54–72. Retrieved from https://norbergfestival.com/anrikningsverket-journal/

Cross, N. (1982). Designerly ways of knowing. *Design Studies, 3*(4), 221–227. doi:10.1016/0142-694X(82)90040-0

Davies, S. R. (2017). *Hackerspaces: Making the maker movement.* Cambridge: Polity.

Di Roma, A., V. Minenna & A. Scarcelli (2017). Fab labs. New hubs for socialization and innovation. *The Design Journal, 20*:sup1, S3152–S3161. doi:10.1080/14606925.2017.1352821.

Dougherty, D. (2011, January). We are makers. Technology entertainment and design. *TED@MotorCity*. Talk presented at 2011 TED conference, Detroit, MI. Retrieved from http://www.ted.com/talks/dale_dougherty_we_are_Makers.html

Dougherty, D. (2012). The maker movement. *Innovations, 7*(3), 11–14.

Foster, E. F. (2017). *Making cultures: Politics of inclusion, accessibility, and empowerment at the margins of the maker movement* (Unpublished doctoral dissertation, Rensselaer Polytechnic Institute, Troy, NY).

Foster, E. F., & Boeva, Y. (2018). Making sense of place: Research reflections from two multi-sited ethnographies. *Making Futures, 5*. Retrieved from www.plymouthart.ac.uk/documents/Ellen_K_Foster_and_Yana_Boeva.pdf

Fox, S., Ulgado, R. R., & Rosner, D. (2015). Hacking culture, not devices: Access and recognition in feminist hackerspaces. In *Proceedings of the 18th ACM Conference on Computer Supported Cooperative Work & Social Computing* (CSCW '15) (pp. 56–68). doi:10.1145/2675133.2675223.

Franz, K. (2005). *Tinkering: Consumers reinvent the early automobile*. Philadelphia, PA: University of Pennsylvania Press.

Gelber, S. M. (1997). Do-it-yourself: Constructing, repairing and maintaining domestic masculinity. *American Quarterly, 49*(1), 66–112.

Gershenfeld, N. (2005). *Fab: The coming revolution on your desktop*. New York, NY: Basic Books.

Gershenfeld, N. (2006, February). Unleash your creativity in a fab lab. *TED2006*. Talk presented at TED conference. Retrieved from www.ted.com/talks/neil_gershenfeld_on_fab_labs?language=en.

Gershenfeld, N. (2012). How to make almost anything: The digital fabrication revolution. *Foreign Affairs, 91*, 43.

Gershenfeld, N., Gershenfeld, A., & Cutcher-Gershenfeld, J. (2017). *Designing reality. How to survive and thrive in the third industrial revolution*. New York, NY: Basic Books.

Hagel, J., Brown, J. S., & Kulasooriya, D. (2013). *A movement in the making*. London: Deloitte Development LLC. Retrieved from https://deloitte.wsj.com/cfo/files/2014/02/MakerMovement.pdf

Haring, K. (2007). *Ham's radio technical culture*. Cambridge, MA: MIT Press.

Hatch, M. (2014). *The maker movement manifesto: Rules for innovation in the new world of crafters, hackers, and tinkerers*. New York, NY: McGraw Hill Education.

Hepp, A. (2018). What makes a maker? Curating a pioneer community through franchising. *NTiK Nordisk Tidsskrift for Informationsvidenskab og Kulturformidling, 7*(2), 3–18.

Hertz, G. (2011). Arduino microcontrollers and the queen's hamlet: Utilitarian and hedonized DIY practices in contemporary electronic culture. In *Proceedings of the 31st Annual Conference of the Association for Computer Aided Design in Architecture* (pp. 44–47).

Hertz, G. (2012). *Critical making*. Hollywood, CA: Telharmonium Press. Retrieved from http://conceptlab.org/criticalmaking/

Hielscher, S. (2017). Experimenting with novel socio-technical configurations. *Digital Culture & Society, 3*(1), 47–72. doi:10.14361/dcs-2017-0104.

Hunsinger, J. (2017). Hacking together globally. *Digital Culture & Society, 3*(1), 95–108. doi:10.14361/dcs-2017-0106.

Irani, L. (2015). Hackathons and the making of entrepreneurial citizenship. *Science, Technology & Human Values, 40*(5), 799–824. doi:10.1177/0162243915578486.

Jeanrenaud, W. (1951). *Inhalt und Methoden der Pionierarbeit. Ost-Probleme, 3*(47), 1457–1463.

Jenkins, T., & Bogost, I. (2015). Escaping the sandbox: Making and its future. In *Proceedings of the Ninth International Conference on Tangible, Embedded, and Embodied Interaction (TEI '15)* (pp. 29–32). doi:10.1145/2677199.2680558

Kline, R., & Pinch, T. J. (1996). Users as agents of technological change: The social construction the of automobile the rural United States. *Technology and Culture, 37*(4), 763–795. doi:10.2307/3107097

Kohtala, C., & Bosqué, C. (2014). The story of MIT-Fablab Norway: Community embedding of peer production. *Journal of Peer Production, 5*. Retrieved from http://peerproduction.net/editsuite/issues/issue-5-shared-machine-shops/peer-reviewed-articles/the-story-of-mit-fablab-norway-community-embedding-of-peer-production/.

Koste, W., Livingston, D., Thiele, J., & Yarbrough, J. (directors). (1960) *American Maker* [motion picture]. USA: Jam Handy.

Kuznetsov, S., & Paulos, E. (2010). Rise of the expert amateur: DIY projects, communities, and cultures. In *Proceedings of the 6th Nordic Conference on Human-Computer Interaction: Extending Boundaries* (NordiCHI '10) (pp. 295–304). doi:10.1145/1868914.1868950.

Latour, B. (1990). Technology is society made durable. *The Sociological Review, 38*(S1), 103–131.

Lindtner, S., Bardzell, S., & Bardzell, J. (2016). Reconstituting the utopian vision of making: HCI after technosolutionism. In *Proceedings of the 2016 CHI Conference on Human Factors in Computing Systems (CHI '16)* (pp. 1390–1402). doi:10.1145/2858036.2858506.

Lindtner, S., Greenspan, A., & Li, D. (2015). Designed in Shenzhen: Shanzhai manufacturing and maker entrepreneurs. *Aarhus Series on Human Centered Computing, 1*(1). doi:10.7146/aahcc.v1i1.21265

Lindtner, S., Hertz, G., & Dourish, P. (2014). Emerging sites of HCI innovation: hackerspaces, hardware startups and incubators. In *Proceedings of the SIGCHI Conference on Human Factors in Computing Systems* (CHI '14) (pp. 439–448). doi:10.1145/2556288.2557132

Lineman, A. (2017). At 3D/DC, gauging Trump's impact on the maker movement. *Shapeways Blog*. Retrieved from www.shapeways.com/blog/archives/29626-3ddc-gauging-trumps-impact-maker-movement.html

Maines, R. (2009). *Hedonizing technologies: Paths to pleasure in hobbies and leisure*. Baltimore, MD: John Hopkins University Press.

Make Media. (2014). *Study of attendance*. Retrieved from http://makermedia.com/wp-content/uploads/2013/01/WMFNY-2014-deck_FOR-PRINT.pdf

Mansuri, G., & Rao, V. (2004). *Community-based and-driven development: A critical review. The World Bank Research Observer, 19*(1), 1–39.

maxigas (2012). *Hacklabs and hackerspaces. Tracing two genealogies. Journal of Peer Production, 2*. Retrieved from http://peerproduction.net/issues/issue-2/peer-reviewed-papers/hacklabs-and-hackerspaces/

Morozov, E. (2014). *Making t. The New Yorker*. Retrieved from www.newyorker.com/magazine/2014/01/13/making-it-2

Noble, D. F. (1999). *The religion of technology: The Divinity of man and the spirit of invention*. London: Penguin Books.

Phillips, R., Silve, S., & Baurley, S. (2013). *The practical maker: Investigating the definitions and requirements of and exploring the motivations behind bespoke making*. European Academy of Design. Paper presented at the 10th European Academy of Design Conference – Crafting the Future, Gothenburg, Sweden. Retrieved from www.trippus.se/eventus/userfiles/39772.pdf

Powell, A. (2012). Democratizing production through open source knowledge: From open software to open hardware. *Media, Culture & Society, 34*(6), 691–708. doi:10.1177/0163443712449497

Raworth, K. (2017). *Doughnut economics. Seven ways to think like a 21st century economist*. White River Junction, VT: Chelsea Green Publishing.

Roedl, D., Bardzell, S., & Bardzell, J. (2015). Sustainable making? Balancing optimism and criticism in HCI discourse. *ACM Transactions on Computer-Human Interaction, 22*(3), 15:1–15:27. doi:10.1145/2699742.

Rosner, D., Shorey, S., Craft, B. & Remick, H. (2018). Making core memory: Design inquiry into gendered legacies of engineering and craftwork. In *Proceedings of the 2018 CHI Conference on Human Factors in Computing Systems* (CHI '18), (pp. 1–13). doi:https://doi.org/10.1145/3173574.3174105.

Saunders, T., & Kingsley, J. (2016). *Made in China – Makerspaces and the search for mass innovation. NESTA Report*, London.

Sennett, R. (2008). *The craftsman*. New Haven, CT; London: Yale University Press.

Sivek, S. C. (2011). "We need a showing of all hands": Technological utopianism in MAKE magazine. *Journal of Communication Inquiry, 35*(3), 187–209. doi:10.1177/0196859911410317.

Smith, A. (2014). Technology networks for socially useful production. *Journal of Peer Production, 5*. Retrieved from http://peerproduction.net/issues/issue-5-shared-machine-shops/peer-reviewed-articles/technology-networks-for-socially-useful-production/

Stangler, D., & Maxwell, K. (2012). DIY producer society. *Innovations, 7*(3), 3–10.

Stein, J. A. (2017). The political imaginaries of 3D printing: Prompting mainstream awareness of design and making. *Design and Culture*, 1–25. doi:10.1080/17547075.2017.1279941.

Tanenbaum, J. G., Williams, A. M., Desjardins, A., & Tanenbaum, K. (2013). Democratizing technology: Pleasure, utility and expressiveness in DIY and maker practice. In *Proceedings of the SIGCHI Conference on Human Factors in Computing Systems (CHI '13)* (pp. 2603–2612). doi:10.1145/2470654.2481360.

Taylor, N., Hurley, U., & Connolly, P. (2016). Making community: The wider role of makerspaces in public life. In *Proceedings of the 2016 CHI Conference on Human Factors in Computing Systems (CHI '16)* (pp. 1415–1425). http://doi.org/10.1145/2858036.2858073

Thoreau, H. D. (1971). *Walden* (ed. J. Lyndon Shanley). Princeton, NJ: Princeton University Press. (Original work published 1854).

Thwaites, T. (2011). *The toaster project: Or a heroic attempt to build a simple electric appliance from scratch*. San Francisco, CA: Chronicle Books.

Toupin, S. (2014). Feminist hackerspaces as safer spaces? *Feminist Journal of Art and Digital Culture 27*. Retrieved from http://dpi.studioxx.org/en/feminist-hackerspaces safer-spaces

Troxler, P. (2014). What is a fab lab for? The philosophy, general and specific purposes. In M. Menichinelli (Ed.), *Fab lab. Revolution field manual* (pp. 51–65). Salenstein: Niggly.

Troxler, P. (2015). *Beyond consenting nerds. Lateral design patterns for new manufacturing.* Rotterdam: Hogeschool Rotterdam Uitgeverij.

Troxler, P. (2017). Making as social fabrication: Towards a new fab commons? In A. Briceño & T. Vivanco (Eds.), *Fabricating society* (pp. 78–88). Ñuñoa-Santiago: Fundacion DID. Retrieved from www.fundaciondid.cl/wp-content/uploads/2017/07/FabricatingSocietyResearchBook.pdf

Troxler, P. & maxigas (2014). Shared machine shops. *Journal of Peer Production, 5.* Retrieved from http://peerproduction.net/issues/issue-5-shared-machine-shops/

Turkle, S. (1984). *The second self: Computers and the human spirit.* Cambridge, MA: MIT Press.

Turner, F. (2018). Millenarian tinkering: The puritan roots of the maker movement. *Technology & Culture, 59*(4S), 160–182.

Zižek, S. (1998). *Cogito and the unconscious: sic 2* (pp. 74–113). Durham, NC: Duke University Press.

Zuboff, S. (2019). *The age of surveillance capitalism: The fight for a human future at the new frontier of power.* London: Profile Books.

18

Blockchain, or, Peer Production Without Guarantees

Pablo Velasco González and Nathaniel Tkacz

1 Introduction

Blockchains are aggregated and distributed databases: shared, chained, and immutable registries that conflate the production of digital tokens with their circulation. At their most basic level, they are technologies for keeping account, or records, of some form of activity, hence they are part of a long lineage of storing data, from clay tablets to bookkeeping. Perhaps their most significant quality is that they enable the distribution and storage of digital data, such as records of transactions, contracts, or other statements, that are independent (or resilient) from centralized points of control. Such resilience is achieved by distributing data among a network of peers who validate the flows of information and store them as sequential blocks: the longer the chain of blocks and the bigger the network of peers, the more secure the transmission of reliable data and the more difficult it is to tamper with. On a technical level, then, blockchains are peer-to-peer (P2P) structures for distributing and storing data. But unlike most other P2P initiatives in which the network serves to share and distribute information and resources, the peers in a blockchain network have the additional function of verifying and validating the data they help distribute.

As is well known, the first implementation of a working blockchain was the Bitcoin cryptocurrency. While Bitcoin has been the subject of wild speculation, both political and financial, it was initially conceived as a way of replacing third-party intermediaries (i.e., banks and card providers). It aimed to enable the secure transmission of currency-type tokens (via a virtual wallet application), with the blockchain simultaneously verifying and recording that a transaction had occurred. By combining the "transaction moment" with the "recording moment" and by making this record public and irreversible, Bitcoin enabled a mode of digital exchange that could bypass the established systems of financial mediation.

During the first few years of Bitcoin's operation (beginning in 2009), Bitcoin was not thought of as an instantiation of blockchain technology; rather, the blockchain was considered a component of Bitcoin. As a digital cryptocurrency, Bitcoin is effectively software and was released under an open source software license. This enabled many early experiments, after which a consciousness of the significance of the blockchain component of Bitcoin emerged. While the early experiments were largely conceived as cryptocurrency alternatives, in time more fundamental evolutions were envisioned, such as Ethereum, which expanded the capacity of blockchains beyond the transmission of units by adding a programming layer on top of the blockchain. This additional layer made it possible to create "smart contracts," agreements in code that have the capacity to self-execute when certain conditions are met.

As an innovation in terms of value transfer (currency) and records (ledger), with additional programmability (smart contracts), the blockchain has inspired the full spectrum of financial and economic thought, from those seeking radical alternatives to global financial capitalism to those hoping to innovate and disrupt economies just enough to get rich quick, to established financial giants seeking to maintain their position by absorbing and disciplining anything that looks like competition. As the collective consciousness of blockchain increased, a whole swath of non-currency implementations

The Handbook of Peer Production, First Edition. Edited by Mathieu O'Neil, Christian Pentzold, and Sophie Toupin.

were experimented with, from real estate title transfers to digital identity systems, political voting, and the creation of wills.

In this chapter, we do not offer the reader a detailed technical explanation of the blockchain or focus on how a blockchain could be used within commons and peer-produced projects. Although we will provide a selected overview of specific projects and at times some technical detail, we do so within the context of an argument about blockchain and peer production. What we are interested in, broadly speaking, is what blockchain means for peer production. How do the two relate to each other? What is the fate of commons-based peer production when it is able to produce software artifacts ostensibly in the service of creating scarcity and market conditions? What kind of peer production is this?

The chapter begins with a historical consideration of the emergence of peer production, including a reevaluation of the work of Yochai Benkler. We aim to show that peer production was given coherence as a model of production by being contrasted with two other modes (hierarchies and markets) and through the lens of Benkler's economic liberalism. Within this framing, what distinguished peer production from these other modes was that it was no longer reliant on price signals (markets) or commands (hierarchies) to organize production. While the earliest blockchain project, Bitcoin, resembled peer production in many ways (as software operating under an open source license), it also reintroduced dynamics that peer production was ostensibly thought to transcend, namely price signals and scarcity. We use this apparent contradiction to reflect on the historical legacy of peer production and blockchain, before proposing a revised approach to peer production; one, following the work of Stuart Hall (1986), that is "without guarantees." The second half of the chapter begins this revised inquiry, focusing on the multifaceted and variegated notion of peers that is present in blockchain projects. We distinguish between four moments or aspects of blockchain initiatives that configure peers in different ways: peer production, peer development, peer governance, and peer exchange. Blockchain is not just another instance of peer production, nor is it straightforwardly a tool for advancing the political project of peer production. It must be taken as an opportunity to revise and rethink what peer production was, and what it might become.

2 Coase's Benkler

Peer to Peer (P2P – a type of network) and the closely-related, "peer production" (a type of collaboration) must be understood in their proper context, as ***historical***. Not in terms of destiny or as a kernel for some possible society to come, but in the somewhat more ordinary sense of emerging at a certain time and within a certain milieu. Consider a simple Google Ngram graph (Figure 18.1) featuring the terms p2p, peer to peer and peer production (not case sensitive). What is evident is that besides a rather modest rise in the occurrence of the tri-gram "peer to peer" beginning around the 1980s, the terms "p2p" and "peer production" were entirely insignificant in published books in English until the late 1990s for "p2p," and the early 2000s for "peer production."

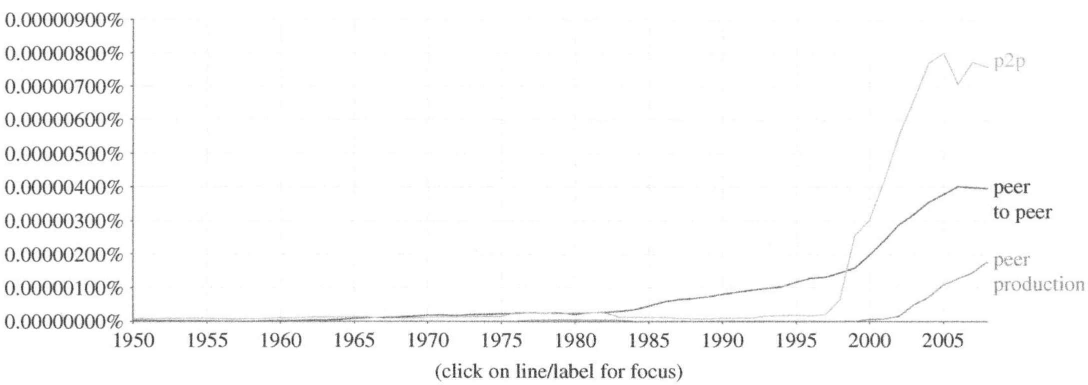

Figure 18.1 Google Ngram query for "p2p," "peer to peer," and "peer production."

It would be naïve and slightly inaccurate (in terms of the Ngram timeline) to suggest that the popularization of P2P – whether conceived as software, infrastructure, practice, or politics – can be attributed to the advent of Napster (1999), but the significant rise of these related terms occurs within something of a *Napster milieu*; a time of technical, political and economic experimentation with network affordances. For its part, "peer production" comes later still, suggesting, perhaps, its status as an *intellectual response* to the 1990s phenomenon of "p2p."

These terms are also historical in the sense that they can be located in relation to discussions around particular projects, especially software and related "open source" and "free culture" projects, and within wider discourses of technological enthusiasm. It was a time of cybercultural experimentation with identity, DIY Indymedia, homepages, chat rooms, and of course, file-sharing. Shortly after, blogs, Wikipedia, YouTube, "social networking sites," and other nascent platforms would mature, providing countless examples of "user generated content." The masses were newly minted as productive and peer production tried to lend political coherency to these highly diverse developments, not a simple feat by any means.

While there were several early influential theorizations of P2P and peer production, and many, many more people involved in promoting and participating in actual peer-produced and related projects, it is difficult to imagine peer production gaining the visibility it has achieved if not for the intellectual work of Yochai Benkler. Benkler lent Ivy-league credentials to thinking seriously about what he coined as "commons-based peer production" and, significantly, he located this notion within a clearly identifiable liberal political-economic framework. His Adam Smith-invoking *Wealth of Networks* (2006) was hugely influential (currently nearing 11,000 citations), but many of the core ideas for this larger monograph can be found in earlier writing and in particular his article "Coase's Penguin, or, Linux and The Nature of the Firm" (Benkler, 2002), where he first makes a case for the significance of peer production as an organizational model. In order to understand something of the historical dimension of peer production, in how it relates to the events of its time and how it was positioned in specific economic and informational ways, we begin with a consideration of Benkler. This weaving of peer production into a broader context of market and non-market production is also precisely what blockchains have done in practice, some years later.

In "Coase's Penguin," Benkler sets up an originary contrast between peer production and two other "models" or "modes of production" (2002, p. 376). To establish the contrast, Benkler draws on Ronald Coase's classic work on firms and the theory of "transaction costs" developed therein (1937). Coase was trying to understand why firms (understood as entities that supersede the price mechanism) exist, given the accepted doctrine within his field that the price mechanism was the superior method of "economic coordination." He sought a rationale for the existence of firms that was largely immanent to, or compatible with this doctrine. That is, he sought a legitimate justification of the existence of firms on (liberal) economic terms. A comprehensive account is not possible or needed here, but Coase proposed that there were costs related to accessing markets (and their coordinating signals) and that in certain circumstances the organization of production can be more efficient if it bypasses or internalizes these signal-exchange relations. As Benkler puts it, "using the price system is costly" (2002, p. 401). Coase's work was further developed by a number of organization theorists (who Benkler also relies on), culminating in a condensed theory of production reducible to two main forms: hierarchies and markets. The key *economic* distinction between them is given in terms of information signals and coordination: hierarchies organize production through *managerial commands*, while markets coordinate production through *price signals*.

While Coase sought to understand firms, he did so from within the market orthodoxy, building explicitly on the (neo-classical) work of Smith, the (neo-liberal) work of Hayek and others, and never challenging the market paradigm fundamentally. His work can be seen as something of a paradigm-native minor corrective, whereby the existence of firms is explained by their capacity to out-coordinate markets in terms of efficiency. Benkler's peer production follows this Coasian trajectory, but instead of just markets, it now contends with market and firms, or markets and hierarchies and their respective signals (managerial commands and price). It is an account of production that presents markets as the default (superior) form. Correspondingly, the main criterion through which to assess a particular form (mode/model) is its capacity to organize or coordinate economic activity through signals and indeed, this is precisely how Benkler understands the unique contribution of peer production.

When peer production is referred to as a "third mode" of production by Benkler and others (Bauwens, 2005; Bauwens, Kostakis, & Pazaitis, 2017; Benkler, 2002, 2006), it must be understood within this Coasian trajectory. Benkler extracted from the political-economic richness and diversity of (FOSS) software development and early web cultures, a specific argument about economic coordination and efficiency within the tradition of Coase: "Peer production emerges, as firms do in Coase's analysis, because it can have lower information opportunity costs under certain technological/economic conditions" (2002, p. 375). And while acknowledging the other important "normative implications" at stake in peer production regarding "freedom and equality," Benkler clearly avoids suggesting that peer production has arrived to replace the other two modes:

> I am not suggesting that peer production will supplant markets or firms. I am not suggesting that it is always the more efficient model of production for information and culture. What I am saying is that this emerging third model is (a) distinct from the other two, and (b) has certain systematic advantages over the other two in identifying and allocating human capital/creativity. (2002, pp. 381–382)

The "systematic advantage" is described in terms of "information gains": the "capacity to collect and process information about human capital" (2002, p. 412).

Whether or not peer production did or does have any kind of systematic advantage over other models is beside the point. By the time one arrives at these technical considerations we are well and truly down the Smith-Hayek-Coase-ian rabbit hole, where peer production is positioned in relation to markets (understood as information processors, ideal forms of economic coordination, and sites of freedom) and firms (which complement markets through internalizing transactions to improve efficiency) and compared in these terms of economic coordination and efficiency. There are empirical and theoretical problems with this orthodoxy (Beckert & Aspers, 2011; Beverungen, Mirowski, Nik-Khah, & Schröter, 2019; Çalişkan, 2010; Ghoshal & Moran, 1996; Mirowski, 2015) and further issues with approaching instances of peer production from within this frame (Terranova, 2009; Tkacz, 2015), although most of these need not concern us. Instead, and to introduce yet another sense in which peer production is *historical* (this time via historical comparison), we turn to the ostensible focus of this *Handbook* entry, blockchains.

3 Coordination Issues

While the perceived potential for blockchain-based initiatives stretches as far as the entrepreneurial imagination, in terms of actual implementations the blockchain space is still dominated by cryptocurrencies, with Bitcoin being the oldest, largest, and most valuable. Acknowledging that there is significant variance among existing blockchains, Bitcoin and related cryptocurrencies introduce new dynamics into the peer production space directly relating to these Coasian questions of transaction costs and production models. We will use Bitcoin as our discussion point below (as one blockchain implementation), before launching into a more expansive consideration of blockchain projects.

If peer production is understood as a third mode of production, how can we understand Bitcoin and blockchain in relation to this mode? The most obvious starting point is at the level of software. Bitcoin and by extension blockchain are open source, with the Bitcoin Core software operating under the MIT License ("MIT License," 2019). In this regard, Bitcoin can be added to the long list of software projects held up as models for peer production. As open source software, it also means that Bitcoin and some other blockchain projects may resemble the development, governance, and organizational dynamics of other open source projects. Beyond this, though, the relationship gets murky, problematic even.

While we (as authors) have never idealized peer production, nor submitted to the idea that there are only two or three ways of organizing production (markets, hierarchies, and peers), some cryptocurrencies (including Bitcoin) and blockchains do resemble other peer production projects in the way they are made. That is, as software, they look like other open source software. But, whether considered money- or currency-like, or as financial assets, exchange infrastructures or distributed accounting systems, blockchains and cryptocurrencies *turn away* from the conditions from which they came into being. That is, they are produced as or within a commons, but turn away from these commons to face (and indeed facilitate) its others. Recall that for Benkler price signals are an artifact of markets and have a cost attached to them. Peer production is precisely that which does not rely on price signals; it coordinates production without such signals. So why the push to recreate market signals and/or related systems of accounting?[1]

There are two possible answers: one that takes us further down the rabbit hole (of economic efficiency) while the other rests on historical contingency and political action. There is an element of truth to both answers, and together they push us to rethink some core elements of peer production. Answer one: while markets are generally the most efficient way to coordinate economic activity, markets themselves must be organized. Such organization is also costly and is typically handled by state and state-regulated actors (i.e., "hierarchies"). Bitcoin and other cryptocurrencies could reduce the costs of facilitating market activity by replacing hierarchical facilitation with peer facilitation. In this argument, the emergence of Bitcoin and blockchains can be explained by the possible inefficiencies of the hierarchical facilitation of markets. Satoshi Nakamoto's early white paper explaining Bitcoin can certainly be read this way (2008). In the title of the white paper, for example, Bitcoin is described as a "Peer-to-Peer Electronic Cash System" – an initial positioning move. The first sentence of the white paper's abstract reads: "A purely peer-to-peer version of electronic cash would allow online payments to be sent directly from one party to another without going through a financial institution" (Nakamoto, 2008, p. 1). "Financial institutions" are presented as an unfortunate but heretofore necessary cost of market activity, especially digital activity where cash is not used. In the opening lines of the introduction, Nakamoto directly locates financial institutions in relation to transaction costs, describing them as third-party intermediaries whereby "[t]he cost of mediation increases transaction costs" (2008, p. 1) and further elaborates that such mediation limits the "minimum practical transaction size" (i.e., micro transactions) and also relies on a system of trust to work (because the time-space of the transaction between parties is not identical to that of the accounts (ledger) where the transaction is recorded).

Libertarians have heralded Bitcoin's claim to disintermediate financial institutions as a kind of market victory over hierarchies; however the white paper makes no reference to this and does not mount a political attack on banks in these terms. Rather, banks are undesirable because they add to transaction costs – even the discussion of "trust" is framed in terms of efficiency.

The rest of the white paper offers a technical description for how Bitcoin works, beginning with a section on transactions. Understood in this way – as part of the Coasian universe of transactions – it's difficult to see Bitcoin as advancing the model of *peer production* in any but the most limited of ways. Instead, it aims to improve the efficiency of *markets* to produce signals by reducing market-based transactions costs. As is well-documented, Bitcoin has not made serious inroads as a currency or money-like entity; it has not been able to produce more efficient price signals and it has by no means spelled the end of financial institutions. However, just because Bitcoin has thus far failed to disintermediate financial institutions does not mean it is not useful to understand it in relation to such aims and actors. As we shall see, blockchains are indeed Coasian technologies; they embody a vision of transactional efficiency that drives experimentation, just not in the way described in the white paper: as disintermediators of financial institutions.

The second answer to the question (of recreating market signals and/or related systems of accounting) lies in relation to the historical event of the 2008 global financial crisis. The publication of Nakamoto's white paper came roughly a month after the collapse of the Lehman Brothers investment bank in September 2008. The significance of this event cannot be underestimated in providing the conditions for Bitcoin and other cryptocurrency experiments to flourish. In a recent interview commenting on this relationship, historian of digital culture Finn Brunton makes a similar observation:

> [Bitcoin] had a really specific genesis, which was introduced and adopted as the world went through an absolute financial catastrophe. We forget how terrifying it was when it looked like we were on the verge of a planet-wide credit freeze, with all of these organizations in free fall and when the tremendous brittleness of the financial infrastructure was all coming due, one piece after another. (Brunton, cited in DuPont, 2019)

The global financial crisis laid many things bare about finance and economy to a generation of non-specialists for the first time: the madness of some financial assets (such as the "collateralized debt obligations" that made their way into popular culture through films such as *The Big Short*); the reality of (housing) bubbles; the interconnectedness of finance companies; the power of governments to intervene via "bail outs"; the interconnectedness of global financial markets (such that the "US" crisis quickly spread); and, the strategic use of such events to make ruthless government spending and investment cuts across many aspects of society (austerity). Through witnessing the crisis unfold, it was possible to see how "everything was connected" in a dark, almost fatalistic way, from an individual housing loan to the viability of large firms and, indeed, the very sovereignty of nations (e.g., Greece). Adding to this connectedness, the financial system seemed arbitrary, highly fragile, and most importantly, a site of power. From this perspective, the rise of Bitcoin (and eventually other blockchains) had less to do with the efficiency of transaction costs, and everything to do with timing. As we have documented previously along with others, Bitcoin is by no means the first attempt to create a digital currency that short-circuits incumbent financial and state actors (De Jong, Tkacz, & Velasco González, 2015; Maurer, 2005).

And while writers such as David Golumbia (2016) are correct in their assessment that a certain right-wing extremism surrounded Bitcoin (and arguably is inscribed in the code), so that a dark current ran through the project from its inception, it is equally true that in the fallout of the crisis many people from a diversity of political leanings saw in Bitcoin an alternative possibility for economic organization. Indeed, we ourselves encountered European anarcho-hackers, Autonomist Marxists, disillusioned bankers, peer producers, and people with no obvious allegiances, alongside the more predictable libertarian and crypto-libertarians in the early days of Bitcoin experimentation (see Lovink, Tkacz, & De Vries, 2015). It may well be that most people were naïve to see in Bitcoin a potential that was other than what was actualized, but the point is Bitcoin was there as a vessel to absorb different visions at a time when the global financial status quo was in tatters.

Of these two possible explanations for the rise of Bitcoin (and later blockchains), the second holds more explanatory power. Indeed, despite being positioned as a technically superior infrastructure for handling transactions – that is, as a technology of market signals – Bitcoin and other cryptocurrencies and blockchains have never managed to operate well in a signal-generating capacity. However, despite Bitcoin (and other blockchains) not living up to their transactional pretensions, the specificity of this "peer-to-peer" cash system is not insignificant. Bitcoin's claim to be able to disintermediate existing financial institutions clearly mattered, as did the "irreversibility" of transactions and the hard-coded limit of possible coins in circulation. These specific qualities spoke directly to the crisis: the corruption of banks; state interventions (bailouts and quantitative easing), and so on. Finally, the technical tone of Nakamoto's paper presented Bitcoin as not overtly political (though one need not dig too deep to find it). When pensions are evaporating and banks are freezing and limiting withdrawals, a technology does not have to do much to appeal.

4 Peer Production After Bitcoin

So far we have dwelled on the early account of peer production as a "third mode" of production, centering on the proposal that it enables forms of economic coordination more efficient than market signals or managerial commands. As theorized by Benkler and others, it was an attempt to lend political coherency to a diversity of projects, happenings, and technologies, and to place these within a liberal economic paradigm. The collaborative dynamics of peer production arose precisely in distinction to signals and commands (and the property and labor relations that underpin then).

While produced as an open source and hence commons-like artifact, the arrival of the first blockchain, Bitcoin, is seemingly at odds with how Benkler first describes the (potential) superiority of commons-based peer production in that it ostensibly attempts to reintroduce dynamics (signals) that peer production was supposed to transcend. It is as if humans were set free and then went about making new chains. We offered two ways to makes sense of this, the first of which is that this "third mode" was never designed to supersede but rather complement the others. This appears to be Benkler's position.

The second way to make sense of this signal-producing effect of peer production is that it seemed like an attractive alternative at a time when the financial status quo was in an especially bad state. That is, the contradictions inherent in Bitcoin were overlooked for pragmatic reasons.

In both cases, what seems pertinent is that peer production offers no clear guarantees in terms of a pathway out of market societies (or authoritarian "command" driven ones). It can be made to fit very well within existing liberal economic ideals (Benkler), and actual instances of peer production can easily be deployed to produce their others, that is, peer technologies can become market technologies. Suffice to say, we do not see in peer production any kind of historical destiny; no kernel for a new society. At worst, there is a potential that it actively contributes to the intensification of market relations via automating elements of bureaucracy and governance; a new kind of protocological neoliberalism.

What we have also suggested thus far, however, is that there is a counter-narrative that runs through peer production and blockchain where both are positioned as historical actors. Peer production gave a name to a cluster of energies that emerged in relation to the web cultures of the late 1990s and early 2000s, while Bitcoin emerged in times of economic crisis. Along these lines, the significant contradictions and political-economic divergences between them perhaps matter less, although are no less easily resolved.

Despite the critical trajectory of this chapter thus far, our critique is not absolute. We are in urgent need of experimenting with different economic models, different relations to production (if that is the term), and different ways of being with each other and the planet. What we think is needed, and what we will begin to articulate in the second half of this chapter, is a critical reflection on existing practices (especially experimental ones) in relation to the terms given to describe them. From this vantage point, the limits and potentials of an emerging political vocabulary can take shape. The fact that we proceed with caution recalls Stuart Hall's position on Marxism some decades back. Hall acknowledged that many aspects of Marxism had been thoroughly "deconstructed" and that the grand narrative of the destiny of the proletariat was not at all given. His Marxism, he declared, was to proceed "without guarantees." This was by no means a dismissal. Rather, if Marxism was to remain relevant it had to be nimble, reinvented around the changing conditions of production. Moreover, he thought many of Marx's ideas still resonated, still got at something, even if they were presented in ways that seemed crude over a century later. This informs our own approach to peer production.

In what follows, we turn to the notion of the "peer" in peer production. The peer is an intriguing figure in a number of ways. It stands in contrast to roles in markets (exchangers, entrepreneurs, debtors, creditors, etc.) and in hierarchies (managers and employees, or "masters" and "servants" in Coase's language). It suggests a notion of equality, with a residual potential for group authority (peer pressure). Among the spaces of blockchain, what exactly constitutes a peer or a "peer relation" is not always clear or immediately discernible. The peer of the blockchain is a complicated and multifaceted figure, politically and even ontologically ambiguous. It is manifested in quite different ways across different aspects or moments of a blockchain implementation. We now pursue our "peer production without guarantees" by reviewing a number of blockchain experiments with the impetus of drawing out the different facets of the peer. We break our analysis into four moments or elements common to all the projects observed (production, development, governance, exchange) and use these to reflect on what it is to be a peer after Bitcoin.

5 Peer Production

A vast sea of blockchain implementations share the common promise of distributed *production*, but not many delve into how this is enacted beyond the lure of signing up for an innovative service. Take for example the ArtID platform ("ArtID Marketplace," n.d.), a service to certify works of art by inscribing information into the Ethereum blockchain and providing its own token ("artid") based on an Ethereum standard (Ecr20).[2] Looking beyond the practical elements of such a system as a different kind of registry, what is significant is the practical absence of user participation within its architecture. Here, the role of the *peers* is equivocal: the platform makes use of Ethereum-derived tokens but only offers these as a service, thus, participants involved in these interactions function straightforwardly as clients, not as producers or active participants of the Ethereum platform. This calls into question what exactly it means to be a peer in terms of production within this architecture.

The peer-producer in blockchains can be understood either as someone benefiting from a network made of peers (such as the ArtID user, benefiting from the Ethereum network), as a participant within a network of peers (enacting a peer-production organizational model), or as an actual producer of items within the organization (in the case of blockchains, a token producer). These three modes are commonly used without distinction, in some cases because they overlap, as this section will show in the case of blockchain "miners," but also because of the ambiguity that comes with the term. Thus, this section looks for a clarification of the blockchain "peer-producer" in relation to its practices, and in particular to the third mode: the producer of tokens.

For the most part, blockchain technology folds two traditional activities of production into one: the process of "mining" is at the same time the distributed strategy for validation of all exchanges taking place through it, and the production of economic units (the "minting" of tokens). The bureaucracies of managerial command and monetization are merged into a strategy to secure highly distributed networks that claim to eschew central command.

This is also the most common scenario for participants to have the possibility of generating capital in the form of tokens with exchange value. While blockchain implementations offer a diversity of schemes regarding capital distribution and access, this is rarely an element that is meant to be available as a common (some cases where this is attempted are mentioned later). All public blockchains use "mining" as a way to produce new assets. In the case of blockchains designed to act as cryptocurrency (e.g., Bitcoin, Litecoin, Dogecoin) this enables a constant influx of exchange-tokens, which are regulated by case-by-case rules. Bitcoin, for example, regulates its supply to guarantee a production of 12.5 tokens approximately every 10 minutes.[3] By design, only 21 million bitcoins will be produced (or *mined*) by the year 2024. Cryptocurrency-oriented blockchains have different reasons to control production, which more or less resonate with their goals: distributed storage, smart contracts (triggered-execution software), registries, and so on, tune different logics of production. However, all blockchains use tokens to execute transactions, and all public blockchains need to use some sort of mining procedure, not only to produce, but also validate transactions. Production in this sense, is framed by the techniques required for token generation, and peer production is enabled (and limited) by the possibility of participating in this manufacturing process.

In the blockchain, the labor to generate numbers, validate transactions, append new blocks to the chain and introduce tokens into circulation is almost fully automated. Miners are machines that verify the signed public keys for each transaction and which validate these into blocks in a public registry (i.e., the blockchain). The job for successfully validating and packing the transactions produces new tokens for the miner, and generates a Proof-of-Work (PoW).[4] The former is the result of a "puzzle," which can be then easily checked by any other machine in the network. Since the design of the system seeks a controlled pace, if the tokens are generated too fast (because there are more and/or stronger miners) the "puzzle" becomes harder (Nakamoto, 2008). The design of the system relies on a form of superabundance (Velasco, 2016) performed by computation, i.e., the puzzle performs an enormous amount of operations in order to work, which serve no other purpose than ensure security, and consumes copious amounts of energy while producing fast cycles of hardware waste. The human part of the miner assemblage is free to modify some variables of this process, but has to ultimately adapt to the rules of the protocol. Bitcoin PoW in this case is a non-human, mechanical kind of labor – "algorithm-made" (Coeckelbergh, 2015) – that produces new tokens. Aside from programming and setting up the machines, barely any human labor is involved in the process. Both programming and setting up the machines are by no means small tasks, and they depend on an assemblage of a huge number of names, discussions, infrastructure, discourses, electricity, investment, and so on. This *programmability*, however, imbues the organization of production with a specific quality: in particular one seeking to flatten hierarchies between producers (of tokens), and to delegate its operation to the network (O'Dwyer, 2014). While the latter is certainly effectively performed in most systems, new hierarchies are also generated even when considering this reduced and technically bounded version of a peer-producer.

Lana Swartz and Kevin Driscoll (2017) explored the *peerness* of Bitcoin mining by setting up a miner and narrating the difficulty to run it in an amateur setting in 2013. In theory, this only required a computer and a continuous stream of electricity and an Internet connection. Their so-called Lonely Old Bitcoin Miner (LOBM) was an anthropomorphization of a machine whose purpose was to exploit the generation of tokens, a "mechanical tin bank." However, the experiment rapidly showed that a one miner operation required a more advanced piece of technology: by 2013 most people invested in mining jumped from powerful GPUs to ASICs (Application Specific Integrated Circuit), dedicated circuits designed uniquely to mine blockchain tokens (Taylor, 2013). This "digital mutualism" – an idealized vision of voluntary peers engaged in the construction of a common that benefits all parties involved (Swartz & Driscoll, 2017) – required a growing investment of time and resources. Since then, mining has become an industrial operation, not only in terms of specialization, but because of the impossibility of running an "artisanal" miner, i.e., a one-machine operation such as the LOBM. Contemporary mining operations run in pools: groups of miners who cooperate to solve puzzles and share earned tokens. This re-centralizes production, as pools are most commonly controlled by one entity, thus undermining one of the main perceived affordances of the technology. However, pools are a good example of collaborative production within these ecosystems, even if here "peer" and "commons" are framed by a very specific profit-driven cooperation.

These for-profit elements raise an obvious dilemma for a commons political economy of cryptocurrencies, but it also applies to other blockchains insofar as the fundamental technological characteristic is distribution through tokens. Pools signal a failure in the original peer (technical) symmetry envisioned by Bitcoin, since mining operations are centralized by the mining pool owner, who manages operations: collaboration is less of a choice and more a necessity to keep production profitable, at the cost of losing the initial symmetry of the peers. Even if we limit our view to the more constrained conceptualization of peer production – the miner as token producer, where the technical affordances of a distributed network allow in theory for a flat hierarchy between the peers – the token generative design is geared towards a managerial organizational model. The figure of the miner is to be expected to leverage production without relying on price or command signals. However, the nature of a token-driven design shows how easily miner pools are reconfigured into hierarchies through the directive of price signals: profitability (or the lack thereof) demands independent miners to collaborate under a managerial system.

This does not undermine other types of collaboration within the frames of peer production, such as the development of the system itself (as shown in the next section), nor mixed alternatives for common governance (as the last section will show), but it positions the limitations that come with a design based on tokens and its permeability to the organizational models of hierarchies and markets.

This collaborative production within asymmetrical peerness also raises the question of how this kind of cooperation would have been different if the goal had been another kind of collective objectives, instead of the *lonely* individual's own advantage (Gianelli & Fumagalli, 2013). The LOBM ultimately shows that the definition of peer within blockchains is not only a problem of policy or governance, but also a problem of technology design and the objectives embodied in it.

6 Peer Development

The conversation regarding the public status of blockchains has been more evident in relation to their development and governance. Both forms of participation provide an opportunity for shareholders to guide the architecture, form, goals, and rules of a system. Either by actively participating in the design of the software, or offering effective input on decision-making, their capacity as peers is first of all dependent on a matter of accessibility: public blockchains would allow an allegedly unrestricted participation, in a similar fashion as other traditional open-development projects. On the other hand, in private, "permissioned," or "incorporated" blockchain, access is restricted by network administrators.[5] However, unlike other open software projects, public blockchain development is disturbed by its token-based value design. This locates blockchains in a complicated *milieu* that is neither publicly (as state owned) or privately owned (as a company), nor explicitly for profit, but simultaneously reliant on a common structure and a market-dependent token economy. Many blockchain projects are based on open models of software production. As mentioned previously, models for open development have a history of commons-like collaboration, where code is a common resource built on the basis of volunteer work. Git[6] repositories are a perfect example of a common-like system that works on the premise that "everyone can use it." Usually working as a meritocratic system, balanced by market signals, changes in open software are free for anyone to propose, as long as these changes are justified, feasible to implement, and validated by a broad community through some agreed-upon protocol. This was and continues to be the case for Bitcoin and Ethereum. Like many open projects, collaboration issues in Bitcoin can be settled via Git-like protocols, but due to its delicate consensus requirement Bitcoin development also adds a second layer of conventions for new suggestions. Bitcoin Improvement Proposal (BIP) followed the tradition of Internet Engineering Task Force's ("The Tao of the IETF," 2012) open calls for improvement. BIPs are essential for changes that are particularly controversial or that introduce changes to core rules of the protocol. A similar process by the name of Ethereum Improvement Proposals (EIP) is followed for Ethereum development.

Open development is an excellent example of peer production in action, as it not only shows the active collaboration in the construction of a common, but also the complex negotiation among peers involved in such tasks. One of the main advantages of working with digital elements (or Bauwens' "immaterial goods") is that, unlike physical goods, they are not constrained by scarcity. The possibility of multiplying projects (along with what these projects have produced) is a common practice known as forking. This practice is usually limited to open projects because it involves a direct duplication, and thus no object subject to copyright can be formally forked (Tkacz, 2011). Since the initial state of the fork includes two projects with the same code, followed by changes in each branch, competition between them is considered a natural part of forking (Raymond, 2008; Weber, 2004). Similar projects to Bitcoin and Ethereum have been generated and ramified from their original source code to produce competing blockchains. Litecoin, for example, the second largest cryptocurrency for a long time by market capitalization, and fifth at the time of this writing, is an example of a competing coin. It behaves as a currency, but uses a different hashing algorithm (scrypt, instead of SHA-256). It also modifies some of the Bitcoin variables, like having a faster block processing and a maximum of 84 million tokens. Litecoin has its own code branch, name and blockchain, which distinguishes it from the Bitcoin chain. It does share, however, the financial asset orientation. Thus, it can be considered a competitor within the cryptocurrencies market but, it remains an independent chain and coexists with Bitcoin and other altcoins.

A few other projects do not even enter in direct competition with Bitcoin because they use tokens merely as mediums for other ends (e.g., distributed DNS servers). Like the competing coins, these blockchain instantiations keep basic characteristics of any blockchain – a shared registry that conflates token circulation and production – but are not identified as currencies or economic elements. Thus, they are neither competing chains nor competing coins. Ethereum ("Ethereum Project," n.d.) is perhaps the best known example; it enhanced the original scope of mere circulation by introducing a programmable and executable aspect to the blockchain. Commonly known as digital contracts, these are arrangements and rules defined between two or more parties and deployed in Ethereum's blockchain. Because "ether" function as tokens to enable smart contracts, they are seen more as "activation" tokens than as currency. However, ether does have a monetary value and, like Bitcoin, is bought and sold in relation to fiat markets. Since all blockchain tokens share these exchange qualities (regardless if they have the intention of being a currency), they are embedded with transactional value. Owing to this, unlike other digital goods, value cannot be split between two forks. Thus, the outcome of the fork represents not only the future of the project, an ideology for or against centralization, the endurance of a group of people, or the restructuring of governance rules, but also a substantial economic interest.

7 Peer Governance

Production and ownership of tokens also correlates directly in some projects that have integrated governance into their technical workings. That is, decision-making is not only being performed through traditional external channels (e.g., forums) but integrated into the infrastructure procedurality. Primavera de Filippi (2015) defines "governance by design," as "the process of online communities increasingly relying on technology in order to organize themselves through novel governance models (designed *by* the community and *for* the community), whose rules are embedded directly into the underlying technology of the platforms they use to operate." This marks an interesting distinction between blockchain projects that not only use governance processes *of* their infrastructure, but also *by* their own infrastructure (De Filippi & Loveluck, 2016). In the former case, to apply changes to the protocols, rules, and general orientation of the project, open source community regulations, like BIP or EIP, are followed.

In this case, expressing arguments, especially when made by a known or respected part of the community, can leverage disagreements. In borderline situations, issues may be settled by a distinctive figure of authority; in the case of the Bitcoin community, for example, the opinion of Gavin Andresen, one of its main developers until 2016, was highly respected and generally accepted as a rational and sometimes definitive voice to settle controversies. In the latter, "governance by infrastructure," a technical consensus must occur to reach agreements. This means, for example, that a majority of computers within the network must implement the proposed changes, by running specific code, in order to successfully adopt them as a standard. The governing rules for this first kind of consensus are embedded directly in the protocol and take into account all the "votes." Some alternatives to the PoW proof algorithm function explicitly as a claim to democratic participation in the governance process. The delegated proof-of-stake (DPoS) matches tokens with decision shares. DPoS, delegated proof-of-stake, gives the shareholders with the majority of shares a major stake in participation in "governance by infrastructure." This is an on-chain governance model, which means that governance is directly integrated in the technical system. Current projects, like Tezos, have made this on-chain capacity their distinctive feature, as a strategy to relocate power dynamics from developers and miners towards shareholders. Both PoS and DPoS carry the risk of monopolies controlling a supposedly distributed system, and even the best on-chain design is vulnerable to meddling by off-chain private interests (Reijers et al., 2018; Schaffers, 2018). However, different projects experiment with tweaks on their protocols to avoid extreme control by accumulation. Peercoin, for example, makes it costly to produce new tokens by PoS, in order to avoid monopolies. Coordination, however, is again dependent on token ownership, thus emphasizing the importance of exchange value within most blockchain systems.

The affordances for peer-governance are not limited to the modification protocols. Some projects use these novel participation possibilities to experiment with democracy and contemporary social issues. An interesting example of this is Civil, a blockchain-based platform for news curation, aimed at confronting the current disinformation crisis of mainstream media and centralized platforms. It has been proposed as a strategy to fund independent journalism as well as a technique to foster high quality and trusted content production. Formed as an alliance of "newsrooms," the platform is in principle open to everyone, but users must go through an application process to be part of the Civil Registry. The peer in this ecosystem is defined by its adherence to a set of in-progress journalistic ethics (Iles, 2018). Like other blockchain-based projects, Civil aims to remove intermediaries – traditional media institutions, but also platforms such as Facebook and Google – and provide a direct channel between journalists and their readers. The project is described as "community owned" because token ownership acts as a share on the platform, and as a mean for voting and for challenging dubious newsrooms. A civil council formed by independent journalists may overrule a community vote, but if a big percentage of the community agrees on a vote, council decisions are overruled, thus enacting peer governance in a similar fashion as the aforementioned Peercoin. Civil describes itself as a "decentralized marketplace," but recognizes that the use of economic incentives does not imply rational choices, and that understanding the motivation of a community is an ongoing process (Vuong, 2018).

Thus, the Civil project is an attempt to exploit distributed ledger technology as a new option for a content market, while recognizing the problematic of a purely market-driven strategy, therefore the stress on the constitutional grounds. Still a work in progress, Civil offers a promising experiment for token-driven content generation for post-truth journalism. And while it has been confronted with the challenges of a novel token-driven economy – failing to reach their 2018 ICO goals, for example – it is a project that uses blockchain technology to experiment for a much-needed sustainable model of online newscasting.

While improved governance is possible outside the mint production, this is not a characteristic that comes with blockchain technology. A distributed infrastructure may be a uniquely useful tool for any attempt to develop the commons, but blockchain is not a groundbreaking technology designed for this goal, and certainly not an inherently collaborative system. What is more, on-chain coordination, is pegged with token exchange-value, which, as we argue is an inherently designed quality of these systems. This clarification is crucial to discuss the limits and affordances of implementing blockchain technology alongside commons-oriented projects.

8 Peer Exchange

It has been argued that the blockchain presents a paradox of social and anti-social qualities (Alqassem, Rahwan, & Svetinovic, 2018). On the one hand it facilitates transactions (and interactions, if we consider the Ethereum blockchain), by virtue of its lack of mediators. On the other hand, part of the community involved in its design concurs with a strong pursuit of privacy and anonymity. The first part of this paradox, the idea that blockchains are "social" due to their transactional resourcefulness, is based on the misled yet common amalgamation of peers as producers and peers as exchangers. Being able to make transactions over a distributed ledger system does not imply the participant is involved with production, and even the peer producers (by means of mint or governance) do not necessarily produce use-value for a community. The exchange use of the blockchain is straightforward, and participation as a peer in this sense is frequent in all projects, due to their token-exchange reliance. However, the discussion regarding participation of peers as producers of communal use-value is significantly more nuanced: some blockchain-based projects consider co-production for specific communities, while others focus on enabling speculative wealth for individuals.

Unlike some of the public blockchains discussed earlier, there is a growing type of private projects that make use of distributed ledgers, but limit production and validation to a selective set of shareholders, while allowing exchange to a broader public. A majority of the investment currently relies on these "incorporative" (Swartz, 2017) blockchain projects. An example is the *Nasdaq Private Market*, a blockchain implementation to manage private shares of technology companies pre-IPOs, or the *Hyperledger*, an open blockchain development project backed by a long list of big institutions and corporations (IBM, the Linux Foundation, Cisco, Intel, JP Morgan, and Wells Fargo, to name only a few), which works to develop a secure blockchain framework for regulated industries. A majority of banks are also developing incorporative blockchains to adapt them to the current economic model. The goal of these kind of projects is "not to disintermediate the financial system, but to determine how to be better intermediaries" (Swartz, 2017, p. 19). Like other private blockchains, the shareholders of the system are limited and deliberately selected. The system design of a distributed ledger remains and some degree of compatibility to other systems remains, but the centralization and ownership differs on each blockchain.

Incorporative blockchains are not the only example of limiting a set of users to exchange-value operations. A recent case to enable exchange-peers for welfare took place with Syrian refugees in Jordan. This was supported by the UN's World Food Program (WFP), and implemented by Parity Technologies, led by Ethereum founder Gavin Wood, and Daterella, a big data and blockchain startup. The WFP project was used to record and authenticate food vouchers given to refugees on the Ethereum blockchain (UN City Copenhagen, 2017). The affordances of using a distributed ledger diminished the costs and risks of using cash or other payment means in areas without a financial infrastructure. The use of this technology also allows for fast reactions, low costs, and manageable implementation, which makes it ideal for emergency situations, unstable geographies, and scarcity of infrastructures. In this context, the use of blockchain technology proved to be useful and this project could most probably be replicated, improved, and enhanced in future iterations. There is an evident advantage in the use of the Ethereum blockchain in this case regarding exchange value: it is an expedite and functional implementation that imbues the users as peer-exchangers. It would be risky, however, to describe all participants in the project as peers participating in a common. The goal of the WFP project in Jordan was aimed primarily at enhancing control, as the system's transparency is tied with its monitoring readiness: the way the funds are used by refugee families is tracked and limited to allowed purchases (such as olive oil, or lentils). The refugees involved in this project had as much participation as any user of a voucher has with the banks behind it. The blockchain in this case serves the production and tracking of the vouchers, but only involves the refugees as endpoints.

A comprehensive role of stakeholders in regards to production, governance, and exchange is rendered more successfully in novel distributed social networks. *Steemit*, a blockchain-based network launched in 2016, offers a blog-like content platform. Pitched as a decentralized and censorship-resistant platform for content generation, the project intends to stimulate a user's production of posts by monetizing them. The broad description "Get paid for good content. Post and upvote articles on Steemit to get your share of the daily rewards pool" ("Steemit," n.d.) summarizes its hybrid economy model. Users can receive micropayments either for producing materials, or for upvoting (similar to Reddit) new "undiscovered" content. The platform has three value forms: STEEM, Steem Power, and Steem Dollars, which pay the role of a cryptocurrency token, a voting token, and a stable-value token. STEEM is a liquid asset, that can be converted into other currencies. It acts as a flexible currency with a clearly identified use-value for upvoting, and high liquidity for exchange value into Steem Power or Steem Dollars. Unlike other mining-based systems, in Steem these tokens are constantly produced for a common pool, available to any user (Thelwall, 2018). In this sense, the platform does provide a common capital for use within the project. Steem Power, on the other hand, is used as an influence tool: the more a user holds, the more weight their votes have. It is however, slow to exchange (or "power down") back into STEEM, so it enhances influence (and governance, in the platform context) at the expense of exchange. Finally, Steem Dollars allow the users to exchange their earning into other currencies, by using a US dollar-pegged system, thus controlling volatility. Authors that are upvoted, or users that identify future high-value posts are rewarded with Steem Dollars. This hybrid use of interchangeable tokens represents different stages of a peer in a blockchain project: the user of the Steem system has an option to produce, and decide upon the content of the platform from a common pool of capital, while also the possibility to exchange earnings either into leverage for decision-making, or into other currencies.

Although Steem has been operational since 2016, and has (until October 2018) handed out US$40 million using its reward token economy model (Kim & Chung, 2019), the platform advertises itself as a beta version. Not only in relation to the open code that runs underneath it, but also highlighting the work-in-progress condition of its token economy model. What experimental projects like Steem show is that, through the architecture of these systems, hybridity is being put at play in different forms: regulated through an algorithmically driven command and control, adopting a shared capital and performing different types of co-production, yet driven by exchange signals, and so, within a market paradigm, but stretching its limits through experimentation.

9 Conclusion

Bitcoin and other blockchain initiatives have been categorized as "Distributed capitalism" (Bauwens et al., 2017), a decentralized but extractive model, supported by P2P infrastructures, but still driven by individual profit. In opposition to the extractive pole stands a generative model, which differentiates itself due to adding "common good" value to communities. What examples like Civil or Steem show is that blockchain-based projects may well be market driven, but not in direct opposition to creating community social value. The evolution of blockchain-based projects shows the interplay of fluid elements that accommodate different parts of their system alongside a broad spectrum of traditional coordinates, even if within a market-born paradigm. The token-based, exchange-guided, and speculation-bound architecture of the blockchain locates its initiatives within Benkler's (and Coase's) mode of production, effectively within the political economy field of markets and firms, with the market as a guiding blueprint. However, the degrees and roles of participation, the definition of a common, and the generative or extractive modes, fluctuate greatly depending on particular undertakings. The variables of development, governance, infrastructure, tokenization, and production are combined, as shown previously, in significantly different ways, pushing our understanding of what constitutes peers, peer production, markets, hierarchies and how these always already troubled notions enter into heady new hybrid relations. The *peer* of blockchains, whether as (lonely old Bitcoin) miner, as token exchanger, Civil or Steem contributor, or proposer of BIPs, is not a consistent identity.

The fascination that currently fuels blockchains is encouraging new forms of experimentation and sparking the imagination of participants. Even considering its architecture constraints and its market paradigms, this technology has pushed the conceivability not only of new types of economies, but of co-designed systems that seek to address current cultural needs, fluid dynamics of exchange, and mixed modes of production, creation, and collaboration. The quantity and diversity of blockchain-based decentralized applications, even if not all become realized, is a testimony to both the need and interest for production within and beyond traditional market and firm paradigms. What is more, the reinvigorated experimentation with blockchains, has encouraged non-blockchain based P2P systems (cf. the DAT protocol, the Interplanetary File System, or the Scuttlebutt protocol, to name a few) which embody different politics and interpretations of what is entailed by co-production, and use blockchains merely as a subsystem, thus folding blockchains into political and economic imaginaries beyond those that have shaped Bitcoin and other blockchains. Perhaps the most interesting quality of blockchains and blockchain-based projects is how they constantly reconfigure and push at the boundaries of peer, market and hierarchical organization; less a "third mode" than an erosion of modal thought. Blockchains are always already hybrid. They force those invested in peer production to proceed with caution, without the certainty of historical destiny or an assumed moral high ground – a peer production without guarantees.

Notes

1 While we have not highlighted questions of property thus far, Benkler also attributes part of the superior efficiency of peer production to its being outside typical property relations, resulting in what he calls "allocation gains": "Peer production relies on making an unbounded set of resources available to an unbounded set of agents, who can apply themselves towards an unbounded set of projects and outcomes" (2002, p. 415). While we are focusing on Bitcoin, currency, and signals here, the non-property dimension of peer production equally raises questions about the accounting dimension of blockchain. Just as we may ask why peer produced artefacts should seek to recreate price signals, we may equally ask why they seek to produce systems of accounts which throughout history have overwhelmingly been used to keep track of property. There are, of course, other uses of ledgers and other uses of accounting. Further exploring what these may be and what potentials they may unleash with regards to blockchain is an intriguing avenue of experimentation, though beyond what we can explore in depth here.

2 This chapter examines different projects, but pays particular attention to the Bitcoin and Ethereum blockchain. As suggested before, despite its shortcomings Bitcoin is the first and still the biggest blockchain by market capitalization, and is especially useful for historical and evolutionary observation of the technology. Ethereum also populates a great part of the state-of-the-art projects. From the 2019 DApps (Decentralized Applications) listed by www.stateofthedapps.com/ website, more than 85% (1898) are based on Ethereum. Also, it must be noted that not all DApps in this landscape share the same degree of activity, as less than 20% show a recent active development.

3 The amount produced is adjusted approximately every four years: starting with 50 in 2009, 25 in 2012, 12.5 in 2016, and 6.25 in 2021.

4 Owing to PoW high energy consumption, other blockchain algorithms have been proposed. The most popular as yet is Proof-of-Stake (PoS). In this proof model, all the token supply is usually created from scratch, and then distributed to the shareholders by network managers, thus threatening the maintenance of a decentralized model. However, these alternative algorithms have been less tested and rarely used by big blockchains. Peercoin, for example, uses a hybrid model, relying on PoW for generation and distribution of new tokens, and on PoS for security and govevrnance. Ethereum discarded the use of PoS due to security reasons, but is still interested in adopting an improved version in a similar hybrid way, in the near future.

5 A third type, Consortium Blockchains, can be defined as permissioned blockchains whose management is shared between multiple organizations.

6 Git is a distributed version-control system, designed for collaborative development and tracking changes in code.

References

Alqassem, I., Rahwan, I., & Svetinovic, D. (2018). The anti-social system properties: Bitcoin network data analysis. *IEEE Transactions on Systems, Man, and Cybernetics: Systems*, 1–11. https://doi.org/10.1109/TSMC.2018.2883678

ArtID Marketplace. (n.d.). Retrieved from https://artid.ch/Site/Index

Bauwens, M. (2005). The political economy of peer production. *CTheory*, *0*(0), 12–1/2005.

Bauwens, M, Kostakis, V., & Pazaitis, A. (2017). *Peer to peer: The commons manifest*. London: University of Westminster Press.

Beckert, J., & Aspers, P. (Eds.). (2011). *The worth of goods: Valuation and pricing in the economy*. New York, NY: Oxford University Press.

Benkler, Y. (2002). Coase's penguin, or, Linux and "the nature of the firm." *The Yale Law Journal*, *112*(3), 369–446. https://doi.org/10.2307/1562247

Benkler, Y. (2006). *The wealth of networks: How social production transforms markets and freedom*. New Haven, CA: Yale University Press.

Beverungen, A., Mirowski, P., Nik-Khah, E., & Schröter, J. (2019). *Markets*. Minneapolis, MN: Meson Press.

Çalişkan, K. (2010). *Market threads: How cotton farmers and traders create a global commodity*. Princeton, NJ: Princeton University Press.

Coase, R. H. (1937). The nature of the firm. *Economica*, *4*(16), 386–405. https://doi.org/10.1111/j.1468-0335.1937.tb00002.x

Coeckelbergh, P. D. M. (2015). *Money machines: Electronic financial technologies, distancing, and responsibility in global finance*. Farnham: Ashgate.

De Filippi, P. (2015, July 31). Commons governance and law with Primavera De Filippi (R. O'Dwyer, interviewer). Retrieved from http://commonstransition.org/commons-centric-law-and-governance-with-primavera-de-filippi/

De Filippi, P., & Loveluck, B. (2016). The invisible politics of Bitcoin: Governance crisis of a decentralised infrastructure. *Internet Policy Review*, *5*(3). https://doi.org/10.14763/2016.3.427

De Jong, E., Tkacz, N., & Velasco González, P. (2015). "Live as friends and count as enemies": On digital cash and the media of payment. In G. Lovink, N. Tkacz, & P. De Vries (Eds.), *MoneyLab reader: An intervention in digital economy* (pp. 257–267). Amsterdam: Institute of Network Cultures.

DuPont. (2019, March 22). New book reveals crypto's radical origins. Retrieved from https://breakermag.com/new-book-reveals-cryptos-radical-origins/

Ethereum Project. (n.d.). Retrieved from www.ethereum.org/

Ghoshal, S., & Moran, P. (1996). Bad for practice: A critique of the transaction cost theory. *Academy of Management Review*, *21*(1), 13–47. https://doi.org/10.5465/amr.1996.9602161563

Gianelli, G., & Fumagalli, A. (2013, December 17). Il fenomeno Bitcoin: moneta alternativa o moneta speculativa? Gianluca Giannelli e Andrea Fumagalli. Retrieved from http://effimera.org/il-fenomeno-bitcoin-moneta-alternativa-o-moneta-speculativa-gianluca-giannelli-e-andrea-fumagalli/

Golumbia, D. (2016). *The politics of Bitcoin: Software as right-wing extremism*. Minneapolis, MN: University of Minnesota Press.

Hall, S. (1986). The problem of ideology – Marxism without guarantees. *Journal of Communication Inquiry*, *10*(2), 28–44. https://doi.org/10.1177/019685998601000203

Iles, M. (2018, May 4). The civil constitution (Beta). Retrieved from https://blog.joincivil.com/the-civil-constitution-beta-64460a181e08

Kim, M. S., & Chung, J. Y. (2019). Sustainable growth and token economy design: The case of Steemit. *Sustainability*, *11*(1), 167. https://doi.org/10.3390/su11010167

Lovink, G., Tkacz, N., & De Vries, P. (Eds.). (2015). MoneyLab reader: An intervention in digital economy. Retrieved from http://networkcultures.org/blog/publication/moneylab-reader-an-intervention-in-digital-economy/

Maurer, B. (2005). *Mutual life, limited: Islamic banking, alternative currencies, lateral reason*. Princeton, NJ: Princeton University Press.

Mirowski, P. (Ed.). (2015). *The road from Mont Pèlerin: The making of the neoliberal thought collective, with a new preface* (new ed.). Cambridge, MA: Harvard University Press.

MIT License. (2019). Retrieved from https://en.wikipedia.org/w/index.php?title=MIT_License&oldid=890561606

Nakamoto, S. (2008). Bitcoin: A peer-to-peer electronic cash system. Retrieved from https://bitcoin.org/bitcoin.pdf

O'Dwyer, R. (2014, September). Other values: Considering digital currency as a fommons. Presented at the Creating Cracks: Value, Commons, and Alternative Economy, London.

Raymond, E. S. (2008). *The cathedral and the bazaar: Musings on Linux and open source by an accidental revolutionary.* Sebastopol, CA: O'Reilly Media.

Reijers, W., Wuisman, I., Mannan, M., De Filippi, P., Wray, C., Rae-Looi, V., . . . Orgad, L. (2018). Now the code runs itself: On-chain and off-chain governance of blockchain technologies. *Topoi.* https://doi.org/10.1007/s11245-018-9626-5

Schaffers, H. (2018). The relevance of blockchain for collaborative networked organizations. In L. M. Camarinha-Matos, H. Afsarmanesh, & Y. Rezgui (Eds.), *Collaborative networks of cognitive systems* (pp. 3–17). New York, NY: Springer International Publishing.

Steemit. (n.d.). Retrieved from https://steemit.com/

Swartz, L. (2017). Blockchain dreams: Imagining techno-economic alternatives after Bitcoin. *Another economy is possible: Culture and economy in a time of crisis* (1st ed.). Malden, MA: Polity.

Swartz, L., & Driscoll, K. (2017, July 19). The lonely old Bitcoin miner touches eternity or, what is a peer? Retrieved from http://kingsreview.co.uk/articles/lonely-old-bitcoin-miner-touches-eternity-peer/

Taylor, M. B. (2013). Bitcoin and the age of bespoke silicon. *Proceedings of the 2013 International Conference on Compilers, Architectures and Synthesis for Embedded Systems,* 16 (pp. 1–16:10). Retrieved from http://dl.acm.org/citation.cfm?id=2555729.2555745

Terranova, T. (2009). Another life: The nature of political economy in Foucault's genealogy of biopolitics. *Theory, Culture & Society, 26*(6), 234–262. https://doi.org/10.1177/0263276409352193

The tao of IETF: A novice's guide to the internet engineering task force. (2012, August). Retrieved from www.ietf.org/tao.html

Thelwall, M. (2018). Can social news websites pay for content and curation? The SteemIt cryptocurrency model. *Journal of Information Science, 44*(6), 736–751. https://doi.org/10.1177/0165551517748290

Tkacz, N. (2011). The politics of forking paths. In G. Lovink & N. Tkacz (Eds.), *Critical point of view: A Wikipedia reader* (pp. 94–109). Retrieved from http://networkcultures.org/wpmu/portal/publications/inc-readers/critical-point-of-view-a-wikipedia-reader/

Tkacz, N. (2015). *Wikipedia and the politics of openness.* Chicago, IL: University of Chicago Press.

UN City Copenhagen, P. D. C. (2017, *May* 30). Blockchain against hunger: Harnessing technology in support of Syrian refugees. Retrieved from http://un.dk/news-and-media/blockchain-against-hunger-harnessing-technology-in-support-of-syrian-refugees

Velasco, P. R. (2016). Superabundant design: From waste to control in Bitcoin mining. *A Peer-Reviewed Journal About, 5*(1). Retrieved from www.aprja.net/?p=2927

Vuong, N. (2018, May 18). The civil registry: A TCR for quality journalism. Retrieved from https://blog.joincivil.com/the-civil-registry-a-tcr-for-quality-journalism-a2452f363665

Weber, S. (2004). *The success of open source.* Cambridge, MA: Harvard University Press.

19

Community Wireless Networks

Gwen Shaffer

1 Introduction

Peer-to-peer networks offer an alternative to corporate Internet service providers (ISPs). This grassroots mode of Internet connectivity enables community members to collectively determine which network infrastructure should be deployed, as well as where and how it should be deployed. On a practical level, peer-to-peer networks provide connectivity. Frequently, however, these projects are defined by their impact on civic engagement and democratic participation. Specifically, stronger community ties, self-reliance and opportunities for democratic deliberation potentially emerge when neighbors share bandwidth. This reflects a theoretical framework that acknowledges the Internet's role as a massive commons. Yochai Benkler (2006) characterizes the commons as "an alternative form of institutional space," where people are unhindered by "the particular restraints required for markets," and where necessary resources are available (p. 144). Not only do community WiFi networks provide access to information, they create a virtual space for participation, creativity, and communication (Holman & McGregor, 2005). The software layer, encompassing non-proprietary protocols such as TCP/IP and open source software, is mutual property of the entire online community (Benkler, 2001; Hofmokl, 2009; Lessig, 2001; Solum & Chung, 2004). The content layer is comprised mostly of information anyone can access (Benkler, 2001; Hofmokl, 2009; Lessig, 2001). Finally, at the structural level, community mesh projects function as an information commons, in the sense that they empower members to create, produce, and distribute knowledge. The value of this commons is derived from the fact that no one owns or controls it – not people, not corporations, not the government (Benkler, 2001; Lessig, 2001).

This chapter explores the history of community wireless networks (CWNs) through case studies, as well as the evolving practices and technologies. Specifically, this chapter describes how peer-to-peer technology works, from a technical perspective; the origins of CWNs as a social movement; motivations for building grassroots broadband networks in various regions throughout the world; how these projects serve as a testbed for related technologies; and the democratizing characteristics of peer-to-peer Internet platforms.

2 Design and Social History of Community Broadband Projects

2.1 Peer-to-Peer Technology

Peer-to-peer networking relies on a method of redistribution that is costless to the giver, in contrast to financial transactions within the corporate ISP market. Mesh networks, created by users themselves, grow virally. Therefore, traditional telecommunication companies – predicated on a central infrastructure that acts as a go-between among nodes and has a finite capacity (M.I.T. Communications Futures Program, 2008) – are much less necessary for one to access the Internet. In a viral communications architecture, components are both "independent" and scalable, and each new node adds capacity to

The Handbook of Peer Production, First Edition. Edited by Mathieu O'Neil, Christian Pentzold, and Sophie Toupin.

the network. A wireless mesh network design includes at least one access point with a direct connection to the Internet – via fiber, cable, or satellite link – and nodes that hop from one device to the next. As the popularity of these networks grow, new users add nodes. As a result, signals have shorter distances to hop and more redundancy is built into the system, ultimately strengthening the network (Rowell, 2007). With conventional WiFi, the access points rely on radio frequency signals. These signals travel through unregulated spectrum to provide computers, tablets, or mobile phones with a connection to the Internet. Because WiFi signals extend for an average distance of 91 meters (300 feet), multiple access points are necessary to cover large areas. A typical citywide project, like the Urban Wireless network in Glasgow, utilizes a hub-and-spoke model in which information beams out of one antenna to an individual computer. Small transmitters placed on utility poles extend the signals. Laptop users logging on from a park bench are likely to receive a reliable signal. However, because WiFi transmitters are low-power and signals do not bend, they lack the ability to penetrate walls. Using the service from inside a home or business required a signal booster, and even using this equipment does not guarantee adequate indoor coverage. Signals in a point-to-multipoint network converge at a single access site. The quality of the radio frequency link between the central access point and the end points determine the reliability of this architecture (Poor, 2003). Additionally, if the central link in a hub-and-spoke system fails, every connection that depends on it is lost.

Dynamic mesh networks, however, operate on a platform that avoids many of the technical challenges vexing hub-and-spoke architecture. In contrast to conventional WiFi, mesh networks work reliably in dense urban areas with many obstructions (BelAir Networks, 2006). Rather than radiating in a straight line from a central antenna, data can hop from individual router to individual router. Therefore, if one link breaks or leaves the network, the rest automatically reconfigure and quickly discover an alternative path for email messages or Web pages to travel. A fishing net has been used as an analogy to explain this ability to "self-heal." Imagine that every knot in the net, where the strings intersect, represents a device (Hatcher, 2005; Priest, 2003). Each device, called a node, is equal. Each of these nodes is also capable of transmitting messages to every node around it. Therefore, when a message is sent, it travels the most efficient path through the nodes to reach its final destination. If the net rips, the message still arrives at its destination by traveling a different path around the hole. Mesh networks work in a similar fashion – they entail self-organizing connections between nodes (computers and routers, for example) and are peer-to-peer based.

Two characteristics are fundamental to mesh networks. First, similar to the Internet itself, mesh networks provide multiple redundant paths for message delivery throughout the network. Second, so-called "plug and play" technology means that these networks do not require manual configuration (Büttrich, 2005; Lippman & Reed, 2003; Meinrath, 2006; Rowell, 2007). As previously mentioned, at least one access point, called the gateway, must be connected to the Internet. In this viral communications architecture, components are both "independent" and "scalable," and each new element adds capacity to the system.

2.2 The Peer-to-Peer WiFi Movement

A confluence of technological advances and ideological convictions influenced the peer networking movement. The underlying concepts associated with WiFi sharing communities can be traced back to 1996, when ICQ launched the first PC-based chat system. ICQ is peer-to-peer because it delegates connection management to individual nodes, as opposed to a central server. This type of connection creates a new temporary IP address that travels with you. A contrasting example is an email address such as @comcast.net or @verizon.net, which disappears if one switches ISPs (Shirky, 2001). Community mesh networks also reportedly have their roots in public safety networks operated by cities and counties in the 1990s. Specifically, police, fire fighters, and other first responders had been using cellular data networks for their mobile communications needs for several decades, sparking interest in wireless networking (Cisco, 2008).

Another major step toward the deployment of CWNs occurred in 1999, when Apple Computer introduced its Airport base station and Linksys released its first home router. These devices helped popularize wireless networking at a time when the typical household was just getting broadband

(Centers, 2018; Powell, 2008; Shaffer, 2010). The inherent versatility of the WiFi standard, coupled with the availability of routers running open source firmware, inspired techies who embraced the challenge of resolving software bugs and modifying hardware features (Shaffer, 2010). Other Internet activists were committed to disrupting the existing telecommunications structure – by shifting it away from corporate control and into community ownership – and joined the CWN movement for broader ideological reasons (Medosch, 2008).

Community broadband projects could not have flourished without the availability of unrestricted spectrum, dubbed "junk bands." Gadgets that do not require a license, such as microwave ovens, baby monitors, and garage door openers all use these frequencies. WiFi routers also live in one of these frequencies – specifically, the 2.4 GHz band, which easily penetrates walls and furniture (IEEE, 2016).

The Champaign-Urbana Wireless Internet Network (CUWiN) pioneered the peer-to-peer broadband movement in the United States. In November 2002, its founders developed software that enabled multi-hop, bandwidth-sharing (Meinrath, 2006). Other early networks included Seattle Wireless in Washington State and Personal Teleco in Portland, Oregon, as well as Berlin's community-driven network, Freifunk, and Vienna's Funkfeuer – all established around 2003.

CUWiN hosted the first National Summit for Community Wireless Networks in 2004. The event brought together more than 200 technology experts, policy analysts, academic researchers, and media activists. Participants believed peer-to-peer technology had the potential to rupture the duopoly market power of cable and phone companies (Peer-to-Peer Foundation, 2011). CWN activists perceived this lack of consumer choice as stifling economic growth and social opportunity. Summit participants hashed out various aspects of peer-to-peer networks, ranging from technology and policy, to challenges associated with organizing volunteers.

Ultimately, leaders of community broadband initiatives drafted a strategic plan for expanding the deployment of networks and lobbying the federal government to create policies that would expand broadband access, and open more unlicensed spectrum. This 2004 summit served as a catalyst for a growing community broadband movement in the United States. By 2007, the "national summit" was rebranded as an "international summit" and more than 300 US communities had planned or built local wireless networks (International Summit for Community Wireless Networks, 2007).

3 Motivations for Deploying Community WiFi Networks

3.1 Commons-Based Ideology

The motivations for building grassroots wireless networks are as diverse as the people and communities involved in these projects. This section explores some of the most common circumstances and ideologies driving CWN deployments. Of course, multiple rationales can be attributed to the emergence of a single community mesh project. This section explains what a CWN is; provides a brief history of the community WiFi movement; identifies the catalysts for specific peer-to-peer networks; and presents case studies of projects.

From the start, two equally powerful notions defined the political significance of CWNs: one characterized by the potential for WiFi to disrupt the existing ISP model, and another focused on the ability for CWNs to provide affordable, or even free, ubiquitous connectivity (Powell, 2008). For participants, CWNs must function as a commons – characterized by bottom-up participation, with minimal reliance on the state. Community mesh networks grounded in a commons-based model of social production include a wide base of volunteers who not only use bandwidth provided by the network, but who also contribute time and skills to sustaining it.

In a commons, communication systems are truly democratic – in the sense that community members themselves determine how the network is designed and deployed. Neither corporations nor policymakers get to influence those decisions. Network participants strive to create an independent entity that stresses creativity and collaboration (Benkler, 2006; Benkler & Nissenbaum, 2006, also this volume; Krowne, 2005; Lessig, 2001). Stronger community ties, self-reliance, and opportunities for democratic deliberation potentially emerge when neighbors share bandwidth. In this sense, WiFi signal sharing is more than a promising "last mile" technology able to reach every home for a fraction of the cost required to lay fiber, DSL, and cable (Martin, 2005). In fact, grassroots mesh projects aim to create "a radically different public sphere" (Burnett, 1999) by situating themselves outside of commercial interests. Typically, one joins, as opposed to subscribes to, the services. As Lippman and Reed (2003, p. 1) observed, "Communications can become something you do rather than something you buy."

3.2 Resiliency as a Catalyst

Centralized structures, such as the traditional Internet, are intrinsically vulnerable. If servers or towers go off line, users can no longer connect. In recent years, entire communities have been cut off from broadband access – the result of circumstances ranging from natural disasters to totalitarian governments shutting down Internet infrastructure. These high-profile "Internet apocalypses" have raised awareness of the weaknesses inherent in a centralized Internet. Hurricane Sandy pummeled the Eastern United States coastline in 2012, causing thousands of residents to lose Internet connectivity. Because Internet users and media outlets across the United States rely on a few critical nodes to connect, Hurricane Sandy had a nationwide impact (Reardon, 2012). Regional ISPs blamed flooded data centers, damaged cables, and electricity outages for the disruption in service (Reardon, 2012). The Red Hook Initiative (2019a), a community broadband project in Brooklyn, had begun deploying a WiFi network the previous year. Although the network had limited reach, it was self-sustaining. In the wake of Hurricane Sandy, the renovated factory that housed the Red Hook Initiative retained electricity, enabling some local residents to connect to the Internet.

When Federal Emergency Management Agency (FEMA) officials checked in on Red Hook residents several days after the storm, they learned about its community WiFi project. They seized the opportunity to attach a satellite link to the network, restoring a main gateway to the Internet. FEMA transformed the site into a base for its own recovery efforts. This enabled Red Hook members to access social media and government assistance earlier than residents of other neighborhoods devastated by the storm (Berwick, 2016). The Red Hook Initiative continues to disaster-proof its network, and to serve as a model for other communities preparing for future floods – a certainty of climate change. As of February 2019, the Red Hook Initiative encompassed 41 access points, including 15 solar-powered "resilient hotspots" (Red Hook Initiative, 2019a).

The Red Hook WiFi project has been transformative for the community in other ways, as well. Young people, between the ages of 16 and 24, spend a year training as "digital stewards." They learn to build a network from scratch and, for many of them, it is their first job. Rather than pouring coffee or babysitting, these young people run cables down the sides of buildings, design network layouts, and learn to code (Red Hook Initiative, 2019b).

Beyond resisting natural disasters, peer-to-peer network design is capable of withstanding attacks from totalitarian governments. This is critical, given the increasing trend of governments throttling or shutting down access to the Internet. With just a handful of companies delivering Internet service in most countries, these providers are accomplices in digital blackouts. By threatening to revoke state licenses to operate, governments easily coerce ISPs into complying with demands to block access. The Open Observatory of Network Interference (OONI) used crowdsourcing and open software to collect data on website censorship and traffic manipulation. OONI performed more than 1 million HTTP requests between 2012 and 2016. "Network anomalies" – specifically, Internet traffic disruptions – occurred in 71 of the 91 countries that OONI researchers tested (United Nations Educational, Scientific and Cultural Organization, 2016). Similarly, UNESCO (2016) researchers identified 56 cases of Internet shutdowns in 18 countries in 2016.

Events in Burma illustrate how governments weaponize Internet infrastructure. A sudden sharp increase in fuel prices spurred an estimated 100,000 Burmese residents to demonstrate in Rangoon in September 2007. Thanks to citizen journalists and bloggers who posted graphic footage and witness accounts – including when the military junta killed about 200 people during a violent crackdown that began on September 26, 2007 – people around the world learned of the political unrest (OpenNet Initiative, 2007). The Burmese government reacted by shutting down Internet access and cell phone service, hoping to stop news about violence from escaping its borders. "When the Internet flows through only a tiny number of central choke points, it doesn't take much ingenuity to figure out a way to strangle it," writes Daniel Jeffries (2016), founder of the Cicada project – a point-to-point, fully encrypted mesh network maintained by individuals and communities. Jeffries (2016) argues that a decentralized network is needed "to survive even if the powers that be kill the Internet." In fact, the Cicada project is grounded in the philosophy that the current centralized Internet infrastructure facilitates abuse of power by authoritarian governments.

Within the past decade, the Syrian government has blocked Internet connectivity between its own residents and the outside world at least a dozen times (Thomson, 2012). In order to counter "common methods of oppression in the digital era," Syrian software engineer Abdul Rahman Alashraf developed FreeCom (2019). Through this open source peer-to-peer protocol, users create their own network by connecting to each other. Freecom (2019) enables smart devices to send encrypted text messages, make calls, receive updates or notifications, and broadcast news even where Internet is unavailable.

In January 2011, Egyptian activists relied on social media and text messaging to mobilize thousands of protesters, who streamed onto the streets of Cairo and Tahrir Square demanding an end to President Hosni Mubarak's 30-year rule. The demonstrators posted images of violence in real-time on Twitter and Facebook – encouraging more people to participate. But on January 25, 2011, the country's Internet and cell phone networks stopped working (Kravetts, 2011; Saleh & El Madany, 2011). Software developers averted a complete communications blackout by patching together a peer-to-peer network that bypassed both the state-controlled Internet and the telecommunications companies that complied with the government order to turn off their networks (Moran, 2012). In the aftermath of this five-day Internet shutdown, Angel investor Shervin Pishevar (2011) launched the OpenMesh Project, aimed at creating and distributing open source software that would allow "citizens of the world" to communicate without telephone or cable companies. "The only line of defense against government filtering and blocking their citizens from freely communicating and coordinating via communication networks was to create a new line of communications technologies that governments would find hard to block: ad hoc wireless mesh networks," wrote Pishevar (2011).

Decentralized infrastructure is also an anecdote for more subtle forms of Internet control widely practiced by totalitarian governments: the censorship of online content and surveillance. Researchers with the OpenNet Initiative (Phillipa et al., 2015) found that countries around the world censor Internet content in multiple ways, such as by tampering with the domain name server; through blockpages; by disrupting HTTP requests; and by triggering TCP resets. The OpenNet Initiative identified Internet censorship in scores of countries, finding the practice common in China, North Korea, Saudi Arabia, Iran, and Egypt (Phillipa et al., 2015).

Genuine need for a broadband network

Often, activists launch peer-to-peer WiFi projects not as alternatives to existing Internet networks, but in the absence of such infrastructure. Several of the most established peer-to-peer networks emerged out of a genuine need for basic Internet connectivity, such as Guifi.net (2019a). In 2004, Spanish incumbent carrier Telefónica offered DSL in some parts of rural Catalonia, Spain – about 75 miles outside of Barcelona. However, the service was expensive and unreliable. Guifi.net boosted the local economy by enabling residents to work from home, and making it possible for rural businesses to run applications ranging from VoIP to surveillance cameras. For the first time, local hog and cattle farmers could transmit test results to veterinarians and accomplish other routine tasks (Shaffer, 2010).

Guifi.net's initial model was to attract entire village governments, as opposed to individuals. Town councils would subscribe to an ISP, then install $100 antennas on street lamps and rooftops throughout their villages. Each of these access points had the capacity to support 30 Internet connections. With an average population of 2,500 residents in each village, it cost just a few thousand euros to deploy nodes throughout an entire village. In order to connect to the signal, individual residents and businesses purchased rooftop antennas (Shaffer, 2010). The network has expanded dramatically in the intervening 15 years. Today, Guifi.net (2019b) has more than 35,200 active nodes, with plans to deploy an additional 18,000 nodes, and covers about 65,000 kilometers. More than 50,000 people get Internet access through the commons-based network, which now includes fiber and wireless infrastructure (Guifi.net, 2019b).

DjurslandS.net in Denmark's rural Djursland Peninsula is another pioneering CWN that emerged out of necessity. In the late 1990s, local residents realized their economy would falter without reliable Internet access. They repeatedly asked incumbent Internet provider Tele Denmark to provide their community with residential DSL service (Shaffer, 2010). When Tele Denmark declined, residents negotiated with 35 smaller Danish ISPs. One after another, these companies insisted that deploying infrastructure in such a sparsely populated community was too expensive (Shaffer, 2010). In 2000, Internet activists purchased wireless access points in bulk for about kr 65,000 (US $10,000), and began connecting the region wirelessly. Meanwhile, the Djursland Peninsula village of Glesborg built a 50-meter high tower and sold it to the Internet activists for a symbolic single krone. In May 2003, omni antennas on the tower communicated with devices on the roof of a sports facility 1.5 km away – creating the network's initial link. By 2009, DjurslandS.net had evolved to include about ten fiber gateways to the Internet and hundreds of strategically located access points in villages throughout the peninsula. Nearly 8,000 households subscribed to the network, paying about one third the price of a corporate broadband subscription in urban Denmark (Shaffer, 2010).

DjurslandS.net ceased function as an umbrella organization in 2009, but 11 regional networks continue to run independently. For instance, VestDjursNet (2019) "aims to create wireless broadband in sparsely populated areas where there are no other options for high-speed Internet." VestDjursNet's (2019) daily operations "are handled by local volunteer technicians" and operational decisions are made by a volunteer board. PrimaNet (2019), another regional network in Djursland with 3,500 subscribers, characterizes its volunteers as "part of an active community that shares the challenges of running, improving and expanding a very large predominantly wireless network."

Two community networks based in African countries also exemplify situations where residents mobilized to create broadband services. Macha is a traditional rural village in Zambia's Southern Province. Most of the 128,000 villagers who live in Macha rely on subsistence farming, as the average villager earns less than US$400 annually (MachaWorks, 2019). The Brethren in Christ church operates a hospital, schools, and community radio station in Macha. But it was the need to connect Macha's medical research institute with the rest of the world that drove villagers to create a community mesh network in 2004 (Mweetwa and Van Stam, 2018). In 2004, they set up a VSAT satellite link – housed in a shipping container – to the Internet and built a local WiFi network. Community members then formed the non-profit cooperative Macha Works to oversee network operations, maintenance, and growth. In the intervening years, the project has expanded dramatically and seven other provinces in Zambia now host community networks (Mweetwa & Van Stam, 2018).

The network transformed daily life in Macha, facilitating access to healthcare, education, and social opportunities. Beyond this, Macha Works operates the LinkNet Information Technology Academy, or LITA. Through this initiative, rural residents earn internationally recognized certificates in computer literacy and ICT engineering. The LITA facility, based in Macha, is a key component "of the overall capacity building objective of the co-operative" (Mudenda & Van Stam, 2012, p. 3).

A community network created out of necessity in South Africa is similarly noteworthy. Mankosi, located in the Eastern Cape Province, is one of the country's poorest communities (Metri, 2017). Nearly all of its 6,000 residents live on less than $2 per day, and are not connected to the electricity grid (Wireless Access Providers Association, 2019). What residents have had since 2012, however, is a solar-powered wireless community network that is owned and operated by a rural cooperative. When it became clear that incumbent ISPs had no interest in providing service to remote Mankosi, researchers from the University of the Western Cape helped start the Zenzeleni Networks project. Zenzeleni means "do it yourself" in the native language isiXhosa. Local residents receive training in how to use the Internet and, significantly, they are taught technical skills to maintain it (Metri, 2017; Tucker, 2017).

The network consists of about a dozen solar-powered mesh network stations mounted on, and inside, homes throughout Mankosi and cover 30 square kilometers. The money collected from network users who purchase data, and voice service for about 20 cents per minute, is re-invested in community projects (Tucker, 2017; WAPA, 2019). Additionally, the excess electricity generated by the solar WiFi stations is used for a mobile phone charging station. The access to a charger changed how residents used their phones, and enabled them to save money they had been forced to spend to charge their phones outside the village (Tucker, 2017). Project organizers believe the network will, eventually, foster economic development that benefits village residents. In fact, Zenzeleni Networks' goal is to expand to support about 25 communities surrounding Mankosi. This would empower about 300,000 people – as well as the schools and hospitals they rely on – to access less expensive voice, data, and phone battery charging (Tucker, 2017).

An alternative to corporate Internet providers

Even in regions across the globe where commercial Internet access is available, a lack of options frequently exists. In the United States, Internet service is practically a duopoly, controlled by Verizon and AT&T, and a handful of cable companies. Three quarters of American homes have no competitive choice, including nearly 20% of Americans who have no service at all (Singer, Naef, & King, 2017). The picture is similarly bleak on the other side of the Atlantic Ocean. As the result of countless mergers and acquisitions, Spain's Telefónica (2018) and Germany's Deutsche Telekom (2018) dominate telecommunications services throughout Europe. British-based Vodafone (2018) and France's Orange S.A. (2018) have also both morphed into multinational Internet companies. This intense concentration in telecommunications markets has real consequences for consumers. When an ISP controls the market, it lacks incentives to invest in infrastructure, to develop innovative products, and to provide quality of service. Not surprisingly, consumers are often dissatisfied with their incumbent telecommunications provider.

Many signal-sharing projects make a deliberate choice to challenge the existing telecommunication model. They treat wireless Internet access as a tool for community empowerment, rather than as a commodity. For example, the Personal Telco Project in Portland launched in 2000, after the so-called "Internet bubble" burst. A group of techies who lost their jobs from the fallout decided to transform their own houses and apartments into wireless hotspots, or nodes. Next, they set about building networks in public locations such as parks and coffee shops. Personal Telco (2019) is dedicated to the idea that users have a central role to play in determining how their communications networks are operated. It mobilizes volunteers to build the infrastructure for the entire community. Currently, about 100 active nodes participate in the project (Personal Telco, 2019). Project volunteers also host hacker nights and trainings so anyone can participate.

NYC Mesh is another community owned network that exists to challenge the status quo. It is grounded in the conviction that Internet access in New York City "is controlled by a cartel of cable companies who overcharge and underdeliver" (NYC Mesh, 2019). The network, which consists of WiFi router nodes spread throughout the city, has no central server and no single ISP providing a gateway to the Internet. All nodes cooperate in the distribution of data, also serving as a stand-alone network in case of an emergency or Internet shutdown. As of February 2019, the network included 282 active nodes, plus 3 "supernodes" and 25 network hubs (NYC Mesh, 2019). As one NYC Mesh participant noted, all traffic traveling over the network bypasses any commercial ISPs. Therefore, network users are not forced to pay arbitrarily high prices to avoid having their connection speeds throttled, and they do not need to worry about ISPs selling their data. "We don't pay into the cable cartel, which means we don't fund their lobbying, advertising, lawyers, executive salaries, bonuses, and dividends," this participant blogged (Ehud, 2018).

Testbed for new hardware and applications

Some community WiFi activists are drawn to the opportunity to tinker with technology. For example, when volunteers began building the Athens Wireless Metropolitan Network (AWMN) in 2002, they certainly hoped to provide an alternative broadband provider. However, the network also immediately functioned as a laboratory for members to design antennas and satellite dish feeders used on backbone nodes. Dozens of techies who joined began refining routing protocols and network management software (Shaffer, 2010). One of the unique aspects of AWMN is that it quickly evolved into a platform for network participants to create new applications and online services exclusively for the Athens wireless community (netCommons, 2016; Shaffer, 2010). As the names of these services imply, they often mirrored sites found on the public Internet: the auction site Wbay; the search engines Woogle and Wahoo; user created content shown on wTube; multiplayer games; dating services; movie and music streaming; a directory of postal codes of Europe; weather reports for each Greek island; and webcams that broadcast traffic; among other services (WiND, 2019). Some members use the network as an opportunity to float ideas for online services they intend to eventually introduce commercially (Shaffer, 2010). As of 2019, AWMN included nearly 2,300 nodes (WiND, 2019).

Democratizing the Internet and bridging the digital divide

A conviction that the Internet should exemplify democratic and inclusive principles serves as another motivation for CWNs. Certainly, this was the driving force behind Freifunk in Berlin. Soldiers from the German Democratic Republic (GDR) began stacking the first concrete blocks of the Berlin Wall on August 13, 1961. Over the next 28 years, East Germany's neglected telecommunications infrastructure fell into disrepair (German Culture, 2009). Following reunification in 1989, the German telephone company ripped out old copper lines. The phone company determined that laying fiber would be prohibitively expensive. But as former East German neighborhoods gentrified, new tech-savvy residents became frustrated at the lack of high-speed Internet access. It is within this cultural and political context that the concept for Freifunk – German for "free radio" – emerged.

In 2002, Berlin resident Juergen Neumann placed antennas on the roof of his building, connecting 35 residents to an ISP. Soon after, he and a handful of friends began setting up wireless nodes in Berlin and sharing bandwidth. They formalized the effort in 2003, with the goal of creating a highly decentralized network with no ownership. Politics and ideology permeate all aspects of the initiative. Freifunk's (2019) goal is to connect neighborhoods, and even entire regions, in an effort to close the digital divide. The hope is to empower people to build and operate their own networks. Volunteers are committed to creating awareness of freedom of information; distributing knowledge and resources; and promoting and supporting both existing and new "social structures" (Freifunk.net, 2019). As noted on Freifunk's (2019) website, its vision "is the democratization of the media through free networks."

As of January 2019, Freifunk (2019) included about 48,000 nodes not only in Berlin, but in communities throughout Germany and neighboring Central European countries. The infrastructure's reach is broad enough to provide access to about 350,000 people (netCommons, 2016). Improved network performance and the falling price of hardware have contributed to the Freifunk's expansion. On a less cheery note, the economic downturn and Germany's refugee crisis also played a role (netCommons, 2016). To host an access point for Freifunk (2019), participants can set up a free wireless router on windows or balconies, automatically connecting to other nodes in the neighborhood. Alternatively, participants can attach routers to their own Internet connections and share bandwidth (each participant determines which percentage of bandwidth to share).

Despite a philosophical commitment to decentralization, in 2013 Freifunk began building a network backbone that interconnects the many wireless "subnetworks" that comprise Freifunk's infrastructure. A $30,000 grant from the local broadcasting corporation Medienanstalt Berlin-Brandenburg (2019) is funding the construction. The backbone enables Freifunk routers to bundle Internet traffic and route it via its own gateways (netCommons, 2016). The network is also "centralized" through its virtual private network (VPN). Specifically, the Freifunk Freedom Fighter Box is a preconfigured access point that sends all data from the public network to Sweden over a VPN connection. This exporting of traffic was deemed necessary because German law states that the owner of an access point is liable for illegal activity using that connection. Therefore, any access point owner who shares his bandwidth with strangers could get caught up in legal trouble (Lehnhardt, 2012), such as a copyright infringement. However, Sweden does not enforce the so-called "liability for interference."

Funkfeuer (2009), a non-commercial mesh network in Vienna open to anybody willing to contribute, is also guided by its commitment to closing the digital divide. As of February 2019, Funkfeuer (2019) included more than 220 active nodes. Funkfeuer is unique among community wireless projects because it owns a 5-gigahertz fiber optic link to the Vienna Internet Exchange.

4 Technological Innovation

Firmware enables routers to connect to other networks; to assign addresses to machines connecting to the router; and to establish a firewall to protect devices using the router; among other essential functions (Kozierok, 2005). In fact, wireless community participants frequently cite the simplicity of installing mesh equipment – sometimes comparing it to plugging in a toaster. "It is so easy, you really don't need to be a nerd to do it," one U.S. community WiFi activist said (Shaffer, 2010). Manufacturers of commercial routers, which determine how messages move through a computer network, pre-install software on these devices. This so-called firmware is analogous to smartphone operating systems such as Android and iOS, or the Windows and OSX operating systems that run computers. By contrast, many CWNs are built with open source routers. This means network designers reconfigure firmware that comes standard on "off-the-shelf" equipment, and rewrite code specifically for their network requirements.

Projects that exclusively develop mesh protocols for CWNs – but do not deploy nodes or connect users – represent a distinct segment of the peer-to-peer networking movement. Router-to-router connections are necessary in order for CWNs to work. However, projects devoted to writing custom firmware with advanced features are, typically, ideologically-driven. Because this firmware is open source – as opposed to proprietary – developers may use, copy, and modify its software code for any network. These coders feel strongly about customizing and controlling management structures for community mesh projects. "We do a lot of experimenting with the technology to see what is most effective . . . so there are no set rules," commented a volunteer in the United States. Other peer-to-peer activists have cited their commitment to open source technology as a key reason for participating in the mesh networking movement. "It was a hack thing, it was great," one technologist said (Shaffer, 2010).

Commotion Wireless (2019a, 2019b) is an example of a community WiFi project that focuses entirely on software development. This open source tool is meant to simplify the process of installing firmware on wireless devices, so that even people who lack technical expertise can build a community wireless network and host local applications. Commotion Wireless is a project of the New America Foundation's Open Technology Institute (2019), which is dedicated to ensuring equitable access to communication technologies. Programmers around the world continually modify and build upon its software platform. Its developers intend for the firmware to support infrastructure that is resistant to both surveillance and disruptions.

A handful of CWNs run on Commotion Wireless firmware, including the Detroit Community Technology Project (2019). The project launched in 2010, when the Detroit Digital Justice Coalition allocated a portion of a US$1.8 million federal grant to launch community wireless networks in several underserved neighborhoods (Allied Media Projects, 2012). A guiding principle of the Detroit Community Technology Project is to "demystify technology" so that residents can create and control their own communications infrastructure (Detroit Digital Justice Coalition, 2010). As of February 2019, "digital stewards" in Detroit had deployed three mesh networks, and four more were planned. The Detroit project is modeled after the Redhook WiFi Initiative in New York, which also runs on Commotion Wireless.

Sudo Mesh in Oakland, California, exemplifies another project devoted to developing technology for community mesh networks. Sudo Mesh (2019) encourages anyone "willing to contribute by reporting bugs, improving software, fixing hardware, setting up a node, installing rooftop nodes. . ." to volunteer. Sudo Mesh participants characterize themselves as "unpaid volunteers" creating open source technology with the potential to connect neighbors; support local businesses; enable community collaboration and cultural production; and serve as a communications tool in the event of a natural disaster or state censorship (People's Open Network, 2019). By combining off-the-shelf hardware and Sudo Mesh's open source software, participants can share bandwidth and create Internet connections in public places and in underserved areas. Sudo Mesh's custom WRT firmware is installed on routers used by the People's Open Network (2019), a CWN also based in Oakland. In October 2018, the Handshake Foundation donated US$100,000 to Sudo Mesh – moving the project closer to achieving its goal of distributing connectivity throughout California's East Bay (People's Open Network, 2018).

FreedomBox (2019) is another open source software tool intended to decentralize the Internet that has been adopted by CWNs for use as WiFi access points. The device sits between a computer and the Internet and is meant to replace a standard wireless router. All aspects of the FreedomBox project rely on community input, from architecture and engineering to administration and funding. Developers around the world contribute to FreedomBox's code, and the software is updated with patches, expanded features, and UI updates about every two weeks (FreedomBox, 2019). As of February 2019, the software stack was at version 49 (Debian, 2019). The developers' goal is to create and preserve personal privacy and anonymity by providing a secure platform for routing, both locally and to the wider Internet. Instead of storing personal information centrally in commercial services, like Facebook or Instagram, each user sets up his or her own FreedomBox server. Small external devices (plug computers or mesh routers) run FreedomBox's software stack, minimizing the possibility of third parties accessing data – legally or illegally (Stilgherrian, 2012).

The FreedomBox initiative is as much about philosophy as it is about technology. Columbia University law professor and free software advocate Eben Moglen founded FreedomBox in February 2010. He laments the privacy violations and security vulnerabilities that have come to define the centralized Web architecture nearly everyone uses for visiting websites, posting to social media, and exchanging email messages. Moglen points out that the Internet was designed as a network of peers without any intrinsic need for hierarchical or structural control, but that things have changed:

> . . . if you have a system which centralizes servers and the servers centralize their logs, then you are creating vast repositories of hierarchically organized data about people at the edges of the network that they do not control . . . All of those decisions architecturally were made without any discussion of the social consequences long-term, part of our general difficulty in talking about the social consequences of technology during the great period of invention of the Internet done by computer scientists who weren't terribly interested in sociology, social psychology, or, with a few shining exceptions – freedom. So we got an architecture which was very subject to misuse. (Moglen, 2010)

Similar concerns about privacy and security motivated the development of PirateBox (2019), which uses free software and off-the-shelf hardware to create offline wireless networks. PirateBox is designed for anonymous file sharing, chatting, message boarding, and media streaming. Users are anonymous – they are never required to sign in – and PirateBox (2019) does not log data. Like CWNs themselves, the technology is about community building. For instance, musicians have used PirateBox to share music during festivals. Teachers have used these devices to distribute and collect electronic assignments from students. Emergency responders, with help from volunteers, have used PirateBox to publish community updates.

While peer-to-peer firmware developers stress the "geek fun" aspect of installing open source software on routers, their practices have real theoretical implications. In fact, these broadband activists are making a political statement by focusing on open source devices that are collectively developed and allow users to maintain control over how they utilize hardware. In this sense, mesh networkers who develop their own routing protocols are challenging key facets of the dominant telecommunications industry. Corporations protect their interests through patents, trademarks, and terms of use restrictions. They lockdown devices like the iPhone and prohibit consumers from copying DVDs for which they already paid. By contrast, open source developers rely on external input from users, and benefit from the viral spread of their technologies.

The open source philosophy mirrors the principles guiding the peer-to-peer networking movement itself. Above all, they privilege collaboration over individual ownership. Community WiFi initiatives – encompassing both mesh networks and open source technology – rely on the combined intellect of "knowledge workers" (Bauwens, 2005), who contribute their time and experience to developing products or networks. They typically do so with no corporate backing and no financial return. In the context of capitalistic economic systems, where decisions tend to be driven by their potential to deliver profits and where intellectual property rights are strictly enforced, this behavior is far outside the mainstream. It is true that open source activists do not reap monetary rewards; what they get instead is the chance to participate in an emerging economy that values "autonomy, democracy, and social justice" (Benkler, 2003, p. 1246). Rather than corporate-owned ISPs controlling which information travels through Internet pipes and at what speed, distributed connectivity provides an opportunity for users themselves to manage content and disseminate it.

5 Conclusion

Peer-to-peer networks in the United States, Europe, and around the globe play a key role in providing broadband for underserved communities. CWNs also highlight the potential of "social innovation at the bottom of the pyramid" (Evolvingwheel, 2008). Specifically, these projects facilitate public participation in technology, as volunteers develop software, hack hardware, and install antennas. The case studies highlighted in this chapter suggest that developing open source applications, infrastructure, and platforms – on which competitive providers may offer content and services – is as important to community wireless projects as connectivity itself.

WiFi signal sharing is more than a promising "last mile" technology. Rather, it has potential to disrupt the telecommunications industry by enabling consumers to take advantage of broadband signals without subscribing to an ISP. As end-users increasingly gain control over technology, they seek alternatives to incumbent ISPs and wireless carriers – both of which exercise their dominance in the marketplace by capping bandwidth levels, centrally managing traffic, and bundling products (i.e., selling Internet and voice services together). The contrasting paradigm of grassroots networks enables participants to operate and maintain their own systems, relying on telecommunications companies solely for backbone connectivity. Peer-to-peer networks face fewer barriers to entry as equipment becomes less expensive, more reliable and easier to install, and as governments allocate additional unlicensed spectrum. These initiatives represent the promise of a transformed communication system, one where access to broadband networks is truly democratic because community members themselves determine how infrastructure is designed and deployed.

References

Allied Media Projects (2012). Learn how we used 1.8 million in stimulus funds to seed a digital justice movement in Detroit. Retrieved from www.alliedmedia.org/news/2012/11/01/learn-how-we-used-18-million-stimulus-funds-seed-digital-justice-movement-detroit.

Bauwens, M. (2005, December 1). The political economy of peer production. *CT Theory*. Retrieved from http://ctheory.net/articles.aspx?id=499

BelAir Networks (2006). *Capacity of wireless mesh networks*. Kanata, ON: BelAirNetworks.

Benkler, Y. (2001). From consumers to users: Shifting the deeper structures of regulation toward sustainable commons and user access. *Federal Communications Law Journal*, 52.

Benkler, Y. (2003). Freedom in the commons: Towards a political economy of information. *Duke Law Journal*, 52, 1245–1276.

Benkler, Y. (2006). *The wealth of networks: How social production transforms markets and freedom*. New Haven, CT: Yale University Press.

Benkler, Y., & Nissenbaum, H. (2006). Commons-based peer production and virtue. *The Journal of Political Philosophy*, 14(4), 394–419.

Berwick, C. (2016, October 3). Yesterday's Internet isn't good enough for tomorrow's cities. *Next City*. Retrieved from https://nextcity.org/features/view/Internet-connection-mesh-networks-resilience

Burnett, R. (1999, Fall). Communications policy and the new public sphere: Towards a new research agenda. *Community Technology Center Review*. Retrieved from www.comtechreview.org/summer-fall-1999/communications_policy_and_the_ne.htm

Büttrich, S. (2005). Mesh networks. ItrainOnline.org. Retrieved from www.itrainonline.org/itrainonline/mmtk/wireless_en/17_Mesh_Networking/17_en_mmtk_wireless_mesh-networking_slides.pdf

Centers, J. (2018, April 27). RIP Apple Airport, 1999–2018. *Tidbits*. Retrieved from https://tidbits.com/2018/04/27/rip-apple-airport-1999-2018/

Cisco (2008). Evolution of municipal wireless networks. Retrieved from www.cisco.com/c/dam/global/en_ca/training-events/networkerssolutions2008/pdfs/5_Wireless_Mesh.pdf

Commotion Wireless (2019). About commotion. Retrieved from https://commotionwireless.net/about/

Debian (2019). FreedomBox release notes. Retrieved from https://wiki.debian.org/FreedomBox/ReleaseNotes

Detroit Digital Justice Coalition (2010). *DDJC 'zine*. Retrieved from www.alliedmedia.org/files/DDJC%20zine%202.pdf

Deutsche Telekom (2018). At a glance. Retrieved from www.telekom.com/en/company/at-a-glance

Ehud (2018, October 31). How we're connecting neighbors. Retrieved from www.nycmesh.net/blog/connecting-neighbors/

Evolvingwheel (2008, January 30). Is "Meraki" bringing social innovation by broadband reselling? *Looking Beyond the Horizon*. Retrieved from https://innovech.wordpress.com/2008/01/30/is-meraki-bringing-social-innovation-by-broadband-reselling/

FreeCom (2019). Our story. Retrieved from https://freecom-info.com/our-story

FreedomBox (2019). Introduction. Retrieved from https://wiki.debian.org/FreedomBox/Introduction

Freifunk (2019). Statistics. Retrieved from https://berlin.freifunk.net/network/stats/

Funkfeuer (2019). How can I participate? Retrieved from www.funkfeuer.at/

German Culture (2009). The reunification of Germany and its aftermath. Retrieved from www.germanculture.com.ua/library/facts/bl_reunification_aftermath.htm

Guifi.net (2019a). *What is Guifi.net?* Retrieved from https://guifi.net/ca/que_es

Guifi.net (2019b). Node statistics. Retrieved from https://guifi.net/guifi/menu/stats/nodes

Hatcher, J. (2005). Mesh networks: A look at the legal future. Social Science Research Network. Retrieved from https://papers.ssrn.com/sol3/papers.cfm?abstract_id=814984

Hofmokl, J. (2009). Towards an eclectic theory of the Internet commons. *International Journal of the Commons*, 4(1), 226–250.

Holman, J., & McGregor, M. (2005). The internet as commons: The issue of access. *Communication Law and Policy*, 10(3), 267–289.

IEEE (2016, June 28). Why Wi-Fi stinks and how to fix it. *IEEE Spectrum*. Retrieved https://spectrum.ieee.org/telecom/wireless/why-wifi-stinksand-how-to-fix-it

International Summit for Community Wireless Networks (2007, April 18). International summit to address future of broadband. Retrieved from https://wetmachine.com/tales-of-the-sausage-factory/a-must-attend-for-community-wireless-networking/

Jeffries, D. (2016, December 12). How we build an unstoppable peer-to-peer Tor with a billion exit nodes. *Hackernoon*. Retrieved from https://hackernoon.com/how-we-build-an-unstoppable-peer-to-peer-tor-with-a-billion-exit-nodes-4b232dac162d

Kozierok, C. (2005, September 30). Overview of key routing protocol concepts: Architectures, protocol types, algorithms and metrics. *The TCP/IP Guide*. Retrieved from www.tcpipguide.com/free/t_OverviewOfKeyRoutingProtocolConceptsArchitecturesP.htm

Kravetts, D. (2011, January 27). Internet down in Egypt, tens of thousands protest in "Friday of Wrath." *Wired*. Retrieved from www.wired.com/2011/01/egypt-Internet-down/

Krowne, A. (2005). The FUD-based encyclopedia: Dismantling the fear, uncertainty and doubt aimed at Wikipedia and other free knowledge sources. *Free Software Magazine*. Retrieved from http://freesoftwaremagazine.com/articles/fud_based_encyclopedia/

Lehnhardt, J. (2012, July 26). Freifunk statt angst. Retrieved from https://writing.jan.io/2012/07/26/freifunk-statt-angst.html

Lessig, L. (2001). *The future of ideas: The fate of the commons in a connected world*. New York, NY: Random House.

Lippman, A., & Reed, D. (2003). Viral communications. M.I.T. Media Laboratory Research. Retrieved from http://dl.media.mit.edu/viral/viral.pdf

MachaWorks (2019). About the organization. Retrieved from www.machaworks.org/en/about-the-organzation.html

Martin, J. (2005, November). Wireless mesh technology: Connecting the new millennium. Ashburn, VA: IJIS Institute.

Medienanstalt Berlin-Brandenburg (2019). Freifunk. Retrieved from www.mabb.de/information/digitale-welt/freifunk.html

Medosch, A. (2008). The construction of the network commons. *Ars Electronica* blog. Retrieved from www.aec.at/en/archives/festival_archive/festival_catalogs/festival_artikel.asp?iProjectID=12949#top05

Meirath, S. (2006, November 1). CUWiN: Wirelessing the revolution with open source mesh wireless technologies. *Government Technology*. Retrieved from www.govtech.com/dc/articles/CUWiN-Wirelessing-the-Revolution-with-Open.html

Metri, N. (2017, April 7). Zenzelini – do it yourself!: How a rural community in South Africa became a telecommunication operator. The Internet Society. Retrieved from www.internetsociety.org/blog/2017/04/zenzeleni-do-it-yourself-how-a-rural-community-in-south-africa-became-a-telecommunication-operator/

MIT Communications Futures Program (2008). Viral communications. Retrieved from http://cfp.mit.edu/groups/viral-comm.shtml

Moglen, E. (2010, February 5). *Freedom in the cloud: Software, freedom, privacy, and security for Web 2.0 and cloud computing*. Speech given at the Internet Society. Retrieved from www.softwarefreedom.org/events/2010/isoc-ny/FreedomInTheCloud-transcript.html

Moran, M. (2012, April 10). From short waves to flash mobs. *Slate*. Retrieved from https://slate.com/news-and-politics/2012/04/revolutionary-technology-facebook-twitter-and-wikileaks-pose-a-challenge-to-governments-everywhere.html

Mudenda, C., & Van Stam, G. (2012). ICT training in rural Zambia: The case of LinkNet Information Technology Academy. Fourth International IEEE EAI conference on e-infrestructure and e-services for developing countries (Africomm 2012).

Mweetwa, F., & Van Stam, G. (2018). Community engagement in community networks in rural Zambia: The case of Macha Works. *Global Information Society Watch: Community Networks* (pp. 256–259). Johannesburg, South Africa: Association for Progressive Communications.

netCommons (2016, December 10). Report on existing community networks and their organization. Network infrastructure as commons project. Retrieved from http://people.ac.upc.edu/leandro/docs/d12.pdf

NYC Mesh (2019). Homepage. Retrieved from https://www.nycmesh.net/

OpenNet Initiative (2007). Pulling the plug: A technical review of the Internet shutdown in Burma. *OpenNet Initiative Bulletin*. Retrieved from https://opennet.net/sites/opennet.net/files/ONI_Bulletin_Burma_2007.pdf

Open Technology Institute (2019). About. Retrieved from www.newamerica.org/oti/about/

Orange S. A. (2018). History. Retrieved www.orange.com/sirius/histoire/en/home/

Peer-to-Peer Foundation (2011, April 16). International Summit of Community Wireless Networks. Retrieved from https://wiki.p2pfoundation.net/International_Summit_of_Community_Wireless_Networks

Peoples Open Network (2019). Development. Retrieved from https://peoplesopen.net/learn/development/

People's Open Network (2018). Press release: Sudo Mesh receives $100K donation from Handshake Foundation. Retrieved from https://peoplesopen.net/blog/press-release-sudo-mesh-receives-100k-donation-from-handshake-foundation/

Personal Telco (2019). Introducing Personal Telco. Retrived from https://personaltelco.net/wiki

Phillipa, G., Crete-Nishihata, M., Dalek, J., Goldberg, S., Senft, A., & Wiseman, G. (2015). Characterizing web censorship worldwide: Another look at the OpenNet initiative data. *ACM Transactions on the Web*, 9(1), Article 4.

PirateBox (2019). PirateBox FAQ. Retrieved from https://piratebox.cc/faq

Pishevar, S. (2011). OpenMesh Project. Retrieved from https://wiki.p2pfoundation.net/OpenMesh_Project

Poor, R. (2003, February). Wireless mesh networks. *Sensors Magazine*. Retrieved from http://archives.sensorsmag.com/articles/0203/38/

Powell, A. (2008). WiFi publics: Producing community and technology. *Information, Communication & Society*, 11(8), 1068–1088.

Priest, J. (2003, June 6). Wireless: From network to community. Paper presented at the Open Cultures Conference, Vienna.

PrimaNet (2019). About the broadband association PrimaNet. Retrieved from www.primanet.dk/index.php?side=foreningen-primanet

Reardon, Marguerite (2012, October 30). Hurricane Sandy disrupts wireless and Internet services. *CNET*. Retrieved from www.cnet.com/news/hurricane-sandy-disrupts-wireless-andInternet-services/

Red Hook Initiative (2019a). Coverage. Retrieved from https://redhookwifi.org/coverage/

Red Hook Initiative (2019b). Digital stewards. Retrieved from http://rhicenter.org/programs-2/youth-development/digital-stewards/

Rowell, L. (2007). Can mesh networks bring low-cost, wireless broadband to the masses? *netWorker*, 11(4), 26–33.

Saleh, Y., & El Madany, S. (2011, January 28). Egypt's Mubarak braced for "Friday of Wrath." Reuters. Retrieved from www.reuters.com/article/uk-egypt/egypts-mubarak-braced-for-friday-of-wrath-idUSLNE70R02B20110128

Shaffer, G. (2010). *Peering into the future: How WiFi signal sharing is impacting digital inclusion efforts* (Unpublished doctoral dissertation). Temple University.

Shirky, C. (2001). Listening to Napster. In A. Oram (Ed.), *Peer-to-peer: Harnassing the power of disruptive technologies*. Sebastopol, CA: O'Reilly.

Singer, H., Naef, E., & King, A. (2017, June). *Assessing the impact of removing regulatory barriers on next generation wireless and wireline broadband infrastructure investment* (based on FCC Form 477 data from June 2016).

Solum, L., & Chung, M. (2004). The layers principle: Internet architecture and the law. *Notre Dame Law Review*, 79(3), 815–948.

Stilgherrian, (2012, January 19). Linux.conf.au 2012: FreedomBox's privacy. Retrieved from www.zdnet.com/article/linux-conf-au-2012-freedomboxs-privacy/

Sudo Mesh (2019). Mesh. Retrieved from https://sudoroom.org/wiki/Mesh.

Telefónica (2018). History. Retrieved from www.telefonica.com/en/web/about_telefonica/history

Thomson, I. (2012). Syria cuts off Internet and mobile communications. *The Register*. Retrieved from www.theregister.co.uk/2012/11/29/syria_Internet_blackout/

Tucker, B. (2017, November 28). How a rural community built South Africa's first ISP owned and run by a cooperative. *The Conversation*. Retrieved from https://theconversation.com/how-a-rural-community-built-south-africas-first-isp-owned-and-run-by-a-cooperative-87448

United Nations Educational, Scientific and Cultural Organization (UNESCO) (2016). *World trends in freedom of expression and media development: Global report 2017/2018*. Retrieved from https://unesdoc.unesco.org/ark:/48223/pf0000261065

VestDjursNet (2019). About VestDjursNet. Retrieved from https://vestdjursnet.dk/om-vestdjursnet/

Vodafone (2018). Our technical network. Retrieved from www.vodafone.com/business/why-vodafone/our-global-network

Wireless Access Providers Association (WAPA) (2019, April 30). The rural community that built a network. Retrieved from https://wapa.org.za/node/263

WiND (2019). Network services. Retrieved from https://wind.awmn.net/?page=services&form_search_services_search=a%3A3%3A%7Bs%3A26%3A%22nodes_services__service_id%22%3Bs%3A0%3A%22%22%3Bs%3A9%3A%22nodes__id%22%3Bs%3A0%3A%22%22%3Bs%3A11%3A%22nodes__name%22%3Bs%3A0%3A%22%22%3B%7D&table_services_showpage=1

20

Commoning the Urban

Nicholas Anastasopoulos

1 Introduction

This chapter seeks to illuminate facets of the urbanized world that most of humanity now lives in, as an introduction to the alternatives to the dominant ways that it functions, that may have existed for the longest time or are being created as we speak. The urban commons are examined as a distinct domain of organizing principles, a category of its own, taking into account cultural traditions and commoning activities in an urban context. Their production, political nature and cultural identity, the dilemmas involving their governance, as well as current trends regarding the impact of creative practices, architecture, and of digital worlds are also examined, and a variety of case studies serve as a reference to these topics.

Just outside of the present dichotomy and the polarizing dilemma of having to choose between private vs. public in the sphere of politics, economy, and the everyday, the commons-based practices and environments offer convincing approaches for the management of urban resources. The private domain is defined by property and is a direct representation of the ownership's legal status and of the owner's will. It is being managed, controlled, designed, and used by individuals and protected from intrusions by various measures and enclosures (Linebaugh, 2014). At the other end of the spectrum, public space is normally controlled and administered by the state. The public domain, in turn, reflects the state's power to exercise control over various types of resources, such as territories, information, knowledge, and services. It is, by definition, the state which governs, manages, controls, designs, and sets the rules. Public spaces may function as the "receptor" of commoning activities and processes, but this does not necessarily identify public spaces as commons. Spaces and territories may become commons by changing their previous status from private or public, and on occasions from sacred to secular, through commoning acts (Agamben, 2007; Linebaugh, 2009, 2014; Negri & Hardt, 2011).

The commons may be understood as the domain of interaction between human communities and collectively managed resources of all kind – natural, territorial, material, and immaterial. Outside of the market/state duopoly, the commons are defined by a community's shared rules over usage and management of any type of territory or resource, by common property or shared vested interest, and by self-governance. The active and ongoing involvement in the sharing of and the caring for those resources are often referred to as commoning (Bollier & Helfrich, 2015).

Commoning and urban commons may be the result of many different processes and circumstances that brought them into being. Sometimes they may be the initiative of an individual, other times a group, or a spontaneous action and a haphazard moment in time. They may result from an urgent and specific need for temporary or longer-term housing for the poor, the homeless, or refugees. They might be driven by libertarian or anarchist ideologies in occupied spaces and squats where creative people live, often collectively, sharing both space and practices, and where long-honored traditions survive and exist beyond political, ideological, financial, and even class distinctions. The spatial and urban commons form a critical part of the history and origins of peer production, yet they curiously remain under-theorized and often misunderstood since they never acquired a recognized status or

The Handbook of Peer Production, First Edition. Edited by Mathieu O'Neil, Christian Pentzold, and Sophie Toupin.

wider and unchallenged acceptance (Bollier, 2014; Moss, 2014). The urban commons tend to develop, survive, and often flourish as a result of traditions, ideology, and need, meaning that their validity in many societies is venerated as a result of their duration in time, while their production and care is the result of a collective need which may be satisfied through some form of commons practices. Yet ideological views also reveal the affinity of the urban commons to social struggles such as access to housing or access to space for collective functions and open areas, such as parks. The urban commons deploy alternative forms of social relations of collective and participatory action and decision-making, and operate within alternative forms of economy, such as the circular economy, barter economy, and solidarity economy. Transactions and activities are here based on reuse, exchange, and gifts respectively. For example, in most of these cases necessary goods are donated, found, or recycled, and labor for producing or constructing objects, as well as maintenance services, are offered by participating individuals. Their resilience over time is a result of many factors. Casualness, ingenuity, conviviality, and creativity may be seen as some of the constant characteristics of the urban commons. Since the urban commons depend on collective participation and care they tend to fluctuate between periods of flourishing and neglect, and because they are most generally not institutionalized they are rarely protected, thus more vulnerable, and so may at times exist in precarious circumstances. It can be observed that various forms of urban commoning tend to be precursors of gentrification processes, thus making us aware of their volatile, nomadic, and impermanent nature.

2 Urban Commons: A Sociospatial Identity

Neither just a locality, nor a social incident, the urban commons invariably possess the dual qualities of being both spatial and social. In order to address the social aspect of commoning as well as its spatial manifestation the term *sociospatial commons* is introduced here, a term that helps illuminate the social dynamics of the commons in an inhabited urban context.

The *sociospatial commons* may be thought of as a hybrid condition, simultaneously and continuously defined both by the social and behavioral aspects of the people involved, and by the physical characteristics of the human-made environments that they occupy. The urban commons may be found in any type of space, within both public and private infrastructures, such as buildings, parks, roads, squares and so on. The coexistence of the two components – social participation and space together with the cultural expression of a community – are inextricably connected in supporting the urban commons.

The terms *Gemeinschaft* and *Gesellschaft* were introduced by German philosopher Ferdinand Tönnies to describe the distinction between two types of human groupings: *Gemeinschaft* refers to the idea of a community based on feelings of togetherness and on mutual bonds, which are cherished and almost synonymous with our understanding of community. *Gesellschaft* refers to groups that are characterized by their members' individual aims and goals (Tönnies, 1887). The distinction between *Gemeinschaft* and *Gesellschaft* corresponds to the different systems pertaining mainly to relatively stable non-urban communities that emphasized mutual obligation and trust vs. the mobile, highly urbanized societies of the 20th century, where the identity of the individual takes precedence and community fades out. The deep transformations in living and massive urbanization processes triggered by the industrial revolution heralded the emergence of *Gesellschaft* with contemporary societies adopting fully private lifestyles in the last two or three generations, imposed largely by the structure of the city itself. These are defined by an individualistic way of life, by confinement in apartments tightly packed next to, and on top of each other, and by being deprived of the privileges and joys of community and collaboration. "Commoning the urban" points to the scattered or organized efforts to recuperate aspects of *Gemeinschaft* in the urban fabric of urbanized societies.

From a societal point of view, the commons involve collective work, maintenance, decision-making, and conflict resolution techniques. Participation and care are human attributes which are dependent not on mechanistic, regulatory, and legal infrastructures, but on a voluntary active role, on community ethics, and emotions. Therefore, in order to be sustainable, collective organizations are in need of members skilled in interpersonal communication and conflict resolution, flexible and at ease in group situations and teamwork, and possessing leadership capabilities without domineering tendencies.

Their spatial identity may be defined by elements such as materials, lighting, noise level, colors, and typology, denoting the various shapes and forms that host urban commons in order to bolster and facilitate social interactions between members of a community. From the point of view of a typological taxonomy, architects, and theorists such as Christopher Alexander and Jan Gehl have been studying both the form, in other words the essential elements that a space consists of, as well as the processes through which public space can be appropriated as urban commons. They also considered the relationship between the urban commons and societal behavior, thus pointing to the often subtle and complex distinctions between *Gemeinschaft* and *Gesellschaft*, the zone in which the urban commons can flourish (Alexander, 1977, 2002; Chermayeff & Alexander, 1965; Gehl, 2011).

The urban commons should be understood as a facet of a larger worldview which encompasses all manifestations of cooperative human activity, hence their affinity to the cooperative economy, alternative education systems, and sustainable food and energy production, all of them focusing on a more human-centered ideology and a nature-friendly use of non-renewable resources and reproduction of renewable ones. Since 2005 the Transition movement has taken the concept of sharing at a whole new level, with the Transition towns' network, which encompasses existing towns and communities and their conscious gradual transitions into reducing dependency on fossil fuels and a more sustainable, communal way of living. The emphasis that the Transition movement places on self-sufficiency and resilience is based on widespread commoning practices, a cooperative economy, and peer production.[1]

The urban commons may drastically vary in scale from small urban gardening lots to metropolitan areas the size of a town. While some of them are more established than others and have developed their own methodologies and know-how, urban commons share practices and ethics.

3 Urban Commons: A Political and Cultural Identity

Urban commons are most often actively supported by people deprived of necessary resources who may, instead of pursuing them individually access them collectively. Alternatively, they are supported by politically oriented individuals inspired by communal or anarchist ideologies, who have a radical sharing vision for the city. Their political identity arises from the social struggles of the underprivileged or dispossessed and revolves around the right to the city, the right to housing, or the contested access to a resource and the claim for the collective right over it (Lefebvre, 1968). Thus, the urban commons are often marginalized as they may exist in direct confrontation with the state or the private domain in the form of squats, occupied spaces, and places of uncertain legal status. Still, the commons characterize a community or society and form part of the collective imaginary as they are rooted in the traditions, common beliefs, and values of a group of people. They are also an expression of a community's or society's social life.

Owing to their cultural identity, various people's understanding and their relationship to the commons change substantially in different cultures and at different historical periods. Former Soviet Union states and communities, for example, have a quite distinct approach towards the commons, hailing from compulsory "communing"[2] routines imposed by the communist regime, which have left an indelible mark on the generations that remember this experience. To some, this is a strong argument in favor of private living, and for rejecting anything communist and compulsory. For noted intellectuals and theorists, such as Slavoj Žižek, R-Urban, Marjetica Potrč, and many others, it serves as an experience upon which they can build a more reflective contemporary and solid critical commoning culture.[3]

Culture, history, and tradition are imbued in practices of the commons in such a way that the land at the edge of the villages, used in common by the peasants in the 16th century, is neither identical to the British commons nor to the pre-Hispanic communal regimes or to the emerging commons we have seen emerging over the past decades (Esteva, 2012).

Words in most languages exhibit a specific cultural affiliation to the commons and lend themselves to different conceptions of urban commons. Thus, the German term *Allmende* indicates a deep-rooted cultural background that indicates common ownership. As such, it has been inspiring projects and initiatives lately, all of them seeking to empower communities against neoliberal investment economics.[4] Traditional communities, such as many indigenous Andean communities, retain strong connections to the commons, thanks to their specific deep-rooted traditions. In Ecuador, for example, an ancestral custom widely known as *Minga*, indicating the active community participation in accomplishing a task which often addresses a community project, such as the maintenance of a square or a school, also permeates current forms of urban life.[5] An *ejido* (from Latin *exitum*) is an area of communal land used for agriculture in which community members have usufruct rights rather than ownership rights to the land. But contemporary urban planning parks in South America may be named *El Ejido*, referring rather to a commons-based land, than to a public park, often reflected in the daily life of the land itself (Figure 20.1).

The same may be argued for several other indigenous or traditional communities in other parts of the world, such as the Maori in New Zealand, the Amish in the United States, or communities of islanders in Greece, whose traditions often survive and permeate the urban condition.[6]

The global squatter movement is another strand of urban commoning, opening empty and derelict buildings to communal use, with emblematic cases in Europe and elsewhere (Germany, Italy, Spain, Greece, Brazil, etc.). Squats exist based on the assumption that they can be challenged and evacuated, as routinely happens at any time, yet in some countries favorable legal and cultural contexts have provided opportunities for converting them into legally recognized entities. Yet another category of urban squatting commons involves creative communities within urban settings such as the case of Teatro Vale in Rome, Macao in Milan, and the Villa Amalia, or Embros Theater in Athens, La Ingobernable in Madrid, or the artistic scenes in Berlin in the 1990s and early 2000s. A different scale of urban commons may be found in the *favelas* and shantytowns of cities in the Global South in South

Figure 20.1 A daily tradition of people engaging in dialogue that takes place in a shallow crater-like circle. El Ejido Park, Quito. 2014. Source: The author.

America, Africa, and Asia, while the Torre David vertical self-organized community in Caracas, Venezuela for a period of time served as an emblematic case of vertical self-organizing commons.

The sense of alertness and care around public space in many northern European countries could be considered an urban commoning behavioral attribute, perhaps a vestige – or even a revival – of traditions in medieval Europe, when commoning practices formed an integral part of the everyday and commons were extensively manifested both spatially and socially (Bravo & Moor, 2008). Traditional and resilient forms of commoning surviving to this day may be found in societies in most parts of the world where they form aspects of a tangible and intangible heritage of humanity. Based on this, several countries have been nurturing forms of urban commons and cooperative practices that respond to the local history, cultural dynamics, and politics in the housing sector, to the creative industries as well as public and social spaces and just about every aspect of human activity. A few examples discussed in the following section demonstrate the variety of scale and form in which the urban commons and commoning practices may appear in the urban fabric.

3.1 Urban Commons and Housing Practices

Commoning practices in the housing category are primarily represented in co-housing systems, which originated in the 1960s in Denmark.[7] Co-housing involves building typologies and models of living based on shared facilities and functions, and sharing ethics and protocols through which residents benefit from mutual support and social bonding. These include common kitchen and living spaces, other facilities such as playgrounds, laundries, exercise rooms, as well as communal gardens, and car sharing. Other common functions and services include shared meals, entertainment and celebrations, child care and elderly care, skill learning, etc. Residents do not own their apartment, rather, they own a share of the entire project, which they may sell whenever they decide to leave. Conversely, applicants do not buy or rent an apartment, but they own a percentage of their rights which therefore could be more flexible in switching spaces as needs change over time.

Germany, Switzerland, Denmark, Sweden, the United States, and eco-community experiments all over the world, including Australia and New Zealand, serve as practical references for the living arrangements that adopt commoning and which realize peer production practices in everyday life. In some cities such as Zurich, the co-housing model is not only highly sought out, but is also supported by the municipality. Kraftwerk1 is a typical example of a co-housing development which exists in Zurich and was finished in 2001. It maintains an active social life and mutual support between residents, who enjoy shared facilities and a relatively cheaper stabilized rent than the average market prices.[8]

3.2 Community Gardens and Community Empowerment

Urban gardening, pocket parks, and citizen-led projects of commoning have been gaining in popularity and numbers in recent years. Their main characteristic is the participants' active role in shaping them and in maintaining the space and the activities hosted. The cases of the Navarinou Park[9] in the neighborhood of Exarheia in Athens and Prinzessinnengarten[10] in the neighborhood of Kreuzberg in Berlin are emblematic of this trend and are discussed here in parallel. Both initiatives have similar trajectories, representing bottom-up collective projects which emerged through the reclaiming of an urban parking lot and a wasteland respectively, and experimented with active popular involvement in the renewal of neglected urban lots into neighborhood parks, where urban commoning and urban gardening were practiced through collective decision-making shaping their use, aesthetics, and world views. Both were the result of collective design, building, and governance, with the anonymous contributions of neighborhood people and citizens as well as (at times) experts such as architects, permaculturists, and botanists. An extraordinary mobilization of human resources occurred in both places and remained active and available for the period of time during which various projects would take place with a visible outcome. The Navarinou Park eventually reached a point of creative saturation: the interest in maintenance chores and participation subsequently dwindled, and as it was part of a neighborhood beset by drug problems, upkeep was never a simple daily routine (Figure 20.2).

Figure 20.2 Mobilization of people in the making of Navarinou Park. Athens, 2009. Source: The author.

Eventually, after an open call to the wider community of supporters, the painful decision was taken to fence the park off and to convert it into a playground, which is its current function. Its counterpart, Prinzessinnengarten eventually succumbed to real estate pressure and gentrification. Ten years down the line, their demise is a reminder that endeavors faced with a hostile environment, real estate speculation, erratic participation, and legal uncertainty remain vulnerable.

In 2006, a competition for the remodeling of the Plaza de la Cebada in Madrid was organized by the Mayor, Alberto Ruiz Gallardón, in the place occupied by the Barley Market and the municipal sports center. In addition to a new market and sports facilities, more green areas and more underground parking spaces were requested. The winning proposal would never be realized due to the economic and financial crisis of 2008 and in 2009 the Madrid City Council decided instead to demolish the sports center. A year later, the site was temporarily occupied as part of the cultural activities linked to the La Noche en Blanco festival, a temporary occupation of the space for various activities. French artist collective Exyzt's shallow outdoor pool, bar, and projection area triggered the involvement of associations of local neighbors, architects, urban planners and activists, and ultimately other Madrileños in a growing movement.[11] A few years later this dynamic may have paved the way for the Indignados who took over the Puerta del Sol in Madrid, in turn serving as an example which was passed on to the Syntagma square movement in Athens, and shortly after to a huge number of cities worldwide.

3.3 Former Airports as Metropolitan Commons

Former airports that have been occupying large land areas, now centrally located as cities have been expanding, have the potential to be transformed into urban commons at a metropolitan scale or a as metropolitan commons, as two cases studies demonstrate. After its shutdown, Tempelhof the former

Berlin airport which operated from 1923 until 2007 was converted into an open green space and has been used consistently as a field hosting commons-based, temporary activities, such as art exhibitions, collective urban gardening, and recreation administered by citizens of the greater Berlin area themselves.[12]

The former Mariscal-Sucre airport (in use 1960–2013) in Quito, the capital of Ecuador, sits in the center of the now expanded metropolitan area. Just three months after its closure it was transformed into "Parque Bicentenario de Quito." With a total area of 125 hectares, this new park is one of the largest in the Metropolitan District of Quito. Since its inception it has been functioning as an urban park available to citizens, and it quickly became a favorite spot for recreational activities such as walking, biking, model airplane flying, picnicking, artistic activities, and so on. In 2015 the Parque Bicentenario hosted the Pope's visit. A huge stage was installed and the vast space of the former airport was filled with hundreds and thousands of people. In a symbolic sense for believers the space was consecrated in the collective imaginary of the people and was elevated to the public sphere (Anastasopoulos, 2015). Meanwhile real estate in combination with future metro access has put high pressure on Parque Bicentenario for development plans.

While speculative discussions about the future of both sites have been raging, their enormous popularity and success became the agent for the emergence of a commons ethics among Berliners and Quiteños, propagating commoning into several more projects.

The common denominator in all these initiatives was the aim to recuperate public space as commons and as community space, and to re-introduce inclusion and positive change, action, and participation in either public space or in private abandoned or misused spaces, ultimately infusing life and empowering communities.

4 Bottom-Up vs. Top-Down, Artists and Architects Initiatives

> Change life! Change society! These ideas lose their meaning completely without producing an appropriate space. (*L'Hypothèse collaborative: Conversation avec les collectifs d'architectes français*, 2018)

Many communities and cities are being eroded today, experiencing internal devaluation and degradation. Neighborhoods, or entire cities, experience cycles of crises reflected in physical space, especially urban space, as neglect and decay take hold of dilapidated infrastructure, inner city neighborhoods, and urban brownfields.[13] What is obscured in this process is the fact that these processes result in the weakening of community as well as of infrastructure, starting a process of enclosure, privatization, and dispossession in order to have easier access to the community's coveted resources. Who should clean up, take care of a park, sidewalk, or playground? Who defines what the needs are for a neighborhood, as well as setting the priorities for the local vs. the large urban scale? How are the rules set, enforced, and administered?

Globalization, population growth, and rapid urbanization are going to be the biggest challenges of the 21st century. Scale remains the reason for the fundamentally different social organization between a community and an urban society and it harbors two major obstacles for a commons-based city structure to adopt Ostrom's (1990) practical principles for managing the urban commons.[14] Governance at a scale that exceeds the size that a community may manage remains a question and a weak point of the commons up to this day. But when we refer to an urban, regional, or larger-scale infrastructure it is still difficult to come up with a viable example other than a centralized one that may address the challenge. Therefore, it is at this level that work should be tested in terms of viability and sustainability and there are examples to which we can resort in terms of a bottom-up approach that is assisted by so-called specialists. These could be professional architects, sociologists, scientists, anthropologists or others. P2P thinkers have also been talking about the partner state, while David Harvey attempts to imagine a confederation of communities coalescing in order to take up the challenge (Bauwens, 2012; Harvey, 2000). This cross-community cooperation should make up for the lack of cohesion and the need for an institutionalized body that will have an overview over infrastructure and the necessary maintenance and governance functions that need to be undertaken.[15] Traditionally, artists tend to pioneer independent cultural centers around de-industrialized sites, abandoned buildings, and brownfields devising urban commons practices. Considerable experience seems to be gained in the cultural domain and the management of such independent centers in Europe and elsewhere.[16]

In a survey of such creative centers, Sandy Fitzgerald asks a number of critical questions about their organization and management, such as:

> . . . Is it grounded enough to distinguish between community and business endeavours, and establish criteria by which the socially more important ones are privileged, even at the cost of conflict or economic risks? Is it courageous enough to outline the policy of responsibility, which, in contrast to the restrictive policies of arranging internal relationships, is based on education and common decision-making? Is it self-reflective enough to clearly see the traps of the cultural industry, and the vicious circle of economisation of culture and culturalisation of economics? And, in light of all the aforementioned specifics, how to prevent it from becoming just some bureaucratic blot, stifling creativity and reactivity with a plentitude of orders and decrees? Where is the line that makes coexistence more human, and where does the speculative economy begin? Is the governance model conceived in a way that allows contemplation on all the above mentioned dilemmas, and the application of findings of such contemplation to practice? (Fitzgerald, 2010)

Taking a position to such questions and actively defending them, while conversely being capable of readjustment if needed, seem to be crucial elements for the sustainability and resilience of related spaces. In recent years, and roughly coinciding with the social movements that erupted as a response to the financial crisis (2008–), revealing a severe socio-political crisis looming underneath the surface (the Occupy movement and the anti-austerity plazas movement in 2011), a generation of young architects started to challenge the definition of architecture and the role of architects themselves by forming architects' collectives and by focusing on participation and participatory practices. Their ideas about the urban commons domain took center stage and ever since connections between architectural theory and practice around urban commons of various forms proliferate. People involved in this movement join forces and exchange knowledge and practices to address both the way they live themselves and understand their professional position, as well as the wider the economy and production by articulating an alternative vision for the society. Redefining their roles as catalysts of social empowerment and as agents of social change, their interventions in public space may be understood as a commoning invitation and a DIY methodology. Their actions take place in disused and abandoned urban land which they strive to endow with a sense of place turning them around as social attractors. This is achieved by means of introducing structures and urban furniture, often constructed through participatory workshops with the community, which encourage congregating, playing, interaction, or simply lingering.

We see conscientious efforts in this direction enacted in various countries and continents as bottom-up, community-led efforts, as initiatives of architects and other specialists (Ferguson, 2014; Rosa et al., 2013; Tonkiss, 2013). Notable examples abound in various cities in Europe and beyond, and case studies include Exyzt and Collectif ETC (FR), Raumlabor (GE), Zuloark (ES), and Rotor (BE),[17] just to mention a few (Figure 20.3). The R-Urban project, for instance, is an architect-led effort to create a community by the Atelier d'Architecture Autogérée (AAA)[18] based in Paris. Their work in the Parisian suburbs (*banlieues*) touches upon issues of commons and commoning which are directly associated with sustainability concerns, investigating food production, and the potential of makerspaces to operate as vehicles of community empowerment (Petrescu et al., 2016).

In recent years several examples have also flourished in Latin America as well.

In this regard, Al Borde[19] is an emblematic architect collective based in Quito, Ecuador which works with participatory methods in disadvantaged communities. It uses local and recycled or discarded materials and community labor to minimize financial cost, thus empowering communities and bypassing the official economy constraints by introducing alternative resources of materials and labor (Figure 20.4). At the heart of their work lies the intention to bring to the foreground capacities already existing within communities to produce, and to come up with creative solutions to the problems they face in a manner which has been coined as "autoconstrucción asistida" (assisted auto-construction). All of these examples experiment with various forms of peer production. They deploy architectural and social strategies for reclaiming public space as urban commons while at the same time they reinvent their internal structure, way of life, and working methods, as well as the architect's role as a consultant rather than an expert.

Figure 20.3 Le PaPoMo (Parlement Populaire Mobile): Creation of a mobile parliament by Collectif ETC for public spaces, that local communities and associations can borrow in order to organize pirate events or symposiums. Marseille, France 2015. Source: Bohn Maxence/Collectif ETC

5 Commoning the Urban: Trends, Gaps, and Opportunities

The Venice Architecture Biennale has hosted three curated shows entitled "People meet in Architecture" (2010), "Common Ground" (2012), and "Freespace" (2018) and one may think that – at least in some architectural circles – there is a trend which currently appreciates the significance of aspects of urban commoning.[20]

With the advent of the Internet and other enabling technologies, contemporary forms of experimentation around sharing indeed proliferate, as becomes evident from various emergent forms of community and hybrid areas of commons and commoning (Bollier & Helfrich, 2012; Hopkins, 2008). Larger-scale networks have also been forming around sharing concepts, where member cities adopt sharing practices, such as urban gardening, car-pooling, makerspaces, and so on. They exchange experience and know-how of principles and practices to explore alternatives to consumerism and individualism. They favor recycling or DIY as well as social, collective practices and habits. But while there is no doubt that various forms of urban commons have been experiencing a renaissance, it also seems that there are many challenges to be met and there is no doubt that there is a severe legal gap to be filled in terms of providing a framework for the urban commons to operate and to thrive.

To this end, the digital sphere can come to serve as a source of inspiration. We can therefore refer to the Internet, a seemingly chaotic sphere which nevertheless is monitored and regulated by a number of nominally independent bodies as helping to generate a new commons territory. Equally complex projects such as the human genome project which was managed by a multitude of stakeholders are examples of successful complexity management. Yet, experiences from some cities such as Bologna or Amsterdam tell a somewhat different story.

When looking for examples of long-standing traditions of urban commons combined with contemporary collaborative practices and innovative actions in the domain of urban governance, observers

Figure 20.4 Paying a visit at the Atucucho neighborhood, Quito, Ecuador, and Al Borde architects collective's intervention in the creation of a communal space with discarded concrete test tubes. August, 2015. Source: The Author.

commonly turn to the city of Bologna. LabGov – the LABoratory for the GOVernance of the City as a Commons – is an international network of theoretical, empirical and applied research platforms engaged in exploring and developing methods, policies, and projects focused on the shared and collaborative management of urban spaces and resources. [21] Co-Bologna is the main field of experimentation of the CO-Cities project, a research project aimed at developing, testing, and adapting the design principles for the governance of the commons identified by the Nobel laureate Elinor Ostrom to the urban context. In recent years, the city has dedicated a lot of efforts to the promotion of a culture of collaboration and has tried to fill a legal gap in terms of the distribution of governance and responsibilities, reducing the distance between the local administration and its citizens and bringing all urban actors to work together around the city's common good. Amsterdam has been following suit with numerous experiments,[22] and one may also cite Naples, Zurich, Berlin, Barcelona, Madrid, among others, for a variety of commoning strategies and case studies. The "right to use," as opposed to the "right to ownership," or the right to self-organize and other new legal statuses, as Christian Iaione discusses in reference to Barcelona and Bologna, are mature enough to be tested in more cities before becoming mainstream (Iaione, 2017).

Yet in cities where there is no specific legal system and substantial experience or infrastructure to act as a support and shield against the whims of hostile economies, the nature of contemporary urban commons in today's complex world makes them all the more vulnerable. In this setting, smartphones and the Internet seem to offer unprecedented options for the exchange of knowledge and communication, networking, and collective planning and action. At the same time, they have been charged with depleting the vitality of the public domain as people become increasingly absorbed in digital exchanges and neglect face-to-face contact during social occasions, even while in the company of others. Undoubtedly, the very complex role that communications and digital technologies play has both disrupted and reinforced what is happening on the ground. Major trends that took societies by storm such as the AirBnB industry, which claims to be part of a "sharing" economy, have negative side effects such as undermining neighborhoods when residents are evicted or driven out by rising housing prices (Srnicek, 2017).[23] Measures need to be taken to that end and it seems that this is beginning to happen.

The social aspects of participation, management, and various interactions include creative processes, collective activities, and rituals. Even jargon found in official policy papers like the New Urban Agenda discussing the need to foster social cohesion, inclusion, and participation could be interpreted as implicit references to various forms of commons and commoning.[24] There, commoning often becomes synonymous with participation and intervention initiatives as strategies for enacting urgently needed urban change. The urban commons do just that.

6 Possible Futures

Could we envision a world in the future governed if not entirely, then at least to a larger percentage by commons? If so, what would it depend upon and what would it look like? The theory around the tragedy of the commons (Hardin, 1968) as seen through the private sphere and the individual's point of view claimed that sustaining a commons-based resource is impossible. This hypothesis has provided fertile ground for neoliberal economics to claim that the market will come up with solutions, even in areas where it should not. But the tragedy of the commons refers to a period in history, particularly in Europe and the West, which coincides with the demise of the commons since the industrial revolution, and more specifically since the middle of the 18th century until recently. During that time, the commons were systematically oppressed, destroyed, or rejected as residues of another era, as inefficient and "backward" forms of organizing. They were dismissed as not serving the purpose of efficient development when it comes to resource management (Bravo & Moor, 2008). Based on this literature, we might conclude that the commons actually embody these contradictions found in the dichotomy between community and society, an agrarian community and city life. But through her community-based research, Elinor Ostrom's groundbreaking theory suggests otherwise and has put together a set of principles for the sustainable governance of the commons. Ostrom was able to challenge the universal validity of Hardin's hypothesis and to open up the discourse about the common's credibility in offering alternative perspectives for the future (Ostrom, 1990).

These scenarios appeal to the emergence of new commons domains, with the proliferation of immaterial commons such as open-source, peer production, and information sharing afforded by the advent of the Internet. In the first part of the 21st century many cities have had to cope with political, social, urban, financial, environmental, migratory, and refugee crises. In the same moment, new forms of community and self-organization have emerged in an effort to achieve viability and resilience through new sorts of solidarity, cooperation, collaborative production, and consumption as well as by way of excluding intermediaries, by emancipation, and genuine forms of commoning (Anastasopoulos, 2013). As a result, previously deprived practices that favored the well-being of a community and a commons ethics are being revived in the context of urban spaces, buildings, and infrastructures to generate new expressions of the commons. The global movements of recent years, from the Occupy movement to the more recent Extinction Rebellion, rely on the viral effect of an idea which ignites similar action massively in geographically disparate societies. Such actions often possess the characteristics of a utopian urban commons-based society in the making.

Yet, while the urban commons provide viable solutions to real needs, we have yet to find convincing answers that allow a balance for their coexistence of various radical ideas stemming from theoretical concepts about alternative social and economic structures within a free market economy and a capitalist and neoliberal world order. Sadly, alternative visions of models of living other than the prevailing Western model of consumerism still fall short of acquiring mainstream recognition. Nevertheless, given the right circumstances, they may be rethought as a realistic option towards a cultural shift for a much-needed change. After all, they may play a key role for survival during the looming transition humanity and the planet are facing in view of climate change, pandemics, and other life-threatening challenges. With appropriate commons-based governance and use of resources, sharing is more economical and therefore more sustainable, and living well in sensible abundance can become a reality.

Notes

1 See https://transitionnetwork.org/

2 The term "commoning" has been popularized by historian Peter Linebaugh, whose book *The Magna Carta Manifesto* shows that the founding document of Anglo-American democracy repeatedly affirms people's right to use the commons to fulfill their basic needs.

3 Slavoj Žižek is a Slovenian philosopher who spoke at Occupy Wall Street, an expression of temporary Urban commons combined with protest against financial elites and capitalism, in Zuccotti Park, NYC in 2011. R-urban is composed of Romanian architects Doina Petrescu and Constantin Petcu who work mainly in France, while Doina is Chair of Architecture and Design Activism at the School of Architecture, University of Sheffield. Among other things, their work is about generating urban commons at the peripheries of cities. Marjetica Potrč is a Slovenian architect who claims that sustainable solutions that are implemented and disseminated by communities serve to empower these communities and help create democratic structures built from below.

4 Allmende-Kontor is another notable urban gardening scheme See www.allmende-kontor.de/

5 Minga is an indigenous term that describes the process of voluntary community teamwork, where the community contributes to the common good.

6 So much so that they have been recorded in diasporic urban living in various continents.

7 Denmark has as well an extensive historical background in cooperative housing, known as the *Andelsbolig*.

8 See www.kraftwerk1.ch/

9 See https://parkingparko.espivblogs.net

10 See http://prinzessinnengarten.net

11 See https://ingobernable.net/

12 A referendum on Tempelhof's future (May 25, 2014) protects the Tempelhof Field (Tempelhofer Feld) from sale, development, and partial privatization and makes it available to the public in its entirety, without any permanent restrictions. The city is investing €100m (£865m) in the headquarters building to turn Tempelhof into Berlin's newest creative quarter in an attempt to match the cultural heritage and the public/common space aspects together through compatible and sustainable uses.

13 Brownfield is a term to describe land previously used for industrial or commercial purposes with known or suspected pollution including soil contamination due to hazardous waste.

14
1. Define clear group boundaries.
2. Match rules governing use of common goods to local needs and conditions.
3. Ensure that those affected by the rules can participate in modifying the rules.
4. Make sure the rule-making rights of community members are respected by outside authorities.
5. Develop a system, carried out by community members, for monitoring members' behavior.
6. Use graduated sanctions for rule violators.
7. Provide accessible, low-cost means for dispute resolution.
8. Build responsibility for governing the common resource in nested tiers from the lowest level up to the entire interconnected system.

15 This dubious hypothesis has been de facto turned down by Bauwens, through his disappointment over the failed experiments of both the FLOK Society Project in Ecuador (2014), and the city of Ghent (2019) in which he was directly implicated.

16 Examples of such centers are: the Ateneu Popular NouBarris in Barcelona, la Ingobernable in Madrid, the Kaapelitehdas in Helsinki, La Friche laBelle de Mai in Marseille, the Melkweg in Amsterdam, to name a few.

17 See www.collectifetc.com, https://raumlabor.net, https://zuloark.com, https://rotordb.org

18 See www.urbantactics.org/

19 See www.albordearq.com

20 In 2012, the US Venice Biennale Pavilion entitled "Spontaneous Interventions for the Common Good" highlighted 124 urban interventions initiated by architects, designers, planners, artists, and everyday citizens.

21 Since 2011, through LabGov and later through CO-Bologna as its pilot application to neighborhoods in the city, the City of Bologna has initiated a policy process to introduce collaboration as a methodology for governing the urban commons. The central regulatory tool of the Bologna Regulation on Public Collaboration for

the Urban Commons is the pact of collaboration, through which the city and citizens (informal groups, NGO's, private entities) agree on an intervention of care and regeneration of an urban commons (green space, abandoned buildings, squares). See http://labgovcity.designforcommons.org/wp-content/uploads/sites/19/bolognaregulation.pdf

22 Such as the NDSM-Werf Shipyard, an emblematic communal work space for artists and other creative professionals, envisioned by a team that joined forces around Eva de Klerk who won the competition in 2000 about the organizational concept that the city had sponsored. See www.evadeklerk.com/en/ndsm-werf/ and https://en.wikipedia.org/wiki/NDSM.

23 See http://insideairbnb.com/

24 See for example the New Urban Agenda (NUA), a policy paper produced during Habitat III under the auspices of the UN, 2016. http://habitat3.org/the-new-urban-agenda/

References

Agamben, G. (2007). *Profanations*. Cambridge, MA: MIT Press.

Alexander, C. (1977). *A pattern language: Towns, buildings, construction*. Oxford: Oxford University Press.

Alexander, C. (2002). *The nature of order: An essay on the art of building and the nature of the universe*. Berkeley: Center for Environmental Structure.

Anastasopoulos, N. (2013). The crisis and the emergence of communal experiments in Greece. In G. S. Meltzer (Ed.), *Communal pathways to sustainable living proceedings* (pp. 350–359). Forres: Findhorn Foundation.

Anastasopoulos, N. (2015, November). Ex airports as metropolitan commons: Challenges, opportunities and contradictions around three case studies. Thematic IASC Conference on Urban Commons, Bologna, Italy. Retrieved from http://hdl.handle.net/10535/9947

Bauwens, M. (2012, July 4). *Blueprint for P2P society: The partner state and ethical economy*. Retrieved from www.shareable.net/blog/blueprint-for-p2p-society-the-partner-state-ethical-economy

Bollier, D. (2014). *Think like a commoner: A short introduction to the life of the commons*. Gabriola Island, BC: New Society Publishers.

Bollier, D., & Helfrich, S. (Eds.). (2012). *The wealth of the commons: A world beyond market and state / David Bollier and Silke Helfrich, editors*. Deal: Levellers Press.

Bollier, D., & Helfrich, S. (Eds.). (2015). *Patterns of commoning*. Deal: Levellers Press.

Bravo, G., & Moor, T. D. (2008). The commons in Europe: From past to future. *International Journal of the Commons, 2*(2), 155–161.

Chermayeff, S., & Alexander, C. (1965). *Community and privacy: Toward a new architecture of humanism*. New York, NY: Doubleday.

Esteva, G. (2012). Hope from the margins. In D. Bollier & S. Helfrich (Eds.), *The wealth of the commons: A world beyond market and state*. Deal: Levellers Press. Retrieved from www.wealthofthecommons.org/essay/hope-margins

Ferguson, F. (2014). *Make_shift city: Renegotiating the urban commons* (bilingual edition). Berlin: JOVIS Verlag.

Fitzgerald, S. (2010). *New times, new models: Investigating the internal governance models and external relations of independent cultural centres in times of change*. Maribor: Pekarna magdalenske mreze.

Gehl, J. (2011). *Life between buildings: Using public space*. Washington, DC: Island Press.

Hardin, G. (1968). The tragedy of the commons. *Science, 162*(3859), 1243–1248.

Harvey, D. (2000). *Spaces of hope*. Edinburgh: Edinburgh University Press.

Hopkins, R. (2008). *The transition handbook: From oil dependency to local resilience*. Cambridge: Green.

Iaione, C. (2017). The right to the Co-City. *Italian Journal of Public Law, 1*, 104–106.

Lefebvre, H. (1968). *1968 Le droit à la ville, Paris: Anthropos* (2nd ed.). Paris: Ed. du Seuil, Collection "Points."

L'Hypothèse collaborative: conversation avec les collectifs d'architectes français. (2018). Paris: Editions Hyperville.

Linebaugh, P. (2009). *The Magna Carta manifesto: Liberties and commons for all*. Berkeley, CA: University of California Press.

Linebaugh, P. (2014). *Stop, thief! The commons, enclosures, and resistance*. Oakland, CA: PM Press.

Moss, T. (2014). Spatiality of the commons. *International Journal of the Commons, 8*, 457–471.

Negri, A., & Hardt, M. (2011). *Commonwealth*. Cambridge, MA: Belknap.

Ostrom, E. (1990). *Governing the commons: The evolution of institutions for collective action*. Cambridge: Cambridge University Press.

Petrescu, D., Petcou, C., & Baibarac, C. (2016). Co-producing commons-based resilience: Lessons from R-Urban. *Building Research & Information*. https://doi.org/10.1080/09613218.2016.1214891

Rosa, M. L., Weiland, U., & Sennett, R. (2013). *Handmade urbanism: From community initiatives to participatory models.* Berlin: Jovis.

Srnicek, N. (2017). *Platform capitalism.* Hoboken, NJ: Wiley.

Tonkiss, F. (2013). Austerity urbanism and the makeshift city. *City, 17*(3), 312–324. https://doi.org/10.1080/13604813.2013.795332

Tönnies, F. (1887). *Gemeinschaft und Gesellschaft.* Essen: Fues's Verlag.

Part V

Conflicts: Peer Production and the World

21

Peer Production and Social Change

Mathieu O'Neil and Sébastien Broca

1 Introduction

This chapter examines claims that peer production can address the social harms of neoliberal capitalism as well as the democratic failures of the centralized state. Our main concern is whether peer production, armed with "do-ocratic" principles such as participatory governance, autonomy in work, technological creativity and free access to information, is an effective tool for achieving social change.

In this chapter we distinguish "political" issues from "economic" ones. This division is of course artificial: self-governance and the commons are recursively intertwined. However, separating terms enables a clearer delineation of recurrent tensions between social change today and tomorrow, between grassroots activism and electoral politics, between the commons for capital and the commons as the germ form for a postcapitalist society. "If the cell form of capitalism is the commodity, the cellular form of a society beyond capital is the common," writes Nick Dyer-Witheford. "A commodity is a good produced for sale, a common is a good produced, or conserved, to be shared" (2007, p. 82). We agree with Dyer-Witheford: the commons transcend the commodity form. Nonetheless digital capitalism has thrived through the commodification of knowledge, but also through *the erosion of the commodity form*. One only needs to consider the free services provided to users by Google or Facebook, or the importance of free and open source software in the Internet economy. In no sense are these "commodities" in the Marxist sense: "a two-sided entity, use-value and exchange-value" (Marx, 1976). They are quite the opposite: use value and exchange value are disconnected, as firms do not extract value directly by selling software or online services enjoyed by users. In fact, many goods "produced, or conserved, to be shared" are now fully integrated in the digital economy. This raises a provocative question: what if the commons now enable the metastasizing of a *capitalism beyond the commodity form* into the heart of peer production projects?

Before developing these ideas, we address the political incarnations of peer production, understood as autonomous and participatory forms of democracy which privilege egalitarian principles, member governance, and the collective resolution of issues. We distinguish distributed and self-governed politics which take place *outside the conventional arena* of parties and elections from those which *directly engage with this conventional arena*. We address each of these forms in turn, dealing only briefly with social movements and protest actors' use of peer production: readers interested in this topic are invited to read our companion chapter on "Peer Production and Collective Action" (Milan, this volume).

2 Self-Governance Outside the Political System: Prefigurative Politics

Prefiguration signifies that ends are embodied in means: actions occurring now reflect the kind of future society sought out by those involved. We touch on social movement actors, then review how different categories of peer producers engage in prefigurative political action.

The Handbook of Peer Production, First Edition. Edited by Mathieu O'Neil, Christian Pentzold, and Sophie Toupin.

2.1 Protest: The Endless Meeting Has Begun

The distributed status of social movements arguably predates the democratization of the Internet in the mid-1990s. World system theorists suggested in the late 1980s that it was no longer necessary for global social movements to be contained by the nation-state; rather, they could be transnationally organized as a network (Arrighi, Hopkins, & Wallerstein, 1989). For Castells (2004), the Internet enabled values such as diversity, decentralization, informality, and grassroots democracy rather than centralization and hierarchy, so that information and communication technologies fit perfectly the ideological and organizational needs of social movements.

Continuing this line of argument, recent social movements have been described as having embraced peer production principles; in particular, the participatory quality observed in offline collective action led some commentators to declare that the "Occupy movement and the *Wikipartido* have explicitly modelled their organization on wikis and other peer production communities in an effort to facilitate deeper participation and democracy" (Shaw & Hill, 2014, p. 202). For a generation of activists and researchers brought up believing that the Internet favours non-hierarchical participation, it was but one step to detect a connection between the networked and leaderless protests that have erupted since 2008 – such as the Arab Spring, Occupy Wall Street, and the countless occupations of city squares around the world – and peer production.

We should however be careful not to project undue influence onto peer production. Francesca Polletta describes in her book *Freedom is an Endless Meeting* (2004) a common attribute connecting 20th-century labor, civil rights, student, feminist, and economic justice movements: participatory democracy. There is a direct thread running from Green politics to the Direct Action Network which challenged the WTO in 1999 and helped to inspire Global Justice, to Occupy Wall Street (Juris et al., 2012). Indeed, many social movements have been organized through decentralized, diffuse, and leaderless networks since the 1960s (Calhoun, 1993; Gerlach & Hine, 1970), with the French *Gilets Jaunes* movement of 2018–2019 representing a notable recent example.

Distributed and anti-authoritarian network structures can be more fair than top-down systems, and they allow more space for autonomous investments by participants. They conform to the prefigurative preeminence of tactics (how shall we best organize direct democracy during this meeting?) over strategy (how can we achieve our long-term goals?). In the age of "mass self-communication" (Castells, 2009), the shift from the DIY Independent Media Centers of the early 2000s to social media results in a loss of autonomous control over the media platforms used to coordinate and report on protests. With "connective action" whereby the network becomes the organization and participants share emotionally resonant content with others (Bennett & Segerberg, 2013), activism ends up being another digital trace to be data-mined and marketed, and we find ourselves very far from participation and transparency. Moreover, the reliance on technology can reinforce existing social divisions: in OWS there was a clear contrast between the informal, fast-paced and individualized modes of participation used by predominantly privileged actors and the more formal grassroots modes of communitarian organization favored by people of color (Juris et al., 2012).

2.2 Peer Projects

Peer production projects such as free and open source software (FOSS) and Wikipedia are also an example of prefigurative politics. Where social movement organizations aim to lessen exploitation and oppression in the offline world, participants in collaborative projects see their online activities as the means to performatively achieve change, by engaging in non-exploitative practice, such as the production or dissemination of non-proprietary commons. This kind of subjective micro-politics challenges traditional categorizations of political action, representing a type of "subactivism" which does not seek conventional political change but rather "personal empowerment, seen as the power of the subject to be the person that they want to be in accordance with his or her reflexively chosen moral and political standards" (Bakardjieva, 2009, p. 96). Traditionally subactivism in peer production has taken the shape of issues which participants agree on, such as "opposition to proprietary software or technology and associated intellectual property rights [. . .] paired with a rejection of surveillance and censorship"

(Delfanti & Söderberg, 2018), Chris Kelty (2008) has argued that these technical and legal assemblages are produced by "recursive publics," that is

> a public that is vitally concerned with the material and practical maintenance and modification of the technical, legal, practical, and conceptual means of its own existence as a public; it is a collective independent of other forms of constituted power and is capable of speaking to existing forms of power through the production of actually existing alternatives. (Kelty, 2008, p 3)

Peer projects have also been defined as "ethical-modular organizations" (O'Neil, 2015) as participants are initially motivated by intrinsic, ethical such as helping humanity, self-fulfillment, and community validation. Peer governance is "distributed" or "modular." All online ethical-modular organizations adopted the computational characteristic of being broken up into distinct autonomous components which can be developed in parallel, allowing asynchronous investments of individuals with varying competencies, who can "define the nature, extent, and timing of their participation in the project" (Benkler, 2006, p. 100).

Since ends are embodied in means, close attention is paid to participatory processes. Combining individual autonomy and the efficient coordination of action has been a challenge for anti-authoritarian groups since the 1970s, when openly debating rules in such groups was shown to be necessary in order to avoid what feminist activist Jo Freeman (1972) called the "tyranny of structurelessness": the absence of explicit rules facilitates power being monopolized by informal cliques which manipulate deliberations and decision-making. A more recent study of several hundred wikis – not including Wikipedia – found that the governance of peer projects led to self-preservation tactics (Shaw & Hill, 2014). The authors suggested that this conformed to the so-called "iron law of oligarchy" (Michels, 1915) which holds that as organizations grow in size, a small group of early entrants increasingly concentrates power in a way that runs contrary to the interests of the majority. Others had made a similar point about minority control as regards Wikipedia (Kittur et al., 2007; Ortega, 2009), leading to negative consequences in terms of justice (O'Neil, 2009) and recruitment (Halfaker et al., 2013).

However, an important distinction is that some forms of peer production are more vulnerable to these harms than others. Concentration of power occurs mostly in projects characterized by vast asymmetries in engagement and rights between participants, by the tolerance for anonymous and/or ill-informed contributions, and where the only costs of poor contribution quality are the time and effort of a limitless army of fact-checkers. In contrast when contributions are not textual and easily reverted but computational and potentially destructive, project participants must be identifiable, and known to be technically competent. Their equal rights in decisions are then guaranteed (for example, all Debian Developers are inducted through the New Member Process and have voting rights in the project) and concentrations of power accrued because of early entry are much less prevalent.

Scholars who examined FOSS communities such as Debian (O'Mahony & Ferraro, 2007) and FreeBSD (Dafermos, 2012) focused on the "succession problem," on the evolution from an informal and centralized form of legitimacy, organized around the figure of the charismatic founder, to a more decentralized and democratic mode. In a book (2009) and articles (2011a, 2011b, 2014) O'Neil also analyzed authority in self-governed, anti-authoritarian projects. He suggested that self-governance in online projects is distinct from the interest or cause-oriented collective identity found in social movements (Ackland & O'Neil, 2011; Melucci, 1995): it depends on a specific type of connection or integration between participants. *Affective* integration connects participants and an exceptional being, in which case the genius of the founder is emphasized; Linus Torvalds remains the archetype. *Democratic* integration unites a community of equals, with no distinction of class, gender, or age, in which case collectively elaborated and implemented procedures are what matter. The hacker-charismatic and procedural modes of authority corresponding to these modes of integration enable controlling actions such as rejections of contributions and exclusions of participants to be made legitimately in the eyes of the community.

In recent years evolving social attitudes, the efforts of feminist FOSS practitioners and the rise of the Maker movement have allowed a better understanding of the intersectionality of domination in FOSS (see the chapters on free and open source software by Couture and on feminist peer production by

Toupin, both in this volume). There is a simple reason for why sexism long went unchallenged in FOSS projects: sexism does not challenge the symbolic integration between participants. This integration is only contradicted when participants appear to be basing expert claims on qualifications issuing from outside the project, or as better informed about important administrative decisions than others. Authority is then perceived to reproduce offline separations and hierarchies, and participants formulate critical reminders, in line with the core principles of the hacker ethic (Levy, 1984). Legitimacy in collaborative projects thus constitutes itself in response to critique: in the realm of knowledge, the critique of external expertise forms the basis for the individualized, hacker-charismatic authority of founders who know the project inside out. In the realm of justice, the critique of opaque deliberations and decisions forms the basis for the collective, procedural authority of administrators who implement community decisions (O'Neil, 2014).

Therein lies autonomous online projects' subactivist potency. Hacker expertise is meritocratic. The only criterion is technical excellence, discussants are equal, and deliberations and critiques are public. This expertise resembles the contempt for non-experts and the cult of efficiency of dominant social actors, but it also constitutes a rejection of the technocratic elite which operates in secret and seldom lets the public interest distract it from advancing industrial and financial concerns. Online peer projects are also governed by procedural authority. Basing authority on practically demonstrated competencies and on the transparency of deliberations and judgments opens the way for a normative critique of the social order. In theory at least, the authority of the better argument and the law being applied irrespective of who one is offer a contrast to wider society, where the "great" laugh at the rules that the "small" observe. That ethical-modular projects need to increase the diversity of their membership does not detract from this achievement.

2.3 Hacktivism: The Perils of Anonymity

When participants in peer-produced online protest activity rely primarily on technology, rather than face-to-face meetings, to coordinate actions that run afoul of the law, they run the risk of seeing their network fatally disrupted by a well-placed node in the network. This is what occurred in Anonymous, where an FBI informant took down a wealth of participants in exchange for reduced sentencing (Coleman, 2014). It has been suggested that since online social movements are easily accessible for research purposes, researchers may be tempted to overstate their importance (Rucht, 2004). Some sectors of Anonymous and other groups such as Telecomix which were motivated by altruistic and progressive principles really did help people rising against dictatorial regimes in Egypt and Tunisia (Coleman, 2014). They also disrupted entities aiming to shut down the filesharing site Wikileaks, for instance through "Operation Avenge Assange" (Firer-Blaess, 2016). However other sectors of Anonymous held fast to the group's original intention of "trolling" for the "lulz," i.e., provoking "laughter at the expense or the misfortune of others" (Coleman, 2014, p. 31). They made fun of "normies," or people holding mainstream values. This ambition was greatly facilitated by the reduced social costs of espousing provocative opinions online. This resulted in Anonymous providing a fertile ground for the "ironic" rejection of progressive values, for subsequent infiltration by ultra-reactionary forces, and for eventual cross-pollination with men's rights, "alt-right," and Nazi ideology (Nagle, 2017). Clearly in liberal democracies in general, and in peer production projects in particular, public debate is greatly diminished by people not having to stand by their opinions.

2.4 Autonomous and Self-Managed Spaces

It would be tempting to call any kind of recent radical protest which does not involve hierarchical social movement organizations "peer production." For example, contemporary protest campaigns are arguably most successful when they are firmly anchored in a physical territory. This was famously the case for the "ZAD" (Zone à Défendre, or Area to Defend) in Notre-Dame-des-Landes in the French Loire-Atlantique department, where the technocratic industrial complex which wanted to build an ecocidal airport was defeated by a coalition of anarchists, environmentalists, and (some) local farmers. By opposing the damaging social and ecological impacts of a hierarchical imposition, a decision made far away by people uninvolved and for the benefit of private capital, this resistance spoke to people across the political spectrum, across ages and lifestyles (Fowler, 2017). However, this campaign lacked a crucial element that would have defined it as "peer production": distributed labor. It constituted an inverted image of fablabs and hackerspaces which do establish self-reliant and self-organizing modes of collective production but remain the province of a small minority (see Boeva & Troxler, this volume) possessing the requisite cultural, economic, and social capital. Another example of restricted participation in a concrete alternative is the Open Source Ecology project, which has made freely available the designs of its "Global Village construction set" which comprises construction plans, tools, and industrial machines necessary for the construction of a "small, sustainable civilization with modern comforts."

The building of autonomous and self-managed spaces echoes the political tradition of libertarian socialism: the ideal of a society based upon collaboration among autonomous individuals, where the role of the market and the state is downplayed. This ideal, which prompted various movements since the 19th century – workers' cooperatives, the kibbutzim, the counterculture of the 1960s, and many others – was revitalized by digital technologies' decentralized affordances. According to this line of thought, the building of a different world will not result from revolutionary actions such as "taking the Winter Palace" (the first act of the Russian Revolution). Emancipation must rather be located deep in the social fabric and consists mainly in the self-organization of civil society apart from the state. The renewal of this utopian vision is arguably what is at stake in the various contemporary forms of peer production (Broca, 2013).

Yet tricky questions remain unanswered. On the one hand, focusing on autonomous enclaves could risk removing active participants from opposition to the system, deterring them from confronting structural inequalities and global power asymmetries, thereby comforting the status quo (Fowler, 2017). On the other hand, the self-sufficiency of these enclaves is debatable to say the least. The state and the market cannot really be discarded, insofar as peer production remains "a proto-mode of production, unable to perpetuate itself on its own outside capitalism" (Bauwens & Kostakis, 2014). In other words participants in peer production projects cannot escape from society, as the vast majority must find their means of subsistence in the dominant market economy.

3 Self-Governance and Engagement in Electoral Politics

In contrast to the constitution of autonomous enclaves outside of conventional politics, a second type of peer-produced politics engages with conventional politics in the form of parties, campaigning, and presenting lists of candidates for election.

3.1 The Pirate Parties

The Pirate Party initially evolved from support for The Pirate Bay file sharing website in Sweden in 2006 around issues such as legalizing file sharing and protecting "freedom" on the Internet (Fredriksson, 2015). From there electoral platforms around digital rights, as well as political parties, were created. At the European Parliament elections in June 2009, the party obtained 7.1% of the votes in Sweden, giving the Pirates two seats in the parliament (Burkart, 2013; Erlingsson & Persson, 2011). Much like the Greens, Pirates adopted a highly decentralized structure designed to emphasize voluntary organization and relying on open access to party meetings, egalitarian decision-making processes as well as "collective forms of leadership, affirmative action, and close links to new social movements" (Miragliotta, 2015, p. 699 ff.).

Tensions arose around electoral participation and leadership. Self-destructive tendencies in the German Pirate Party were described in a 2014 article in *Die Zeit*: ingrained distrust towards traditional politics meant that all forms of political representation were viewed with suspicion, even those originating from the party itself. Pirate members elected to four state legislatures were vilified for benefiting from the system (Pham, cited in Fredriksson, 2016). In the Party's early days, the lack of formal processes leading to informal concentrations of power in the hands of cliques also came to the fore

(Fredriksson, 2015). Somewhat surprisingly, this issue resurfaced when the participatory decision-making "Liquid Democracy" model was introduced. This digital voting platform was supposed to enhance direct democracy within the party but operated much in the same way that administrative power is concentrated on Wikipedia (O'Neil, 2009): a minority of active participants uses its mastery of discursive and administrative codes to effectively concentrate power (Cammaerts, 2015).

3.2 Radical Municipalism and Civic Tech

Another form of peer-produced electoral engagement hinges around the idea that the municipality is the organizational scale best able to achieve a transition to more concretely democratic ways of living. Whereas mobilizations such as Occupy Wall Street, Nuit Debout and the Greek *aganaktismenoi* (anti-austerity) movement petered out or lost their radical edge, the Spanish anti-austerity activism of Movimiento 15M/Indignados gained power. In the May 2015 municipal elections, the Mayors of Madrid, Barcelona, Zaragoza, Valencia, A Coruña, Cadiz, Pamplona, and Santiago de Compostela were elected through "citizen platforms" (García, 2017: 463). These platforms are distinct from political parties in that they use offline neighborhood assemblies and online democracy platforms (aka "civic tech") to "decide everything from their policy agenda to their organizational structure" (Baird, 2015). *Decidim* in Barcelona and *CONSUL* in Madrid, the two most used platforms, follow the FOSS development model in that they are free to use and modify, and are frequently improved by global and local volunteers. This municipalist movement is attempting to create radically democratic-grassroots political processes whose distinctive characteristics include co-design, co-ownership, trans-local collaboration, and open access. This is meant to enable a different type of governmentality in which local activists, organizations, councils, and citizens benefit from being connected to the global commons movement and other local councils to create participatory and collaborative democratic tools (Husain et al., 2019).

Two frequent references amongst many actors in this space are North American social anarchist Murray Bookchin's (1992) proposition that a city should operate as a self-governing commune, and the Kurdish autonomist region of Rojava, also known as the Democratic Federation of Northern Syria (Arafat, 2016). According to proponents, the nationalist struggle for a Kurdish homeland has been surpassed by a municipalist struggle not only for the Kurdish identity but for the overcoming of all religious, sectarian, ethnic, economic, political, gender, national and ecological oppression (Fowler, 2017). Spanish radical municipalists lost power at the 2019 elections, but peer politics continue to play an important role in other countries such as Iceland where Birgitta Jonsdottir and Smari McCarthy's Pirate Party won seats at the Althing in the 2013, 2016, and 2017 elections.

4 The Economics of the Commons and Labor

The following section considers the economic aspects of peer production. It first explains why many radical thinkers have considered peer production as the germ of a postcapitalist future. It then critically engages with this vision, arguing that peer production projects are often intertwined with neoliberal digital capitalism, and that commoners regularly fail to address the questions raised by unpaid voluntary labor.

4.1 The Digital Commons Against Informational Capitalism

Elinor Ostrom's Nobel Prize for economics in 2009 contributed to spread the idea that there were alternatives to managing resources only through exclusive private ownership of land and knowledge (Ostrom, 1990; Ostrom & Hess, 2007). Ostrom helped to make people realize that Hardin's (1968) so-called "tragedy of the commons," whereby self-interest by participants leads to the over-use and destruction of common goods, was in fact a tragedy of ill-managed open access. The main bulk of Ostrom's work was applicable to similar commons as Hardin's pasture, that is to say finite natural resources (such as fish stocks in a river) administered by a local community. Such rival resources raise issues linked to their long-time preservation; in contrast digital commons raise issues concerning their production and enrichment (Coriat, 2011). Further, most digital commons are globally available to everyone with an Internet connection, whereas physical commons usually imply that different stakeholders are given different "bundles of rights" (Ostrom & Schlager, 1996).

Free software is the most famous example of a digital commons (Benkler, 2006). In the 1980s and 1990s, even if it was not explicitly anti-capitalist, the world of free software was positioned as external to capitalist accumulation because of the adoption of non-proprietary licenses such as the GPL. In a knowledge economy that was at the time epitomized by Microsoft, i.e., globally built around the enclosure of information through intellectual property, free software represented a dangerous alternative. It was a radical threat to the dominant logic of capital accumulation, not only in the software industry but in all economic sectors whose business models depend on strong intellectual property rights: cultural industries, pharmaceutical companies, agribusiness, and so on. Indeed, insofar as most people working on FOSS viewed it as an end in itself (Dafermos & Söderberg, 2009) and were willing to abjurate their exclusive property rights over the resource they had created, they could be portrayed as fierce critics of informational capitalism (Aigrain, 2005). These characteristics led some radical thinkers to theorize that peer production was heralding a postcapitalist society (Söderberg, 2008). André Gorz presented free software hackers as embodying "a practical negation of capitalist social relations" (2003, p. 93). McKenzie Wark described informational capitalism as the dispossession of the collaborative production of the "hacker class" by the communication channels-possessing "vectoral class" through the use of patents and copyright (Wark, 2004).

Writers in the Italian autonomist tradition were also active in this intellectual space. They described a form of cognitive capitalism where value is placed on what cannot be automated, in the capacity to resolve problems (Moulier Boutang, 2007; Virno, 1996). Intelligence, relationships, emotions, creativity are what matter. This new accumulation system is, however, rife with contradictions. The extension of intellectual property rights, on which accumulation depends, collides with the need to be able to access existing knowledge in order to build new knowledge; the "immaterial" nature of work rarely fits into the institutional logic of waged labor; and standard economic tools become unable to correctly measure the production of economic value (Moulier Boutang, 2007). In a nutshell, a productive "multitude" whose free or waged immaterial labor collectively produces immaterial products is being expropriated by capital (Hardt & Negri, 2004). All is not lost however, as "through the increasing centrality of the common in capitalist production – the production of ideas, affects, social relations, and forms of life – are emerging the conditions and weapons for a communist project. Capital, in other words, is creating its own gravediggers" (Hardt, 2010, p. 355).

In contrast to the multifaceted multitude, German activists in the Oekonux (*Linux Oekonomy*) network were clear: it is peer producers who are the vanguard actors in a historical process leading to the disappearance of capitalism and hierarchy. The non-alienated labor of workers in ethical-modular organizations foretells a society where everyone will be free to work at will (Meretz, 2012; Merten & Richardson 2001; Rigi, 2013). Oekonuxers attempted to bring together the online and the offline, the old and the new, green, and social concerns (Euler, 2016). They defined the exchange value of digital commons as close to zero as they are not a form of property or possession but are defined by their actual use in a social relationship; usage is not founded on reciprocity but contribution (Siefkes, 2009). Revisiting Marx's *Critique of the Gotha Programme* (2008) in which the new capitalism emerges from the old feudal model, Meretz (2012) described peer production as a "germ form" which needs a strong capitalism in order to develop. Common goods represent concrete examples of the proprietary logic of exclusion slowly giving ground to the commons' logic of inclusion. Though the commons depend on capital, the new logic is incompatible with the old logic; it cannot be entirely absorbed (Meretz, 2013): open source software can only be produced openly. At some point, after several stages, the new logic will supersede the old, and become dominant: network effects will start occurring, and people will stop being materially dependent on the old mode of production (Euler, 2016).

Oekonux describes a future society without coercion, money, and hierarchy unfolding almost mechanically, through the law of history. It also idealizes the power of cooperative commons and foregoes discussions of conflicts, politics, and law. Siefkes (2009) suggests that in a commons-based society "when effort is distributed, there will probably be a few tasks that nobody [...] wants to do, say because they are annoying, dirty, dangerous, or just plain boring." One way to resolve this would be to weigh such tasks higher, "i.e., to count short times of doing such a task as equivalent to longer times of doing other tasks" (Siefkes, 2009). The question of what happens if people refuse or abuse the system is not considered.

All these radical perspectives – be they that of André Gorz, *Autonomia*-inspired, or issuing from the German left – see today's peer production projects as harbingers of global social change. They often imbue these projects with a decisive political significance which is not always shared by participants themselves, whose involvement can be explained by many factors and is often oriented towards more modest goals (Broca, 2013). Radical thinkers usually define peer production as important because of its prefigurative role. Peer production projects, goes this argument, demonstrate that there is a deep antagonism between the new source of economic value – social collaboration through information technologies – and the institutions of contemporary capitalism (Moulier Boutang, 2007). The free circulation of information necessary for immaterial production is the force that will eventually displace private appropriation; this is a vision that clearly harkens back to the classic Marxist view that the conflict between productive forces and relations of production is the main driver of social change.

However, if one considers the available data on peer production and labor in the digital economy, one may doubt whether there is "an antagonism between the networked productive forces and the class relations of digital production" (Fuchs, 2020). It is not a given that, with digital technologies, labor power is reappropriating the means of production, thus becoming more autonomous and less subject to the capitalist forms of discipline and control that dominated the Fordist era (Hardt & Negri, 2018). Instead, it appears that the idea of a parasitic or extractive capitalism, simply expropriating the commons "from the outside," is overly simplistic, and that immaterial production can also be organized *within* capitalism, providing firms make certain adjustments (Dardot & Laval, 2014). Instead of discussing whether the commons are on the brink of superseding capitalism, a more relevant question might be: at what point has the threshold for capitalist firms to have thoroughly digested and retooled the commons been reached? In other words, have the digital commons mainly become commons for capital?

4.2 The Commons for Capital

The notion, popular in the 2000s, that cyberspace was a cornucopia brimming over with free or pirated content in effect created in consumers the *need* to purchase the requisite hardware and bandwidth; and it told them to do so in the name of rebelling against the power of evil corporations, who were intent on protecting their private intellectual property (O'Neil, 2006). As the Apple Computer slogan once infamously had it: "Rip. Mix. Burn. It's your music." Parallel trends can be observed in some parts of the IT sector, where it eventually dawned on firms that maintaining a death grip on their intellectual property at all costs was hindering innovation. An *informational* capitalism strongly relying on intellectual property, Microsoft's capitalism, was thus gradually superseded by a new *digital* capitalism, Google's capitalism (Broca, 2018), ushering in the new age of the so-called "Openness Ideology" (Lund & Zukerfeld, 2020) which is dominated by firms such as Google, Facebook, Red Hat, Coursera and Elsevier. Informational capitalism's profits stemmed from the commodification of information, whereas digital capitalism's innovative business model taught itself to take advantage of free and open resources. In the computing world this process began in 1998, after the Freeware summit in Palo Alto, hosted by O'Reilly & Associates, during which many key players in the free software community (Linus Torvalds, Larry Wall, Guido Van Rossum and many others) decided to use the term "open source" in order to eliminate the anti-business overtones of the term "free software" favored by Richard Stallman.[1]

The subsequent integration of peer production into the capitalist economy has taken several forms. New firms such as Red Hat began producing FOSS whilst offering legal guarantees and services. Red Hat combines waged and volunteer labor in the Fedora project (Birkinbine, 2017). Products are created at a much lower cost than a fully-waged workforce would entail, and Dafermos and Söderberg (2009) argued that a possible after-effect of this business model would be the advent of a race towards the bottom in the software industry. In fact, this downward pressure on wages and working conditions in the computer sector has not materialized. At this point it becomes necessary to reflect on whether traditional Marxist schemata still apply: firms are not appropriating FOSS code, as it is accessible to all. This is quite different from the situation of proletarians who are dispossessed from the fruits of their labor (Broca, 2013; O'Neil, 2015). In this case the exchange seems to be mutually profitable, even if the goods being swapped – economic profits for the firms, self-realization for the developers – are different. It could even be argued that Red Hat creates an environment where developers can play with passion (Lessig, 2008).

Existing IT firms such as IBM also began adopting Linux, abandoning traditional protection strategies in which value is safeguarded through copyright. FOSS has now become the "roads and bridges" (Egbahl, 2016) of the digital economy, a "digital infrastructure" that is essential for contemporary commercial activity. From the developer's perspective, participation facilitates integration into the labor market: experience in FOSS helps with recruitment by IT firms and FOSS is a source of lifelong learning and support for employees. Many firms no longer sell information goods but generate behavioral data for marketing or algorithmic intermediating between providers and consumers of goods and services (Broca, 2018). Recent examples of firm investments include Google adopting Debian as its internal operating system in 2017, Microsoft buying the FOSS GitHub repository (where 28m FOSS developers uploaded code in 2018) for US$7.5b, and IBM buying Red Hat the same year for US$34m. These striking numbers clarify the enormous economic value of open digital infrastructures (Alcaras, 2019).

Further, 85% of Linux code is now written by firm employees (Linux Kernel, 2019) and the largest block of uploads to GitHub are made by firm employees (O'Neil et al., 2020b). This type of capitalist engagement – firms paying the salaries of developers to produce FOSS – is significant, as it reaches deep into projects and raises the issue of the subsistence of peer producers (Broca, 2018; O'Neil et al., 2020a). It also challenges conventional understandings of labor (Marx, 1976). In the present case, labor appears to be hybrid, being both "alienated" since it is sold to firms, yet "not alienated" or "communal" as workers work cooperatively and democratically to produce commons (O'Neil, 2015). Maurel (2018) argues that projects such as Linux becoming enmeshed with large IT firms potentially compromises project self-governance, as employee-developers are formally subordinated to their employers. In contrast for Mansell and Berdou (2010) workers being paid by firms to contribute to the commons does not affect the "cooperative spirit" of projects; nor can these workers prevent the results of labor from being socialized into commons. Indeed, Debian Developers paid wages by firms to develop FOSS systematically professed a stronger allegiance to the project, which they had been associated with longer than any employer (O'Neil et al., 2020a).

4.3 Accounting for Unpaid Voluntary Labor

For Richard Stallman, whose opinions have historically been highly influential in this debate, FOSS produced by free labor will always be preferable to non-free software produced by decently paid waged labor: the defense of the "four freedoms" (to use, study and change, redistribute, and distribute modified versions of software) matters more than the fair distribution of profits stemming from software development (Broca, 2018). This normative stance was made very early in the history of free software. In the *GNU Manifesto*, R. Stallman thus contends that "there is nothing wrong with wanting pay for work, or seeking to maximize one's income," but only "as long as one does not use means that are destructive." Indeed, "extracting money from users of a program by restricting their use of it is destructive" (Stallman, 1985). Over the years, every time he has been asked to comment on the valuation of open source companies and the fact that they benefit from unpaid voluntary labor, R. Stallman has given the same answer: these issues are secondary. They are mainly "a distraction from what really matters: that these programs (e.g., free software) are available for everyone to use in freedom and community" (Stallman, 2018). In other words, the free software movement should consider software as *resources* upon which users have certain *rights*, not as products of a *labor* that deserves *monetary retribution*. Hence, free riding that is not based on enclosing code but on free labor is not a major concern for R. Stallman.

Yochai Benkler's (2006, 2013) near-canonical accounts of peer production similarly insist on the altruistic or passionate intrinsic motivations of peer producers, who will need to find means of subsistence outside peer production. A few voices have also risen from within the FOSS community against generating funds to compensate developers for their labor. The simplicity and elegance of FOSS code is born of necessity, goes this argument; the introduction of money means developers will start embellishing code to fill the amount raised, resulting in "consultant ware," needlessly complex software which requires expert advice to run; if intrinsic motivations are replaced by the marketplace, the "magic and beauty" of FOSS will be lost (Heinemier Hansson, 2013).

In contrast, other actors in the FOSS community have claimed that "open source has a working-for-free problem," with some launching firms with the explicit aim of providing independent open source maintainers with a reliable income (Pennington, 2019). In the same vein a candidate for the 2019 Debian Project Leader election proposed in his platform that Debian Developers should be able to make Debian their careers if they chose to, thanks to increased firm involvement and grants (Michlmayr, 2019). A few months later the issue was revived on Debian's "Project" email list where it generated a robust debate, with some Developers arguing that introducing financial rewards within the project risked distorting Debian's emergent and self-directed means of determining the relative value of project goals (debian-project, 2019).

Such perspectives acknowledge the reality of firm support for FOSS but argue that the highly distributed nature of this support, as well as the ethical norms of FOSS, guarantee project autonomy. Yet the rejection of financial rewards inside the Debian project fails to address the social stratification which results in 1.5% of FOSS developers being female, against 28% for proprietary software (David et al., 2003; Dryden, 2013). The relationship of ethical-modular organizations to existing social hierarchies is seldom evoked in terms of class, with gender often framed as the most significant source of power imbalances (Nafus, 2012). That voluntary labor in ethical-modular organizations is only accessible to a minority is clear. However, it is not only women who do not have the disposable income, cultural capital, or family support to engage in unpaid labor (Huws, 2013).

Not only does the rejection of financial rewards reproduce class and gender inequalities, it also hinders the evolution of the commons into an autonomous mode of production. If peer production is to transform society and not simply provide digital capitalism with a pool of free labor, it cannot depend only on volunteer work, but must invent institutions and metrics to compensate peer labor and measure value. This does not mean that voluntary unpaid labor should be abolished. In the foreseeable future, it will probably remain a necessary component of numerous peer production projects. But peer production thinkers and practitioners need to devise viable mechanisms to make the commons economy more robust and more autonomous, in order to challenge the structural imbalance between what digital capitalism obtains from the commons, and what it gives back.

5 Conclusion: Towards a Sustainable Contribution Society?

Neoliberal capitalism persists because of its hegemony, because of the neutralization of its politics into common sense, and because of its infiltration of the perceptions and imaginations of people who inhabit it (Fowler, 2017). Its rationality, efficiency, or ability to provide well-being do not keep it alive; capitalism survives thanks to its economic and political power and the lack of a coherent, compelling, and believable alternative and counter-narrative.

This chapter has argued that there are reasons to believe that peer production and the commons could be part of such an alternative. However social change is neither a simple effect of technological innovation, nor is it the mechanical outcome of the contradiction between productive forces and relations of production. The capacity of peer production to inform a believable and meaningful alternative thus depends on peer producers' ability to find new allies in civil society, unions, and government, as well as on their capacity to invent and keep building new practices, organizations, economies, and political strategies. These and related questions are explored in this *Handbook*'s final chapter.

Note

1 It has recently come to light that the term 'open source' was introduced by the Foresight Institute's Christine Peterson (Peterson, 2018).

References

Ackland, R., & O'Neil, M. (2011). Online collective identity: The case of the environmental movement. *Social Networks, 33*(3), 177–190.

Aigrain, P. (2005). *Cause commune*. Paris: Fayard.

Alcaras, G. (2019). *Le travail du code source – une socio-histoire de git, de 1995 à nos jours*. (Thèse de sociologie en cours), EHESS.

Arafat, H. (2016). "Rojava" no longer exists, "Northern Syria" adopted instead. *Kurdistan24*. Retrieved from www.kurdistan24.net/en/news/51940fb9-3aff-4e51-bcf8-b1629af00299/-rojava--no-longer-exists---northern-syria--adopted-instead-

Arrighi, G., Hopkins, T. K., & Wallerstein, I. (1989). *Anti-systemic movements*. London: Verso.

Baird, K.S. (2015). Rebel cities: the citizen platforms in power. *Red Pepper*. Retrieved from www.redpepper.org.uk/rebel-cities-the-citizen-platforms-in-power/

Bakardjieva, M. (2009). Subactivism: Lifeworld and politics in the age of the Internet. *The Information Society, 25*, 91–104.

Baldwin, C., & Clark, K. (2000). *Design rules: Vol. I. The power of modularity*. Cambridge, MA: MIT Press.

Bauwens. M., & Kostakis, V. (2014). From the communism of capital to capital for the commons: Towards an open co-operativism. *Triple C, 12*(1). Retrieved from www.triple-c.at/index.php/tripleC/article/view/561

Benkler, Y. (2006). *The wealth of networks: How social production transforms markets and freedom*. New Haven, CT: Yale University Press.

Benkler, Y. (2013). Practical anarchism: Peer mutualism, market power, and the fallible state. *Politics & Society, 41*(2), 213–251.

Bennett, L. (2004). Communicating global activism. In W. van de Donk, B. Loader, P. G. Nixon, & D. Rucht (Eds.), *Cyberprotest. New media, citizens and social movements* (pp. 109–128). London: Routledge.

Bennett, L., & Segerberg, A. (2013). *The logic of connective action: Digital media and the personalization of contentious politics*. New York, NY: Cambridge University Press.

Birkinbine, B. (2017). From the commons to capital: Red Hat, Inc. and the business of free software. *Journal of Peer Production, 10*. Retrieved from http://peerproduction.net/issues/issue-10-peer-production-and-work/from-the-commons-to-capital/

Boltanski, L. (2011). *On critique. A sociology of emancipation*. Cambridge: Polity.

Bookchin, M. (1992). *From urbanization to cities, toward a new politics of citizenship*. Guildford: Cassell.

Broca, S. (2013). *Utopie du logiciel libre. Du bricolage technologique à la réinvention sociale*. Neuvy-en-Champagne: Le passager clandestin.

Broca, S. (2018). Du modèle du logiciel libre au modèle productif des communs. Les licences pair à pair contre le free software? *Working Paper 9*, Projet EnCommuns.

Burkart, P. (2013). *Pirate politics*. Cambridge, MA: MIT Press.

Calhoun, C. (1993). "New Social Movements" of the early nineteenth century. *Social Science History, 17*(3), 385–427.

Cammaerts, B. (2015). Pirates on the liquid shores of liberal democracy: Movement frames of European Pirate Parties. *Javnost – The Public, 22*(1), 19–26.

Castells, M. (2004). *The power of identity. The information age: Economy, society and culture* (vol. 2, 2nd ed.). London: Blackwell.

Castells, M. (2009). *Communication power*. Oxford: Oxford University Press.

Coleman, G. (2014). *Hacker, hoaxer, whistleblower, spy: The many faces of anonymous*. London: Verso Books.

Coriat, B. (2011). From natural-resources commons to knowledge commons. Common traits and differences. WP 2011-4. Retrieved from www.mshparisnord.fr/ANR-PROPICE/documents.html

Dafermos, G. (2012). Authority in peer production: The emergence of governance in the FreeBSD project. *Journal of Peer Production, 1*. Retrieved from http://peerproduction.net/issues/issue-1/peer-reviewed-papers/authority-in-peer-production/

Dafermos, G., & Söderberg, J. (2009). The hacker movement as a continuation of labour struggle. *Capital & Class*, *33*(1), 53–72.

Dardot, P., & Laval, C. (2014). *Commun*. Paris: La Découverte.

David, P., Waterman A., & Arora, S. (2003). The free/libre/open source software survey for 2003. Stanford University. Retrieved from www.stanford.edu/group/floss-us/

debian-project (2019). Realizing good ideas with debian money. May–June. Retrieved from https://lists.debian.org/debian-project/2019/05/threads.html

Delfanti, A., & Söderberg, J. (2018). Repurposing the hacker. Three cycles of recuperation in the evolution of hacking and capitalism. *ephemera*, *18*(3), 457–476.

Dryden, A. (2013). The ethics of unpaid labor and the OSS Community, 13 November. Retrieved from www.ashedryden.com/blog/the-ethics-of-unpaid-labor-and-the-oss-community

Dyer-Witheford, N. (2007). Commonism. *Turbulence*, 1. Retrieved from http://turbulence.org.uk/turbulence-1/commonism/

Eghbal, N. (2017). *Roads and bridges: The unseen labor behind our digital infrastructure*. New York, NY: Ford Foundation.

Erlingsson, G. O., & Persson, M. (2011). The Swedish Pirate Party and the 2009 European Parliament election: Protest or issue voting? *Politics*, *31*(3), 121–128.

Euler, J. (2016). Commons-creating society: On the radical German commons discourse. *Review of Radical Political Economics*, *48*(1), 93–110.

Firer-Blaess, S. (2016). *The collective identity of anonymous: Web of meanings in a digitally enabled movement* (Unpublished doctoral dissertation). University of Uppsala. http://urn.kb.se/resolve?urn=urn:nbn:se:uu:diva-292734</div>

Fowler, K. (2017). *Tessellating dissensus: Resistance, autonomy and radical democracy – can transnational municipalism constitute a counterpower to liberate society from neoliberal capitalist hegemony? (Unpublished master's thesis)*. Schumacher College.

Fredriksson, M. (2015). The Pirate Party and the politics of communication. *International Journal of Communication*, *9*, 909–924.

Fredriksson, M. (2016). Pirate politics between protest movement and the parliament. *Ephemera*, *16*(2), 97–114.

Freeman, J. (1972). The tyranny of structurelessness. *Berkeley Journal of Sociology*, *17*, 151–165.

Fuchs, C. (2020). Communicative socialism/digital socialism. *TripleC*, *18*(1), 1–31.

García, B. (2017). New citizenship in Spain: From social cooperation to self-government. *Citizenship Studies*, *21*(4), 455–467.

Gerlach, L. R., & Hine, V.H. (1970). *People, power, change: Movements of social transformation*. Indianapolis, IN: Bobbs Merrill.

Gitlin, T. (1980). *The whole world is watching: Mass media in the making and unmaking of the new left*. Berkeley, CA: University of California Press.

Gorz, A. (2003). *L'immatériel: Connaissance, valeur et capital*. Paris: Galilée.

Halfaker, A., Geiger, R. S., Morgan, J. T., & Riedl, J. (2013). The rise and decline of an open collaboration system: How Wikipedia's reaction to popularity is causing its decline. *American Behavioral Scientist*, *57*, 664–688.

Hardin, G. (1968). The tragedy of the commons, *Science*, *162*, 1243–1248.

Hardt, M. (2010). The common in communism. *Rethinking Marxism*, *22*(3), 346–356.

Hardt, M., & Negri, T. (2004). *Multitude*. New York, NY: Random House.

Hardt, M., & Negri, T. (2018). The powers of the exploited and the social ontology of praxis. *TripleC*, *16*(2), 415–423.

Heinemier Hansson, D. (2013). The perils of mixing open source and money, 12 November. Retrieved from https://dhh.dk/2013/the-perils-of-mixing-open-source-and-money.html

Huws, U. (2013) The underpinnings of class in the digital age: Living, labour and value. *Socialist Register*, *50*, 80–107.

Husain, O. S., Franklin, A., & Roep, D. (2019). Decentralising geographies of political action. Civic tech and place-based municipalism. *Journal of Peer Production*, 13. Retrieved from http://peerproduction.net/issues/issue-13-open/peer-reviewed-papers/decentralising-geographies-of-political-action/

Juris, J. S., Ronayne, M., Shokooh-Valle, F., & Wengronowitz, R. (2012). Negotiating power and difference within the 99%. *Social Movement Studies*, *11*(3–4), 434–440.

Kelty, C. (2008). *Two bits: The cultural significance of free software*. Durham, NC: Duke University Press.

Kittur, A., Chi, E., Suh, B., & Mytkowicz, T. (2007). Power of the few vs. wisdom of the crowd: Wikipedia and the rise of the bourgeoisie. *Proceedings of Alt. CHI*. New York, NY: ACM.

Kreiss, D., Finn, M., & Turner, F. (2011). The limits of peer production: Some reminders from Max Weber for the network society. *New Media & Society, 13*, 243–259.
Lessig, L. (2008). *Remix: Making art and commerce thrive in the hybrid economy.* New York, NY: Penguin.
Levy, S. (1984). *Hackers: Heroes of the computer revolution.* New York, NY: Doubleday.
Linux Kernel Patch Statistic (2019). Retrieved form www.remword.com/kps_result/
Lund, A., & Zukerfeld, M. (2020) *Corporate capitalism's use of openness.* Basingstoke: Palgrave Macmillan.
Mansell, R., & Berdou, E. (2010). Political economy, the Internet and FL/OSS development. In J. Hunsinger, M. Allen, & L. Klastrup (Eds.), *International handbook of Internet research* (pp. 341–362). Amsterdam: Springer.
Marx, K. (1973). *Grundrisse. Harmondsworth*: Penguin.
Marx. K. (1976). *Capital: Vol. 1.* London: Pelican.
Marx, K. (2008). *Critique of the Gotha programme.* London: Wildside Press.
Maurel, L. (2018). Les communs numériques sont-ils condamnés à devenir des communs du capital? *Scinfolex*, 24 June. Retrieved from https://scinfolex.com/2018/06/24/les-communs-numeriques-sont-il-condamnes-a-devenir-des-communs-du-capital/
Melucci, A. (1995). The process of collective identity. In H. Johnston & B. Klandemans (Eds.), *Social movements and culture* (pp. 41–63). Minneapolis, MN: University of Minnesota Press.
Meretz, S. (2012). Ten patterns developed by the Oekonux movement. *Journal of Peer Production, 1.* Retrievedfromhttp://peerproduction.net/issues/issue-1/debate-societal-transformation/ten-patterns-developed-by-the-oekonux-project/
Meretz, S. (2013). *Commonismus begreifen, Oder: Die bestimmte negation des kapitalismus.* Transformationsseminar. Hiddinghausen. Retrieved from http://keimform.de/2013/commonismus-begreifen
Merten, S. & Richardson, J. (2001). Free software & GPL society. *Subsol.* Retrieved from http://subsol.c3.hu/subsol_2/contributors0/mertentext.html
Michels, R. (1915). *Political parties: A sociological study of the oligarchical tendencies of modern democracy.* New York, NY: Hearst.
Michlmayr, M. (2019). Platform. Retrieved from www.debian.org/vote/2019/platforms/tbm
Miragliotta, N. (2015). Minor organizational change in Green parties: An Australian case study. *Party Politics, 21*(5), 699–711.
Moulier Boutang, Y. (2007). *Le capitalisme cognitif: La nouvelle grande transformation.* Paris: Éditions Amsterdam.
Nafus, D. (2012). "Patches don't have gender": What is not open in open source software. *New Media & Society, 14*(4), 669–683.
Nagle, A. (2017). *Kill all normies: Online culture wars from 4chan and Tumblr to Trump and the Alt-Right.* New York, NY: Zero Books.
O'Mahony, S., & Ferraro, F. (2007). The emergence of governance in an open source community. *Academy of Management Journal, 50*, 1079–1106.
O'Neil, M. (2006). Rebels for the system? Virus writers, general intellect, cyberpunk and criminal capitalism. *Continuum: Journal of Media & Cultural Studies, 20*(2), 225–241.
O'Neil, M. (2009). *Cyberchiefs: Autonomy and authority in online tribes.* London: Pluto Press.
O'Neil, M. (2011a). Wikipedia and authority. In Lovink, G., & Tkacz, N. (Eds.), *Critical point of view reader* (pp. 309–324). Amsterdam: Institute of Network Cultures.
O'Neil, M. (2011b). The sociology of critique in Wikipedia. *Journal of Peer Production*, 0. Retrieved from http://peerproduction.net/issues/issue-0/peer-reviewed-papers/sociology-of-critique/
O'Neil, M. (2014). Hacking Weber: Legitimacy, critique and trust in peer production. *Information, Communication & Society, 17*(7), 872–888.
O'Neil, M. (2015). Labour out of control. The political economy of capitalist and ethical organizations. *Organization Studies, 36*(12), 1627–1647.
O'Neil, M., Muselli, L., Raissi, M., & Zacchiroli, S. (2020a). "Open source has won and lost the war": Justifying commercial-communal hybridisation in a FOSS project. *New Media & Society.* https://doi.org/10.1177/1461444820907022
O'Neil, M., Cai, X., Muselli, L., & Zacchiroli, S. (2020b). Mapping open source capitalism: The firm-volunteer project co-production network and its media representation. ASA *Annual Meeting*, San Francisco.
Ortega, F. (2009). *Wikipedia: A quantitative analysis* (Unpublished doctoral dissertation) Universidad Rey Juan Carlos, Madrid.
Ostrom, E. (1990). *Governing the commons: The evolution of institutions for collective action.* New York, NY: Cambridge University Press.
Ostrom, E., & Hess, C. (2007). *Understanding knowledge as a commons.* Cambridge, MA: MIT Press.

Ostrom, E., & Schlager, E. (1996). The formation of property rights. In S. Hanna, C. Folke, C., & K.-G. Mäler (Eds.), *Rights to nature* (pp. 127–156). Washington, DC: Island Press.

Pennington, H. (2019). Open source has a working-for-free problem. Retrieved from https://blog.tidelift.com/open-source-has-a-working-for-free-problem

Peterson, C. (2018). How I coined the term "open source". Feb 1. Retrieved from https://opensource.com/article/18/2/coining-term-open-source-software

Polletta, F. (2004). *Freedom is an endless meeting: Democracy in American social movements*. Chicago, IL: University of Chicago Press.

Rigi, J. (2013). Peer production and Marxian communism: Contours of a new emerging mode of production. *Capital & Class*, *37*, 397–416.

Rucht, D. (2004). The quadruple "A": Media strategies of protest movements since the 1960s. In W. van de Donk, B. Loader, P. G. Nixon, & D. Rucht (Eds.), *Cyber-protest. New media, citizens and social movements*. (pp. 25–48) London: Routledge.

Söderberg, J. (2008). *Hacking capitalism: The free and open source software movement*. London: Routledge.

Shaw, A., & Hill, B. M. (2014). Laboratories of oligarchy? How the iron law extends to peer production. *Journal of Communication*, *64*(2), 215–238.

Siefkes, C. (2009). The commons of the future: Building blocks for a commons-based society. *The Commoner*. Retrieved from www.commoner.org.uk/wp-content/uploads/2009/03/siefkes_future-commons.pdf

Stallman. R. M. (1985). *The GNU Manifesto*. Retrieved from www.gnu.org/gnu/manifesto.en.html

Stallman. R. M. (2018). Private conversation with Rob Lucas. Posted on the mailing list Nettime with permission. Retrieved from https://nettime.org/Lists-Archives/nettime-l-1811/msg00049.html

Virno, P. (1996). Notes on the general intellect. In S. Makdisi, C. Casarino, & R. Carl (Eds.), *Marxism beyond Marxism* (pp. 265–272) London: Routledge.

Wark, M. (2004). *A hacker manifesto: Version 4.0*. Retrieved from http://subsol.c3.hu/subsol_2/contributors0/warktext.html

22

Peer Production and Collective Action

Stefania Milan

1 Introduction

Over the last decade, a number of progressive social movements around the world have embraced peer-production principles such as collaboration, co-production, and self-organization. Think for example of Anonymous, the informal community of "hacktivists" engaging in disruptive online action in support of freedom of expression on and beyond the web (Coleman, 2014). Anonymous' distributed denial of service attacks and its video messages filled with references to digital and alternative culture (e.g., the Guy Fawkes mask) mobilized hundreds of people who had never met each other in person. Launching unconventional actions and online campaigns that often went viral, Anonymous showcases the power of peer-production mechanisms, and of Internet-mediated collaboration in particular. But the dynamics of peer production also characterize protest movements that are not obviously and/or primarily "digital" and yet employ peer production as a core practice and value system. The Spanish 15M, or *Indignad@s* (the outraged), is a case in point. Emerging in Madrid on May 15, 2010 – hence the name – the 15M leveraged the values and repertoires of the free culture and digital commons movement. It expressed a strong preference for "open and collaborative formats of knowledge creation" and developed "an information and knowledge policy favoring public domain and access" (Fuster Morell, 2012, pp. 386–390). Similarly, the Occupy Wall Street (#Occupy) mobilization, which sprung up in New York in 2012, popularized narrative frames (e.g., the well-known slogan "We are the 99%") later appropriated worldwide by the protest wave it inspired. #Occupy protests saw thousands of people taking over city squares to question austerity policies and call for equality. The mobilization also resulted in numerous Internet memes, that is to say "multimodal artifacts remixed by countless participants, employing popular culture for public commentary" (Milner, 2013, p. 2357). These memes played a key role in producing a "polyvocal public discourse" (2013, p. 2357) able to involve bystanders of various cultures and backgrounds, resulting in #Occupy protests in as many as 70 countries worldwide (*Occupy Directory*, n.d.). But peer-production mechanisms are increasingly popular also amongst the so-called Alt-Right, a reactionary online political movement composed of "teenage gamers, pseudonymous swastika-posting anime lovers, ironic South Park conservatives, anti-feminist pranksters, nerdish harassers and meme-making trolls" (Nagle, 2017, p. 2). The Alt-Right "culture war," too, feeds on the values of open culture that were once the flagship of the progressive left.

As these introductory examples illustrate, a variety of social actors across the political spectrum are now taking advantage of peer-production mechanisms, creating models and symbolic references for ensuing protests to appropriate. They do so for a number of reasons, including the creation of shared normative references and the development of collective action frames able to sustain long-term mobilization. Peer production also functions as a key resource to reach out to novel audiences and inspire other like-minded individuals to mobilize. By facilitating the appropriation of effective collective action frames by third parties, peer production promotes emulation. The slogans, memes, and visuals of Anonymous, the 15M and #Occupy, among others, have rapidly become part of a "repertoire of

The Handbook of Peer Production, First Edition. Edited by Mathieu O'Neil, Christian Pentzold, and Sophie Toupin.

communication" (Mattoni, 2013) shared by a number of contemporary campaigns, movements, and coalitions across the globe. On the one hand, these mobilizations have in common a faith in horizontal modes of production and the power of communal cultural references. On the other, they share the rejection of formal leadership in favor of a central role attributed to the individual as the main producer and carrier of meaning. But these traits, far from merely empowering innovative forms of collective action and peer production, may also contribute to create tensions within groups and networks.

This chapter investigates the consequences of peer production for social protest, looking at how peer production reshuffles and "remediates" (cf. Bolter & Grusin, 1998) social change activism today. It explores the convergences and tensions between peer networks and contemporary social movements ranging from informal coalitions and amorphous groupings to old-school social movement organizations. Although peer-production mechanisms today appear to unite the left and the right of the political spectrum, the chapter is illustrated with examples from progressive movements working for social change.

The chapter is structured as follows. First, it traces the historical trajectory of peer production as it has come to permeate the progressive social movements of the last three decades, linking distinct approaches to organizing to technological innovation. Second, it reflects on the so-called social affordances (and constraints) of digital infrastructure and their role in fostering specific modes of creativity and convergence apt to support protest actors. Third, it explores three types of consequences of peer production for social movements, namely cultural production and norm change, collective identity, and the commons. The chapter then examines three tensions that might emerge in the process of embedding peer-production mechanisms and values in instances of collective action, namely: individual vs. collective engagements, peer networks vs. social movement organizations, and self-organized vs. commercial infrastructure.

2 Peer Production in Context: A Brief History

With the advent of Internet-enabled products and services, social movements have undergone major changes. The engagement with peer production has transformed, too. We can distinguish at least three stages in the evolution of organized collective action since the 1990s to the present. Each stage is characterized by a specific ideal type of organizational form ***modeled after*** – that is, reflecting the features and modes of interaction typical of – the digital infrastructure available at each time, with distinct consequences on peer production (see also Milan, 2015a). It is worth noting that the boundaries between the stages and ideal types are porous and artificial, set by the researcher to capture complex socio-political and techno-cultural developments. However, thinking in stages and ideal types can help us to make sense of the role of the digital milieu in "constructing" specific modes of peer production. It illustrates how digital technology often provides not only the digital backbone for people to organize, but also specific imaginaries of empowerment able to stimulate the emergence of specific social structures at the grassroots level (Milan, 2013).

The first stage we can identify roughly covers the period between the 1990s and the mid-2000s and is characterized by the diffusion of the World Wide Web – the "network of networks" (Berners-Lee, 1999). Although horizontal participatory politics pre-existed the advent of the Internet as we know it (see also O'Neil & Broca, this volume), the World Wide Web supported the proliferation of informal groupings, networks, and global coalitions mimicking the features of their digital backbone – thus distributed, rooted in nodes which are relatively independent from each other, and borderless. It inspired progressive social movements to prefer horizontality and "leaderless" participation over the traditional hierarchical ways of organizing (Juris, 2008). The global justice movement, which thrived in Western democracies since the late-1990s, is a paradigmatic example of these informal networks (Della Porta, 2007). Nicknamed "the movement of movements" with a clear reference to the "network of networks," it advocated for social justice and organized around a variety of local nodes that were independent yet connected to each other also across national borders (Tarrow, 2005). Composed of both traditional social movement organizations and informal groupings and networks, the movement gave rise to a number of epic street protests. It managed to shut down the World Trade Organization (WTO) summit in Seattle (1999) and mobilized against the meetings of inter-governmental forums like the Group of Eight (G8), considered the driving force as well as the epitome of the neoliberal globalization hindering social justice globally. Interestingly, the cultural and symbolic production of the global justice movement unfolded through a plethora of news-oriented peer-production projects such as the Independent Media Center, or Indymedia. Nicknamed "the mother of all blogs" by virtue of its innovative "open publishing" software (Milan, 2010, p. 89), Indymedia allowed street protesters to partake for the first time in news production in real time, bypassing the gatekeeping role of journalists and webmasters. Thanks to ready-to-use free and open source software (FOSS) enabling collaboration, an editorial process easy to copycat, and the attractive collective identity of the "media activist," Indymedia nodes mushroomed rapidly since its appearance at the Seattle anti-WTO protests: three years after its foundation, the network already counted 89 nodes distributed in six continents (Kidd, 2003).

In the second stage, spanning approximately the mid-2000s to the mid-2010s, the introduction and swift popularity of social networking services like Facebook and Twitter promoted a more central role of the individual in collective action (Juris, 2012). Non-mediated and "personalized" action was then saluted as one of the main outcomes of the diffusion of social media (Bennett & Segerberg, 2013). Social media made the involvement of individuals possible without the filter of traditional movement organizations, as showcased by, among others, #Occupy. For the first time, this concerned not only the ability to customize individual participation, but also to directly feed and shape the movement's cultural production (e.g., collective action frames). Individuals were able to produce and share on social media meanings associated with collective action and social change, from video footage to hashtags. In this way, "the soft resources necessary for collective action become broadly and readily available to potential and engaged participants alike, and know-how necessary for mobilization 'travels' across borders and protest waves" (Milan, 2015b, p. 894). In addition, activists could easily customize their involvement in collective action and the narrative associated with it (2015b). Leveraging the collaborative features of social media, novel peer-production communities organized around the principle of "open collaboration" and engaged in "crowd-enabled action," whereby participants jointly developed, reproduced, and customized "action frames and participatory ethos (e.g., 'we are the 99%')" (Bennett et al., 2014, p. 233). The specific "structuring conditions" of digital contexts allowed for a "modular and granular" involvement of distinct individuals, resulting in self-selection and decentralization as core organizing principles (2014, p. 233). These forms of collective action exacerbated the centrality of the subjective and private experience of the individual, emphasizing the role of shared experience and the specific "grammars of embodiment" (McDonald, 2006, p. 37) of social media.

The third stage identified for the purposes of our analysis began approximately in the mid-2010s and is still ongoing. Alongside with the ever-rising popularity of social media, this stage is characterized by the diffusion of instant messaging smartphone applications such as WhatsApp, FireChat, and Telegram, which became the main channel for the production and mass circulation of peer-produced protest-related content. Messaging services support microcoordination (Ling & Lai, 2016) and enable a specific form of sociality characterized by a "felt-life of being together" which results from the "togetherness and intimacy [. . .] enacted through small, continuous traces of narrative, of tellings and tidbits, noticings and thoughts, shared images and lingering pauses" (O'Hara et al., 2014, p. 1131). Accessible via a deeply intimate technology like the smartphone (cf. Farman, 2012), instant messaging has brought personalized action to the next level (Milan & Barbosa, 2020). Facilitating a constant, always-on engagement and fostering novel "relationship 'doings'" (O'Hara et al., 2014, p. 1131), it has furthered what we might call the "mundane-ization of politics." Inaugurated by social media but exponentially bolstered by chat apps, the mundane-ization of politics stems from content of a political nature becoming immersed in and part of ordinary exchanges with peers and family, e.g., being served to users alongside leisure-related content. In fact, private group chats have become the preferred channel for sharing peer-produced and/or alternative content of a political nature, especially in countries where social networks are used to surveil and curb activism, such as Iran (Alimardani & Milan, 2018). Videos, hashtags, and memes related to collective action flow through group chats in near-real time following a live streaming logic. The *#vomitaço* protest, which targeted the interim president of Brazil Michel Temer (2016–2019), is an example of peer-produced creative protest circulating on chat apps. the The *#vomitaço* (Portuguese for vomiting) action consisted in multiple users simultaneously posting a "vomit" emoji to signal their disgust towards Temer's inauguration as president and his tenure. "How to" files detailing how to start a *#vomitaço* were frequently circulated in chat channels as a call to action (Milan & Barbosa, 2020). They have an expressive and a mobilizing function, and contribute to coalesce flexible, multi-headed communities of interest based on emotion- and content sharing. They simultaneously support group-making activities and individual self-directed action, including one-to-one exchanges (by means of, e.g., the "direct reply" function allowing users in a group to engage directly with one another).

Far from arguing that "[e]very major display of social unrest these days seems to come with a game-changing technological accompaniment" (Bland, 2014), this brief historical perspective did aim to show how peer production is deeply engrained in technological innovation. Not only does it evolve at pace, but it also takes inspiration from the very same design of its supportive technology. In the next section, we look at what it is, in digital services and their contexts, that makes peer production possible and/or prominent.

3 Enabling Infrastructure and its Social Affordances

We can understand World Wide Web, social media, and instant messaging applications as the enabling infrastructure supporting the diffusion of protest-oriented peer production, its mechanisms, and values. This enabling infrastructure embodies its own specific "social affordances" – in other words, the properties of the digital infrastructure are conducive to a certain range of social actions (Bucher & Helmond, 2017). If we look at social media and instant messaging applications, for example, these properties include but are not limited to: the immediacy of communications; the intimacy of a device of quotidian and personal use such as the smartphone; the effortlessness of familiar, targeted content as served to each user by personalization algorithms, and the ease of forging ties with others (e.g., "friends" and "followers").

Interestingly for our analysis, these "mechanisms built into platforms," including "filtering, recommendation, and reputation systems," help users to "navigate the many challenges of concerted action in conventional peer-production contexts" (Bennett et al., 2014, p. 235). For example, technological innovation has contributed to lower the costs of participation and mobilization by empowering node-to-node exchanges and fostering connections between individuals regardless of their location (see, e.g., Shirky, 2008). But it has also consequences when it comes to the cultural and normative production of protest movements. Leveraging the affordances of the hashtag, for example, allows users to connect (and facilitate the retrieval of) discourses and repertoires of action, forging complex narratives and assembling novel, composite publics worldwide.

Taking a closer look at #Occupy, we can see how social media have offered protesters not only the backbone to novel ways of social organizing that take peer production to heart, but also the "imaginaries" (Wyatt, 2004) upon which to mold collection action. Activists, in fact, have been found to "echo" and reproduce familiar modes of interaction propagated by the digital services they use most, namely social media platforms. Think of the "human microphone" pioneered on a large scale by #Occupy activists worldwide: in the eventuality of a lack of sound amplifiers, the individuals located closer to speakers addressing a large group would repeat their sentences one by one, to reach those sitting further away from the center of action (Costanza-Chock, 2012). This mechanism, widely adopted in the so-called "general assemblies" of the movement's local chapters, allowed attendees of what was otherwise a "dispersed and individualized constituency" (Gerbaudo, 2012, p. 5) to afford visibility and have a say in group decisions. But they also reiterated the "reverberating" mechanism, emotion-laden exchange, and multi-author discourse so characteristic of social media (Milan, 2015a). These imaginaries of grassroots organizing typically travel well across geographies and cultural preferences as they build on the shared experience of commercial social media services. Think of the 2014 Hong Kong protest wave that became known as the "umbrella movement," where social media worked as a catalyzing agent (Agur & Frisch, 2019).

However, while empowering (a certain breed of) collective action, the digital infrastructure enabling protest today might also impose specific limitations on users. Personalization algorithms on social media platforms, to name just one shortcoming, present users only with select content based on their past behavior on the platform (cf. Pasquale, 2015), allegedly contributing to envelop users in a "filter bubble" (Pariser, 2011). The business model specific to social media platforms – constructed to compete for user attention, time and monetizable "clicks" on ads and "sponsored content" – influences the design of user interactions (van Dijck et al., 2018). In other words, every such infrastructure embodies specific "politics of the platform" (Gillespie, 2010, p. 347) able to shape user possibilities by structuring modes of interaction and relationships, with both positive and negative consequences. It is in this framework – rigid and scripted yet described as empowering and liberating (Diamond, 2010) – that peer production emerges today, and is successfully appropriated for social protest. Not without tensions, as we shall see.

4 The Consequences of Peer Production for Collective Action

With its emphasis on the non-commercial and non-proprietary mechanisms supporting shared production, the notion of peer production is imbued with potentialities for social change (Benkler, 2006). We can identify three types of consequences of peer production for collective action, namely: on cultural production, on identity-making, and on protest replicability and diffusion.

4.1 Cultural Production and Norm Change

According to Melucci (1996), collective action is rooted on the combination of shared cultural production, shared meanings and, as we will see, a shared collective identity. Shared cultural production today is primarily made visible in the array of social media platforms and group chat apps that contribute to spread and popularize digital content, allowing many to "share" and "like" the utterances of their "friends" or to express a range of emotions such as surprise, rage or distaste. On these services, content travels very rapidly, with a high potential of "going viral." But sharing on the web is typically also a gateway to novel content, as other individuals and groups might appropriate pictures, footage, and text to repurpose them at will, in an array of creative "mash-ups" (Lievrouw, 2011).

This variably selfless cultural production results in new "codes" sustaining collective action (Melucci, 1996) – that is to say, novel ways of thinking about and "feeling out" a problem. It can help to turn a concern shared by only a subset of the population into the concern of many, contributing to changing the public perception of an issue and even influencing agenda setting at the governmental level. The mobilization of hundreds of thousands of youngsters worldwide, seeking with the #Fridays4Future protests and civil disobedience actions to draw public attention to the climate emergency, is a case in point.

Social movements' cultural production is a key ingredient of norm change (Rochon, 1998). Norm change refers to a significant shift in the way a certain problem is perceived by the population and/or policymakers. It is one of the primary long-term goals of social movements (Finnemore & Sikkink, 1998). Think of historical examples like the civil rights, anti-apartheid, and "Black" movement, lately evoked by the Black Lives Matter campaign (Carney, 2016). Far from remaining confined to the online public sphere, norm change facilitated by peer production is able to fuel and sustain offline action (Freelon et al., 2016), with real-life consequences, including policy decisions.

The features of peer production that most contribute to cultural production and norm change are the cooperative discursive interventions, the low cost of participation, the emphasis on non-commercial and non-proprietary content that enable repurposing, and the appeal to participation that peer-produced mechanisms harbor. Combined with the social affordances of platforms, such as the hashtag, these features empower peer production to forge "affective publics" activated by feelings of belonging and solidarity (Papacharissi, 2015) that become the driving force behind norm change. The #metoo hashtag, for instance, sought to make visible misogyny and sexual harassment in a variety of sectors, from the movie industry onwards (Mendes et al., 2018). It was used to harness public attention across national borders. Emotions and affect played a crucial role in this process, for two reasons: because they become constitutive of publics coalesced around values (see also Kelty, 2008), and even more so because they appeal to the personal sphere of individuals as well as highly contested issues.

But what makes the #metoo mobilization so representative of the power of peer production as it encounters social media is the ability of today's digital infrastructure to knit together the personal and the collective. As Clark observed, "a hashtag's narrative logic – its ability to produce and connect individual stories – fuels its political growth. The online telling and connecting of personal stories distinguish hashtag feminism from earlier forms of feminist personal politics" (Clark, 2016, p. 789). The weaving of the personal and the collective on the one hand, and the personal and the political on the other, promotes "empowerment through empathy" (Suk et al., 2019) – that is to say, allows others, including potential victims, to recognize themselves in the situation, and in the collective actor sustaining the mobilization. The big numbers associated with social media-facilitated peer production in turn generate discursive change. "[O]rganizing movements on the basis of empathy" (Rodino-Colocino, 2018, p. 96) allows participants to simultaneously appeal to bystanders to turn a problem into a contentious issue that resonates with others, and to challenge the status quo.

In using hashtags strategically, activists promote their values and their vision of reality, to the extent of changing the way people perceive(d) sexual harassment and rape in society – all the way to taking down offenders and changing established social norms (e.g., racial segregation, rape culture) and even legislation. However, the short life span and the extraordinary volatility of peer-produced content as it travels on social media and messaging apps might hinder the activists' ability to sustain long-term mobilization, for which a shared collective identity is key, as we analyze next.

4.2 Collective Identity

Today, peer culture constitutes one of the building blocks of collective organizing, and of collective emotional experiences and collective identity in particular. The question of collective identity is acknowledged to be at the core of organizing and of the ability of collective action to survive the test of time (Flesher Fominaya, 2010). Collective identity is the process through which distinct social actors recognize each other as part of the same struggle and develop emotional attachment to their fellows (Polletta & Jasper, 2001). Distinct individuals "dissolve" into a group and learn to identify themselves in a real or imagined "we-ness" that stands in for collective action (Snow, 2001). Identity-building occurs at the interplay of the individual and the collective dimension of action. It is an "interactive and shared" process in which "elements are constructed and negotiated through a recurrent process of activation of the relations that bind actors together" (Melucci, 1996, p. 70). While there is no agreement on the relevance of collective identity in the age of personalized activism (see, e.g., Bennett & Segerberg, 2013), collective identity is indubitably a key building block of concerted political activism (Gerbaudo & Treré, 2015).

Peer production supports the creation and sustainability over time of a movement's shared identity. The features of peer production that most enhance the development of a successful collective identity are its horizontal relationships and social structures and the malleability of the discursive production, whereby content is easy to adapt to one's own perspective and preferences. We can see these mechanisms at work in the adoption of protest avatars, or digital images that act as collective symbols for a movement (Gerbaudo, 2015). In the aftermath of the terrorist attack on the French satirical magazine Charlie Hebdo in January 2015 and of the wave of attacks in Parisian bars and concert venues in November 2015, social media users changed *en masse* their social media profile picture adopting the colors of the French flag. Profile pictures thus work as "memetic signifiers": vague and inclusive, they easily go viral, allowing participants to "experience a collective fusion in an online crowd" (Gerbaudo, 2015, p. 916). Furthermore, the associated hashtag #JeSuisCharlie was a three-word identity statement in its own right. It leveraged the "affective intensity of the here-and-now" which "creates dividing lines of evaluation and identification" (Giaxoglou, 2018, p. 13). By deploying #JeSuisCharlie, the user became a "teller," as opposed to a passive spectator of events (Giaxoglou, 2018, p. 18), witnessing and taking sides, while signaling others what side to be on.

These forms of identification into a collective actor are particularly transient and might not necessarily coalesce into a social movement. But they promote collective identification by "frame alignment," whereby a specific social change narrative put forward by a movement resonates with the lived experiences or values of bystanders (Snow & Benford, 1988). This might result in real-life actions as in the case of the #NeverAgain/#MarchForOurLives student campaign that followed the mass shooting in a high school in Parkland, Florida in 2018 (Maxwell, 2019). Giving voice to emotions and allowing recognition of like-minded individuals, social media and group chat apps signal interest in the subject matter and tell individuals that "they are not alone" in the face of a situation of distress. In a nutshell, social media allow the rapid communication and reinforcement of a sense of belonging, boosting confidence. A successful collective identity is in fact rooted in shared emotions (Goodwin et al., 2001), which speaks to the importance of "moments of digital enthusiasm" expedited by, e.g., Facebook pages during the Arab Springs protests (Gerbaudo, 2016).

4.3 Protest Replicability and Diffusion

One of the most recurrent features of recent social movements is the replicability of collective action to a vast array of distinct socio-cultural settings, showcased by the #Fridays4Future protests. In an era in which co-presence is no longer a *sine qua non* condition of joint action (Earl & Kimport, 2011), movements are known to diffuse transnationally when their instances, claims, and demands resonate with other populations in distinct localities (Bennett et al., 2014; Kramer et al., 2014; Tarrow, 2005).

The peer-production mechanisms that most contribute to the replicability of social action and thus protest diffusion across national and socio-cultural boundaries are the availability of self-selecting content that exists in the realm of the commons, and the ease of retrieving and adapting this content to local contexts and preferences. When slogans and memes, but also action repertoires travel rapidly on social networking services, crossing linguistic barriers, they can be appropriated by other concerned individuals, who can easily adapt them to their local contexts. It is the case of the encampments and the slogans promoted by #Occupy, which inspired a global wave of protest, with protest sites in over 1,500 locations worldwide (*Occupy Directory*, n.d.). However, typically the diffusion is not as seamless as it might appear, and cultural resonance is the result of "laborious dynamics of cultural translation and domestication" (Gerbaudo, 2013) seeking to bring a narrative closer to its audience.

The following section explores three important tensions that can originate when peer production is mobilized for social change purposes.

5 Three Unresolved Tensions

We have reviewed the contribution of peer-production mechanisms and values to progressive social movements – from norm change to the replicability of collective action across distinct social-cultural geographies. But this appropriation is not without friction. We can distinguish at least three points of pressure, which are detailed next.

5.1 Individual vs. Collective

The advent of social media has promoted horizontal leadership, whereby traditional social movement organizations have been supplanted by individuals acting as "soft choreographers" of collective action (Gerbaudo, 2012). Personalized communication has resulted in personalized action frames (Bennett, 2012), and flexible identities are now built around minimum common denominators which have visibility at the core (Milan, 2015b).

Marshaling the individual to the core of collective action, however, has at least two potential negative consequences on group dynamics. First, the resulting dispersed but "leaderful" (Costanza-Chock, 2012) groupings are typically characterized by influential individuals such as Facebook Page or WhatsApp group admins. These rapidly emerge as "soft leaders" (Gerbaudo, 2012) who define the boundaries of the collective experience in ways that are less than transparent and have to do with visibility and high availability on social media. These dynamics are made more complicated by the very architecture of platforms, which contribute to forge "connective leaders" (Poell et al., 2016). Second, some individuals are afforded much higher levels of visibility on social media than the majority of users, in an entanglement of fame, influence, and agency which is typical of social media (Marshall, 2019). Celebrities and influencers might enjoy a disproportionate power in setting the discourse. As research on the 15M illustrated, "better connected users, that is, the broadcasters or celebrities acting as the hubs of the network, led the diffusion process, triggering a snowball effect that quickly reached global proportions thanks to their larger personal networks" (González-Bailón et al., 2013, p. 952). Finally, the over-reliance on self-standing individuals does not impose the same degree of responsibility towards fellow activists that real-life groups do, which might have "dramatic effects on [the] duration, sustainability, identity production" of social movements, and "on how robust the political trajectory can become" (Leistert, 2013).

5.2 Peer Networks vs. Social Movement Organizations

The preference for horizontality, leaderless participation, and non-mediated individual action might create tensions within action networks. Historically, social movements have been characterized by the presence of social movement organizations acting as "entrepreneurs" of collective actions and identities (McCarthy & Zald, 1977). The Internet first, and social network services subsequently, have dramatically downsized the role of these traditional entrepreneurs in masterminding protest events and networks. With the complicity of social media services, peer networks have made even resourceful nongovernmental organizations lose traction and relevance. As a result, nongovernmental organization have adapted to a role of service providers and/or evidence producers, while trying to stay afloat in the discursive flows of social media.

These tensions are often unspoken, but they bear consequences for the social movement ecology. First, the notion that movements today are (allegedly) leaderless does not sit well with news media, which seeks representatives to operate as heroes or villains – think of Greta Thunberg, the Swedish youngster named *Time*'s 2019 Person of the Year for her vocal role in the climate change debate. The hyper-identification of a discourse with one individual may contribute to over-simplify complex issues and homogenize the discourse. It may also expose the (unwilling) spokesperson to mockery and attacks aimed at downplaying the goals of the movement as a whole. Second, though the lack of formal structure facilitates "connective action" whereby the network becomes the organization, it may also cause burnout, and hinder the movement in the long run (Cox, 2010). Third, the amorphous shape of today's movement networks may facilitate infiltration by informants and saboteurs, with potentially high costs to movement viability.

5.3 Self-Organized vs. Commercial Infrastructure

One last friction that emerges in the appropriation of peer production for protest and collective action concerns the key supporting role of digital infrastructure. Curiously, while recent history is awash with examples of small-scale to mid-range infrastructure such as wireless community networks translating into practice the ideals of peer production (see, e.g., Crabu et al. 2015; Csikszentmihalyi et al. 2018; De Filippi & Tréguer 2014; Maxigas 2012; Musiani 2016; Shaffer, this volume; Trudel & Tréguer 2016), the question of *self-organized* infrastructure appears to be consistently overlooked by contemporary progressive activists, whose critical attitude is not applied to commercial digital infrastructure of the likes of Facebook and Google.

While digital infrastructure is indeed a resource of utmost value and running independent infrastructures is costly and requires a certain degree of technical expertise, relying exclusively on commercial platforms is known to have chilling effects on free speech (Penney, 2017) and to expose activists to surveillance and repression (Duffy & Chan, 2018). In addition, the operating values of commercial platforms – centralized and profit-driven – have little to do with "the direct, participatory, collective and autonomous nature of the emerging social movement(s)" (Milan, 2010, pp. 88–89) and even less with peer-production principles. Although a number of scandals have recently brought under the spotlight the role of social media and messaging app services and service providers in facilitating large-scale surveillance (Andrejevic, 2012; Nadir, 2012, p. 201; Trottier & Lyon, 2012) as well as behavioral change through microtargeting (Borgesius et al., 2018), many progressive activists, including those sensitive to the commons and/or autonomous action, appear to ignore this tension. Others are acutely aware of it but are unable to act upon these concerns; others again prefer to mobilize where their potential audiences are, hence on commercial platforms. All in all, the drive to self-determination that characterized, among others, Indymedia, seems to have lost traction, with potentially negative consequences for peer-production communities.

6 Conclusions

Technological innovation has radically altered the dynamics of organized collective action, promoting novel ways to engage in peer production and to mobilize peer production for social change. And we cannot imagine social change today without evoking the collaboration of many distinct, self-directed individuals contributing to the collective production and dissemination of digital artifacts of various kinds – from symbols to slogans, from policy documents to video footage of protest events. This chapter has analyzed the consequences of peer production for contemporary progressive activists and networks. It distinguished three stages in the evolution of organized collective action over the last three decades, each characterized by specific organizational modes modeled after the available digital infrastructure. It posited digital infrastructure as, on the one hand, the enabling infrastructure of collective action, with its specific social affordances and constraints, and on the other hand, as a "metaphor" or the organizing model inspiring specific forms of coming together around agendas of a political nature. It identified three realms in which peer production exerts its influence, namely: cultural production and norm change, collective identity, and protest replicability and diffusion. Finally, it reflected on three lingering, unresolved tensions: individual vs. collective, peer networks vs. social movement organizations, and self-organized vs. commercial infrastructure. These tensions, often unspoken, are likely to linger on for the years to come, but they also offer possibilities for activists to creatively navigate them and strategically leverage some of these points of pressure in support of their causes.

In conclusion, what does the future yield for peer production and progressive collective action? As we have seen in the cases of the World Wide Web, social media, and messaging apps, technological innovation is likely to be appropriated by grassroots movements to further their agenda, organize, and pursue new audiences. It will certainly harbor new possibilities for innovative protests: it is worth asking, for instance, how the Internet of Things might change street protests as well as the associated cultural and normative production (see, e.g., Puri, 2017). But technological innovation might also turn out to harbor novel Achilles heels for social movements – see, for example, the surveillance potential of Artificial Intelligence applications like facial recognition, as feared by Hong Kong protesters who in summer 2019 took down the "smart lampposts" allegedly spying on them (Binder, 2019). What we know for now is that the creative possibilities unleashed by peer production are here to stay. They are likely to contribute to foster global action networks able to alter the way we think of the complex issues of our time and might even empower protesters to reach the control rooms of decision-makers.

References

Agur, C., & Frisch, N. (2019). Digital disobedience and the limits of persuasion: Social media activism in Hong Kong's 2014 Umbrella Movement. *Social Media + Society*, *5*(1). https://doi.org/10.1177/2056305119827002

Alimardani, M., & Milan, S. (2018). The Internet as a global/local site of contestation: The case of Iran. In *Global Cultures of Contestation* (pp. 171–192). New York, NY: Springer.

Andrejevic, M. (2012). Ubiquitous surveillance. In D. Lyon, K. D. Haggerty, & K. S. Ball (Eds.), *Routledge Handbook of Surveillance Studies* (pp. 91–98). Abingdon: Routledge.

Benkler, Y. (2006). *The wealth of networks: How social production transforms markets and freedom*. New Haven, CT: Yale University Press.

Bennett, L. W. (2012). The personalization of politics: Identity, social media, and changing patterns of participation. *The Annals of the American Academy*, *644*, 20–39.

Bennett, L. W., & Segerberg, A. (2013). *The logic of connective action digital media and the personalization of contentious politics*. Cambridge: Cambridge University Press.

Bennett, L. W., Segerberg, A., & Walker, S. (2014). Organization in the crowd: Peer production in large-scale networked protests. *Information, Communication & Society, 17*(2), 232–260.

Berners-Lee, T. (1999). *Weaving the web: The original design and ultimate destiny of the World Wide Web by its inventor*. London: Harper Collins.

Binder, E. (2019, August 30). Hong Kong protesters spy a new enemy: Lampposts. *The Wall Street Journal*. Retrieved from www.wsj.com/articles/hong-kong-protesters-spy-a-new-enemy-lampposts-11567161002

Bland, A. (2014, September 29). FireChat – the messaging app that's powering the Hong Kong protests. *The Guardian*. Retrieved from www.theguardian.com/world/2014/sep/29/firechat-messaging-app-powering-hong-kong-protests

Bolter, J. D., & Grusin, R. (1998). *Remediation: Understanding new media*. Cambridge, MA: MIT Press.

Borgesius, F. J. Z., Möller, J., Kruikemeier, S., Ó Fathaigh, R., Irion, K., Dobber, T., Balazs, B., & de Vreese, C. (2018). Online political microtargeting: Promises and threats for democracy. *Utrecht Law Review, 14*(1), 82–96. https://doi.org/10.18352/ulr.420

Bucher, T., & Helmond, A. (2017). The affordances of social media platforms. In J. Burgess, T. Poell, & A. Marwick (Eds.), *The Sage handbook of social media*. Thousand Oaks, CA: Sage.

Carney, N. (2016). All lives matter, but so does race: Black lives matter and the evolving role of social media. *Humanity & Society, 40*(2), 180–199. https://doi.org/10.1177/0160597616643868

Clark, R. (2016). "Hope in a hashtag": The discursive activism of #WhyIStayed. *Feminist Media Studies, 16*(5), 788–804.

Coleman, G. (2014). *Hacker, hoaxer, whistleblower, spy: The many faces of anonymous*. London: Verso.

Costanza-Chock, S. (2012). Mic check! Media cultures and the Occupy Movement. *Social Movement Studies: Journal of Social, Cultural and Political Protest, 11*(3–4), 375–385.

Cox, L. (2010). How do we keep going? Activist burnout and personal sustainability in social movements. Retrieved from http://mural.maynoothuniversity.ie/2815/

Crabu, S., Magaudda, P., Giovanella, F., & Maccari, L. (2015). A transdisciplinary gaze on wireless community networks. *Tecnoscienza. Italian Journal of Science & Technology Studies, 6*(2), 113–134.

Csikszentmihalyi, C., Mukundane, J., Rodrigues, G. F., Mwesigwa, D., & Kasprzak, M. (2018). The space of possibilities: Political economies of technology innovation in Sub-Saharan Africa. Proceedings of the 2018 CHI Conference on Human Factors in Computing Systems (CHI '18), 306 (pp. 1–13).

De Filippi, P., & Tréguer, F. (2014). Expanding the Internet commons: The subversive potential of wireless community networks. *Journal of Peer Production, 6*. Retrieved from http://peerproduction.net/wp-content/uploads/2015/01/De-Filippi-Tréguer-Expanding-the-Internet-Commons-with-Community-Networks.pdf

Della Porta, D. (Ed.). (2007). *The global justice movement. Cross national and transnational perspectives*. Boulder, CO: Paradigm.

Diamond, L. (2010). Liberation technology. *Journal of Democracy, 3*, 69–83.

Duffy, B. E., & Chan, N. K. (2018). "You never really know who's looking": Imagined surveillance across social media platforms. *New Media & Society, 21*(1), 119–138. https://doi.org/10.1177/1461444818791318

Earl, J., & Kimport, K. (2011). *Digitally enabled social change. Activism in the Internet age*. Cambridge, MA: MIT Press.

Farman, J. (2012). *Mobile interface theory. Embodied space and locative media*. Abingdon: Routledge.

Finnemore, M., & Sikkink, K. (1998). International norm dynamics and political change. *International Organization, 52*, 887–917.

Flesher Fominaya, C. (2010). Collective identity in social movements: Central concepts and debates. *Sociology Compass, 4*(6), 393–404.

Freelon, D., McIlwain, C. D., & Clark, M. D. (2016). *Beyond the hashtags: #Ferguson, #Blacklivesmatter, and the online struggle for offline justice* (Center for Media and Social Impact). School of Communication, American University. Retrieved from http://cmsimpact.org/resource/beyond-hashtags-ferguson-blacklivesmatter-online-struggle-offline-justice/

Fuster Morell, M. (2012). The free culture and 15M movements in Spain: Composition, social networks and synergies. *Social Movement Studies, 11*(3–4), 386–392. https://doi.org/10.1080/14742837.2012.710323

Gerbaudo, P. (2012). *Tweets and the streets social media and contemporary activism*. London: Pluto.

Gerbaudo, P. (2013). Protest diffusion and cultural resonance in the 2011 protest wave. *The International Spectator, 48*(4), 86–101. https://doi.org/10.1080/03932729.2013.847689

Gerbaudo, P. (2015). Protest avatars as memetic signifiers: Political profile pictures and the construction of collective identity on social media in the 2011 protest wave. *Information, Communication & Society, 18*(8), 916–929. https://doi.org/10.1080/1369118X.2015.1043316

Gerbaudo, P. (2016). Rousing the Facebook crowd: Digital enthusiasm and emotional contagion in the 2011 protests in Egypt and Spain. *International Journal of Communication, 10*, 254–273.
Gerbaudo, P., & Treré, E. (2015). In search of the "we" of social media activism: Introduction to the special issue on social media and protest identities. *Information, Communication & Society, 18*(8), 865–871. https://doi.org/10.1080/1369118X.2015.1043319
Giaxoglou, K. (2018). #JeSuisCharlie? Hashtags as narrative resources in contexts of ecstatic sharing. *Discourse, Context & Media, 22*, 13–20. https://doi.org/10.1016/j.dcm.2017.07.006
Gillespie, T. (2010). The politics of platforms. *New Media & Society, 12*(3), 347–364.
González-Bailón, S., Borge-Holthoefer, J., & Moreno, Y. (2013). Broadcasters and hidden influentials in online protest diffusion. *American Behavioral Scientist, 57*(7), 943–965. https://doi.org/10.1177/0002764213479371
Goodwin, J., Jasper, J. M., & Polletta, F. (Eds.). (2001). *Passionate politics: Emotions and social movements.* Chicago, IL: Chicago University Press.
Juris, J. S. (2008). *Networking futures. The movements against corporate globalisation.* Durham, NC: Duke University Press.
Juris, J. S. (2012). Reflections on #Occupy Everywhere: Social media, public space, and emerging logics of aggregation. *American Ethnologist, 39*(2), 259–279.
Kelty, C. M. (2008). *Two bits: The cultural significance of free software.* Durham, NC: Duke University Press.
Kidd, D. (2003). The Independent Media Center: A new Mmodel. *Media Development, IV.* Retrieved from https://docs.indymedia.org/pub/Global/PDFsOfIndymediaGuide/brochurepages.pdf
Kramer, A. D. I., Guillory, J. E., & Hancock, J. T. (2014). Experimental evidence of massive-scale emotional contagion through social networks. *Proceedings of the National Academy of Sciences of the United States of America, 111*(24), 8788–8790. https://doi.org/10.1073/pnas.1320040111
Leistert, O. (2013). Smell the fish: Digital Disneyland and the right to oblivion. *First Monday, 18*(3). Retrieved from http://firstmonday.org/ojs/index.php/fm/rt/printerFriendly/4619/3422
Lievrouw, L. A. (2011). *Alternative and activist new media.* Bristol: Policy Press.
Ling, R., & Lai, C.-H. (2016). Microcoordination 2.0: Social coordination in the age of smartphones and messaging apps. *Journal of Communication, 66*(5), 834–856. https://doi.org/10.1111/jcom.12251
Marshall, P. D. (2019). Celebrity, politics, and new media: An essay on the implications of pandemic fame and persona. *International Journal of Politics, Culture, and Society.* https://doi.org/10.1007/s10767-018-9311-0
Mattoni, A. (2013). Repertoires of communication in social movement processes. In B. Cammaerts, A. Mattoni, & P. McCurdy (Eds.), *Mediation and protest movements* (pp. 39–56). Bristol: Intellect.
Maxigas. (2012). Hacklabs and hackerspaces – tracing two genealogies. *Journal of Peer Production, 2*, June. Retrieved from http://peerproduction.net/issues/issue-2/peer-reviewed-papers/hacklabs-and-hackerspaces/
Maxwell, P. J. (2019, September 12). Online social movements: Is it me, you, or us? What the frame says. *2nd International Conference on Research in Humanities & Social Sciences*, Rome, Italy. Retrieved from www.dpublication.com/wp-content/uploads/2019/09/S-F484.pdf
McCarthy, J. D., & Zald, M. N. (1977). Resource mobilization and social movements: A partial theory. *American Journal of Sociology, 82*, 1212–1241.
McDonald, K. (2006). *Global movements: Action and culture.* Oxford: Blackwell.
Melucci, A. (1996). *Challenging codes. Collective action in the information age.* Cambridge: Cambridge University Press.
Mendes, K., Ringrose, J., & Keller, J. (2018). #MeToo and the promise and pitfalls of challenging rape culture through digital feminist activism. *European Journal of Women's Studies, 25*(2), 236–246. https://doi.org/10.1177/1350506818765318
Milan, S. (2010). The way is the goal: Interview with Maqui, Indymedia London / IMC-UK network activist. *International Journal of E-Politics, 1*(1), 88–91.
Milan, S. (2013). *Social movements and their technologies: Wiring social change.* Basingstoke: Palgrave Macmillan.
Milan, S. (2015a). When algorithms shape collective action: Social media and the dynamics of cloud protesting. *Social Media + Society.* https://doi.org/10.1177/2056305115622481
Milan, S. (2015b). From social movements to cloud protesting: The evolution of collective identity. *Information, Communication & Society, 18*(8), 887–900. https://doi.org/10.1080/1369118X.2015.1043135
Milan, S., & Barbosa, S. (2020). Enter the WhatsApper: Reinventing digital activism at the time of chat apps. *First Monday, 25*(1). https://doi.org/10.5210/fm.v25i12.10414
Milner, R. M. (2013). Pop polyvocality: Internet memes, public participation, and the Occupy Wall Street Movement. *International Journal of Communication, 7*, 2357–2390. https://doi.org/1932–8036/20130005
Musiani, F. (2016). Alternative technologies as alternative institutions: The case of the domain name system. In F. Musiani, D. L. Cogburn, L. DeNardis, & N. S. Levinson (Eds.), *The turn to infrastructure in Internet governance* (pp. 73–86). Basingstoke: Palgrave Macmillan.

Nadir. (2012). *We need to talk about Facebook* [Nadir.org]. Retrieved from www.nadir.org/txt/We_need_to_talk_about_Facebook.html

Nagle, A. (2017). *Kill all normies: Online culture wars from 4Chan and Tumblr to Trump and the Alt-Right.* London: Zero Books.

Occupy Directory. (n.d.). Retrieved from http://directory.occupy.net

O'Hara, K. P., Massimi, M., Harper, R., Rubens, S., & Morris, J. (2014). Everyday dwelling with WhatsApp. *Proceedings of the 17th ACM Conference on Computer Supported Cooperative Work & Social Computing* (pp. 1131–1143). Retrieved from https://dl.acm.org/citation.cfm?id=2531679

Papacharissi, Z. (2015). *Affective publics sentiment, technology, and politics.* Oxford: Oxford University Press.

Pariser, E. (2011). *The filter bubble: What the Internet is hiding from you.* London: Penguin.

Pasquale, F. (2015). *The Black Box society: The secret algorithms that control money and information.* Cambridge, MA: Harvard University Press.

Penney, J. W. (2017). Internet surveillance, regulation, and chilling effects online: A comparative case study. *Internet Policy Review, 6*(2), 1–38. https://doi.org/10.14763/2017.2.692

Poell, T., Abdulla, R., Rieder, B., Woltering, R., & Zack, L. (2016). Protest leadership in the age of social media. *Information, Communication & Society, 19*(7), 994–1014.

Polletta, F., & Jasper, J. M. (2001). Collective identity and social movements. *American Review of Sociology, 27,* 283–305.

Puri, D. (2017, October 10). How the Internet of Things can help protect both our safety and right to free speech. *Network World.* Retrieved from www.networkworld.com/article/3231984/how-the-internet-of-things-can-help-protect-both-our-safety-and-right-to-free-speech.html

Rochon, T. R. (1998). *Culture moves. Ideas, activism, and changing values.* Princeton, NJ: Princeton University Press.

Rodino-Colocino, M. (2018). Me too, #MeToo: Countering cruelty with empathy. *Communication and Critical/Cultural Studies, 15*(1), 96–100. https://doi.org/10.1080/14791420.2018.1435083

Shirky, C. (2008). *Here comes everybody. The power of organizing without organizations.* London: Penguin.

Snow, D. A., & Benford, R. D. (1988). Ideology, frame resonance, and participant mobilization. *International Social Movement Research, 1,* 197–217.

Snow, D. A. (2001). Collective identity. In N. J. Smelsner & P. B. Baltes (Eds.), *International encyclopedia of the social and behavioral sciences* (pp. 2212–2219). Elsevier.

Suk, J., Abhishek, A., Zhang, Y., Ahn, S. Y., Correa, T., Garlough, C., & Shah, D. V. (2019). #MeToo, Networked Acknowledgment, and Connective Action: How "Empowerment Through Empathy" Launched a Social Movement. *Social Science Computer Review.* https://doi.org/10.1177/0894439319864882

Tarrow, S. (2005). *The new transnational activism.* Cambridge: Cambridge University Press.

Trottier, D., & Lyon, D. (2012). *Key features of social media surveillance. In C. Fuchs, K. Broersma, A. Albrechtslund, & M. Sandoval (Eds.), Internet and surveillance: The challenges of Web 2.0 and social media* (pp. 89–105). Abingdon: Routledge.

Trudel, D., & Tréguer, F. (2016). *Alternative communications networks throughout history* [NetCommons Deliverable number: D5.1]. Retrieved from https://halshs.archives-ouvertes.fr/halshs-01418826

van Dijck, J., Poell, T., & de Waal, M. (2018). *The platform society. Public values in a connective world.* Oxford: Oxford University Press.

Wyatt, S. (2004). Danger! Metaphors at work in economics, geophysiology, and the Internet. *Science, Technology & Human Values, 29*(2), 242–261.

23

Feminist Peer Production

Sophie Toupin

1 Introduction

Over the years, feminists have developed specific digital forms of collective practices that effectively illustrate what feminist peer production is. In this chapter, I examine some of the practices which have rendered visible the analytical category of gender – and at times other categories such as race, class, and caste, which are often absent from the analysis of mainstream peer-production scholarship and practice. The move to use feminist perspectives allows for a broadening of understandings of peer production and in turn enables the following questions: What is feminist peer production? Why has it emerged and how is it done? These apparently simple questions contain complexity, the exploration of which will be the work of this chapter.

To begin, I define two concepts: peer production and feminism. I follow Benkler's (2006) notion of peer production where he suggests that the "networked environment makes possible a new modality of organizing production: radically decentralized, collaborative, and nonproprietary; based on sharing resources and outputs among widely distributed, loosely connected individuals who cooperate with each other without relying on either market signals or managerial commands" (p. 60). When I use the framework of feminism to qualify peer production, I am influenced by bell hooks' definition. She understands feminism as "the movement to end sexism, sexual exploitation and sexual oppression" (2000, p. 33). She goes further in adding that feminism cannot be separated from racism and from colonial histories, including slavery, as integral to the development of capitalism and the enrichment of the Global North (Ahmed, 2017; hooks, 2000). Rooting feminist peer production in bell hooks' understanding makes visible the intersectional and decolonial aspects of feminism which are crucial for many feminists today. From those definitions, it is possible to get a sense of what is feminist peer production and highlight the gendered aspect of the social relations which form peer production.

This chapter will be divided into three sections. First, to understand feminist peer production, it is imperative to focus on feminist objectivity. Feminist objectivity is one of the main features of feminist science and technology studies and is integral to understanding feminist peer production. I draw from the debates on objectivity that were prevalent in the 1970s, 1980s, and 1990s. These debates will offer the reader a much better understanding of how feminist objectivity has influenced the emergence of and shaped feminist peer production practices. Second, I focus on feminist critiques of peer production to show the main reasons why this practice has emerged and how it has grown over the years. The invisibility of feminist peer production is part of a long history of erasing the contributions and specificities of women in technology. Third, I examine the ways in which feminist peer productions are produced through a process of struggle. To do so, I draw from a number of case studies of feminist peer production, which reflect on the plurality of these practices. I conclude with some thoughts on the future of feminist peer production.

This research uses mixed methods. First, it is grounded in ethnographic fieldwork with a variety of feminist technologists involved in peer production projects. I have myself been taking part in a number of feminist projects, which has led me to discuss, observe, and come to the conclusion that feminist

The Handbook of Peer Production, First Edition. Edited by Mathieu O'Neil, Christian Pentzold, and Sophie Toupin.

peer production is grounded in distinct practices. While commonalities exist between traditional forms of peer production projects and feminist ones, the practice of feminist peer production remains distinct, especially as it is attentive to gender. I have followed a number of these feminist peer production projects as a participant observer either online or in face-to-face gatherings. Second, this research is rooted in a classical literature review of some of the work produced in the field of science and technology studies.

Finally, I want to stress that the picture of feminist peer production that I am about to draw will be incomplete. This chapter is meant as an introduction to what feminist peer production looks like and how it differs from mainstream peer production. It, however, does not account for all the types of feminist peer production projects and practices that exist. I have in particular been attentive to some of the practices in which I am rooted in or have witnessed in addition to those that focus on the Global South since such practices make visible what would otherwise likely be hidden from view. Haraway's (1997) metaphor of the game of cat's cradle, a game that kids play involving creating various string figures with the hands, illustrates what I am about to do. The metaphor of the cat's cradle gestures towards the different types of articulations, formations, and entanglements that emerge while the game is being played, which involves humans and non-humans. Despite the emergent formations created through the game of strings, gaps and holes remain within every formation.

2 Feminist Peer Production as Situated Knowledge

One of the main principles behind feminist peer production is the fact that it uses situated knowledge to understand and act in the world. As is outlined below, the ground-breaking work of feminist science and technology scholars such as Donna Haraway (1988, 1997, 2004) and others (Harding, 1986, 1993, 2011; Wajcman, 1991) have been essential to re-imagine the ways in which we see and understand the world that surrounds us and our relationship to science and technology. Such research has been crucial to the emergence of the practice of feminist peer production.

In the 1970s, 1980s, and 1990s a debate about objectivity became of great importance for feminist scholars of science and technology studies. The question that was under scrutiny pertained to how scientific knowledge was made: Was an insider's perspective, within scientific knowledge, a valid one? A well-known trope of feminist analysis of scientific and technological knowledge (but clearly not the only one) can be located in the White[1] American women's health movement of the 1970s (Murphy, 2012). The White women's liberation movement had identified the body as a way to gain knowledge and control over women's sexuality and fertility – an early example of non-digital feminist peer production. Women would gather in a living room to show one another how to examine their bodies and in turn produce what Michelle Murphy (2012) calls a feminist protocol. As I will show below, their practice would influence the emergence of feminist biohacklabs. As part of heterosexual family planning techniques, they practiced menstrual extraction and as such gained expert knowledge through peer-to-peer knowledge exchange. This knowledge had generally been in the hands of gynecologists who were White men. It is in a context of the re-appropriation of their bodies that women could develop new knowledge and skills collectively and with their bodies that had a direct impact on their lives (Wajcman, 1991). They recognized science, and scientific knowledge and practices, as patriarchal, as it was based on a masculine project grounded in so-called objectivity, reason, and truth. To disrupt what they saw as a biased understanding of science and technology (or the portmanteau term technoscience), they needed to develop their own knowledge based on their specific experiences and praxis.

During the 1980s, a scholarly body of literature started to criticize Western science as patriarchal. Sandra Harding's influential 1986 book entitled *The Science Question in Feminism* aimed at shifting the script by asking the science question in feminism rather than the woman question in science. At the core of the feminist questioning of science and reproductive technologies were the questions of what counts as knowledge and what produces better knowledge.

The endeavor of these feminist scholars was about the needs of women (often with White women in mind) to create and access knowledge for women, because in science and technology the default human was a (White) man. This approach allowed scholars to have a more complete understanding of dominant institutions and structures. By mapping the dominant groups and their practices, feminist scholars were able to show the patterns of domination in the relationship between gender, science, and technology. Sandra Harding's position was that knowledge production based on experiences was not just situated opinions: "Men's dominating position in social life results in partial and perverse understanding, whereas women's subjugated position provides the possibility of more complete and less perverse understandings" (as cited in Wajcman, 1991, p. 10). While Judy Wajcman (1991) recognized that Harding's standpoint theory was based on the universal features of women's experience, she was nonetheless critical of this project: the fractured identities of women along lines of class, race, and culture limited the standpoint framework. Was there a need, she asked, for other kinds of feminist science and technology studies rooted in experiences of, for example, Black women, Asian women, lesbian women, Indigenous women?

In 1988, in her article "Situated Knowledges: The Science Question in Feminism and the Privilege of Partial Perspective," Haraway proposed another way of framing the objectivity question, thereby rethinking the standpoint. She suggested that all knowledge production – not only subjugated knowledge – was in fact situated. Even so-called objective and rational science was situated because "all science is already embedded in matrices of capital, subjectivity and history" (Haraway as cited in Murphy, 2012, p. 99). To talk about feminist objectivity, Haraway preferred to adopt the metaphor of the vision: "The moral is simple: only partial perspective promises objective vision" (Haraway, 1988, p. 583). However, she warned feminists about the dangers of romanticizing or appropriating a vision from below. Seeing from below or from the margin is not an innocent position, she argued. While it remains preferred because it seems to provide a better account of the world, it requires enormous skill to be mobilized well. It requires passionate detachment and the recognition that vision is always a question of power (Haraway, 1988, p. 585).

Situated knowledge is about challenging "the god-trick of seeing everything from nowhere" (Haraway, 1988, p. 581) and in turn makes visible other ways of seeing from other positionalities. Situated knowledge is important for thinking about and with science and technology as it is at the basis of many feminist peer production practices. Feminist peer production practices do challenge the default positions of who participates in peer production projects and how, and the reasons why some peer production projects or understandings are more visible than others.

Moreover, taking an expansive view on technology is fundamental to capturing feminist peer production practices. The way technology is often thought of is as things or artifacts as opposed to processes or practices (Slack & Wise, 2015). An expansive view considers technology not only as an artifact with particular values thereby revealing the situatedness embedded in such position, but also its social and cultural implications. Wajcman (1991) clearly illustrates this situated expression in relation to the history of technology when she says that "women's exclusion from, and rejection of, technology is made more explicable by an analysis of technology as culture that expresses and consolidates relations amongst men. Technologies bear the imprint of the people and social context in which they developed" (p. 22). Wajcman's reference ought to be taken in the context of an era in which the rejection of science and technology by many White feminists in the Global North came about during the Cold War and the Vietnam War. Nuclear power, weapon systems, and the earlier Manhattan project were the types of scientific and technological systems that were front and center for feminists' rejection of military technologies.

So, what does peer production bring to situated knowledge? I suggest that it is not only about ways of seeing, but also about feminist ways of doing (Rentschler, 2019; Toupin & SpiderAlex, 2018). One feminist scholar who brings this relationship of seeing and doing together in the realm of science and technology studies is Michelle Murphy (2012) in her book *Seizing the Means of Reproduction: Entanglements of Feminism, Health, and Technoscience*. For Murphy, ***doing*** feminist technoscience is about an understanding of knowledge that is produced in the process of struggle. Her book shows that the proliferation of diverse forms of participatory science and the politicization of collective feminist health projects in the late 20th century were a reaction to a dissatisfaction with patriarchal and commercialized forms of gynecology and of biopolitics (such as the control of population through forced sterilization of Indigenous women and women in the Global South) it supported and constrained. Feminists wanted to create their own feminist biopolitics and allow women to seize their means of reproduction. For Murphy, the feminist strategy of developing non-commoditized health techniques through exchange practices of sharing information and doing together resonates with contemporary practices around free and open source software (FOSS) development.

Participatory feminist science is still alive today in the Gynepunk Lab, a feminist biohacking space (see chapter on Biohacking by Morgan Meyer) in Spain, which aims at bringing back knowledge of reproductive justice in a context where it is threatened. In this feminist biohacking lab, the Gynepunks aim at regaining control of their fertility through menstrual extraction techniques and processes, and have developed DIY emergency gynecological kits with 3D printers that any woman can use safely if needs be.

Using Murphy's understanding and applying it to feminist peer production highlights the fact that feminist peer production practices are rooted in feminist everyday struggles. Such an understanding adds to Benkler's (2002) conceptualization of the commons-based peer-production model by making visible the gender and race aspects of peer production. For feminists, peer production is about the importance of considering race and gender as integral to peer production.

2.1 What Are the Feminist Critiques?

I now highlight some of the critiques of peer production from feminist perspectives. These critiques will help to distinguish practices that make up feminist peer production, to show that there are different forms of feminist peer production practices and that tensions exist within and between these projects. The four feminist criticisms I examine are (1) the lack of feminist content and/or women's participation in some of the best-known peer production projects (Wikipedia and FOSS), (2) the trend towards eclipsing women in technology, (3) the ongoing debate about who can take part in feminist peer production projects, and (4) the invisibility of the oppressive cycle of technological production in terms of labor and the origins of the material that constitutes our technologies. Focusing on these critiques helps answer the question of how, as a result of feminist critiques, feminists have performed feminist peer production.

2.2 The "Culture" Within Peer Production Projects

Since Wikipedia has long been dominated by the participation of White men (Sichler & Prommer, 2014) often from the Global North and in English, feminists have organized themselves to create feminist Wikipedia edit-a-thons (Wikimedia Foundation, 2018). The goal is to add feminist content within the online encyclopedia. Such activities have been organized by feminist academics, librarians, artists, activists, and technologists in many parts of the world. The feminist Wikipedia edit-a-thons are designed to mitigate the lack of feminist content and the low participation of women as Wikipedia contributors (Bear & Collier, 2016), and also as a means to teach other women how to edit within Wikipedia. The hope was that women who participated in the one-time Wikipedia edit-a-thon would become regular contributors (Evans, Mabey, & Mandiberg, 2015). This do-it-together methodology has been recognized as a feminist approach to learning, particularly when it relates to technology, in order to shy away from technological fears (Toupin, 2017). Despite initiatives to widen feminist content in languages other than English and in geographical location such as the Global South, it remains that the core activities have been dominated by White women from the Global North.

Within free and open source software projects, the percentage of women is very low (Geek Feminism Wiki, n.d.). In 2001, Ghosh, Glott, Krieger, and Robles' (2002) research showed that only 1.1% of women were active in the development of free and open source software (see chapter on FOSS by Stéphane Couture, in this volume). To start changing this trend, women coders have tried recruiting other women with coding skills. Moreover, some have started organizing training gatherings where they show other women how they can contribute to FOSS projects, dispelling the myth that one absolutely needs to know how to code to participate in such projects (Haralanova, 2010). In fact, these gatherings aim at highlighting that, for those who want to get involved and learn, many tasks do not require coding knowledge but rather demand translation skills and being able to identify errors or bugs within the code (sometimes spelling mistakes), among other skills. The intention behind these projects being that more women with coding skills and women with skills other than coding can get involved.

Why is it important to focus on "culture" to shed light on the lack of women within mainstream peer production projects? In her work ***Feminism Confronts Technology***, Wajcman (1991) stresses the relationship between technology and culture by articulating the mutual shaping of technology and gender relations. She argues that the fact that technology was recognized as masculine culture was the result of the historical and cultural construction of gender (Wajcman, 1991, p. 135). It was an ideological position, the ideology of masculinity, which created a bond between men and machines. In her discussion about technology and masculinity she writes, "treating technology as a culture has enabled us to see the way in which technology is expressive of masculinity and how, in turn, men characteristically view themselves in relation to machines" (Wajcman, 1991, p. 149). Further, Wajcman and other feminist scholars (Balsamo, 1995; Faulkner, 2001; Hicks, 2017) insist that there are a lot of cultural factors – including associated symbols, images, and representations of technology – that play a role in lowering the interest of girls and women in technology.

Wajcman's (1991) research brings important aspects to understanding the lack of visibility on feminist peer production practices. More recent scholarship has shown that the contributions of women to the histories of technology have been invisibilized particularly when it comes to the development of computing (Ensmenger, 2010; Hicks, 2017). Thanks in part to new declassified archival material and the decentering of the inventor or the machine as objects of study, it has been demonstrated that women were the first computer programmers in both the United Kingdom and the United States. In the UK, thousands of women were hired at Bletchley Park during World War II to do cryptanalytic work, and helped with programming the first computer, Colossus (Hicks, 2017). There were also women in the United States, mostly after the war, who programmed the Electronic Numerical Integrator and Computer (ENIAC), a computer based on the Turing machine. Those two government-led computing projects followed a division of labor where men were responsible for plans and analysis, while women operated the machines. This division of labor resulted in rendering women's contribution to the history of computing invisible for nearly 50 years, as they were not seen as programmers but rather involved in the low clerical task of "simply" pressing buttons (Hicks, 2017).

In both cases, research has documented the ways in which women were excluded from the field of computing as it began to take prominence whether through governmental or corporate policies as well as in terms of representation in advertising. But more so, Mary Hicks (2017) argues that computerization has been designed as a project built on a type of labor organization that reproduces a sexualized division of labor. She concludes her book wondering whether the increased number of women and girls with programming skills will simply succeed in flooding the market and creating a feminized sphere of computer labor rather than changing social and economic inequalities (Hicks, 2017). Her conclusion shares similarities with Christina Dunbar-Hester's (2014) reflection on social movements and their technologies where she asks whether it is necessary to have the attainment of technical expertise as a universal goal? She observed that participation in technology is no guarantee towards empowered social conditions, even among a group of activists (Dunbar-Hester, 2014). Such critiques are of great value when trying to understand, map, and make sense of feminist peer production practices and their limits.

2.3 Who Takes Part in Feminist Peer Production Practices?

The group composition of who can ascribe to feminist peer production practices varies and depends on the communities that are associated with these projects. Certain projects feel that the identity of the members of the group must be limited to women (most include queer and trans people), while others rather recognize that anyone independent of biological body at birth can be a feminist. Feminist peer production practices are thus attuned to how the community itself understands feminism. This relationship implies that there might be disagreements among the different groups that carry out feminist peer production practices. However, those disagreements do not necessarily preclude participation in broader feminist peer production projects or ideas. Feminist peer production practices and groups are, however, not static. Depending on the composition of the group and their political orientation, they can go from one understanding (e.g., women only groups) to another (e.g., recognizing that anyone can be a feminist).

2.4 Invisibility of Feminist Peer Production Practices

There are a lot of feminist peer production projects out there, but many are not well known or might not be thought of as peer production. Most of these projects do not articulate what they do as feminist peer production. They might instead be using terms such as do-it-together, collective work, or co-production.

There are many reasons why feminist peer production projects are relatively invisible. First, because many of these projects address issues that affect mostly women, trans, and queer people they are considered projects that tackle "minority" issues. The understanding that gender issues concern women, trans people, and queer people only erases the responsibility of other actors in gender and intersectional dynamics. As a case in point, since street and online harassment is common for most people who are not White, heterosexual, and male, many of the peer production platforms that are feminist in nature concern gender-based violence. Some examples which will be further explained below include HarassMap and Hollaback!, platforms which allow feminist activists to crowd-source information on street harassment and have it available for others to see.

Second, in the past few years, feminist makers and hackers involved in peer-production practices have foregrounded the importance of the politics of visibility. SSL Nagbot (2016) articulates this type of political orientation as follows: "Th[e] combination of visualization with emancipatory alterity demonstrates the ways that feminism in hacking is largely based on a politics of visibility; that is, hacking and making serve the broader objectives of bringing to light the invisible infra/structures of power that render technological achievement possible" (p. 1). Making visible the type of work necessary for feminist peer production practices to exist illuminates why some feminists simply cannot take part in them or, if they do so, why their time is limited. Women are still predominantly responsible for taking care of children and elders as well as reproductive tasks (cleaning, cooking, etc.), giving them less time for leisure. Finally, some feminist peer production projects are invisible because they want to remain as such. Certain projects that aim at building feminist servers or other types of feminist technologies benefit from staying under the radar to protect themselves and their projects from online or offline violence and harassment.

2.5 Cycle of Technology Production

Part of the analysis of feminist peer production is about understanding the production cycle of technologies that allows peer production to happen. This is usually a forgotten dimension in most peer production projects with the exception of such projects such as the Fairphone (Haucke, 2018), Precious Plastic, and a handful of others. Recognizing that the cycle of production is rife with inequality – from the production of technology to its access, uptake, development, and governance, until its end cycle –feminists make visible aspects which are often invisible. This feminist approach to peer production sheds light on forms of violence intrinsic to resource extraction, the factory, and assembly line, a type of work that is highly gendered and raced (Toupin & Haché, 2015).

3 Three Case Studies

Having examined some of the main criticisms of peer production projects from the standpoint of feminist perspectives, I now turn to three case studies of feminist peer production. This discussion is in an attempt to make visible feminist peer production practices and better understand how they are put in practice. The three case studies are the following: (1) the building and maintaining of feminist spaces, collectives, and servers; (2) projects that address gender-based violence through peer production; and (3) the ways in which speculative thinking is being mobilized by feminists to rethink technologies and their imaginaries.

3.1 Feminist Peer Production and Space

In the past couple of years, feminist tech collectives from Brazil, Mexico, the Netherlands, and Spain, among others, have come together to set up feminist hackerspaces/hacklabs (Toupin, 2014, 2017) (see chapter on Makerspaces by Kat Braybrooke and Adrian Smith, in this volume), discussion lists, and bots and maintain what they call feminist servers. These feminist projects are a way to host feminist content, combat violence, say no to the commercialization of the internet's infrastructure, learn better in a do-it-together environment, and make visible the development of feminist projects. These projects follow feminist principles and employ peer production to build, maintain, and repair them.

Their do-it-together methodology is a form of peer production that is feminist in its pedagogy and allows other feminists to learn to set up or maintain a given technology (how to build a bot, a server, etc.). What they do is meet online or have face-to-face gatherings where they learn together and produce collectively. The documentation of their meetings and gatherings is then archived on a feminist server and circulated on feminist discussion lists. The ways in which these projects have seen the light are twofold: feminist techies have either impressed upon feminist groups the importance of hosting their content on feminist technological infrastructures, or feminist groups have come to feminist techies to have them build the needed physical or virtual space.

Through such a case study, we come to understand that feminist peer production is an intentional feminist political act. This practice aims at (re)gaining control and consent over one's technological infrastructure, be it software, hardware, or data produced. Moreover, it is about creating new practices and new expectations regarding technologies and its associated services, in this case being provided by feminist tech collectives.

Feminist peer production is the product of collective reflections, interactions, dialogue, and co-production of knowledge and practices on what it means to be building, maintaining, and repairing feminist technological infrastructures and being a feminist. Recognizing this history of co-production (or peer production) is all about taking a feminist approach that credits feminists.

3.2 Feminist Peer Production and the Fight against Sexual Harassment

Feminist peer production has been used to map instances of sexual harassment in the streets and to make (unofficial) lists of sexual harassers public. When it comes to gender-based violence, feminist peer production projects have brought to the fore the intrinsic relationship between its online and offline dimensions. In making visible the systematic culture of violence Black women, queer, and trans people experience daily in the public sphere and virtually, feminist peer production has shown that this is a worldwide and structural phenomenon.

Hollaback! (Keller, Mendes, & Ringrose, 2018) was established in New York City in 2005 and later became a platform where feminists could geolocate and map details of the street harassment they experienced or witnessed. The aim of this feminist-powered platform is no less than to end street harassment. The platform is available in 21 cities and 16 countries in both the Global North and South.

Following instances of harassment in Egypt during the Arab Spring in 2010, a group set up HarassMap to map and spatially locate instances of violence against women in Egypt (Abdelmonem & Galan, 2017). The platform is designed in such a way that with a cell phone a woman can easily anonymously report the street harassment she has been subjected to. She can send a report through social media, SMS, email, or by clicking on one of the two main buttons appearing on site: Report an incident. The other button available is to report an intervention, which means that you can describe the kind of assistance or response that was offered to the harassed woman. HarassMap verifies the report and geolocates it on a Google map of Egypt.

Another feminist peer production project called HeartMob was set up in the United States by Hollaback! to provide yet another essential service: care from the community to reduce trauma to those who suffer harassment. Their premise is to fight fire with water. The app HeartMob, designed by a feminist workers collective, can be installed on a cell phone and send secure and private messages of care to those who have experienced harassment (Blackwell, Diamond, Schoenebeck, & Lampe, 2017).

These three projects are in the tradition of feminist peer production as they are volunteer-based with a feminist vision of ending sexual harassment in society and tackling their consequences. But more so, these examples show that feminist peer production is more than participating voluntarily in a peer production project. It stresses the need for the bystander to act and say no to a violent practice both online and offline (Rentschler, 2017). The peer is not just a volunteer who participates for free in a project that she believes in, but a feminist political actor who helps transform the society in which we live. Furthermore, the socio-technical imaginary behind this project is the pre-figuration of a world where gender-based violence, patriarchy, and racism come to an end.

More recently, a new type of feminist peer production has surfaced online to make (unofficial) lists of harassers in and around university campuses. In a digital context, two examples are worth mentioning, in South Africa and India. In 2016, an anonymous Facebook post appeared on the Rhodes University (RU) Queer Confessions and Crushes Facebook page. Rhodes University (or the University Currently Known as Rhodes, as it bears the name of a White supremacist, Cecil Rhodes, who pillaged the resources of the southern African continent) is located in the southern part of South Africa. The post made a public peer produced "reference list" (which was to trend on Twitter in South Africa the next day under the hashtag #RUReferenceList) consisting of 11 names of people from Rhodes University who were believed to be sexual offenders (Seddon, 2016). The list called on the university administration to take immediate action and investigate the alleged offenders. The list circulated widely and forced yet another debate about sexual violence and rape culture in South Africa generally and more specifically in university settings. This action by South African queer, trans, and feminist students succeeded in putting the topic of rape culture back in public discourse. South Africa has terrible statistics regarding the rape and murder of Black women, queer, and trans people, making it one of the worst places in the world to be a woman (Gqola, 2015). A similar example of the constitution of a list of sexual harassers came to the fore on Facebook in an Indian and diaspora context. The online list, called Raya Sarkar's list or later LoSHA (List of Sexual Harassers in Academia), was released by an Indian female law student at the University of California (Davis). Raya Sarkar, a 24-year-old Indian citizen studying law in the United States, decided in 2017 to publish the name of dozens of men who had allegedly been accused of sexual harassment by women (Chowdury & Deep, 2017). Her list was then expanded to 75 names after other women contacted her on Facebook and WhatsApp to add more names. It was possible to consult the list on a non-editable Google document and identify the names of Indian male academics who had allegedly been harassers in India, the United States, and elsewhere. The list, which was peer produced, was extremely controversial and polarized Indian feminist movements. The controversy surrounded the ways in which the list was produced and circulated, as well as its accuracy (Menon, 2017). Independent of this controversy, the list has been named as a catalyst for an organized #MeToo movement[2] in India and within the Indian diaspora (Kumar, 2018). Sarkar understands this form of feminist peer production as a form of dissent in a world where sexual harassment is rife and where female identified students in particular have a right to know if their professors have a history of harassment (Prashad, 2018).

The peer production of lists by feminist, queer, and trans people who face violence and harassment seems to be a tactic that has emerged of late. While the ways in which feminist peer production is activated varies, "lists" seem to represent a feminist means of leaking peer produced information relatively anonymously using corporate platforms and as one of the multiple ways to fight against gender-based violence and harassment. The paradox with the latter practice is that on the one hand it can be said to depoliticize peer production by using corporate and proprietary tools, but politicizes peer production through its feminist strategy regarding its content and the importance of reaching out to a large audience. All and all, the two types of feminist peer production practices highlighted above show differences: one type aims at building FOSS feminist platforms and crowd-sources its content (as with Holloback!, HarassMap, and HeartMob) in line with Benkler's definition while the other has the goal to constitute crowd-sourced lists using corporate platforms such as Facebook and WhatsApp.

3.3 Feminist Peer Production and Imaginaries

> it matters what ideas we use to think other ideas (with). It matters what matters we use to think other matters with; it matters what stories we tell to tell other stories with; it matters what knots knot knots, what thoughts think thoughts, what descriptions describe descriptions, what ties tie ties. It matters what stories make worlds, what worlds make stories. (Haraway, 2016, p. 29)

Donna Haraway's (2016) citation signals the importance of stories, and how we tell them. This point seems particularly fitting when engaging with feminist peer production and imaginaries. Research on peer production is generally not a field that acknowledges that it tells stories and that in part these stories help shape future technologies and their infrastructures. The imaginary around peer production is mainly a post-capitalist one (Bauwens, Kostakis, & Pazaitis, 2019; Mason, 2015; Srnicek & Williams, 2016) where the struggle against patriarchy and racism is not specifically acknowledged. The fact that feminists are mobilizing imaginaries to (re)activate their peer production practices gestures towards rendering visible particular sensibilities, cultures, and entanglements.

Regarding socio-technical imaginaries, science and technology scholar Sheila Jasanoff (2015) states that they are "collectively held, institutionally stabilized, and publicly performed visions of desirable futures, animated by shared understandings of forms of social life and social order attainable through, and supportive of, advances in science and technology" (p. 4). Although in this definition an emphasis is placed on the institutional aspect, socio-technical imaginaries have also gained support outside of institutions and within Black feminist movements in particular as a way to engage with current problems whether they are sexual harassment, the climate crisis, or the racism that can blatantly surface with technologies (Benjamin, 2019; Eubanks, 2018; Noble, 2018). The relationship between technologies and imaginaries has a long and contested history. Some believed it has had and still has a direct influence on technologists (the term *cyberspace* being coined by William Gibson's 1984 novel *Neuromancer*) while others say that there is no direct link. To re-imagine technologies and their peer production practices, feminists have designed a way to collectively write stories on and about feminist technologies and the infrastructures that sustain them (data centers, the submarine cables and the boats that repair them, the electricity necessary to power our devices, the minerals that power them, etc.). These stories are a way to reclaim, through peer produced narratives, feminist technologies and practices. As a case in point, workshops have used a feminist peer production methodology to produce new narratives or what SpiderAlex calls feminist futurotopias (Radio Cargo & SpiderAlex, 2018). These workshops are a way to prefigure the future that feminists want, in addition to going back in time to better understand the present. These workshops take, though not explicitly, a feminist approach *à la* bell hooks, recognizing the patriarchal, colonial, and imperial past is shaping our present and future. The workshops are about collectively conceiving imaginaries that are alternatives to existing oppressive or dystopian technological models.

3.4 The Future of Feminist Peer Production

As a practice feminist peer production seems to be here to stay. The examples that were briefly showcased in this chapter exemplify this assertion. While the term never had much traction, the principles behind the practice remain widespread. Feminists involved in such practices have preferred terms such as do-it-together (DiT), co-production, or collective work. What underlines these terms are what feminist peer production is about: the social transformation of society, especially the collective fight against gender-based violence, the fact that the personal is political, making visible projects and practices that mobilize feminist peer production practices, the belief and acknowledgment that all postures and practices are rooted in situated knowledge, and shedding light on the cycle of production of technologies and their impact on the environment. Moreover, feminist peer production adds to our understanding of the commons-based peer-production model through the gendered and racial aspects too often invisibilized.

These feminist peer production practices are not homogenous and often differ widely. But feminist peer producers are often able to work together despite varying feminist ideological positions. Friction remains front and center, but it does not prevent or close the possibility of collaboration. Feminists involved in peer production practices have criticized certain ways of doing things, either of other feminists or peer producers, not for the sake of critique, but rather to make visible the invisible and the irreconcilable contradictions that open up new possibilities. The imagination that has been deployed by feminist peer producers up to now offers hope for the future. A feminist position that is antiracist and which attends to the colonial and the imperial in its practice is a welcome approach in the landscape of feminist peer production. Feminists from the Global South and women of color living in the Global North have much to do with this turn. Their work and advocacy to highlight the blind spots of White feminism have been essential for this shift.

Notes

1 I have decided to capitalize the B in Black and the W in White to affirm that being Black and being White are races and that race is a social construct. While this gesture is to call attention to how Whiteness functions in our societies, I condemn the use of the capital W in White when it is being used by White supremacists to evoke racial domination and violence.
2 #MeToo is a concept first used by Black community organizer Tarana Burke.

References

Abdelmonem, A., & Galan, S. (2017). Action-oriented responses to sexual harassment in Egypt: The cases of HarassMap and wendo. *Journal of Middle East Women's Studies. 13*(1), 154–167. doi:10.1215/15525864-3728767

Ahmed, S. (2017). *Living a feminist life*. Durham, NC: Duke University Press.

Balsamo, A. (1995). *Technologies of the gendered body: Reading cyborg women*. Durham, NC: Duke University Press.

Bauwens, M., Kostakis, V., & Pazaitis, A. (2019). *Peer to peer. The commons manifesto*. London: University of Westminster Press.

Bear, J. B., & Collier, B. (2016). Where are the women in Wikipedia? Understanding the different psychological experiences of men and women in Wikipedia. *Sex Roles: A Journal of Research, 74*(5–6): 254–265.

Benjamin, R. (2019). *Race after technology: Abolitionist tools for the new Jim Code*. Cambridge: Polity.

Benkler, Y. (2002). Coase's penguin, or, Linux and "The nature of the firm." *The Yale Law Journal, 112*(3), 369–446. https://doi.org/10.2307/1562247

Benkler, Y. (2006). *The wealth of networks: How social production transforms markets and freedom*. New Haven CT: Yale University Press.

Blackwell, J., Diamond, J., Schoenebeck, S., & Lampe, C. (2017). Classification and its consequences for online harassment: Design insights from HeartMob. *Proceedings of the Acm on Human-Computer Interaction, 1*(Cscw) (pp. 1–19). doi:10.1145/3134659.

Chowdury, S. R. & Deep, A. (2017, Nov. 4). Universities respond to Raya Sarkar's list of alleged sexual predators:Mostlysilence,somedenials.Retrievedfromhttps://scroll.in/article/856589/universities-respond-to-raya-sarkars-list-of-alleged-sexual-predators-mostly-silence-some-denials

Dunbar-Hester, C. (2014). *Low power to the people: Pirates, protest, and politics in FM radio activism*. Cambridge, MA: MIT Press.

Ensmenger, N. (2010). *The computer boys take over: Computers, programmers, and the politics of technical expertise*. Cambridge, MA: MIT Press.

Geek Feminism Wiki. (n.d.). FLOSS. Retrieved from http://geekfeminism.wikia.com/wiki/FLOSS

Eubanks, V. (2018). *Automating inequality: How high-tech tools profile, police, and punish the poor*. London: St-Martin's Press.

Evans, S., Mabey, J., & Mandiberg, M. (2015). Editing for equality: The outcomes of the art feminism Wikipedia edit-a-thons. *Art Documentation: Journal of the Art Libraries Society of North America, 34*(2), 194–203.

Faulkner, W. (2001). The technology question in feminism: A view from feminist technology studies. *Women's Studies International Forum, 24*(1): 79–95.

Ghosh, R. A., Glott, R., Krieger, B., & Robles, G. (2002). *Free/libre and open source software: Survey and study* (No. V). Retrieved from www.infonomics.nl/FLOSS/report/Final4.htm.

Gqola, P. (2015). *Rape: A South African nightmare*. Auckland Park: MF Books Joburg.

Haralanova, K. (2010). L'apport des femmes dans le développement du logiciel libre (Master's thesis). Retrieved from https://www.ludost.org/sites/www.ludost.org/files/memhark_complete_1.pdf

Haraway, D. (1988). Situated knowledges: The science question in feminism and the privilege of partial perspective. *Feminist Studies, 14*(3), 575–599.

Haraway, D. (1997). *Modest_witness@second_millennium: Femaleman_meets_oncoMouse: Feminism and technoscience*. New York, NY: Routledge.

Haraway, D. (2004). A manifesto for cyborgs: Science, technology and socialist feminism in the 1980s. In D. Haraway (Ed.), *The Haraway reader* (pp. 7–45). New York, NY: Routledge.

Haraway, D. J. (2016). *Staying with the trouble: Making kin in the Chthulucene*. Durham, NC: Duke University Press.

Harding, S. (1986). *The science question in feminism*. Ithaca, NY: Cornell University Press.

Harding, S. (1993). Rethinking standpoint epistemology: "What is strong objectivity"? In L. Alcoff & E. Potter (Eds.), *Feminist epistemologies* (pp. 49–82). New York, NY: Routledge.
Harding, S. (ed.) (2011). *The postcolonial science and technology studies reader*. Durham, NC: Duke University Press.
Haucke, F. (2018). Smartphone-enabled social change: Evidence from the Fairphone case? *Journal of Cleaner Production, 197*, 1719–1730. doi:10.1016/j.jclepro.2017.07.014
Hicks, M. (2017). *Programmed inequality: How Britain discarded women technologists and lost its edge in computing*. Cambridge: MIT Press.
hooks, b. (2000). *Feminism is for everybody: Passionate politics*. Boston, MA: South End Press
Jasanoff, S. (2015). Future imperfect: Science, technology, and the imaginations of modernity. In S. Jasanoff & S.-H. Kim (Eds.), *Dreamscapes of modernity: Sociotechnical imaginaries and the fabrication of power* (pp. 1–47). Chicago, IL: University of Chicago Press.
Keller, J., Mendes, K., & Ringrose, J. (2018). Speaking "unspeakable things": Documenting digital feminist responses to rape culture. *Journal of Gender Studies, 27*(1), 22–36. doi:10.1080/09589236.2016.1211511
Kumar, R. (2018, Dec. 7). Why the "Me too" movement in India is succeeding at last. Retrieved from www.opendemocracy.net/en/5050/me-too-india-succeeding-at-last/
Mason, P. (2015). *PostCapitalism: A guide to our future*. New York, NY: Farrar Straus and Giroux Publishers.
Menon, N. (2017, Oct. 24). Statement by feminists on Facebook campaign to "NAME AND SHAME." Retrieved from https://kafila.online/2017/10/24/statement-by-feminists-on-facebook-campaign-to-name-and-shame/.
Murphy, M. (2012). *Seizing the means of reproduction: Entanglements of feminism, health, and technoscience*. Durham, NC: Duke University Press.
Noble, S. U. (2018). *Algorithms of oppression: How search engines reinforce racism*. New York, NY: New York University Press.
Prashad, V. (2018, Oct. 29). India's Me too movement: An interview with Raya Sarkar. Retrieved from www.counterpunch.org/2018/10/29/indias-me-too-movement-an-interview-with-raya-sarkar/
Radio Cargo and SpiderAlex. (2018, Nov. 16). Wild waves, feminist futurotopies. Rencontres bandits-images & European media art platfrom. Retrieved from http://rencontres2018.bandits-mages.com/ondes_sauvages/?lang=en
Rentschler, C. (2017). Bystander intervention, feminist hashtag activism, and the anti-carceral politics of care. *Feminist Media Studies*. doi: 10.1080/14680777.2017.1326556.
Rentschler, C. (2019). Making culture and doing feminism. In T. Oren & A. Press (Eds.), *The Routledge handbook of contemporary feminism* (pp. 127–147). London: Routledge.
Seddon, D. (2016, June). "We will not be silenced": Rape culture, #RUReferencelist, and the university currentlyknownasRhodes.Retrievedfromwww.dailymaverick.co.za/opinionista/2016-06-01-we-will-not-be-silenced-rape-culture-rureferencelist-and-the-university-currently-known-as-rhodes/
Sichler, A., & Prommer, E. (2014). Gender differences within the German-language Wikipedia. *Essachess, 7.2*(14), 77–93. Retrieved from www.essachess.com/index.php/jcs/article/view/250/281
Slack, J. D., & Wise, M. J. (2015). *Culture + technology: A primer* (2nd ed.). New York, NY: Peter Lang.
Srnicek, N., & Williams, A. (2016). *Inventing the future: Postcapitalism and a world without work*. New York, NY: Verso Books.
SSL Nagbot. (2016). Feminist hacking/making: Exploring new gender horizons of possibility. *Journal of Peer Production, 8*. Retrieved from http://peerproduction.net/issues/issue-8-feminism-and-unhacking/
Toupin, S. (2014). Feminist hackerspaces: The synthesis of feminist and hacker cultures. *Journal of Peer Production, 5*, 1–9.
Toupin, S. (2017). Le hacking féministe: La résistance par la spatialité. In M. Bonenfant, F. Dumain & G. Trépanier-Jobin (Eds.), *Les pratiques transformatrices de l'espace socionumérique: appropriation, résistance et expérimentation* (pp. 163–182). Montreal: Presses de l'Université du Québec à Montréal.
Toupin, S., & Haché, A. (2015). Feminist autonomous infrastructure. Global information society watch: Sexual rights and the internet. Retrieved from https://giswatch.org/sites/default/files/gw2015-hache.pdf
Toupin, S., & Spideralex (2018). Radical feminist storytelling and speculative fiction: Creating new worlds by re-imagining hacking. *Ada: A Journal of Gender, New Media, and Technology, 13*. 10.5399/uo/ada.2018.13.1 Retrieved from https://adanewmedia.org/2018/05/issue13-toupin-spideralex/
Wajcman, J. (1991). *Feminism confronts technology*. Cambridge: Polity Press.
Wikimedia Foundation. (2018). Advancing gender equity. Retrieved from https://meta.wikimedia.org/wiki/Gender_equity_report_2018

24

Postcolonial Peer Production

Maitrayee Deka

1 Introduction

Peer-production communities have come a long way since Yochai Benkler published his celebrated work, *The Wealth of Networks* in 2006. Since then, mediatized networks have become a global phenomenon while at the same time getting embedded in local cultures in unique ways. Some of these networks carry radical potentialities, while others have been co-opted under a capitalist regime. In the present day, it is the latter phenomenon that has become dominant. Many scholars are critical about the metamorphosis of decentralized networks into rent based capitalist structures, particularly the dominance of Airbnb and Uber style actors in the sharing economy (Srnicek, 2017; Zuboff, 2019). In spite of recent developments, there are communities and activists that have kept the spirit of a participatory community alive. This resistance has gained further momentum through sustainable blockchain based technology, freeing organizations and members from the clasp of big financial actors and laying out a workable structure of mass scale peer-to-peer exchanges. We see this happening at the level of national governments as well as in small communities, a movement that offers a real possibility of an alternative future (Bauwens, 2009; Bennett, Segerberg, & Walker, 2014).

Even when activists-led discourses have been highly critical about the monopolistic nature of digital platforms, the literature has not been able to grasp the realities of people operating within a postcolonial context. The oversight is hardly intentional and has more to do with the complex histories of postcolonial nations. More often than not, the overthrow of an imperial power created another elite structure, not very different from the one that colonizers had left behind. Whenever a Southern story (Connell, 2007) has been highlighted, it is the story of technological elites and not that of the majority of the people who generally operate in the interstices of law and civic rights. In a country like India for instance, most developmental programs, including technological ones were fashioned by and for the educated middle- and upper-class citizens. This led a large section of the population to jump onto the technology bandwagon through semi-legal channels, and partake in bazaar style innovation with particular groups located outside networks of corporate elites and the professional middle class. Gordon Mathews et al. (2012) have defined the non-elite networks of semi-legal goods and services as "globalization from below," where street-level markets sell cheap and cheerful products aided by ubiquitous Internet connectivity and smartphones. The chaotic bazaar networks create a subterranean modernity for non-elites in the Middle East, Asia, and Africa and also parts in Europe (Pine, 2012). China became an important node in lower-end globalization as low-cost products connected peddlers in Laos, Delhi, Hong Kong, and Naples. Adjacent to factories that produced branded smartphones in China were small units that assembled components for knock-offs such as "hi-phone" that reached nooks and crannies of the global popular markets through trade nodes such as Hong Kong's Chunking Mansion and MKT in Bangkok.

Interestingly, many of these street-level markets and bazaars became part of an innovation and knowledge culture that displays some of the classic characteristics of peer-to-peer communities: sharing and collaboration. Although not plugged into any of the rhetoric and practices of dominant versions

The Handbook of Peer Production, First Edition. Edited by Mathieu O'Neil, Christian Pentzold, and Sophie Toupin.

of peer production, of high skilled labor and technology, in practice street-level markets exhibited sharing cultures with unique relations to knowledge acquisition and transfer. This was made possible by the nature of bazaar transaction that traditionally depended on informal networks of people and popular knowledge in the absence of formalized institution of trust. A place for commerce was a place for sociality, where actors shared their skill and tools in the backdrop of small exchanges.

This chapter will focus on India to give an overview of postcolonial peer production (PCPP) and show how the bazaar model provides an alternative perspective to peer production. It does not provide an exhaustive survey of the various discussions and practices around peer production. Instead, the focus of the chapter is on the important strands that make a historical study and contemporary analysis of PCPP possible. To this effect, the chapter presents bazaar-level innovation and organizational structure as a unique bottom-up model of peer production. It situates bazaars in the debates and literature of peer production in two ways: (1) Bazaar knowledge as part of a long tradition of popular knowledge, and (2) Engagement of the bazaar model with dominant Euro-American versions of peer production, particularly by investigating the value-virtue relationship in both these models. Throughout the chapter "bazaar" and PCPP will be used interchangeably as the primary empirical and theoretical focus.

The chapter starts with a brief political history of the Indian context to show why a particular type of PCPP emerged. It develops traditions of popular wisdom and embodied knowledge as open repositories for non-elites to learn and tinker with technological products. The chapter moves into a more specific discussion on peer communities, particularly through a focus on money, creativity, and modes of organization. WeMake, a fablab based in Milan will be used as an empirical example of the Global North model to situate PCPP in a comparative context.

It is important to highlight that the PCPP model discussed here comprises men within the 19 to 55 age group. Delhi's electronic bazaars are masculine spaces, which means that the forms of sociality and innovation cultures described in the chapter do not extend to women. Women have been part and parcel of the technological landscape in India, however they have been more active in the elite networks of knowledge production and call center-based global work. Within the bazaar networks, women have been active in sectors other than electronics, which might be very well stem from the marketplaces starting as male dominated spaces or from the type of commodity that did not attract the attention of women vendors. One could also connect the absence of women in the electronic marketplaces to patterns of interaction in public spaces that have traditionally been masculine spaces, although there has been no systematic study of this relationship in the electronic marketplaces in Delhi.

2 A Brief History of PCPP: An Indian Perspective

The Indian technological question is rooted in its colonial past to the extent that the approach to science and technology became a way to break away from the erstwhile colonial rulers as well as to pose a distinct identity for an independent nation. India was a British colony for more than four centuries, starting under the tutelage of East India from the early sixteenth century to the mid-nineteenth century, after which the country came directly under the British crown until it gained independence in 1947. Initially, the discourse around science and technological development became a way to give credit to "native science" which under colonial rule was relegated to museums and a pseudoscience status. Noteworthy amongst these voices were that of leaders such as Mahatma Gandhi and Ram Manohar Lohia; both in their ways resisted an out and out model of Global North-centric industrial development. Instead, their work gave prominence to ground-level realities; India at that time being an agrarian country entrenched in local-level caste and class politics. In contrast to Gandhi, who was keen on strengthening cottage industries, Lohia was not entirely against machinery and electrical power. Instead, his concern was more about finding "appropriate technology, one that accorded priority to equality over productivity, and encouraged decentralized governance and autonomous, connected villages" (Basole, 2010, p. 106).

In the end, the vision of Jawaharlal Nehru, independent India's first Prime Minister, prevailed. Nehru set the country on the path to modern scientific and technological development. India went through a series of five-year plans, each plan focusing on the development of critical sectors, a significant part of which was the advancement of modern scientific and technological knowledge. Nehru's friendship with many leading industrialists and technocrats of that time helped shape the premier institutions including the Indian Statistical Institute, the Bhabha Atomic Research Centre, and the Tata Institute of Fundamental Research (Subramanian, 2014). Emphasis on the public over private enterprises together with India's non-aligned status during the Cold War led to the growth of both USSR- and US-style technological institutes, the latter having a more lasting impact on the technological landscape of the country. The first specialized technological institute, the Indian Institute of Technology Kharagpur was modeled after the Massachusetts Institute of Technology (Raina, 2017; Sharma, 2015).

A further push towards Global North-centric science and technological development occurred in the early 1990s when the Indian government liberalized its economy, opening it to a slew of global companies and commodities. Moving into the 21st century, the state-sponsored knowledge economy promoted technological institutes, import substitution measures, and IT parks. Colliding with these changes was the rise of a "new middle class" of white-collar urban professionals together with the emergence of a "corporate" class (Chatterjee, 2008). Around this time, there was a shift in the power structure as the traditional elites of rural gentry, politicians, and civil servants (Kaviraj, 1988) accommodated a new class of business elites. The corporate class became the central actor designing the technological landscape of the country. This situation is not unique to India, and is symptomatic of a global trend where Silicon Valley-style entrepreneurs and venture capitalists increasingly invest and design urban spaces and amenities.

A parallel development to the state-supported and increasingly big business-led knowledge environment was the emergence of a group of activists and programmers. The growth of technological institutes and educated graduates created an alternative climate of knowledge production that did not follow the principles and expectations of a corporate model. Chris Kelty (2008) observed that Indian programmers were participating in the early web-based newsletters like *Tasty Bits for the Technological Front* about the same time that these platforms were becoming popular in the Global North. Many of the early bulletin board members would go on to become active members of the FOSS community. Amongst the various projects, that practitioners contributed were government organized Hackathon for an open and transparent data culture.

In both official and critical practices, a large part of the Indian population is left out. This large set of the population lies outside the network of specialized knowledge and global creative culture. They are further marginalized by the high prices of consumer goods alongside the dominance of a branded aesthetic. In order for popular classes to enjoy consumer products at an affordable price, pirate networks have sprung up. A certain type of pirate media urbanism in India started in the 1970s and 1980s. One important event was the ninth Asian Games held in Delhi in 1982, after which television reached Indian households. There were transnational video and film networks that brought in pirated media content from countries such as Pakistan and made it part of a domestic network through informal cable network providers. What Peter Manuel (1993) calls the "cassette culture" further provided a broader base for a popular music industry as Hindi movie songs were remixed and recorded by new singers. The growth of the pirate music scene aligned with the emergence of a distributive network of media products: the so-called "suitcase entrepreneurs" smuggled various products from countries such as Hong Kong, Dubai, and Thailand by striking informal deals with customs officials (Deka, 2016).

Despite piracy generating employment and access to consumer goods, both the government and critical movements have been wary of it. A serious concern raised by a series of activists was the use of piracy for commercial purposes. Postcolonial media theorists such as Ravi Sundaram (2010) regarded the commercial lens as narrow, unable to gaze at the multidimensional life and politics of piracy, especially if we consider issues of access and informal networks of governance. Instead, the pirate is a "precarious citizen" who is not bound by the same legal rules and responsibilities as the elites, for the simple reason that non-elites do not enjoy additional privileges that follow a liberal citizen (Krikorian & Kapczynski, 2010). As such, the non-elites create their channels of access by trespassing legal requirements and the aesthetics of "authorized" consumption (Nakassis, 2016).

Even though until recently anthropologists and media theorists were the ones focusing on the disruptive nature of pirate cultures, "bazaars" have not gone entirely unnoticed by scholars and activists from the Global North. Already in the 1990s, Eric S. Raymond wrote his essay *The Cathedral and the*

Bazaar, where he compared the gilded medieval cathedral to the restricted development model of the Free Software Foundation and the open atmosphere of the bazaars to that of Linux. Raymond (1999, p. 21) referred to Linux software as "a great babbling bazaar of different agendas and approaches." Lori Byrd Phillips (2013) argues that museums should embrace the spirit of the bazaars, making archives public and open. Interestingly, for all these authors, the term "bazaars" has become a metaphor, standing for something other than their actual historical or contemporary existence. Even when real marketplaces have appeared, they are mostly exotic bazaars of the Orient; unfamiliar and irrational.

There has been no detailed account of physical marketplaces to understand how traditional economic systems, while continuing their network-based knowledge, have upgraded themselves to innovate and sell advanced technological products. This oversight has been one of the reasons why a people-centric peer production has so far been outside the scope of the literature of peer production. In this chapter, the focus is on bazaar-based peer production, to account for one of the most extensive networks of technological knowledge in a postcolonial context. Ironically, the dominant paradigm, the IT industry that put India on the global map, seldom addressed the demands of the lower section of its population. To put it differently, the inability of government (as the primary utility provider) to gaze at everyday use and desire for technology have forced popular classes to come up with their systems of knowledge. The hard truth remained that many of the official plans and policies, including for instance the Delhi Development Plan of 1962, excluded the popular classes from the official map, because of their involvement in so-called irrational and "hazardous" employment (Sundaram, 2010). Interestingly the bazaar model did not emerge because of the successes of government or critical projects, but despite them, as a way for ordinary people to claim the rhetoric and practices of a networked society.

PCPP develops from the collective amnesia and neglect of elite actors regarding the needs, constraints, and aspirations of the popular classes. The primary driving point of PCPP is not creativity and knowledge per se, but has to do with developing channels of access to existing technology that have been exclusionary because of the absence of a political will to change the status quo. Whenever certain access to available technology is unavailable, bazaar-level practitioners innovate to create alternatives, or if need be, new products with comparable features. The sociability of the bazaars (emerging mainly through bargaining practices) makes it amenable to a diverse range of consumers, as a result of which they build customer-friendly alternatives at affordable prices.

Broadly, the chapter discusses the primary features of bazaar-based peer production, and how they differ or extend the tenets of peer production. The first part analyzes the bazaars side by side with traditional bases of popular knowledge, as a way to carve out historical roots and bring out their essence. The rest of the chapter situates bazaar knowledge and innovation in the dominant literature of peer production. The empirical material comes from ongoing fieldwork with traders of video games in Lajpat Rai market and Palika Bazaar, and street vendors in Nehru Place in Delhi. Lajpat Rai is primarily a wholesale market of electronics in North Delhi. Palika Bazaar, situated in central Delhi, is a retail market of a variety of consumer items, a large part of which are electronic products. Nehru Place is a market of computers and computer accessories in South Delhi. I have interviewed 50 street vendors and 38 traders in the three marketplaces as part of a year-long ethnography from September 2012 to September 2013. Since then, I have continued to study these markets; the most recent visits have been in July 2018. Semi-structured interviews, observation, and conversations constitute the central part of data collection. The empirical material on WeMake, Milan is part of a five-month long study in 2015 that I conducted as part of my postdoctoral research with the European Union funded FP7 project, Peer-to-Peer (P2P) Value. In the course of the fieldwork, I interviewed 15 core, middle, and periphery members of the community and attended various workshops and meet-ups. WeMake is a fab lab specializing in 3D printing, wearables, and laser cut technology.

3 Traditions of Popular Knowledge

In the second edition of *A Handbook on the Use of Bazaar Medicines of India*, Santappa Chetty (1900) of the Mysore government museum curated a long list of medicines for various ailments. What was interesting in this list was that none of the cures were factory-manufactured medicines but were rather an amalgamation of everyday herbs and spices. Out of the many household remedies, there was onion and garlic prescribed for phlegmatic fever and elephant creeper root for rheumatism. What is visible in the field of bazaar medicine is present elsewhere: alternative and different usage for everyday objects and resources. This attitude in the scholarly and journalistic literature has appeared as a non-western approach to the external world including science, particularly when compared with the "scientific" approach of the colonial rulers (Nandy, 1980). In contemporary times, informal knowledge finds its most durable base in the popular classes, who have limited access to cultural and economic capital. Their way to stay on the map has been to grasp fragments of resources, in whatever form they may come: knowledge, objects, and people in order to carve livelihoods in profoundly changing political and economic context.

Take for example, Amar Dhand's (2009) account of heroin addicts in Yamuna Bazar and how their first education about drug use, as well as treatment, is provided by "street doctor" and "peer-based systems of practical medicine." If things went wrong, then heroin addicts' first contact of care came in the form of seasoned addicts who applied herbal medicine and grounded pills to the wounds. A different anatomical knowledge develops because of peer knowledge and the sharing of care responsibilities.

Perhaps the most vivid example of a frugal "human technical assemblage" is the Indian phenomenon of *jugaad* (Rai, 2015). Radjou et al. (2012) in their book, *Jugaad innovation* attach six main attributes to such practices: seek opportunity in adversity, do more with less, think and act flexibly, keep it simple, include the margins, and follow your heart (p. 19–20). A lot has been written about *jugaad*, hailing it as India's answer to innovation for a popular base. Amongst the din of celebratory calls, one acknowledges the inability of the government to provide safety nets, and public services, as a result of which, the non-elites have to resort to ad hoc measures. As Manu Joseph (2018) writes in a popular e-newspaper, "The existence of 'jugaad' is evidence that the circumstances of a society are so bad that its smart people are doing what smart people in other civilisations do not have to do."

Even considering the unequal power and resource structure that push ordinary people to *jugaad* style fixes, we cannot imagine the extent of marginalization that would follow if such ad hoc measures were unavailable. Without street-level innovation, quick fixes and smart tactics, non-elites would have remained outside contemporary experiences or would have entered these worlds at a much later stage. For instance, the Indian government does have many ICT (Information Communication Technology) programs in schools, and there is an increasing push towards making computer education a compulsory feature of all government schools. However, those who are outside a school system would not be able to avail these training programs. Furthermore, there is no real estimation of what the non-elite demand for technological education could be. In many cases, the schools offered basic computer literacy while many of the students might have niche needs and skills. Technology in the hands of the non-elites does not have to prepare them for formal employment alone but also fits around entrepreneurial skills.

Historically the inability of elites to provide an inclusive public infrastructure has pushed the popular classes to create their unique tradition of knowledge and innovation. Bazaar knowledge is as much a story about survival and inclusion, as about connectedness between a peculiar range of objects and people. In John Zubrzycki's (2018) book, *Jadoowallas, Jugglers and Jinns: A Magical History of India*, the popular classes' pursuit of employment defines the creation of different strands of knowledge. They were excluded from the Brahminical traditions of scriptural knowledge, and their way of making ends meet was to create an audience for their traditional practices. In the world of street magicians and jugglers, the novel movements that their body conjured up often guaranteed a livable income. The street magicians and their acclaimed "Indian rope trick" became so well known that magicians would travel to England to put up a show. There was not a dull moment in these shows: a monkey and a boy climbed a rope and disappeared in the sky and fell down a few seconds later.

Such stories of street performers, jokers, and jugglers are not unique to India. These marginalized figures are present in Bakhtin's (1984) account of marketplaces in Europe, where the carnivalesque spirit portrayed the coarse humor of Rabelais and other marginalized figures such as the jester and the prostitute. Peter Stallybrass and Allon White (1986) described a 19th-century village fair in England as, "Even the smallest fair juxtaposed people and objects which were normally kept separate and thus provided a taste of life beyond the narrow horizon of town or village" (p. 30). The fair "turned the world inside out" in its mercantilist aspect just as much if not more than it "turned the world upside down in its popular rituals." There was something about marketplaces that carried within them endless possibilities to entertain bystanders. More important for the argument here is the aspect that historically commerce accommodated a large pool of "unskilled" and marginal labor.

In fact, it is nomadic "Romany gypsies and Irish travellers" that popularized the term "tinker." The Romas' use of scarce resources, moving from house to house mending pots and pans were one of the first modern examples of popular classes' commerce keeping alive a culture of merging random objects and people (MacRitchie, 1894). If we bring the argument to contemporary times and focus on the bazaar knowledge in Delhi, we see that the traders tinker with advanced electronics with the aid of everyday objects. Scottish tinkers used clay to repair pots and pans, from which the phrase "tinker's dam" originates. Similarly, bazaar traders soldered chips to the circuit board of gaming consoles. For both Scottish and bazaar tinkers, inexpensive technology and finding opportunities where others did not enabled them to survive on the margins.

Popular knowledge does not come from one single source; it arrives as half-baked ideas. The informal nature of popular knowledge is what the novelist Arvind Adiga speaks of through his protagonist Balram in *The White Tiger*:

> Me, and a thousand others in this country like me, are half-baked because we were never allowed to complete our schooling. Open our skulls, look in with a penlight, and you'll find an odd museum of ideas: sentences of history or mathematics remembered from school textbooks . . . sentences about politics read in a newspaper . . . triangles and pyramids seen on the torn pages of the old geometry textbook which every tea shop in the country uses to wrap its snacks in . . . Entrepreneurs are made from half-baked clay. (Adiga, 2008, p. 10)

It is perhaps this disjointed learning process, a certain irreverence towards formal methods, which creates a vibrant innovation culture at the bottom of the society. A fundamental way to bring together the different traditions of popular knowledge, as I have argued elsewhere (Deka, 2018), is the prominence placed on the physical body. Without falling into the trap of the modernist binary of the body and the mind, of which scientific thought is a manifestation, the centrality of the body shows a flexible approach to knowledge. Unlike a textual tradition that is continually borrowing and building a field of expertise, bodily knowledge is chaotic. There is a remembrance of past knowledge, but also unlearning and new learning that happens on the spot. Depending on the ways one's fingers worked on the machine, new knowledge about the product can be unearthed. Or the time that one spent working laboriously "modding" consoles brought the traders one step closer to time-saving and economic solutions.

Marginal figures throughout history employed what was at their disposal, bodily knowledge when dominant knowledge and privileges were enclosed within selective groups. The reliance placed on bodily knowledge meant that practitioners could enter the creative process at any point without formal education, often training the body through a trial-and-error process. It further allowed for a flexible culture of creativity, as most of the times, an accidental sleight of hand can lead to a new fix. A repairperson in Palika Bazaar can spend hours with a broken console trying to figure out what the problem was with it. Some of the routine procedures included plugging the console to a switchboard and see if it reprised at the spur of the moment. Other tactics included connecting the console to the TV monitor at the shop. If the screen turned blank, then the repairperson opened the motherboard and tested various chips with the soldering wire. Once they were relatively satisfied with the state of the console, they would connect it to a monitor. Sometimes, the console worked for a short period of time and went blank thereafter. With every attempt made by the trader, they gained a better perspective on how to tackle the problem. They fixed the console on the same day if no other tasks needed their immediate attention. Otherwise, they worked on it another day, giving the impression to the consumer that it was a difficult repair. Contrast this situation to the digital commons of our times. Almost all communities presume expertise in programming language. This created an unspoken hierarchy between software programmers and other roles. Rozas et al. (2018) observed in the free and open source Drupal platform that affective labor does not get as much visibility as writing code, even though it is equally important, if not more important for the community.

In practice, bazaar-level repair had no fixed rules of identifying the value of knowledge and the correct methods as long as the problem got solved. It had a flexible approach when it came to following one's instinct and rhythms of the body. Embodied knowledge encouraged a different culture of sharing. Most knowledge developed by observing what others were doing. In the subsequent sections, we go deeper into aspects of sharing and creativity through the value-virtue paradigm that has been so consequential in building the ideology and organizational framework of dominant Euro-American peer production.

4 Money and Creativity in PCPP

Ever since the publication of Karl Polanyi's (1944) *The Great Transformation*, there has been a thorough revision of the devastating outcome of a modern economy. Polanyi related the development of a market society to a gradual erosion of economic practices from social relations subsequently creating self-interested market actors. Polanyi's premise has encouraged a stream of alternative ideas and thoughts, principally looking at how to re-embed the economy in superior moral values, or what Benkler and Nissenbaum (2006, also in this volume) call "virtue." They analyzed the capacity of commons-based peer production (CBPP) to create meaningful social action. Benkler (2004) has also argued elsewhere, that social motivations like "sharing nicely" is what drives actors to take part in productive CBPP communities whether it is the FOSS movement or other public and community-based projects. In his work, CBPP carries within it a holistic promise – the fulfilling aspect of being creative, greater autonomy, and the joy of contributing to decentralized projects. In such renditions, money or economic pursuits become problematic.

The general distrust that peer-production practitioners show towards money springs from an understanding that the survival instinct is not very different from a capitalist one, where individuals can go to any extent to procure their interest. In practitioners' opinion, a collective vision is missing in such a survival-based society (Himanen, 2001). After all, if one is concerned with one's bread and butter, then critical network communities do not see within this model, the time or the inclination to build a better society. The main ideological concerns of any peer-production communities such as fablabs and collaborative platforms are future-oriented (Arvidsson, 2018) working towards a fairer society, wherein in the best possible scenario nobody has to worry about livelihood concerns.

In fact, empirical studies on Euro-American peer communities highlight their concerns for building future communities. In WeMake, a fablab in Milan where I conducted a six-month-long empirical research project in 2015, I observed that many of the participants framed their rationale for being part of such a community through the possibility of creating an alternative future. X, one of the founders of WeMake problematized the current state of mass production, where a company like IKEA controlled an entire industry selling low-quality products. Instead, in their words fab labs provided initial steps for developing small-scale productive units through 3D printing. Fab labs exist as a way to imagine and produce a more emancipatory and participatory future. In such a narrative, survival concerns drag people into narrow and parochial concerns when there is a real possibility to build an inclusive society. This genuine responsibility that peer-production practitioners feel towards future generations legitimizes a certain discourse of creativity, as a catalyst for change. However, we observe that in reality, it is not just acts of creativity that are celebrated; there also emerges a cult of the creative person.

G, a peripheral member at WeMake, fills her time at the community working with laser cut technology. Her real ambition is to create something that can go up in a gallery or an art exhibition. G says after her name gets attached to her creation, she does not mind sharing it with anyone or uploading the prototype on GitHub. Creativity becomes a meaningful way to create a reputation for individual members in peer-production communities which in turn guarantees rewards for their valuable contribution. Interestingly for many of them, reputation affirms status in the group, and inadvertently categorizes certain acts and people as creative (Gandini, 2016; Hassan & Filippi, 2016). There is a sidelining of the role of chance, errors, and accidents at the altar of a creator.

The discourses around money and creativity emerge in a completely different fashion under PCPP than what is outlined by dominant versions of peer production. First, money in the bazaar does not get conflated with capitalism. It is a scarce good that one earns after a day's work. Further money acquires new meaning based on the context it is used and the motivations behind earning it, the primary one being taking care of the survival needs of the family. The embedded understanding of money is not new to the social sciences. Anthropologists and economic sociologists have argued for a long time that money circulates in a web of human interaction and social meanings (Zelizer, 1995). In most societies, money or monies are part of face-to-face market exchange and social interactions. Braudel (1983) in *Wheels of Commerce* gives us a glimpse of this world of everyday trade of bread and wood that sellers bring to the market from different producers. This dense interaction is the sphere of petty producers, traders, and people; money passing hands in densely populated haats, bazaars, and marketplaces (Bayly, 1983). They operate under the radar of capitalism, but that does not imply that these popular marketplaces follow the same ethic and principle of capitalist accumulation. In small-scale economies, the forms of collaboration and innovation paradigms develop a particular relationship to money. Here, money becomes a way to earn a living, pay rent, and sustain a family. As the basic level resource is precarious, this results in everyday innovation getting co-opted within a survival paradigm. Even if there is a steady income for a particular period, the fear remains that income might dwindle in the near future; this predicament ties the petty traders to a survival paradigm.

Similar to the world of money, the attitude of bazaar-level actors towards creativity is complex. On the one hand, there is no real estimation of what they do as creative. On the other hand, there is an understanding that creativity exists in a more formal environment. Harish in Lajpat Rai market keeps abreast with different technical repairs. After a certain amount of time passes working with a new problem, he develops time and resource-saving solutions. For example, in the beginning, he was "modding" PlayStation 3 consoles by downloading the hack from a website and then re-writing it on a customer's console. The problem arose when the code did not work. One day he learned from a trader in China (with whom he communicated through an e-messaging app) that there were producers who were making chips to mod consoles. Harish imported a few chips and was happy with the result; he did not have to struggle with a corrupt code anymore. There was more time and space for him to concentrate on niche repair work or come up with new products for the market. After Harish found out that "modding" PlayStation need not be time-consuming, he came up with the idea of selling pen drives loaded with pirated games. When the consumers came for their recycled consoles, he offered them the pre-installed pen drive at an additional price. The product became popular as gamers could now play many more franchises without individually paying for them.

Harish did not consider any of the activities in the bazaar as creative. In his words, they were an act of *jugaad*, desperate and careless measures to fix a problem. He understood creativity as something to do with an original idea and its implementation more often than not requiring expensive and sophisticated technologies. Like Harish, most traders in Delhi's market had a complicated relationship with the notion of creativity. For one, they all felt alienated from a creative process they saw flourishing in an enabling environment of advanced technologies, and fat salaries. Many of the traders believed that they never had the chance of becoming a creative person, as they do not hold educational certificates and "degrees."

The opinions about money and creativity bring us back to the question of *jugaad*, and whether it is in essence uncreative. Or, instead, we need to get used to a different definition of creativity that does not attach creativity to a talented individual or a systematic mode of thought. This radical outlook of analyzing knowledge as necessarily fragmented and emerging out of an ad hoc human-technological assemblage is not entirely a new frame of thought. Bruno Latour (1993) has described how Boyle's experiment on the vacuum pump was a long series of trials and errors, tinkering with wayward objects and tools. Even then, the scientific world, as well as that of peer production, celebrates to an extent the Renaissance figure of the individual genius such as Leonardo da Vinci, and their ability to create.

The zeitgeist does affect the bazaar-level traders conclusively. Everything they do becomes a way to create economic value. In this way, they exist outside the scope of a Euro-American vision of creating communities around virtues of creativity, autonomy, and passion. To include bazaar-level innovation as creative, we would need a complete rethinking of both our relation to the surrounding objects as well as dense market relations from a capitalist one. PCPP provides a pragmatic approach to value and virtue, where economic value becomes a way to survive in the margins. What emerges as a virtuous action is strongly tied to a lived ethical universe, an aspect that is developed in the next section.

5 The Organizational Structure of PCPP

One of the most celebrated aspects of peer production is its organizational logic. Many of these communities developed out of the initial reservations about corporations' gatekeeping knowledge production and circulation. This awareness has led to the unique organization principle of peer-production communities, many of them existing as decentralized bodies with particular mechanisms of collaboration and cooperation. Siobhan O'Mahony (2006) observes, in the Debian project, that decision-making occurs through a democratic process, where the addition of a new feature has to be vetted by each member. However, such a community for everyday operation also creates bureaucratic structures to efficiently assign different tasks to different members.

The organizational logic of many peer production communities comes from a very well-thought-out plan, often times communicated to a wider public through social media or a manifesto-like document. In the bazaars instead, the organizational logic develops from traditional practices that have historically governed such marketplaces. Bazaars have been open places, as it is to their advantage to attract a diverse range of consumers and people to counter the classic dilemma of information asymmetry. Also in contemporary times, vis-à-vis other places of commerce like shopping malls, supermarkets, and online markets, bazaars have literally to scrape the bottom of the barrel to hold a consumer base.

The openness of the bazaar comes out immediately through the aesthetic of such places; they are crowded and full of people at any moment of the day. Not all of them are buyers, most are loiterers, tourists, and romantic couples on dates. This openness also allowed a range of marginal figures to find their footing in a new city, which came out for instance in Aman Sethi's (2012) *A free man*, where the protagonist Mohammed Ashraf navigates a host of popular commercial places including the wholesale market Sadar Bazaar and the underbelly of Delhi's organ market to survive. There is of course a deep-seated deprivation that is part of bazaar-like places, where extreme forms of corruption and desperation come together. However, bazaars are also sites for camaraderie and a last resort for those who do not have other ways to make ends meet. Many young men that I met in the three marketplaces started as migrants in a new city from the hinterlands of Delhi and some of them were slum dwellers, as was the case with most of the street vendors in Nehru Place. Their stories about gaining access to the marketplace followed the trajectory of how they stumbled upon the market in one of their aimless wanderings. The markets fascinated the young men because of the activities surrounding them; crowded with people and objects. They would come back and repeat these day-long adventures until someone at the shop or the street said, "as you visit the market so often, why don't you come and work for me?" The simple suggestion opened for many young men an earning opportunity in the marketplace.

If we focus on trade secrets, traditionally a bastion of enclosed knowledge, one observes that rather than a particular trader perennially protecting trade secrets, there are situations when holding on to knowledge helps. For example, if a trader develops an innovation such as selling secondhand DVDs, there is a small window when the new product draws exclusive profit for him. Very soon other traders will copy his innovation, a common practice given that traders do not think of their creative process as sacred. They are aware that their ideas are constantly coming from multiple sources, often as half-baked information from YouTube comments and learning from observing others in the marketplace (Deka, 2018).

A general tolerance towards copying does not mean that hierarchies do not exist within PCPP. One of the hierarchies that develop with regards to the organizational structure of bazaar knowledge is between people using knowledge gainfully and others who are merely sellers. In Nehru Place, the seasoned traders and repairpersons regarded the street vendors who sold pirated software and games as wasting away their potential. Their main complaint was about how these young men ran after quick money. Skilled technicians saw alcohol and smack addiction as the worst habit afflicting young street vendors. There is an existing ethical universe that places work and money in a stream of responsibilities: (in no particular order) to look after one's ailing parents, to start a family of one's own, have savings and, offer money in temples. As the street vendors failed to accomplish many of these responsibilities, they did not appear as an ethical market actor to senior traders in the marketplaces.

Amongst the repairpersons and traders, there is another hierarchy, arising from customary ties that have continued from the old ***karkhana*** (factory) model, guild-like structures of the 18th and 19th centuries. Tirthankar Roy (2008) has extensively written about the guilds in modern India and how during the Mughal rule, they gained prominence under the patronage of merchants and nobility. One of the relationships that were specific to these institutions was that of the ***ustad*** (teacher) and his ***karigar*** (workman), much like the master and apprenticeship relationship of medieval Europe. Even though contemporary bazaars do not have the same organizational and commercial role that guilds played in the past, some traditional nomenclature continued. Principally the relationship developed between the ***ustad*** and the novice. The ***ustad*** is the trader under whose exegesis a newcomer learns the trick of the trade. Lambu, a shop assistant in Palika Bazaar recalls how in the initial days his ***ustad*** asked him, "What are you doing standing outside the shop, if you keep up this behavior, you will not learn anything new! Observe what I am doing, you will slowly pick up things."

Interestingly, the ***ustad*** is not painted as a universally generous figure. There are stories of how a few teachers would hide their knowledge, making it difficult for newcomers to follow them. Pankaj, a street vendor in Nehru Place, remembers that whenever he went to his mentor in the market with a broken computer, suddenly at a crucial juncture, whether by keeping his hand at a particular angle or by working very quickly, the mentor would obscure the repairing process. Lalit moved from Uttar Pradesh as a young man, and trained for a couple of years under his ***ustad*** in Lajpat Rai market. The shop appeared like a laboratory where about ten young men would gather every morning to learn hardware repair. Lalit mentioned that their ***ustad*** was very strict, would often exhaust the young men before they could get any new information out of him. To this day, they were not sure if their ***ustad***'s strict attitude was because he was reluctant to share trade secrets or because he wanted his students to develop their unique style. While many others dropped out, Lalit went on to become a skilled technician and rented his repair shop at Palika Bazaar. Lalit and his ***ustad*** to date have a cordial relationship. It would have been a problem if Lalit went to work for a competitor, someone with equal respect and economic hold as his ***ustad***.

The lack of background training and embodied nature of knowledge create an open structure in the bazaars, where new entrants to the city, as well as those exiting in the margins, found means of livelihood. There is a flexible framework in the bazaars, whereby if someone wanted, they could build their knowledge and skills by simply observing what others are doing, or by training under someone. However, there are specific hierarchies that distribute people within the bazaar system. These hierarchies emerge on the basis of prevalent ethical tropes, whether it is do with religious sanctions, familial roles, or customary ties. PCPP organizes people based on their ethical responsibilities.

6 Conclusion

To include PCPP within the literature and practice of Euro-American peer-production is as much a descriptive act as it is political. Undoubtedly, the dominant peer-production communities have strong roots and justification when they talk about breaking from unequal power and economic structures. However, somewhere within this narrative, people-centric postcolonial models get left out. This has to do with many of the dominant models focusing on building emancipatory communities in the future. While these imaginations are necessary, a sole focus on a "coming community" disables engagement with practices in the "capitalist ruins." Perhaps a subverted gaze would allow for collaboration with radical practices in the present day as we embark on a journey for a participatory future.

As this chapter outlined, some of that solidarity can emerge along the lines of including embodied knowledge of marginal figures, and a creative culture that values ad hoc techno-social arrangements as much as a creative genius. Moreover, in this evolved framework, the embedded ethical universe has to be on par with external virtues. This would mean a radical rethinking of the present as we continue to work on alternative ideas. To end with the words of Donna Haraway, "Staying with the trouble does not require such a relationship to times called the future. In fact, staying with the trouble requires learning to be truly present, not as a vanishing pivot between awful or edenic pasts and apocalyptic or salvific futures, but as mortal critters entwined in myriad unfinished configurations of places, times, matters, meaning" (Haraway, 2016, p. 1).

References

Adiga, A. (2008). *The white tiger*. New Delhi: Harper Collins.

Arvidsson, A. (2018). Value and virtue in the sharing economy. *The Sociological Review*, *66*(2), 289–301.

Bakhtin, M. (1984). *Rabelais and his world*. Bloomington, IN: Indiana University Press.

Basole, A. (2010). The technology question in Lohia. *Economic and Political Weekly*, *45*(44–45), 106–111.

Bauwens, M. (2009). Class and capital in peer production. *Capital & Class*, *33*(1), 121–141.

Bayly, C. A. (1983). *Rulers, townsmen and bazaars: North Indian society in the age of British expansion, 1770–1870*. Cambridge: Cambridge University Press.

Bennett, W. L., Segerberg, A., & Walker, S. (2014). Organization in the crowd: Peer production in large-scale networked protests. *Information, Communication & Society*, *17*(2), 232–260.

Benkler, Y. (2004). "Sharing nicely": On shareable goods and the emergence of sharing as a modality of economic production. *The Yale Law Journal*, *114*, 273–358.

Benkler, Y. (2006). *The wealth of networks: How social production transforms markets and freedom*. New Haven, NJ: Yale University Press.

Benkler, Y., & Nissenbaum, H. (2006). Commons-based peer production and virtue. *Journal of Political Philosophy*, *14*(4), 394–419.

Braudel, F. (1983). *Civilization and capitalism 15th–18th century: The wheels of commerce*. London: Collins.

Chatterjee, P. (2008). Democracy and economic transformation in India. *Economic and Political Weekly*, *43*(16), 53–62.

Chetty, B. S. (1900). *A handbook on the use of the bazaar medicines of India* (2nd ed.). Bangalore: Mysore Government Museum.

Connell, R. (2007). *Southern theory: The global dynamics of knowledge in social science*. Cambridge: Polity.

Deka, M. (2016). Bazaars and video games in India. *Bioscope*, *7*(2), 172–188.

Deka, M. (2018). Embodied commons: Knowledge and sharing in Delhi's electronic bazaars. *The Sociological Review*, *66*(2), 365–380.

Dhand, A. (2009). Street "doctory" among a group of heroin addicts in India: Naturalistic peer learning. *Ethnography and Education*, *4*(1), 101–116.

Gandini, A. (2016). *The reputation economy: Understanding knowledge work in digital society*. Basingstoke: Palgrave Macmillan.

Haraway, D. J. (2016). *Staying with the trouble: Making kin in the Chthulucene*. Durham, NC: Duke University Press.

Hassan, S., & Filippi, P. (2016, February 1) Reputation and quality indicators to improve community governance. Retrieved from https://papers.ssrn.com/sol3/papers.cfm?abstract_id=2725369

Himanen, P. (2001). *The hacker ethic and the spirit of the information age*. London: Secker &Warburg.

Joseph, M. (2018, August 18). "Jugaad," India's most overrated idea. Livemint. Retrieved from www.livemint.com/Leisure/2c3sntdHfJ8Py2tWxEqgcN/Jugaad-Indias-most-overrated-idea.html

Kaviraj, S. (1988). A critique of the passive revolution. *Economic and Political Weekly*, *23*(45/47), 2429–2444.

Kelty, C. M. (2008). *Two bits: The cultural significance of free software*. Durham, NC: Duke University Press.

Krikorian, G., & Kapczynski, A. (Eds.). (2010). *Access to knowledge in the age of intellectual property*. New York, NY: Zone Books.

Latour, B. (1993). *We have never been modern*. Cambridge, MA: Harvard University Press.

MacRitchie, D. (1894). *Scottish gypsies under the Stewarts*. Edinburgh: D. Douglas.

Manuel, P. (1993). *Cassette culture: Popular music and technology in north India*. Chicago, IL: Chicago University Press.

Mathews, G., Ribeiro, G. L., & Vega, C. A. (Eds.). (2012). *Globalization from below: The world's other economy*. Abingdon: Routledge.

Nakassis, C. V. (2016). *Doing style: Youth and mass mediation in south India*. Chicago, IL: Chicago University Press.

Nandy, A. (1980). *Alternative sciences*: New Delhi: Allied Publishers.

O'Mahony, S. (2006). Developing community software in a commodity world. In M. S. Fisher &G. Downey (Eds.), *Frontiers of capital: Ethnographic reflections on the new economy* (pp. 237–266). Durham, NC: Duke University Press.

Phillips, L. B. (2013). The temple and the bazaar: Wikipedia as a platform for open authority in museums. *Curator: The Museum Journal*, *56*(2), 219–235.

Pine, J. (2012). *The art of making do in Naples*. Minneapolis, MN: University of Minnesota Press.

Polanyi, K. (1944). *The great transformation: The political and economic origins of our time*. Boston, MA: Beacon Press.

Rai, A. S. (2015). The affect of Jugaad: Frugal innovation and postcolonial practice in India's mobile phone ecology. *Environment and Planning D: Society and Space, 33*(6), 985–1002.

Raina, D. (2017). Engineering science education and the Indian Institutes of Technology: Reframing the context of the "cold war and science" (1950–1970). *Contemporary Education Dialogue, 14*(1), 49–70.

Radjou, N., Prabhu, J., & Ahuja, S. (2012). *Jugaad: A frugal and flexible approach to innovation for the 21st century*. London: Random House.

Raymond, E. S. (1999). *The cathedral and the bazaar: Musings on Linux and open source by an accidental revolutionary*. Sebastopol, CA: O'Reilly Media.

Roy, T. (2008). The guild in modern South Asia. *International Review of Social History, 53*(16), 95–120.

Rozas, D., Gilbert, N., Hodkinson, P., & Hassan, S. (2018, November 7) Talk is silver, code is gold: Beyond "object" centric notions of contribution in peer production. SSRN Electronic Journal. Retrieved from https://papers.ssrn.com/sol3/papers.cfm?abstract_id=3272342

Sethi, A. (2012). *A free man: A true story of life and death in Delhi*. New Delhi: Penguin.

Sharma, D. C. (2015). *The outsourcer: The story of India's IT revolution*. Cambridge, MA: MIT Press.

Srnicek, N. (2017). *Platform capitalism*. London: Polity.

Stallybrass, P., & White, A. (1986). *The politics and poetics of transgression*. Ithaca, NY: Cornell University Press.

Subramanian, R. (2014). *India and information technology: A historical and critical perspective. Journal of Global Information Technology Management, 9*(4), 28–46.

Sundaram, R. (2010). *Pirate modernity: Delhi's media urbanism*. New Delhi: Routledge.

Zelizer, V. A. (1995). *The social meaning of money: Pin money, paychecks, poor relief, and other currencies*. New York, NY: Basic Books.

Zuboff, S. (2019). *The age of surveillance capitalism: The fight for a human future at the new frontier of power*. London: Profile Books.

Zubrzycki, J. (2018). *Jadoowallahs, jugglers and jinns: A magical history of India*. New Delhi: Picador India.

25

Gaps in Peer Design

Francesca Musiani

1 Introduction

In the history of Internet-based services, the concurring pushes towards different types of design choices, in particular peer-based designs versus ones based on a client/server architecture,[1] has been a long-standing source of compromises and tensions – social, technical, political, economic, and legal. In a client/server architecture, having a server separate from the user's machine, possibly managed by a third party, greatly complicates resistance to censorship, because it provides an obvious control point (DeNardis, 2014) for authorities. On the other hand, without such a server, communication between users who are not permanently connected becomes much more difficult. This is why fundamental Internet services such as email most often resort to intermediaries, able to ensure the ongoing functioning of the service, but which can also potentially stop, limit, or block this service, as well as read what passes through their servers (on the liability of Internet intermediaries in this respect see Riordan, 2016).

Efforts aimed at developing decentralized systems date back to the early Internet (Minar & Hedlund, 2001). They were generally built as ad hoc strategic responses to specific threats of shutdown. The file-sharing system BitTorrent, for example, was developed as a response to the shutdown of Napster, in order to make legal prosecution for breaches of IP in file-sharing networks much more complicated (Izal et al., 2004). A system such as Napster, which had a peer-to-peer component in file search, but was in fact being run by a small group of people with associated mandatory passage points for all exchanges, means the system is fully dependent on these nodes, technically as well as legally. As a consequence, it can be effectively neutralized by turning off servers or seizing the equipment or the people running it (Ku, 2002). In contrast, decentralized systems do not have any central servers, and the functioning of the system involves many peers (people, and the computing resources at their disposal) who do not, or may not, even know each other. If any particular node leaves the system, it continues to run regardless. Thus, the reason for the creation of decentralized systems is that such systems are designed to be resilient to targeting by authorities, and are accordingly often deemed to be superior to proprietary, closed, more centralized systems, because they value long-term robustness over cost-effective commercial expedience (Oram, 2001).

While decentralized systems have been subject to "waves" of interest in the last 20 years, starting with the early 2000s file-sharing frenzy and the hailing of p2p as a "disruptive technology" (Oram, 2001), in recent years there has been an even greater interest in, and uptake of, decentralization. In the process, the motivations for adopting decentralized technologies have broadened from a particular strategy of opposition to specific companies or pieces of regulation, to proposals of an alternative "vision" of what corporate, legal, and state institutions should be. Two main dynamics have driven this tendency: the first was the emergence of blockchain technology (in particular with Bitcoin technology, as a response to the 2008 financial crisis; see Campbell-Verduyn, 2017 and Brunton, 2019), and the second was spurred by Edward Snowden's revelations of mass surveillance operations facilitated by a number of telecommunications companies and Silicon Valley giants, on behalf of the US

The Handbook of Peer Production, First Edition. Edited by Mathieu O'Neil, Christian Pentzold, and Sophie Toupin.

National Security Agency (see Pohle & Van Audenhove, 2017). These events greatly raised the general public's awareness of the surveillance-based, and personal data-based, business models of near-monopoly tech companies, and their "dangerous liaisons" with state security agencies (Musiani, 2013).

As a consequence of these dynamics, "both decentralization and the notion of authority took on broader meaning and decentralization became a technical, political, economic and social aim in and of itself, reaching outside the 'hacker' circles of the early p2p systems" (Brekke & Isakiidis, 2019). However, this wider appreciation of decentralization as a principle and a vision is not devoid of side effects; most notably, oftentimes, decentralization has become an objective in and of itself, with little understanding of intent or assessment of actual effects. As information studies scholar and Internet pioneer Philip Agre said in 2003, "architecture is politics, but should not be understood as a substitute for politics"; decentralized protocols are too readily assumed, because of their technical qualities, to bring about decentralized political, social, and economic outcomes.

In this context, it is important to assess not only the qualities, but also the potential problems and difficulties of peer-to-peer and decentralized design, and to assess the extent to which economic, social, legal (and of course, technical) factors may get in the way of the linear "translation" from a peer-to-peer technical architecture to a successful decentralized socio-economic system, able to compete with more proprietary and centralized models. A number of factors may be the cause of this phenomenon, including the difficulty in providing proper quality of service in the early phases of a system that relies exclusively on users' contributions; the volunteer development model, which oftentimes lacks incentives for performing routine tasks; the difficulty in equaling user-friendly, sometimes non-technical aspects of proprietary systems, such as "cool" design; or the difficulty in finding a straightforward business model that can successfully be merged with decentralization.

The issue of the (de-)centralization of networked structures and organizations has been a central concern for both social and human sciences and computer science when studying the Internet and its numerous social, legal, and economic consequences. Beyond specific discussions of technical architectures and their implications, decentralization connects to broader notions of openness and access in informatics, computing, and networking – at the different layers of physical infrastructure, protocols, and content – and of information freedom. The emergent field of peer production studies has built upon, and is currently contributing to, discussions around these concepts, drawing from works in anthropology, law, sociology, and cultural studies. While this is not the place to review them extensively,[2] it is important to mention their relationship to the present work as they constitute the common foundations on which this chapter, and most chapters in this volume, rest upon.

In the legal studies field, drawing on pioneering works such as those of Yochai Benkler on sharing as a paradigm of economic production in its own right (2004) and of Lawrence Lessig on "code as law" (2002), Niva Elkin-Koren understands architecture as a dynamic parameter and explores how the law does not merely respond to new technologies, but also co-shapes them and may affect their design (Elkin-Koren, e.g., 2006). In the field of economics, Elinor Ostrom (in particular in an edited volume with Charlotte Hess, 2007) demonstrates that collective action and institutional design play as large a part in the shaping of information as do law- and market-based instruments, thereby laying the foundations for new ways of looking at knowledge as a resource shared by a group of people subject to social dilemmas and necessities of choice. From a cultural studies/anthropology perspective, Christopher Kelty has examined the politics of FOSS by asking about openness the same questions outlined above for decentralization: whether it can be considered as a good in itself, or should be seen as a means to achieve something else, such as interoperability or integration (Kelty, 2008); within FOSS practices, Gabriella Coleman has examined the role of hacker culture, exploring what hackers mean by freedom and how they enact it, as a form of self-determination that considers unrestricted access to knowledge as a necessary precondition for the evolution of their "technical art" (Coleman, 2005). In a more engaged and explicitly activist perspective, practitioners of the FOSS community, most prominently Richard Stallman, have also outlined the tenets of (networked) information freedom and the consequences of choosing particular sets of technical features, legal licenses, and business models (e.g., Stallman, 2006).

Within this broader scenario, this chapter focuses on the issue of the "gaps" in the peer-based design of the technical architecture of Internet-based services; although net architecture will be our primary focus, we will also consider how dynamics of motivation/incentives to participate in peer-based systems, and their attractiveness/usability, are fundamentally linked to the constraints and opportunities of different architectural designs. The chapter builds upon empirical material derived from two case studies[3] of particular events in the development processes of decentralized systems. The first one, the Faroo peer-to-peer search engine, made the key socio-technical choice of placing users and their computing equipment at the core of the system; as a consequence, several strategies were required to remedy technical conundrums and maintain users' enrollment in the system. The second case, the Wuala distributed file-storage platform, made, at a particular point in its history, the choice of reverting to a centralized model, so as to "simplify" the technical development process and the related business model. This revealed a number of technical and social features that pioneer users were attached to, but that developers were unable to make self-sustainable in order to maintain the peer-to-peer nature of the system.

Via these two case studies, the chapter shows that one of the reasons why these gaps are particularly problematic is that they are not merely "technical" or "social" or "economic," but a mix of the three. The lessons of these two case studies are also useful to partly explain a widely-known phenomenon related to decentralized architectures, i.e., that while user-controlled, decentralized alternatives to Internet-based services have been regularly put forward as an alternative to Internet giants, their developers have generally been unable to compete with proprietary market leaders.

2 Faroo, or the Socio-Economic Implications of a Peer Design Choice

Faroo[4] was an Anglo-German start-up that, during the late 2000s–early 2010s, proposed a search engine meant to be decentralized at several levels of its technical architecture, based on principles of affinity and common preferences among users. Born in 2007, Faroo was one of the few projects to build a peer-to-peer search engine able to go beyond the stage of thought experiment or university research project. It was led by a small interdisciplinary team including three engineers, a psychologist and a linguist, and assembled a community of about 2 million users/peers. Its developers, in the early stages of the service, could count on a core of pioneer users, particularly interested in the "full decentralization" model and the "social research"[5] principles underpinning the search engine. A peculiar characteristic of this service, meant to be p2p, was the choice by its developers to make it proprietary software – which they "justified" by the fact that the specificity of the system was in the technology they developed, not the costs of its development, thus, the open source release of the software would be giving the advantage to the competition.[6]

Faroo proposed a "six-times distributed" search, the six levels being indexation, information discovery, search, classification, bootstrapping and software updates, with the aim of building a decentralized network able to free itself from the client-server dichotomy. Pairs were mobilized as components of a socio-technical system entangling users, algorithms, and the available technical resources. The fact of not considering peers as merely a support for an intrinsically centralized system,[7] but of "taking seriously" their status of equals within the network and undertaking all development work according to this vision, became a problematic issue in the early development stages of the system. As technical blogger Brad McCarthy commented (2010),

> Peer-to-peer search has some huge advantages. Not being reliant on a single (or even multiple) data center can clearly free up a lot of concern. Socially relevant results do indeed have their place for my own use, and I'm sure they would for others as well. [... At the same time, Faroo] is not being marketed toward anyone. This leaves the entire future of the project (at this moment) in the hands of the user. [...] if you're relying on the world at large to be your network, and nearly refusing to entertain the idea of hybrid technology, you may be on the way to kill your company's shot. (McCarthy, B. (2010). "A look at Faroo's P2P Search. Will this one make it?", The Next Web, May 25. http://thenextweb.com/apps/2010/05/25/p2p-search-is-here-but-who-will-use-it/.)

Thus, the crucial importance of a peer-centered design took shape at once as the best opportunity and the greatest risk for the Faroo system. Among the opportunities was the fact that the system did not

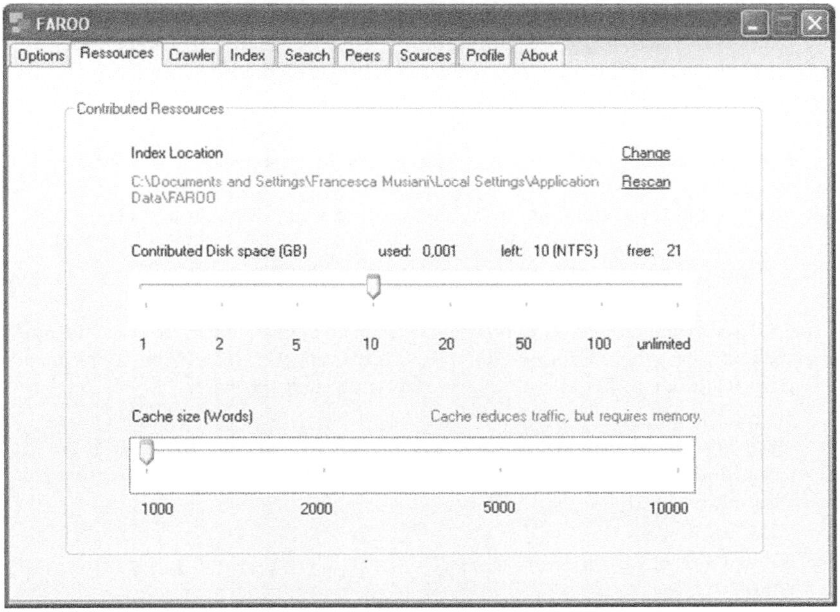

Figure 25.1 Screen shot of Faroo's control panel, allowing users to monitor and supervise their contributions to the distributed search engine. Source: Faroo control.

entail significant initial investments of resources by the firm, and was in principle sufficiently flexible to work at once in stable and unstable situations. The system was built to reconfigure autonomously after a potentially disruptive event; in the absence of a central entity, the "surviving" peers were designed to find one another again and ensure, in short delays, the reconfiguration of a distributed entity that could still perform its functions.[8]

At the same time, the centrality of users and of the material and computing resources they would bring to the system could also be problematic for its sustainability, particularly in its initial phases. Faroo "relie(d) on the entire world as a network," as McCarthy (2010) put it. This positioning could be risky, as being able to raise the interest of an important number of users was not only a matter, for Faroo, of reclaiming its niche in the market of search engines, but it was vital for the system to function correctly from a technical standpoint, and for its growth and stability over time: no users meant no resources, in the most fundamental way (Figure 25.1).

Instead, more users meant not only more material resources (memory, CPU, bandwidth) available to the network as a whole, but also a progressive refining of the variety and trustworthiness of the results obtained by each of the peers, and a better user experience for the network as a whole. This tension was at the core of several points of discussion and controversy during Faroo's first years of existence.

2.1 "The Egg-Index and the Chicken-Users": Which Comes First?

The relevance and the precision of the search results that could be obtained during the bootstrapping phase of the system – thus, their "quality," in the view of Faroo's developers – was the most debated point among the team members: the quantity, and, as a direct consequence, the quality of the p2p search index depended on the quantity of people/peers using Faroo. During the early beta phase, in 2007, pioneer users considered the quality of the index to be the core problem – to the point of compromising, in some cases, the user experience:

> the P2P search index scales with the number of people using FAROO. It looks like a show-stopper. As the number of searchers is very low right now, people are likely to try it, get lousy results and forget about it. (McCarthy, 2010)

The fact that the indexation of web sites was limited to pages that had actually been visited previously by one or more peers was a way to limit spam and improve the quality of the results – but with a low number of "starting" users, the few selected results oftentimes led to disappointing results or even an absence of results. Said for example User 1:

> When you go to perform a search, Faroo will only search sites that real people have visited. This is both the service's strength and its weakness. You don't get a lot of junk sites this way. But since Faroo is in private beta with a limited user pool, you don't get many sites at all that way. A major letdown.

User 2 summed up this dilemma as the evergreen "chicken and egg problem":

> As with any social/user-generated software, there's a chicken and egg problem. It's hard to market a product that has a small user base, because it just doesn't perform very well yet. At the same time, it's hard to attract new users with a product that doesn't deliver right away all its promises.[9]

ReviewSaurus, a web site displaying technical evaluations of emerging Internet services, proposed a possible solution, highlighting however that it should be temporary if the innovative potential of the p2p index had to be maintained:

> Mahalo[10] displays search results from different search engines. Until they don't have many users logged in at one time, why not?[11]

Facing the possibility that a majority of pioneer users rapidly lose all interest for the project due to lack of proper incentives and motivation to engage in active participation, the Faroo team revised its immediate priorities. The main objective in the phase of "recruitment" of a critical mass of users had to be the scalability of the index, for a rapidly visible improvement of the quality of results, which would have triggered a virtuous circle of motivation and improved technical performances. Temporarily, the technical solution that could be implemented most rapidly was to aggregate, and show to the user, results obtained by other search engines in addition to the results obtained by Faroo. Lead developer Georg underlined that

> (such a solution) is not by itself a big deal and used by other alternative engines, and I wasn't thrilled. But we had to get stuff that worked immediately, if we wanted to climb on the shoulders of our users, so to speak – and, by getting some results, make the next results better. So, we made a tool with which at some point you could get some results no matter what you were looking for, and you were more willing to continue to use the software. In the meantime, our index was scaling faster.

The temporary solution produced some beneficial effects for the motivation of the users, and their "recruiting" improved; in March 2008, technical commentators (e.g., on SeaDVD) concluded that Faroo developers have indeed implemented remedial measures, or even that

> they took our suggestion and have started to display the search results from Yahoo or Microsoft Live, Wikia and a few others if Faroo can't deliver the search result on its own.[12]

The design choice led to other "side effects" when it came to the motivation of the users: some of them appreciated that the team of developers was in close connection with pioneer users ("This also shows that they listen to their customers and are ready to customize the application as their user wants"), while others explicitly invited potential users, in forums or in the comment section of specialized blogs, to join the system.

Interestingly, an argument supporting this "call for users" was the fact that a peer could join with different degrees of engagement, at different times, all the while helping the system to bootstrap. Indeed, from the very start, and even as a passive user (i.e., installing the Faroo p2p client even without actively launching search requests), a peer already contributed to bring resources to a system that was in great need of "globalizing" itself:

> a suggestion for all of you too, go ahead and install Faroo and use it as much as you can, it's high time and we need an alternative of Google and you can help in doing that by just running this client.

Thus, it was possible – if the user was unsatisfied by the performance of the search engine once the client was installed – to let it run in the background for a few months and then come back to it, when more peers had joined the system and as a consequence, the index had grown in size and complexity. According to the technical blog ReviewSaurus, referenced above, this was "worth a try," especially as the quantity of resources mobilized by Faroo did not reduce the reactivity of the user's terminal, a problem that is often caused by software functioning "locally," such as p2p clients. This feature became, in the eyes of the users, a gesture of "technical politeness" which prompted them to recruit other users. SeaDVD's review of Faroo remarked that

> the client doesn't seem to eat much RAM, so you could always install it now, go about your business and check out the search engine feature in a few weeks or months.

While the episode shows that some strategies can be implemented to temporarily remedy the "chicken and egg" problem of users and their resources, it is indeed an important hurdle that peer-to-peer based design has to face.

2.2 Downloading a P2P Client: A Barrier to Entry?

A second point of controversy in the Faroo case which is illustrative of a problematic aspect in peer-to-peer design concerns the status of the p2p client,[13] that Faroo users needed to install as a precondition to becoming a part of the system. In Faroo's early development phases, the p2p client was for the developers a central artifact, both materially and symbolically: the concrete "place" and technical device where the core operations of indexation and search happened, and the symbol of the "distributed centrality" of users in the system. However, at the moment when the closed beta version (a relatively stabilized version, accessible by developer invitation) of Faroo was released in early 2008, users issued frequent objections to the use of client software which was to be installed on their terminals, citing two recent (at the time) tendencies in the market of Internet services: the rise of cloud computing, and an increasing number of users worldwide shifting to exclusively mobile Internet connections. Said a user on a Faroo-related discussion thread:

> Faroo, being P2P, requires a client download. That can be an incredibly big hurdle to adoption. At first I thought your best shot was to build on top of existing P2P services such as LimeWire or Gnutella. But your strategy appears to be different as you are building on top of .Net. . .. Oops there go all the Mac and Linux heads. . . bye. This is potentially a serious issue as lots of early adopters tend to be on the Mac. If I am a Mac user, then I may not use Faroo without borrowing somebody's PC.[14] (From Lunn, B. "Faroo: Could P2P Search Change the Game?", ReadWriteWeb (now ReadWrite), Web technology blog formerly at http://www.readwriteweb.com/archives/faroo_could_p2p_search_change_the_game.php.)

Furthermore, the fact that the p2p client was developed on NET (an app-building kit of tools and libraries developed by Microsoft, thus, strictly linked to the Windows environment) seemed to be a deal-breaker for some pioneer users. The most concise one, Fabian, summed up "No Mac, No Linux? Then not for me." Another user, Samidh, in a less strict but equally critical fashion, remarked that the multiplication of devices linked to the Internet that are now owned by the average user, and the current tendency to make terminal-installed applications increasingly "light" and mobile, could cause supplementary problems for the diffusion of an application that implies a mandatory client download:

> I think the biggest obstacle is that they require a download. Startups need to realize that the age of the download is dead. Skype was the end of an era. People are far too mobile and use too many computers to worry about specific downloads anymore. Everybody now needs to keep repeating the mantra, "The Internet is the Platform."

And danah boyd, in her then-capacity as a Microsoft researcher,[15] contributed to present the obligation to install a p2p client as a "weight" on the shoulder of potential users, rather than emphasizing potential advantages such as resilience:

> I'm also curious about the future of peer-to-peer systems in light of the move towards the cloud, but I'm not convinced that decentralization is a panacea to all of our contemporary woes. Realistically, I don't think that most users around the globe will find a peer-to-peer solution *worth the hassle*. The cost/benefit analysis isn't in their favor.[16]

Thus, the p2p client, symbol of the articulation of the global and the local dimensions of a p2p system, and "sold" by the Faroo team as one of the innovative and original features of the system, became, through the feedback of several users and technical commentators, an obstacle and a barrier to entry – something that prevented an immediate and comfortable appropriation of the system by the users. The Faroo developers were aware of the issue: one of them, Jiechi, placed the difficulties the service faced within a broader landscape of similar problems faced by other p2p systems:

> Indeed, it is not easy even for successful ones, look at Joost[17] – they've been forced to move to a pure browser model because nobody took the download plunge. The only people who can successfully get downloads distributed are people who already have wide distribution.

However, the p2p client download was, in lead developer Georg's vision, one of the distinctive, and "deal-breaking" features of a decentralized search engine, having to do with how users could monitor the incoming and outgoing traffic of its p2p client – traffic which was an indicator of two things: that the search query was being indexed by the peer that the client represents, and that this peer was in the process of answering queries. Thus, he engaged in a "translation" attempt, and redefined the p2p client as a benefit for the user:

> With the bandwidth rise downloads are not that uncomfortable anymore. [. . .] "The Internet is the Platform". So do we [try to], referring to the decentralized structure of the internet, include users as part of the internet, not only reduce them to a bunch of central servers;-) I believe that people will jump over this "download hurdle" if they get something in return, which they can't get in a centralized system. . .

This point of view was shared and clearly communicated by Bernard Lunn, a technical commentator who would later become a Faroo consultant:

> The download hurdle is a common reservation. It has not proved a hurdle when the payoff is big enough – think Skype and Spotify. And more recently the iPhone and Android app stores have totally changed the mindset around downloading software. The key issue I think is what you mention – "hassle." It is a hassle-free experience on iPhone but people have found the experience less than hassle-free on the PC.[18]

Thus, from a narrative of gaps, hurdles, and problems, another one sought to emerge, describing the download of a client that could be hosted on one's terminal, assuring the p2p connections, as an economic opportunity and a market niche, the symbol of another way to conceive information search on the Web and maybe, eventually, the Internet at large. Indeed, at stake was the interoperability of p2p applications with the rest of the network:

> The first P2P application was filesharing, the next Skype, now we are building P2P search. What could be the next? Instead of another isolated P2P application, we would like to see P2P built into the OS and Internet stack in a standardized manner.[19]

In the end, Georg summed up, the download and local running of the p2p client could be framed as either a gap/problem, or an opportunity. What would tip the scales on one side or the other was the added value that the decentralized system would be able to provide, both in terms of business model and user experience: ". . . this will be privacy protected personalization, attention based ranking, and revenue sharing based on the saved infrastructure costs." But once again, the controversy around Faroo's client download demonstrates the technical and socio-economic costs of making the peer design choice, and the need for Faroo's developers to engage in a number of strategies in order to keep the product attractive for users with respect to centralized platforms.

3 Wuala, or Ultimately Renouncing Peer Design

Wuala[20] (2007–2015) was a start-up born in Switzerland, later acquired by a leading French manufacturer of storage hardware, offering a distributed file-storage system that also included social networking features. Created in a late-2000s landscape replete with online storage applications based on "classic" servers or cloud platforms, Wuala's team of developers – fifteen at the time of my initial fieldwork in 2010, all computer engineers – worked for a number of years on a hybrid storage platform, combining centralization and decentralization, and advocating a sophisticated system of "bartering" disk space in exchange for permanent storage space on the user network. Wuala was an interesting example of a "compromise" between an approach largely dependent on a server platform, and a distributed and decentralized architecture approach. It proposed a decentralized version of an online storage platform, of which Dropbox currently is the leading centralized representative. Furthermore, it possessed a relatively large user base actively engaged, as in Faroo's case, in the discussion and shaping of the decentralized aspects of the system.

I analyze here a specific moment of "gap" in the p2p-based design of Wuala, the time at which – on September 28, 2011 – Wuala developers chose to interrupt one of their defining features, the "barter" of disk space described above. This decision *de facto* meant the "re-centralization" of the system at the level of the technological architecture, depriving it of its fundamental distributed feature – the feature that made the user's computer an active part of the archive system, rather than the mere "starting point" for files which end up in central servers. The decision also meant ending what was universally regarded as the main specificity of Wuala compared to similar services available on the market. The developers were not insensitive to the controversy that this might cause, and tried to prevent it through the following post on their main user forum,[21] emphasizing the positive implications of the choice not only from their point of view, but in terms of advantages for users:

> Why is trading not supported anymore?
> With the introduction of our new storage system, the whole file storage scheme has changed. There are some major improvements coming with this new storage system, such as incremental upload, higher redundancy, multiple down- and uploads and increased file size limit (up to 100 GB). Also, we want to free up engineering resources so we can focus on features we consider more important. (Gianluca, Wuala developer, September 28, 11:53)

The discussion that follows is an exhaustive testimony of how the re-centralization, and to some extent the failure, of the peer-to-peer project in Wuala – which seemed to betray the system's original intent and "reform it" according to the dominant model, so much so that it was for many a deal-breaker – not only responded to a variety of motivations from the developers, but was understood in different ways by the user community, and affected their perception of whatever incentives were left to them for participating in the system. On the basis of the issue of re-centralization, the discussion about a specific system and its evolution became a broader one about the pros and cons of peer-based design – from "practical" and technical aspects to issues of principles, of community, of organization of the Net.

3.1 The Many Facets of Discontinuing a Peer-Based Feature

First of all, the end of the peer-based feature was framed in terms of the loyalty that developers owed to users in exchange for their active contribution to product improvement. In fact, as always happens in the initial phases of the development of a totally or partially open source software, the first Wuala users and their computers served as a "test-base" for the system, highlighting, through their practice of it, a series of aspects that could be improved. The re-centralization therefore signified the failure of the trust that users placed in the developers – trust that they were contributing to something of value, different from most of the solutions proposed on the market, and whose creation process really "took seriously" the community on which, socially and technically, the system was based:

> As a user from day one it makes me really sad and angry to see this change.During the last months Wuala has already stopped listening to its loyal users and with every update there were missing features, incomprehensible design decisions, etc.
>
> And now – booom! [. . .] You tricked the community in believing they have a saying in your decisions. I want the time and effort I spent with testing, *commenting*, suggesting and bug tracking over the last years back from you!! (joe12, October 3, 5:15pm)

> My guess is that a lot of users will feel kind of mis-used [. . .] And as fast as new products and services can appear in the internet they can also disappear. I would think that the impact on wuala's image is much bigger than it looks right know: one key essential when using cloud-based services is trust. Changing the game in a way that it looks solely money-driven will disappoint the community that also supported wuala a lot in the past. (Doug0915, October 3, 5:55pm)

A related issue, but with its specificity, concerned the possible economic models for peer-to-peer. Various users seemed to suggest that the barter system was a victim of the impossibility of continuing to maintain a peer-based model when it comes to profit. Was Wuala's story that of many originally peer-based systems, that were born peer-to-peer to grab some "loyal" users seduced by the model, and that ended up succumbing to the temptation of improving their economic performance by introducing a simpler model technically, but which made it easier to enter mechanisms that could be sources of income for the company, such as advertising or premium solutions?

> No trading = no wuala for me. Thanks for the run – this was a great idea that seems to have been ruined by a corporation. I understand that everyone is just in this game to make money, but it is sad to see a truly useful feature disappear from this product. (VenMen, October 3, 5:09pm)

> I understand you want the money, but imo this change will reduce long-run income (no more advertising from guys like me who is/was in love with the system). Of course it will give more money in a short-term. Probably that's your goal, maybe you are not planning to live long. (Dimps, October 3, 5:48pm)

One of the fundamental aspects that seemed to fail with the re-centralization was the attribution of independence, autonomy, and control to the users allowed by p2p design, with the strong "local" dimension of the storage and the sharing of each user's disk space with other users. Many had chosen Wuala precisely for this distributed character, which allowed to free oneself as much as possible from the service provider and therefore to be as independent as possible from servers and the terms of use they imposed, by replacing them with a network of peers. A user went even further, explicitly mentioning the failure of the company's "social experiment," which therefore invalidated in the eyes of users any solution that could be proposed later.

> [The reason] that I selected Wuala is gone for me. The whole idea of having data stored on several peers (which improves speed, etc.), and sharing out one's own disk accordingly, was what appealed to me and what I had been looking for all since this idea originated. [. . .] The very strong argument of having the data distributed across many computers in the net made Wuala so appealing, giving the impression of independence and not rely on a single company for storing all the data fragments. (flexgrin, October 3, 6:12pm)

> [Now you could at least] release a Wuala based client [. . .] using the old scheme as open-source, so that new clouds of **social** storage can emerge using it. Not doing so basically declares the attempt to implement a social file storage system as a failure, and if it was a failure, why trust the company behind it? (Tor, October 3, 5:23pm)

And for others still, renouncing peer design was a failure because the peer-to-peer model embodied various forms of "technological activism" – environmental protection (using free and underutilized disk space on their computers instead of feeding the data centers with content duplicates), being part of a community interested in the social value of sharing hardware and software resources, or even referring to the original (and idealistic) principles of the Net such as cooperation, symmetry, equality:

> Maybe I have been operating under a happy delusion for the past couple of years, but I thought that by sharing my storage with other Wuala users, I was, in some small way, contributing to the health of our planet. I thought I was helping to reduce power consumption by making it possible for your company to store data on underutilized storage space on my PC and notebook computers. (droom, October, 6:09pm)

> There was always this "promise" that one can be part of the storage community by sharing space. Many wondered whether this could sustain but the answers I remember have always been confirming the commitment to sharing space as being one of the core principles of Wuala. I am disappointed because I thought the trading option was a very nice social experiment marred by a flawed implementation. I am a strong believer in distributed systems at the endpoints of the net. (Milic, October 3, 6:45pm)

4 Conclusions

This chapter has addressed the issue of the "gaps" in peer-based design of the technical architecture of Internet-based services, by building on two case studies of particular events in the development processes of decentralized systems. Via these two case studies, the chapter has sought to address the "mix" of technical, social, and economic factors that lead to dilemmas and problematic issues in peer design.

With the case of Faroo, we saw how the fact of considering peers not as mere outposts of a centralized system, but taking "seriously" their status of equals at the network's periphery, and having this design as foundation of all further work, represented an extremely delicate issue for developers at the early stages of system's expansion. We also saw how the conception of the user's "centrality" (and the resources that it brings to the whole system) resulting from this distribution approach emerged as both the greatest opportunity and the major risk for the system. Among the advantages, such a system did not require large investments of resources from the company, and could work in stable and unstable situations. It was also possible for it to recover autonomously after an event had changed its structure, thanks to the "surviving" peers that guaranteed the readjustment of a distributed entity that could still perform its functions. At the same time, the centrality of users and their computing resources could also be problematic for the sustainability of the system, particularly in its initial phases. To rely on no other resources than those provided by users, "to rely on the world to be its network," also meant, for developers, to take a huge risk, betting on the fact that they would manage to "enroll" (Callon, 1984) a number of users large enough to ensure the proper technical operation of the system itself, and to allow its growth and stability over time. More users meant not only more material resources (memory, CPU, bandwidth) available to the whole, but also, as we have seen, a progressive refinement of the variety and reliability of the results obtained by each of them, and a better user experience for the entire network. We have followed how this tension took shape around two points of controversy within the team, and between the team and some users, during the first years of Faroo's existence.

The first of these cases, concerning the quality of the results during the bootstrapping phase, showed how – given that the quantity and quality of the p2p search index depended on the amount of people/peers using Faroo – the shortage of users proved to be the main problem, both human and technical, for early adopters, a situation which forced developers to temporary technical fallback solutions. The second case followed discussions focused on the p2p client, which Faroo users had to install as a precondition to entry into the system. The symbol of the articulation between the local and global dimensions of the system, understood and "sold" by the Faroo team as one of the innovative and original points of their service, was transformed, by the reactions of some users and commentators, into an obstacle, a barrier to entry, forcing the Faroo team to renegotiate the p2p client as a market opportunity, symbol of another way of conceiving the search for information on the Web and perhaps the Internet as a whole.

In the case of Wuala, we have witnessed a moment in the history of a p2p system in which its developers decided to renounce decentralization, and discontinue peer design. We have seen how changes in the design of the technological architecture of a communication network has both causes and consequences that are economic, political, and social; when networking models, the management of data flows, their archiving and technical treatment are discussed, the peers no-longer-to-be were in fact talking about these broader consequences, their perception as to whether these consequences were more or less socially and economically fair, and more or less successful. The ability of the network (and its developers) to sustain peer design became the core notion around which dynamics of equality, of "justice," of community took shape. The re-centralization of the system was the focus of the negotiations between innovators and pioneer users, and it showed the reconfiguration and re-appropriation of peer-based innovation, creating multiple definitions of what constitutes the technical, economic, and political "failure" of a p2p system that is not such anymore.

This process also shows that the "barter" feature in the Wuala system had, by the time it was discontinued, assumed the quality of "distributed governance," implemented through peer-to-peer technology. When the feature was suspended, users felt deprived of this management model by an act of politico-economic governance, giving predominance to a simpler business model, which is perceived as more dictatorial and oblivious to the wishes expressed by users, as well as the care they have displayed to make the previous system a better technical artifact, driven by practical and principled motivations. The choice to discard peer design becomes a symbol of this evolution of power and control relations, as well as trust. The evaluation of the soundness of the different architectural solutions – the previous one and the new proposal – with respect to Wuala's business model, which the developers thought they could make exclusively based on technical and economic considerations, became charged with other meanings and eventually caused the system's demise.[22]

In conclusion, assessing the advantages and disadvantages of peer design is a foremost example of a long-standing heavily discussed issue: whether a technical architecture should be centralized or decentralized, its design based on equal interaction among peers or on a client/server architecture. It is important to underline that, when it comes to discussing the degree of (de-)centralization, protocols are not intrinsically good or bad. The two cases we have examined – the gaps and shortcomings they show when it comes to peer design, and the strategies that developers and users display to counter them – ultimately illustrate that choices related to the design of technical architectures "come with" a number of social, organizational, and at-large political dynamics, and lead to stories of compromises and trials.[23]

However, they also reinforce Philip Agre's seminal argument that while architectures are politics, they should not be assumed to be a substitute of politics (Agre, 2003). When it comes to privacy, user autonomy, and freedom of expression, Silicon Valley giants and their centralized silos may arguably not be the best option, but having a system with intermediaries and obligatory passage points may actually prove more useful, in some cases, than a completely egalitarian and peer-based design, and in particular, it can be understood by developers-turned-start-up-leaders as a simpler way to go for elaborating a business model. And despite an accompanying rhetoric of openness and freedom that presents them as the ideal solution to all problems, including censorship, decentralized architectures also bear or can bear issues.

This is not only because of socio-technical and socio-economic constraints, but also – being potentially subject to issues of data poisoning, net neutrality, display of metadata such as IP addresses, and difficulties to enforce identity verification – from the standpoint of civil liberties and human rights. While the present chapter has focused more on the former aspects, and what they mean for the capability of peer-based systems to compete with centralized ones, we should not neglect the latter as they are, as well, an integral part of how Internet governance is increasingly conducted via its infrastructure. Together, all these implications of architectural design choices inform what we have called "the turn to infrastructure in Internet governance" (Musiani et al., 2016), and as a consequence, the kind of Internet user that we can and want to be.

Notes

1 Many Internet protocols are client/server, which means that the machines that communicate are not equivalent: one is a server, permanently on and waiting for connections, the others are clients, who connect when they have something to ask. This is a logical mode of operation, for example, in the case of the Web: when you visit a website, you are a reader, and the entity which manages the website produces the content you seek to read. But not all uses of the Internet fit into this model; they include direct sending of messages, or file exchange – not a one-way communication but a peer-to-peer one, with two machines or two humans communicating directly.

2 The editors' introduction to this volume provides a general overview of infrastructure; we prefer to focus in this chapter on more specific instances related to the "problem" of decentralization and the consequences of the peer-based design of the technical architecture.

3 The fieldwork on which the two cases are based was conducted between 2009 and 2011 and then again in 2014. Results have never been published in English before, but some discussion of them was published in French (Musiani, 2015) and Italian (Musiani, 2018). The names of developers used throughout the chapter are pseudonyms, although "Faroo" and "Wuala" are not.

4 www.faroo.com, has now evolved into SeekStorm, a "Search as a Service" project to whose website it is re-directed.

5 A number of protocols taking advantage of the dynamics of proximity and affinity between users, notably "gossip" and "epidemic" protocols (see https://en.wikipedia.org/wiki/Gossip_protocol).

6 The Faroo FAQs, now archived (previously at www.faroo.com/hp/p2p/faq.html#opensource), read: "open source is perfect when competing by a cost advantage with a commercial product on the same technological level (Linux, OpenOffice). But it's not a good idea to hand over your technological advantage to a monopoly, when competing with its free service with enormous brand power."

7 Some users consider that Skype, in its early phase of peer-to-peer configurations, was actually a system of this kind: "a new node in Skype goes through 2 big steps before it fully joins the network. The first is that it has to authenticate to the central servers, and the second is that it has to find a good 'supernode' to initialize/bootstrap and maintain connectivity to the central servers. [. . .] due to the synchronized restart of clients, when all the nodes started banging on the central servers, the supernodes started losing connectivity and dropped out of the 'supernode' mode. Creating a chicken and egg problem and making the situation worse than before. Is it possible that the real reason why the Skype network failed was because of its dependence on central servers?" See comments section of "Skype Failed the Boot Scalability Test: Is P2P Fundamentally Flawed?," High Scalability, http://highscalability.com/skype-failed-boot-scalability-test-p2p-fundamentally-flawed

8 In a white paper published in 2007 (formerly available at www.faroo.com/hp/p2p/whitepaper.html), lead developer Georg used a biology-grounded metaphor to account for peer design: "In biology organisms naturally deal with the rise and falls of its cells; simple elements form superior systems. We believe that evolution works in search too, and that the future belongs to multicellular organisms/systems" (p. 22).

9 The highlights of user feedback in this paragraph are issued of a discussion during Faroo's private beta phase, conducted on SeaDVD, a former website where beta-phase software was evaluated (http://beta.seadvd.com/faroo-follow-up-p2p-search-engine-will-only-work-with-more-users/).

10 A now-defunct web directory and human search engine, formerly at www.mahalo.com/

11 ReviewSaurus, "Faroo: P2P Web search engine, impressive idea!," January 2008, formerly at www.reviewsaurus.com/web-applications-reviews/faroo-p2p-based-search-engine-impressive-idea/

12 ReviewSaurus, "Faroo opens for public and takes our suggestion!," March 2008, formerly at www.reviewsaurus.com/software-reviews/faroo-opens-for-public-and-takes-our-suggestion/. The following two citations are also from this webpage.

13 So as to clarify all semantic ambiguity, it should be pointed out that throughout this section the word "client" is used to indicate the piece of software that users need to install on their machine so that it can become a node in the p2p system.

14 User feedback in the commentary section of Lunn, B. "Faroo: Could P2P Search Change the Game?," ReadWriteWeb (now ReadWrite), Web technology blog formerly at www.readwriteweb.com/archives/faroo_could_p2p_search_change_the_game.php. The following two interventions are also hosted on this thread.

15 Thus, it should be noted, not a completely neutral position in the centralization/decentralization debate...

16 danah boyd's contribution on the P2P Foundation mailing list, May 27, 2010, formerly at http://listcultures.org/pipermail/p2presearch_listcultures.org/2010-June/008941.html.

17 Joost was an Internet TV service created by Niklas Zennström and Janus Friis (previous founders of Skype and Kazaa), which during a period of its development used peer-to-peer television technology for content distribution.

18 Bernard Lunn's contribution on the P2P Foundation mailing list, May 27, 2010, as an answer to danah boyd's previously cited remarks.
19 For this and the following citation, see Faroo white paper at p. 30.
20 Formerly at www.wuala.com/
21 Wuala's user forum was hosted on the start-up's website, www.Wuala.com, and is no longer accessible since November 15, 2015, when the service was shut down.
22 On August 17, 2015, Wuala announced that it was discontinuing its service and that all stored data would be deleted on November 15, 2015.
23 In the sense of *épreuve* (see e.g., Martuccelli, 2015).

References

Agre, P. (2003). Peer-to-peer and the promise of Internet equality. *Communications of the ACM, 46*(2), 39–42.

Benkler, Y. (2004). Sharing nicely: On shareable goods and the emergence of sharing as a modality of economic production. *The Yale Law Journal, 114*(2), 273–358.

Brekke, J. K., & Isakidis, M. (2019). Principles for decentralized protocol design. *NEXTLEAP Project Deliverable 2.5.*

Brunton, F. (2019). *Digital cash: The unknown history of the anarchists, utopians, and technologists who created cryptocurrency.* Princeton, NJ: Princeton University Press.

Callon, M. (1984). Some elements of a sociology of translation: Domestication of the scallops and the fishermen of St Brieuc Bay. *The Sociological Review, 32*(1-suppl), 196–233.

Campbell-Verduyn, M. (Ed.). (2017). *Bitcoin and beyond: Cryptocurrencies, blockchains, and global governance. Abingdon*: Routledge.

Coleman, E. G. (2005). *The social construction of freedom in free and open source software: Hackers, ethics, and the liberal tradition* (Unpublished doctoral dissertation), University of Chicago.

DeNardis, L. (2014). *The global war for Internet governance.* New Haven, CT: Yale University Press.

Elkin-Koren, N. (2006). Making technology visible: Liability of Internet service providers for peer-to-peer traffic. *New York University Journal of Legislation & Public Policy, 9*(15), 15–76.

Izal, M., Urvoy-Keller, G., Biersack, E. W., Felber, P. A., Al Hamra, A., & Garces-Erice, L. (2004). *Dissecting* BitTorrent: Five months in a torrent's lifetime. *International Workshop on Passive and Active Network Measurement (pp. 1–11).* Berlin: Springer.

Kelty, C. M. (2008). *Two bits: The cultural significance of free software.* Durham, NC: Duke University Press.

Ku, R. S. R. (2002). *The creative destruction of copyright: Napster and the new economics of digital technology.* The University of Chicago Law Review, *69*, 263–324.

Lessig, L. (2002). *The future of ideas.* New York, NY: Vintage Books.

Martuccelli, D. (2015). Les deux voies de la notion d'épreuve en sociologie. *Sociologie, 6*(1), 43–60.

McCarthy, B. (2010). "A look at Faroo's P2P search. Will this one make it?" The Next Web, May 25. Retrieved from http://thenextweb.com/apps/2010/05/25/p2p-search-is-here-but-who-will-use-it/

Minar, N., & Hedlund, M. (2001). A network of peers – peer-to-peer models through the history of the Internet. In A. Oram (Ed.), *Peer-to-peer: Harnessing the power of disruptive Technologies* (pp. 9–20). Sebastopol, CA: O'Reilly.

Musiani, F. (2013). Dangerous liaisons? Governments, companies and Internet governance. *Internet Policy Review, 2*(1). Retrieved from https://policyreview.info/articles/analysis/dangerous-liaisons-governments-companies-and-internet-governance

Musiani, F. (2015). *Nains sans géants: Architecture décentralisée et services Internet.* Paris: Presses des Mines.

Musiani, F. (2018). "Quando il peer-to-peer si 'ricentralizza': Vincolo sociotecnico, spinta di mercato o fallimento?" In G. Balbi & P. Magaudda (Eds.), *Fallimenti digitali: Un'archeologia dei nuovi media.* Bologna: Unicopli.

Musiani, F., Cogburn, D. L., DeNardis, L., & Levinson, N. S. (Eds.). (2016). *The turn to infrastructure in Internet governance.* New York, NY: Springer.

Oram, A. (Ed.) (2001). *Peer-to-peer: Harnessing the power of disruptive technologies.* Sebastopol, CA: O'Reilly.

Ostrom, E., & Hess, C. (2007). *Understanding knowledge as a commons: From theory to practice.* Cambridge, MA: MIT Press.

Pohle, J., & Van Audenhove, L. (eds.) (2017). Post-Snowden Internet policy. *Media and Communication, 5*(1). Retrieved from https://www.cogitatiopress.com/mediaandcommunication/issue/view/59

Riordan, J. (2016). *The liability of Internet intermediaries.* Oxford: Oxford University Press.

Stallman, R. (2002). *Free software, free society: Selected essays of Richard M. Stallman.* Retrieved from Lulu.com.

26

Makerspaces and Peer Production
Spaces of Possibility, Tension, Post-Automation, or Liberation?

Kat Braybrooke and Adrian Smith

1 Introduction

Makerspaces are open community workshops for peer production which provide people with technical tools and training to experiment with making, learning, and hands-on participation around material cultures. These workshops come in a variety of forms, and they are called by many names, including shared machine shops, hackerspaces, fab labs, digital studios and many other labels – including makerspace, which we will use as an umbrella term to keep things simple. What they have in common, however, is a commitment to providing people with the skills and means needed to access versatile design and fabrication technologies, and to fostering communities that share an open and collaborative ethos regarding the possibilities that democratized design and fabrication technologies might offer personally, socially, economically, and culturally. In this chapter, we discuss the diverse dynamics of makerspaces, and how encounters between makerspaces and institutional interests in particular are shaping what is possible. We also ask what is gained from the radical redistribution of prototyping capabilities in societies that makerspaces represent, and what is diminished.

Makerspace communities engage with a versatile selection of design and fabrication tools – frequently with special emphasis on digital tools for fabrication, in which code and designs are shared over networks, but also including more traditional hand and craft tools as well, that enable people to engage in material participation, pursue new livelihoods, and simply have fun making cool projects (O'Donovan & Smith, 2020). Tools are maintained and accessed through peer networks of makerspace members, who collectively manage spaces, share skills and mentor incoming makers, and who design, prototype and make anything from street furniture to wearable electronics to hydroponic systems, from open source beehives to 3D printed prosthetic limbs using recycled materials, and from environmental monitoring systems to musical instruments. Members might call themselves hackers, makers, tinkerers, activists, entrepreneurs, or simply creators. A typical makerspace like Gearbox in Nairobi, Kenya[1] or Chaihuo Makerspace in Shenzen, China,[2] for example, provides *tools* (such as 3D printers, computer-numerical-controlled machining (CNC) machines, laser cutters, sewing machines, soldering equipment, and micro-electronics), *community services* (such as pedagogical materials for digital making, training sessions for educators, and support for start-ups and small businesses), and *membership* to its users (with plans ranging from daily to life-long in length).

When makerspaces are also founded with the ideologies and non-proprietary gift economies of "commons-based peer production" (Benkler, 2002; Wolf et al., 2014) in mind, their *practices* (such as co-designing), their *modes of production* (like 3D printers), and their *outputs* (like an open source windmill or a prosthetic arm built from open hardware designs) are collaborative, non-hierarchical, decentralized, and openly shared, with the aim being to encourage future innovations amongst members, who will then share their contributions in turn. Other makerspaces are founded with a more entrepreneurial focus, and act as small business incubators, where outputs are privately owned by their

The Handbook of Peer Production, First Edition. Edited by Mathieu O'Neil, Christian Pentzold, and Sophie Toupin.

creators. So, makerspace practices may include commons-based peer production or personal and proprietary projects – and many sites encourage all of these.

However, whilst they are important sites for prototyping peer production through material engagement, makerspaces are other things beside that. Their versatility has caught the imaginations of a wide variety of institutions, who see in them an exciting buzz of organized possibilities. This speaks to a growing body of research literature that analyses the wide range of socio-technical practices that are experimented within makerspaces, and the multiple issues that arise around these (Hielscher & Smith, 2014; SSL Nagbot, 2016). This chapter makes use of that literature, and it also takes a broader perspective by considering makerspaces as situated in relation to social institutions. By social institutions, we mean both broad norms and routines in societies, such as conventions for property in a market-based economy, and also more specific institutional actors such as museums, schools, and innovation agencies, amongst whom there has been interest and support towards makerspaces.

What kinds of encounters do makerspaces have with these institutions, and how do these interactions enrich our understanding of the critical implications of peer production? We address this question by first looking at makerspaces as spaces of possibility, by introducing some of the innovative practices that have emerged from the prototyping of alternative socio-technical configurations in makerspace settings. We then consider makerspaces as spaces of tension, examining the contradictions and contestations they manifest, and the ways they interpret, reinforce, and challenge prevailing socio-technical logics. Here, it is important to understand how the power relations of institutional encounters shape the terrain of makerspace possibilities and limitations. We engage in a historical analysis of the deeper conflicts of open community workshops and the institutional tendencies that have structured them, discussing how makerspaces can be conceived of as spaces for "post-automation," constituting more human-centered alternatives for commons-based technology as compared to the approaches of typical automation theories, represented most recently in Industry 4.0 discourse. And we conclude by discussing the emergence of even more liberatory visions of makerspaces as sites of radical socio-technical reconfiguration. In doing so, our aim with this chapter is to capture some of the ways institutional encounters in makerspaces anticipate and shape peer production futures – while leaving space for the many contingencies that give each makerspace its own unique flavor.

2 Spaces of Possibility

We opened this chapter by introducing makerspaces as open community workshops which offer people the tools and mentors to engage with materially-engaged making and learning. This basic template has caught the imagination of a wide variety of actors who are intrigued by the possibilities of such spaces, from computer hackers and schoolchildren to global corporations. It has also meant that makerspaces have become ciphers for the broader socio-technical aims and imaginations of those who encounter them – and as such they are described as innovators of education, cradles for entrepreneurship, studios for digital creativity, catalysts of social change, prototyping shops for local manufacturers, 21st-century libraries, laboratories for smart urbanism, galleries for digital explorations of material culture, and so on – not forgetting, of course, spaces simply for hanging out and having fun.

More radical depictions of makerspaces have anticipated them becoming not only workshops for making and learning, but also the "occupied factories of peer production theory" (Troxler & Maxigas, 2014), by providing people with alternatives to mass production and business-as-usual (Arvidsson, 2020). Such conceptions of makerspaces foresee their potential to interrogate more complicated socio-technical dynamics through material experimentation that elicits wider socio-cultural transformations. In this framing, makerspaces can be understood not only as educational environments where people can learn digital fabrication skills, but also as "real-life laboratories" (Dickel et al., 2014) where new ways of working and interacting within civil society can be experimented through what the craftivist Sarah Corbett has called "slow and mindful activism" (Corbett, 2017, p. 2). These possibilities have led to some pundits going so far as to laud makerspaces as harbingers of a "fourth industrial revolution" (Anderson, 2012) which is claimed to have originated from the examples of a few influential ventures in North America, like the newly-defunct *Make Magazine*.

With aspirations for global sustainable development like that of the circular economy[3] starting to redefine the norms and rules by which products and services are produced, consumed, and remanufactured, meanwhile, makerspaces are also increasingly being cast as actors within even broader imaginaries of systems-level transformations (Prendeville et al., 2017). Their tools and technical possibilities are also being used to rethink how global circulations of design and machining can be harnessed for more sustainable and local fabrication (Smith & Light, 2017). The concept of redistributed manufacturing (RDM) in particular has become alluring for a number of institutional agendas, which look to makerspaces as pioneers in prototyping systems and practices aimed at shifting the organization of manufacturing through local supply chains (Corbin & Stewart, 2018).

The realities of current-day makerspaces, however, are even more heterogeneous, and humble, than the grand visions of fourth industrial revolutions and circular economies. When makerspaces do engage with such visions, their strength rests in conceptualizing their practical requirements on a day-to-day level (Kohtala & Hyysalo, 2015). Makerspace possibilities have always been grounded by the many contingencies of the cultural and historical contexts within which they are situated. Makerspaces also come from a long tradition of hacking, crafting, fixing, remaking, circuit bending, and other innovative socio-technical practices that have taken many different forms around the world, especially in nations on the periphery (Braybrooke & Jordan, 2017; Smith et al., 2016; Turner, 2018).

In India, for example, unofficial makerspaces have flourished for years through the assorted activities of "jugaad" culture, a colloquial Hindi and Punjabi word meaning an innovative fix rooted in necessity, the practices of which have been described as "community-led combinations of making-do, hacking, and frugal engineering" (Ray Murray & Hand, 2013; see also Deka, this volume). Jugaad projects, which typically re-use recycled and discarded materials from digital consumer cultures, range from Frankenstein bikes made up from discarded bike parts to walls built with recycled bottles. As an inextricable part of Indian life and cultural rhetoric, jugaad is a visible postcolonial practice in many different kinds of environments, from repair shops to homes and roadside cafés (Rai, 2015).

Spaces for making and hacking in China, meanwhile, have long been associated with copyleft or "shanzai" cultures of technological disobedience, as well as China's extensive history of digital fabrication as a mode of mass manufacturing (Arvidsson, 2020; Braybrooke, 2019; Lindtner & Li, 2012). Makerspaces have more recently become a core focus of economic development for the national government, which has introduced policies that emphasize digital making as a way of transforming the next generation of Chinese young people into entrepreneurs to address concerns over future unemployment (Justice, 2012; Shea & Gu, 2018). The specific circumstances that fostered the creation of today's government-funded makerspaces in China started to emerge around 1988, when the Ministry of Science and Technology launched its first "Torch Programme" to help foster new R&D spaces across the nation to promote an "innovation-oriented China" such as science and technology industrial parks – a conceptualization of material participation as economic development that was also introduced in the UK around this period (Massey & Wield, 1992; Wang, 2016). This history has meant that peer production is typically viewed in China not as a creative hobby, but instead as good business, and government subsidies have helped over 7,500 makerspaces and tech hubs open across the nation – a higher concentration of such spaces, China claims, than anywhere else in the world (The People's Republic of China, 2017).

3 Spaces of Tension

Because makerspaces emerge from ever-evolving and contingent contexts like the above, a deeper examination of their impacts conjures up many contradictions as well as possibilities. Some makerspaces have been criticized for perpetuating contrasting ideologies of egalitarian openness on one hand, while at the same time maintaining an apolitical stance towards issues of race, class, and gender, which perpetuate privileged and hierarchical or "pushyocratic" (Toupin, 2014) cultures. In North American and European makerspaces in particular, members remain predominantly white, male, educated, and middle class (Davies, 2018; Nesta, 2015). Ideologically-oriented makerspaces, such as the feminist hackspace Mz Balthazar's Laboratory in Vienna, attempt to address these issues by experimenting not only with new technologies, but also with new ways of working that tackle these asymmetries.

Other kinds of spaces, however, have become synonymous with neoliberal business-as-usual, where a kind of entrepreneurial citizenship is prototyped through the exploitation of precarious labor by businesses and institutions alike, with producers blocked from the rights to their own creations (Irani, 2015; Lindtner, 2017; Smith et al., 2017). Critics argue this kind of enactment turns the makerspace model into yet another site of consumption, in which designs, materials, tools and cool lifestyle projects are marketed to hobbyists, with little of the social awareness or transformational aspirations of peer production (Hertz, 2012).

There is also an increasing proliferation of makerspaces that have opened with institutional associations, situated within libraries, museums, universities, and companies. The emergence of these hybrid spaces provides a new set of opportunities and tensions. The first fab lab, for example, was opened through partnerships at the MIT Center for Bits and Atoms in the United States in 2001, and the model soon spread to other regions, who opened their own fab labs enacting the same fabrication tools and modes of knowledge production as the original fab lab. There are now nearly 2,000 fab labs active in over 40 nations, many in partnership with private and institutional actors (Fab City Research Lab, 2019). A network of city and regional authorities, from Barcelona and Plymouth to Kerala and Seoul, is now using the idea of designing globally and making locally to promote a "Fab City" concept, which situates fab lab districts as the basis for prototyping new urban metabolisms.[4] The idea is to seed a transformation to more circular economies by creating city-scale infrastructures that are networked with other cities, so that capabilities expand for "designing globally and making locally" (Diez, 2012; Kostakis et al., 2015).

Increased institutional attention towards makerspaces means they are subject to a plurality of developments and demands, which can result in a loss of autonomy when they conform to the ways of working of their institutional hosts (Braybrooke & Smith, 2018). However, more enlightened institutions are finding the experimental approaches of makerspaces are of value for reforming their own practices. Noteworthy here are the Citizen Innovation Laboratories pioneered by Medialab-Prado in Madrid and the regional government of Aragon in Spain, and now being adapted across Latin America. These laboratories convene citizens in the co-production of new institutional forms for resolving development needs and aspirations, which are framed collaboratively, and with the ensuing, peer-produced social innovations treated as a commons (Pascale, 2018). The critical factor here will be whether the conditions under which these kinds of initiatives are cultivated are sufficient to carry forth their prototypes and turn them into instruments for wider institutional reforms (Smith, 2018), under which a more radically open and democratic commons can be instituted.

Here, institutional theory can come in handy, as it seeks to explain the kinds of settled (and unsettled) social environments in which organizations operate, and the effects that different kinds of organizations have on each other's development. Institutions can be very broad and cultural, such as those concerning property, or they can be instrumental and specific, like a government agency, the formation of a law, or a particular regulation. The increased institutionalization of makerspaces means they may experience powerful pressures to conform – an example being isomorphism, a process which occurs when organizations encounter each other's practices over time in such a way that they inevitably come to resemble one another (DiMaggio & Powell, 2000). By definition, institutions also seek to normalize and routinize and, when challenged, tend to adapt and elaborate rather than transform and liberate (Willmott, 2014).

Criticisms of institutional theory argue that it can be overly static and conservative (Munir 2015), prompting perspectives that view institutions more dynamically, and propose the creation of new ***kinds*** of institutions that transform social environments. Institutional collaborators can transform institutions, for example, by exploiting social movements and shifts in social discourse, which undermines the legitimacy of incumbent institutions and opens space for the development of alternatives (Levy & Scully, 2007; Zietsma & Lawrence, 2010). Whilst institutions do constitute a powerful pressure for conformity, meanwhile, there is nevertheless scope for strategic maneuvers by organizations encountering these pressures. Research has found, for example, that when grassroots innovation movements encounter mainstream institutions, alternative pathways to innovation are made possible through the synthesis of their different modes of engagement (Fressoli et al., 2014). Institutional environments have also been depicted as hybridized and complex, impacted by multiple logics (or belief systems that structure behavior) and demands (Pache & Santos, 2012). These ever-shifting agencies, it is argued, can enable shifts or "cracks" in the power relations that otherwise maintain institutional control, opening up new spaces for the development of alternative ways of doing and interacting (Fuenfschilling & Truffer, 2014; Hirsch & Lounsbury, 2015; Zietsma & Lawrence, 2010). Such situations open possibilities for institutional entrepreneurs and social movements to press for institutional redesigns and reforms.

Research on "collections makerspaces" based in museums in London, UK, for example, has found that they foster institutional change by providing friendly environments where digital tools enable publics to manipulate, remix, and play not only with a set of cultural artifacts, but also with their perceptions of the museum that acquired them in the first place (Braybrooke, 2018). This kind of encounter enables collections makerspaces to be activated by their users in ways that can foster critical inquiry into museums themselves, and the conventional logics of culture and privilege that they precipitate. These logics do not disappear as a result of encounters between collections makerspaces, and museums, but they do evolve, and enable the redistribution of power and agency from museum headquarters to publics. Studies of institutional makerspaces in South Africa, meanwhile, have illustrated how a co-existence of informal and institutional practices can be possible. The grassroots engagements of maker communities are being formalized by virtue of their institutional partners, it is argued, while at the same time a commitment to informality and open collaboration is increasingly valued amongst those same partners (Armstrong et al., 2018).

In other cases, the attraction to makerspaces is derived from their presumed ability to help institutions reach out and connect beneficially with large informal sectors, which themselves already have cultures of creativity and innovativeness, by strengthening that potential economically and socially. In São Paulo, the city authorities opened public fab labs in a variety of districts, including the disadvantaged Cidade Tiradentes on the margins of the city (literally and figuratively). In this case, aspirations to seed inclusive developments within a community depend upon makerspaces connecting with the Brazilian culture of improvisation and making-do known as *gambiarra*, as well as learning from earlier programs for social technology aimed at emancipating people from poverty through other kinds of participatory technology programs (Fonseca, 2015; Thomas, 2013). What is striking in these spaces, and familiar to public support for makerspaces in other cities, is how makerspaces are seen as an instrument that permits a new "script" to be written for technology and local development, compared to the norms and routines of more institutional approaches to innovation (Akrich, 1992). However, the critical factor here is whether the intended beneficiaries recognize this potential also, and can develop their *own* scripts for local development with makerspace technologies, skills, and prototyping possibilities. What is important for initiatives like São Paulo, and others like it, is the processes by which people can bring their own scripts into technology developments in makerspaces – as well as narratives about the communities in which they are situated, and what they would like them to become (Dias & Smith, 2018).

Institutional encounters with makerspaces are not restricted only to specific programs, however. Recalling that institutions also precipitate broader norms in society, such as property relations, or norms for how innovation should operate, these norms can also be conceived of as pervasive within makerspaces. Makerspaces have been found to intersect with institutions in ways that generate expectations amongst their participants in relation to the impacts of those social institutions – sometimes as an Other to be challenged (such as intellectual property rights used by commons-based maker activists) and sometimes as an unquestioned reality (such as the practices of start-up culture). This means that when a society more generally seeks to change its institutions, its makerspaces are presented with key opportunities to be in the vanguard of defining what those aspirations can mean in practice.

4 Spaces for Post-Automation?

The tensions above demonstrate in particular how the ways digital technologies are interpreted and used in makerspaces may serve to advance or challenge the institutions of the societies within which they are located. The relations of makerspaces, like those of cafés, bookstores, libraries, parks and other "third places" (Oldenburg, 1989) in between home and work where people gather for less institutionalized purposes, provide spaces for considering the forces at play within the dominant socio-technical regimes that constitute our worlds. Makerspaces show us that technologies are adaptable, and that their impacts depend upon the environments within which they are produced and interacted with – which may be hegemonic, or hackable, as circumstances permit. At their best, makerspaces provide people with capabilities to question prevailing assumptions, and to do things differently, according to social values that may diverge from economic orthodoxy (Ratto & Boler, 2014).

One orthodoxy whose foundations need to be interrogated through practical material activity as a matter of urgency, is, arguably, automation in society. The experience of building and using community-run and managed routers and laser cutters in a neighborhood woodworking project dedicated to reclaiming a public space with upcycled street furniture, for example, is obviously a different social phenomenon from using similar digital technologies to fabricate on-demand flat-pack furniture in an automated factory, where the operative has no control over the outputs of their labor, and is alienated from the process via fabrication robots, mass marketing, and shipping between warehouses. The socio-technical configurations of these two different interactions with similar fabrication technologies are radically different, as are the networks of technological and non-technological actors that are deployed, and the social and private values produced. By redefining value chains and inserting new (social) values, geographies, and directions into digital fabrication, makerspaces enable the kinds of socio-technical experiments that can challenge prevailing industrial relations.

Many of the digital fabrication tools in current-day makerspaces hold within them particular historical ironies that only grow deeper upon examination. The introduction of computer-aided-design (CAD), computer-numerical-controlled machining (CNC), and computer-integrated-manufacturing (CIM) equipment in the 1970s and 1980s North America and Europe, for example, threatened the livelihoods and identities of local manufacturing communities of the time (Noble, 1984). In the current day, however, their more accessible technological descendants have been widely celebrated, for enabling new opportunities for learning and agency amongst maker communities (Gauntlett, 2013).

Indeed, it is a curious feature of makerspaces that these sites of creativity make use of digital technologies whose origins can be traced back to the automation of design and manufacturing. Where previous generations had experienced these technologies as automation artifacts which deskilled and robbed operatives of livelihoods – something that continues with Industry 4.0 today – in makerspaces we see people cultivating new skills, finding collaborative and commons-based forms of working with peers, and exploring potentially sustainable livelihoods for their futures (O'Donovan & Smith, 2020). Whilst still marginal to the latest wave of automation being promoted by capital and installed by management consultants under the rubric of Industry 4.0, makerspaces nevertheless demonstrate in a vivid and practical way something that technology studies scholars have long known: there is nothing automatic about automation.

Industry 4.0 provides a seamless vision of global "cyber-physical systems" for manufacturing, logistics, and markets, whose revolutionary advances in production and consumption will be made through the proprietary deployment of cloud computing, the Internet of Things, robotics, 3D printing, Big Data, machine learning, genetics, material science and other technologies of communication, computation, and control (Schwab, 2017). Advocates like the World Economic Forum, leading business consultancies, and, increasingly, national governments anticipate a cornucopia of opportunities by actively investing in these new technology systems; and which differ from previous automations of specific processes and organizations by now seeking the automation of entire systems of production and consumption. However, not everyone is so sanguine about these opportunities. Waves of enthusiasm for automation by capital have always been accompanied by undertows of social anxiety about the consequences; whether for the future of work, the consumption of our finite planet, concentrations of political and economic power, or the fact, as illustrated in the makerspace examples from India, China, and Brazil above, that visions of universal technological frontiers seem to be blind to the uneven geographies and technological realities on the ground.

Socializing control over the means of automation is a longstanding response to some of these anxieties, by reassuring citizens that the benefits (and risks) will be spread fairly by the state (Ad Hoc Committee, 1964). State coordination and planning is advocated as vital to translating the labor productivity gains of computer technologies into reduced work for all, fair distribution of technological abundances in goods and services, and often backed by a universal basic income for citizens unable to retrain into the new high-tech jobs. Fully Automated Luxury Communism (FALC) is the most recent articulation of these ideas, updated with arguments from ecological modernity anticipating technology-fixes for environmental problems (Bastani, 2019; Srnicek & Williams, 2015; cf. Benanav, 2019). These "automation theorists" seek to advance automation as widely and profoundly as possible, but adapted to public good rather than capital accumulation by socializing ownership of the technology.

Because arguments like these are centered on the social control of seemingly inevitable automation, the technologies of automation themselves are all too often treated as black boxes whose ownership and/or application is disputed, rather than being opened up to scrutiny as sites of political struggle which frame the socio-technical configuration of society (Benanav, 2019). Here, makerspaces become especially relevant, because many of their particular socio-technical experiments point to different development directions for digital technologies, which imply a "post-automation" space whose assumptions, values, and practices of care contest the universalist logics of control in automation theory. Where automation theory pursues a logic of ever-increasing labor productivity, tightening management control, and capital accumulation, makerspaces can open up digital configurations to other kinds of developments that foster durable, locally sensitive material cultures, that have human creativity and collaboration at their core, and whose organization brings together networked technologies and platforms with skillful contributors to constitute a new kind of commons. Such experiments in post-automation are already taking place, and perhaps even finding roots, in spaces that exist beyond conventional industrial settings and global chains of design, production, consumption, and disposal (Smith et al., 2020).

Of course, the design entrepreneurialism and product prototyping that takes place in many makerspaces can just as easily feed into the global value chains of Industry 4.0: a creative input into business-as-usual. But when post-automation initiatives are accompanied by concerted efforts to develop other kinds of institutional logics such as commons-based peer production, the prototypes and experiments of makerspaces may succeed at fostering alternative socio-technical futures. Given how heavily-instituted Industry 4.0 is becoming, constituting but the latest wave of a long history of automation (with advances in machine learning only beginning to become dominant), so arguments for post-automation spaces appear utopian. But, given the diverse geographic realities of technology use today, and the imperatives for sustainable development that are now challenging many of the foundational assumptions in industrial society, perhaps post-automation activity is no more naïve than the presumptions of automation theorists?

5 Spaces of Liberation?

This being said, it is important to question exactly how makerspaces might enable the technologies of digital design and fabrication to escape their origins and become democratic infrastructure for commons-based economic activity. Just how open to radical socio-technical reconfiguration *are* these technologies? Whilst primitivist anarchists like John Zerzan might argue that any historical turnaround in the significance of automating technologies is a mirage, and that activity today is still based in an inherently technological (and therefore oppressive) society, social anarchists like Murray Bookchin might be more hopeful and enthusiastic regarding their alternative technological possibilities. Fifty years ago, Bookchin, like other activists, welcomed a post-scarcity future in which technological progress would give collectives the opportunity to own tools and organize production non-hierarchically and sustainably, harnessing "liberatory technologies" for socially useful purposes (Bookchin, 1967). In this view, as Janet Biehl (2007) has stated, technological innovation need not always lead to a continued alienation from the means of opening up the "black box" of technological production, and can instead provide new opportunities to enact the concepts of commons-based peer production through these socio-technical experiments.

These are enduring challenges. In a different setting, organized workers in Scandinavia, Germany and the UK collaborated with leftist researchers in the 1980s for the introduction of human-centered computer technologies into workplaces, in ways that would democratize the labor process (Ehn, 1988; Rasmussen, 2007). These initiatives were inspired in part by a movement in the UK for socially useful production, whose iconic expression originated in the "Lucas Plan."[5] Launched by shop-floor trades unionists at Lucas Aerospace in 1976, and emulated by workers in other factories, the Lucas Plan presented an alternative approach to technology development, industrial production, and socialized consumption that was ecologically-aware (Smith, 2014). Workers at Lucas Aerospace – a firm heavily dependent upon government defense contracts, and whose jobs were threatened by industrial restructuring and new technologies – prototyped over 150 alternatives for socially useful products which diverged from military applications, ranging from wind turbines to medical equipment, from devices for people with disabilities to public transport. These practical prototypes served also as emblematic illustrations for a more profound proposition: democratic re-organization of the workplace; manufacturing planned in consultation with local communities and for the common good; computer technologies that enhanced and extended operator skills rather than displacing their labor; and socialized market based in use value (Wainwright & Elliott, 1982) – prototyping as "technological agit-prop" (Cooley, 1987).

Inspired by these developments, a movement of labor activists in London worked with the municipal government in 1982 to open a series of open prototyping workshops called Community Technology Networks (Smith, 2014). These networks enabled citizens to access new technologies to design socially useful products, with the intention of linking them to local authority schemes for creating cooperative manufacturing businesses (amongst the start-ups launched by the Greater London Enterprise Board were a re-manufacturing enterprise and a women's computing cooperative). Commenting on these moves, the sociologist Donald Mackenzie noted, "Whatever the eventual success or failure of these efforts to alter the nature of technology, our understanding of how technology changes can only profit from them. For, by making contingency and choice actual rather than merely hypothetical, they throw into ever-sharper light the ways in which social relations shape technical development" (Mackenzie, 1984, p. 502).

The experiments of these open community workshops sat in stark contrast to the automated "workerless factory" that owners, management, and consultants were promoting in the 1980s (Kaplinsky, 1984). The workers' alternatives were ignored by management, investors, government, and even the senior trade union hierarchy, who were more comfortable negotiating redundancy packages and uneasy about radical workplace democracy. Ultimately, the creativity arising from the worker initiatives generated little lasting impact because the coalitions that were formed and the spaces they created were swept aside by the rise of Thatcherism in the UK, and the installation of neoliberal ideologies in public institutions. Notions of social purpose in technology and production were overwhelmed by the ideology of free markets. Organized labor became contained within restrictive legislation. The demonstration of prototypes proved insufficient in the absence of a political economic program that challenged economic restructuring and relocation by capital – and that closed factories, redefined jobs, and put pressures on communities.

Ironically, in pursuing a socio-technical alternative, the activists of open community workshops did succeed in pioneering methods in participatory technology design whose appropriation by business is widely practiced today – though for user-centered and customer-loyalty purposes rather than democratizing production (Asaro, 2000). But the alignment of institutions to the movement's more radical political and economic visions proved beyond its agency at the time. This is a salutary lesson today for the kinds of spaces that makerspaces might be able to sustain in the absence of wider social mobilizations around political and economic programs for peer production (see for example Birkinbine; O'Neil & Broca; and Pazaitis & Drechsler, all in this volume).

These lessons from history do not mean that the social significance of makerspaces' more radical aspirations should be ignored, however, nor should we discount their potential to be enacted as sites for experimenting with post-automation. As we have written elsewhere: "Timing and contingency are always significant in the social shaping of technology, but the ready provision of plural possibilities is a never-ending requirement. Even if alternative pathways are by definition more open and less powerfully articulated than conventional institutions for innovation, they nevertheless cultivate ideas and practices that sometimes resonate through time, and can have real material consequences when the moment is right . . . [The movement for socially useful production] pioneered ideas and activities for a more constructive and democratic relationship with technology development in society [. . .] Insisting upon democratic technology developments, and attempting to advance this practically, was probably the most socially useful product of the movement" (Smith et al., 2017, p. 55).

And today, it is precisely because automation trajectories have become so powerfully entrenched institutionally that we need makerspaces, because they provide publics with valuable opportunities to experiment with, and question, our socio-technical interactions by engaging in peer production through communities of support and solidarity.

6 Conclusion

The discussions of this chapter illustrate the value of situating the dynamics of current-day makerspaces within the historical and cultural contingencies that are interwoven into their relationships with institutions. This enables us to cultivate a more nuanced appreciation of the many potentials, and limits, of these spaces, and the wider socio-technical configurations and power relations that they both facilitate and challenge. Throughout this chapter, we have examined the many encounters that can emerge between makerspaces and other actors (from grassroots communities and entrepreneurs to flows of global capital), asking how these relations produce makerspaces as sites of possibility, tension, post-automation, and liberation. We have also asked how these relations impact on the ways that the ideologies, structures and mechanisms of peer production can be experienced in practical ways, through material participation. As spaces of not only making and learning, but also of social experimentation with new ways of working and interacting, makerspaces reflect the tensions, understandings, and visions of the diverse actors who engage with them. In this sense, makerspaces can be perceived of as kaleidoscopes – ever-shifting, diversely colored prisms through which makers, institutions, and societies more generally can reflect upon the different possibilities of flux and disruption that can shift the conditions of our socio-technical engagements.

We have seen, for example, how some makerspaces are being enacted as spaces of rearticulation, where new ways of working and being-in-the-world are fostered, while others are produced in ways that reinforce the logics of incumbent institutions. We have also seen how the practical implementation of peer-production practices in makerspaces remains a privilege, which is more accessible to some groups and cultures than to others, and which remains structured by institutional and systemic biases that need to be more proactively countered. We used the historical experiences of automation in design and fabrication, and corresponding movements for socially useful production, to cast the limits faced by these kinds of spaces in a stark, political economic light.

The important takeaway here, though, is that makerspaces also show us how such limits *can* be countered through material participation. Even when they are captured by institutions, makerspaces can enable further transformations by regrouping, restating, and renegotiating the terms of such encounters. We have seen how institutional interest in makerspaces is sometimes derived from their perceived potential to deliver longstanding agendas in novel ways, which may or may not involve peer production. In other cases, makerspaces attract interest because they anticipate new institutional possibilities, and offer incumbents responses to pressure for change. Thus, the extent to which peer production within makerspaces can inform the development of institutions (and society) to engage in more peer production will depend upon wider social pressures and aspirations for the latter. As the examples above illustrate, there are complex mixes of both currents within the kaleidoscope: existing institutional agendas moving into makerspaces, new institutional possibilities emerging out of makerspace experimentation.

So, makerspaces are subjects in a plurality of institutional advances and developments. There is a contradiction at the heart of these encounters that drives their dynamics: makerspaces are about experimentation, improvisation, and unruliness; and yet institutions promote regularity, certainty, and orderliness. This often uneasy co-existence illustrates the many challenges of translating emancipatory ideas into mundane realities. It is important to recall here that while makerspace users and facilitators are, in many contexts, energized by the possibilities of commons-based peer production, these motivations are not necessarily shared across other kinds of makerspaces, or other cultures. The ideals of peer production, and the promotion of its advancement in society at large, must thus be adapted to the particular kaleidoscope of each makerspace in turn, with a sensitivity to the community histories of control, emancipation, and transformation which structure the shape of its relations. Most importantly, the contingent and multi-layered interactions of makerspaces and institutions show us that local geographies and experiences matter. These dynamics are the socio-technical anchors with which peer production ceases to be an ideal, and starts to become a new way of being.

Notes

1 Located on the web at www.gearbox.co.ke/
2 Located on the web at www.chaihuo.org/home/index
3 The circular economy is a framework that correlates economic growth with sustainable development by replacing the current "take-make-waste" industrial model with one where waste and pollution are designed out of the system, by keeping materials in use, transitioning to renewable resources and regenerating natural systems (Ellen MacArthur Foundation 2018).
4 https://fab.city/
5 For a recent documentary about the Lucas Plan, see www.theplandocumentary.com

References

Ad Hoc Committee (1964). The triple revolution. *International Socialist Review, 24*(3), 85–89.

Akrich, M. (1992). The de-scription of technical objects. In W. E. Bijker & J. Law (Eds.), *Shaping technology / building society: Studies in sociotechnical change*. Cambridge, MA: MIT Press.

Anderson, C. (2012). *Makers: The new industrial revolution*. New York, NY: Random House.

Armstrong, C., de Beer, J., Kraemer-Mbula, E., & Ellis, M. (2018). Institutionalisation and informal innovation in South African Maker communities. *Journal of Peer Production, 12*, 1–23.

Arvidsson, A. (2020). Capitalism and the commons. *Theory, Culture & Society, 37*(2), 3–30. doi:10.1177/0263276419868838.

Asaro, P. M. (2000). Transforming society by transforming technology: The science and politics of participatory design. *Accounting Management and Information Technologies, 10*, 257–290.

Bastani, A. (2019). *Fully automated luxury communism*. London: Verso.

Benanav, A. (2019). Automation and the future of work – 2. *New Left Review, 120*, 117–146.

Benkler, Y. (2002). Coase's penguin, or, Linux and "the nature of the firm." *The Yale Law Journal, 112*(3), 369–446.

Biehl, J. (2007). Bookchin breaks with anarchism. *Communalism: International Journal for a Rational Society*, 1–20.

Bookchin, M. (1967). Towards a liberatory technology. *Anarchy, 78*, 107–138.

Braybrooke, K. (2018). Hacking the museum? Practices and power geometries at collections makerspaces in London. *Journal of Peer Production, 12*, 40–59.

Braybrooke, K. (2019). "Placeless" making? Reframing the power-geometries of digital platforms in China through tactical co-creation. In L. Bogers & L. Chiappini (Eds.), *The critical makers reader: (Un)learning technology* (pp. 258–269). Amsterdam: Institute of Networked Cultures.

Braybrooke, K., & Jordan, T. (2017). Genealogy, culture and technomyth: Decolonizing Western information technologies, from open source to the maker movement. *Digital Culture & Society, 3*(1), 25–46.

Braybrooke, K., & Smith, A. (2018). Liberatory technologies for whom? Analysing a new generation of makerspaces defined by institutional encounters. *Journal of Peer Production, 12*, 2–12.

Cooley, M. (1987). *Architect or bee? The human price of technology* (2nd ed.). London: Hogarth Press.

Corbett, S. (2017). *How to be a craftivist: The art of gentle protest*. London: Unbound Publishing.

Davies, S. R. (2018). Characterizing hacking: Mundane engagement in U.S. hacker and makerspaces. *Science, Technology, & Human Values, 43*(2), 171–197.

Dias, R., & Smith, A. (2018). Making in Brazil: can we make it work for social inclusion? *Journal of Peer Production, 12*. Retrieved from http://peerproduction.net/wp-content/uploads/2018/07/jopp_issue12_dias_smith.pdf

Dickel, S., Ferdinand, J.-P., & Petschow, U. (2014). Shared machine shops as real-life laboratories. *Journal of Peer Production*, *5*, 1–9.

Diez, T. (2012). Personal fabrication: Fab labs as platforms for citizen-based innovation, from microcontrollers to cities. *Nexus Network Journal*, *14*(3), 457–468.

DiMaggio, P. J., & Powell, W. W. (2000). The iron cage revisited: Institutional isomorphism and collective rationality in organizational fields. *Advances in Strategic Management*, *17*, 143–166. https://doi.org/10.1016/S0742-3322(00)17011-1

Ehn, P. (1988). *Work-oriented design of computer artifacts.* Stockholm: Arbetslivscentrum.

Ellen MacArthur Foundation. (2018). What is The circular economy? Retrieved from www.ellenmacarthurfoundation.org/circular-economy/what-is-the-circular-economy

Fab City Research Lab. (2019). "Fab Lab List." FabLabs.Io – The Fab Lab Network. Retrieved from www.fablabs.io/labs

Fonseca, F. (2015). Gambiarra repair culture by Felipe Fonseca. *Makery: Media for Labs* (March), 1–15. Retrieved from www.makery.info/2015/03/31/gambiarra-la-culture-de-la-reparation/?lang=en.

Fressoli, M., Arond, E., Abrol, D., Smith, A., Ely, A., & Dias, R. (2014). When grassroots innovation movements encounter mainstream institutions: Implications for models of inclusive innovation. *Innovation and Development*, *4*(2), 277–292. https://doi.org/10.1080/2157930X.2014.921354

Fuenfschilling, L., & Truffer, B. (2014). The structuration of socio-technical regimes — Conceptual foundations from institutional theory. *Research Policy*, *43*(4), 772–791.

Gauntlett, D. (2013). *Making is connecting.* Oxford: Polity Press.

Hertz, G. (2012). *Critical making.* Retrieved from www.conceptlab.com/criticalmaking/

Hielscher, S., & Smith, A. (2014). *Community-based digital fabrication workshops: A review of the research literature.* SPRU Working Paper 2014/08. Brighton.

Hirsch, P., & Lounsbury, M. (2015) Toward a more critical and "powerful" institutionalism. *Journal of Management Inquiry*, *24*(1), 96–99.

Irani, L. (2015). Hackathons and the making of entrepreneurial citizenship. *Science, Technology, & Human Values*, *40*(5), 799–824.

Justice, L. (2012). *China's Design Revolution.* Cambridge, MA: MIT Press.

Kaplinsky, R. (1984). *Automation: The technology and society.* Harlow: Longman.

Kohtala, C., & Hyysalo, S. (2015). Anticipated environmental sustainability of personal fabrication. *Journal of Cleaner Production*, *99*, 333–344. doi:10.1016/j.jclepro.2015.02.093.

Kostakis, V. et al. (2015). Design global, manufacture local : Exploring the contours of an emerging productive model. *Futures*, *73*, 126–135. doi:10.1016/j.futures.2015.09.001.

Levy, D., & Scully, M. (2007). The institutional entrepreneur as modern prince: The strategic face of power in contested fields. *Organization Studies*, *28*(7), 971–991.

Lindtner, S. (2017). Laboratory of the precarious: Prototyping entrepreneurial living in Shenzhen. *WSQ: Women's Studies Quarterly*, *45*(3), 287–305.

Lindtner, S., & Li, D. (2012). Created in China: The makings of China's hackerspace community. *Interactions*, *19*(6), 18–22.

Mackenzie, D. (1984). Marx and the machine. *Technology & Culture*, *25*(3), 473–502.

Massey, D., & Wield, D. (1992). Science parks: A concept in science, society, and "space" (a realist tale). *Environment and Planning D: Society and Space*, *10*(4), 411–422. https://doi.org/10.1068/d100411

Munir, K. A. (2015). A loss of power in institutional theory. *Journal of Management Inquiry*, *24*(1), 90–92. doi: 10.1177/1056492614545302.

Nesta. (2015). *Open dataset of UK makerspaces.* Retrieved from www.nesta.org.uk/sites/default/files/ukmakerspacesdata_0.pdf

Noble, D. (1984). *Forces of production: A social history of industrial automation.* New York, NY: Knopf.

O'Donovan, C., & Smith, A. (2020). Technology and human capabilities in UK makerspaces. *Journal of Human Development and Capabilities*, *21*(1), 63–83.

Oldenburg, R. (1989). *The great good place: Cafés, coffee shops, bookstores, bars, hair salons, and other hangouts at the heart of a community.* Trowbridge: Paragon House.

Pache, A.-C., & Santos, F. (2012). Inside the hybrid organization: Selective coupling as a response to competing institutional logics. *Academy of Management Journal*, *56*(4), 972–1001. https://doi.org/10.5465/amj.2011.0405

Pascale, P. (2018). Laboratorios de innovación ciudadana: Nueva institucionalidad para un futuro sostenible. *Pensamiento Iberoamericano*, *6*, 63–72.

Prendeville, S. et al. (2017). Circular makerspaces: The founder's view. *International Journal of Sustainable Engineering*, *10*(4–5), 272–288. doi: 10.1080/19397038.2017.1317876.

Rai, A. S. (2015). The affect of Jugaad: Frugal innovation and postcolonial practice in India's mobile phone ecology. *Environment and Planning D: Society and Space*, *33*(6), 985–1002.

Rasmussen, L. B. (2007) From human-centred to human-context centred approach: looking back over "the hills," what has been gained and lost? *Ai & Society*, *21*(4), 471–495. doi:10.1007/s00146-007-0088-3.

Ratto, M., & Boler, M. (2014). *DIY citizenship: Critical making and social media*. Cambridge, MA: MIT Press.

Ray Murray, P., & Hand, C. (2013). Making culture: Locating the digital humanities in India. *Visible Language Journal*, *49*(3), 140–155.

Schwab, K. (2017). *The fourth industrial revolution*. Geneva: World Economic Forum.

Shea, P., & Gu, X. (2018). Makerspaces and urban ideology: The institutional shaping of fab labs in China and Northern Ireland. *Journal of Peer Production*, *12*, 1–16.

Smith, A. (2014). Socially useful production. STEPS Centre Working Paper 58. Brighton. Retrieved from http://steps-centre.org/publication/socially-useful-production/.

Smith, A. (2018). *Forking the SDGs: How prototypes cold transform the new global goals, STEPS Centre Blog*. Retrieved from https://steps-centre.org/blog/forking-sdgs-prototypes-transform-new-global-goals/

Smith, A., Ely, A., Fressoli, M., Abrol, D., & Arond, E. (2017). *Grassroots innovation movements*. Abingdon: Routledge.

Smith, A., Fressoli, M., Galdos, M., & A. Moreira (2020). Post-automation: Report from an international workshop. Science Policy Research Unit, Brighton.

Smith, A., & Light, A. (2017). How to cultivate sustainable developments with makerspaces. *Liinc em revista*, *13*(1), 162–174. Retrieved from http://revista.ibict.br/liinc/article/view/3900

Srnicek, N., & Williams, A. (2015). *Inventing the future: Postcapitalism and a world without work*. London: Verso.

SSL Nagbot (2016). Feminist hacking/making: exploring new gender horizons of possibility. *Journal of Peer Production*, *8*, 1–10.

The People's Republic of China. (2017, September 10). China tops the world in incubators, makerspaces. *The People's Republic of China Press Releases*. Retrieved from http://english.gov.cn/state_council/ministries/2017/09/19/content_281475870146634.htm

Thomas, H. (2013). Tecnologías para la inclusión social en America Latina: de las tecnologías apropiadas a los sistemas tecnológicas sociales. Problemas conceptuales y soluciones estratégicas. In H. Thomas, M. Fressoli, & G. Santos (Eds.). *Tecnología, desarrollo y democracia: nueve estudios sobre dinámicas socio-técnicas de exclusión/inclusión social*. Córdoba, Argentina: Ministerio de Ciencia, Tecnología y Innovación Productiva.

Toupin, S. (2014). Feminist hackerspaces: The synthesis of feminist and hacker cultures. *Journal of Peer Production*, *5*, 1–5.

Troxler, P., & Maxigas. (2014). Editorial note: We now have the means of production, but where is my revolution? *The Journal of Peer Production*. Retrieved from /issues/issue-5-shared-machine-shops/editorial-section/editorial-note-we-now-have-the-means-of-production-but-where-is-my-revolution/

Turner, F. (2018). Millenarian tinkering: The Puritan roots of the Maker Movement. *Technology and Culture*, *59*(5), S160–S182. https://doi.org/10.1353/tech.2018.0153

Wainwright, H., & Elliott, D. (1982). *The Lucas plan. A new trade unionism in the making?* London: Allison & Busby.

Wang, J. (2016). The makers are coming! China's long tail revolution. In M. Keane (Ed.), *Handbook of cultural and creative industries in China* (pp. 46–63). Cheltenham: Edward Elgar.

Willmott, H. (2014). Why institutional theory cannot be critical. *Journal of Management Inquiry*, *24*(1), 105–111.

Wolf, P., Troxler, P., Kocher, P.-Y., Harboe, J., & Gaudenz, U. (2014). Sharing is sparing: Open knowledge sharing in fablabs. *Journal of Peer Production*, *5*, 1–11.

Zietsma, C., & Lawrence, T. B. (2010). Institutional work in the transformation of an organizational field: The interplay of boundary work and practice work. *Administrative Science Quarterly*, *55*, 189–221.

27

Peer Production and State Theory
Envisioning a Cooperative Partner State

Alex Pazaitis[1] and Wolfgang Drechsler

1 Introduction

Anything regarding the state will be immediately regarded with suspicion by most of the protagonists of peer production and, indeed, of the cooperative movement. For both share an implied distrust and opposition against hierarchy, bureaucracy, or anything above and beyond the people, including or pertaining to the state. So, to make a case for a positive, let alone ***necessary*** role for the state, however transformed and reduced for peer production, even at the level of discussion, would be an uphill battle.

As we explain later, across a broad spectrum of diverse political interpretations and ambiguities of peer production, one thing that appears to be common is this almost ingrained view of the state as antagonistic, repressive, or obsolete. However, it is exactly the absence of state-like structures or institutions – in the sense of a regulating, ordering principle outside of the immediate economic sphere of peer production – that seems to inevitably lead to all kinds of disasters, both theoretically and in experience. From the Facebook Cambridge Analytica scandal[2] to the precariousness of Uber drivers (Scholz, 2016) and the population displacement in cities evoked by AirBnB,[3] and from the discourse on the ecological costs of Bitcoin (Krause & Tolaymat, 2018) or the failure of "The DAO" experiment (DuPond, 2017; Mehar et al., 2017) to the Deletionists vs. Inclusionists dispute in Wikipedia (Kostakis, 2010) or the issue of gender discrimination in free and open source software communities (Terrell et al., 2017), numerous examples illustrate this tension. Hence, though this may be an uphill battle, it is most probably one we cannot avoid.

Yet, contradictory as this may seem, peer production is far from being the only case of state support to alternative or emergent economic structures. Many historical instances come to mind, including the role of the state in the early Israeli kibbutz movement and the subsequent formation of the State of Israel (Tsuk, 2000); the formation of the Italian cooperative network in the Emilia Romagna region (Adeler, 2014; Corcoran & Wilson, 2010; Zamagni, 2006); or the contemporary policies for the social and solidarity economy amidst the ongoing degradation of the welfare state in Greece (Adam, 2012, 2016; Kalogeraki et al., 2018; Vathakou, 2015). In all these cases, state-led policies have consciously created emancipatory environments for civic action, which in turn influenced the political agenda and even state reform. Likewise, peer production evinces new forms of economic and social organization, accompanied by a shared morality coalesced around the P2P administration of social affairs.

But how can the contemporary state support peer production? To answer this question, we interrogate the notion of the partner state, as a new form of symbiosis between state and civil society, based on the principles and practices of peer production. Recent experimentations such as Ecuador's Open Knowledge Society project (Bauwens et al., 2015), and the urban commons policies in cities like Barcelona, Bologna, and Naples (Bauwens & Niaros, 2018), allow us to sketch out the prefigurable contours of the partner state through current practice and informed speculation. This chapter's aim is to examine and discuss these early theoretical and empirical foundations through the lens of State

The Handbook of Peer Production, First Edition. Edited by Mathieu O'Neil, Christian Pentzold, and Sophie Toupin.

Theory. Our main objective is to ***understand*** and ***explain*** the ontological and functional foundations of the partner state, beyond the myth and in the context of current social and economic transformations.

A tentative glimpse of these elements may be offered through a synergy of peer production with the principles of the cooperative movement, which has been proposed with the concept of "open cooperativism" (Bauwens & Kostakis, 2014; Conaty & Bollier, 2014; Pazaitis et al., 2017). An open cooperative is legally and statutorily dedicated to the creation of commons and shared resources. It adopts multi-stakeholder forms of governance to internalize negative externalities and organizes, socially and politically, around global concerns, while operating locally (Bauwens & Kostakis, 2016; Bauwens et al., 2019).

Acknowledging the common line of critique on cooperativism (Luxemburg, 1899), open cooperatives are presented as an opportunity to revitalize cooperativism in the digital era. But what is particularly important for our inquiry is that they also provide an analytical basis for state theory to approach peer production. They illustrate hybrid configurations incorporating elements from reciprocity, as well as commons-based organization. These proto-institutions gradually create a playing field on the micro and meso level that breeds the democratic means for commoners to pursue their own livelihood, but also broader social and political transformation. Open cooperatives uphold peer production as a learning process of bottom-up political action and influence to forward state reform.

In the following sections we briefly explore the economic dynamics of peer production and their socio-political implications. Afterwards, we provide an overview of the state as the agent for social reform and change, drawing from the philosophy of G. W. F. Hegel and the critical analysis of Antonio Gramsci, to conceptualize the role of the partner state in enabling and strengthening peer production within and beyond the current system. We then follow with an exposition of these transformative elements through open cooperativism. Finally, we discuss a tentative political agenda towards a commons-centric society.

2 With or Without the State: Peer Production from the Economic Sphere to the Political, and Back Again

Stemming from the enhanced capabilities of Information and Communication Technologies and the rapid expansion of the Internet, peer production embodies the diversities and ambiguities of its socio-technological background. We, thus, have different political objectives and interpretations pursued through various socio-technological frameworks of peer production (Bauwens et al., 2019).

On one side of the spectrum, we have generative, civic-driven economic forms of localized commons, such as transition towns or ecovillages, and global digital commons, like free and open source software and Wikipedia. Despite the deviance in scale or impact, both local and global commons-based communities are celebrating self-management and autonomy over control and coercion; decentralization over concentration of power; and ad-hocracy over planning and execution, which goes almost by definition against most contemporary views of state institutions. Moreover, the main theoretical underpinnings around the commons, from Ostrom (1990) to Harvey (2011), Hardt and Negri (2011), and De Angelis (2017), to Bollier and Helfrich (2012, 2019), would rather speak for a dynamic shift contra to the state or, at least, away from– and around it.

In addition, many practitioners and activists of commons-based peer production, often with a Marxist or Anarchist background, are socialized into thinking that the state must and, eventually, will perish, having always been more of a problem than an answer to social questions. Likewise, those with a green, urban, peace-oriented or other activist background have personally experienced the state as the physical enemy, in the form of riot police or state prosecutors. In this perspective, under the current system, peer production will be doomed to remain subject to massive co-optation, providing yet another medium for precariousness and exploitation to feed capitalist growth. And admittedly, these critical views have merit, but taken as absolutes, they risk, as we will see, fighting a paper tiger.

On the other side of the spectrum, the dynamics of peer production are exploited by centralized platforms such as Facebook and Uber, or distributed rent-driven networks like Bitcoin (Bauwens et al., 2019). Here a widespread techno-enthusiasm, from Silicon Valley to the crypto-economy would hold any type of state intervention, regulation, or oversight as an impediment to innovation and progress. Technology and the state are deemed statutorily antagonistic and technological progress is often assumed to advance towards a condition that would simply engineer the state away. Hence, if the state delays the advance of technology, it is seen as something bad or, at best, inefficient rather than as a creative friction in human affairs that opens a space for discussion and negotiation (Drechsler & Kostakis, 2014).

At the extreme of this outlook, there is often the idea that the state, often simply conflated with the nation state, would simply wither away via new forms of technologically-enabled sovereignty (e.g., Manski & Manski, 2018). The discussion around the so-called blockchain or distributed governance has since the introduction of Bitcoin (Nakamoto, 2008) reached the size of a fashion item of global intellectual significance, into which peer production can seamlessly be situated. A highly economized understanding of politics is also at play here, where a distributed form of money issuance surpassing the need for a central bank – other than of course financing everything related to the existence of the said distributed system – is equivalent to state-less freedom.

Finally, peer production is often conflated with the promises of distributed production and a post-scarcity or post-capitalist society (e.g., Mason, 2015; Rifkin, 2014; Tapscott & Williams, 2006). Yet these views lack a serious analysis of the social and political ambivalences of such a transition, holding strongly utopian views of societies that would perhaps be nice to have, but to which very few bridges are leading. After all, can a change of production alone really transform society and politics? Conversely, peer production has a strongly reformist element, though often this is not admitted. A lot of the discussion is based on the exploration of seed forms that are transforming what we currently have, working within the system. Even more, conflicts and division very often burst out between groups, where one would normally expect unity and complementarity.

And this is so, because, in the end, peer production is an economic theory, as the name already implies. It is often the case that state, legal, and political-philosophical matters are relatively downplayed and many things merely assumed. But the relationship between economics and politics is not linear, i.e., a radical change in the former is not necessarily followed by corresponding ones in the latter. Things can go many ways and, as Werner Sombart (1932) already pointed out, the question of which economy to choose is primarily a political rather than an economic one.

For instance, in its early stages industrial capitalism and the state were enemies, as the latter was feudalistic-agrarian in all early-capitalist systems. It was in fact only relatively late in the former's development that they became aligned, with their synthesis best embodied in the form of the welfare state (Reinert, 2019). And that is a key point: the classic welfare state, in the form of the social market economy, assumes that the market can efficiently allocate scarce resources and this is, for significant parts of the economy, a sensible *modus operandi*. Incidentally, the welfare state also seems to be the main context for any form of real-existing peer production today, i.e., peer production really exists, to the extent that it does, mostly within systems that are social market economies.

However, the results of economic operations are not necessarily congruent with what society in a given context prioritizes and desires (Drechsler, 1997). In peer production, if we assume an absolute situation, where market-based operations are absent or heavily reduced, this question will still remain. Even under the best of circumstances, even optimally and with the best of well-minded people, what reason do we have to assume that peer production will lead to a society that is as we desire? What about those who cannot co-produce? What about those who do not do it well? What about, indeed, those who do not want to? If this is not managed ad hoc and arbitrarily, we need someone or something to generalize the commonweal in an institutional, predictable, and systematic way. And even if we call this committees, colleges, or assemblies, of course that is the state.

Moreover, peer production discourse, by and large, assumes that humans are naturally better than they actually are. Already William Morris (1890) conceded that his classic, indeed archetypical, commons-based, peer-productive utopia *News from Nowhere* needed a big cataclysm, a new human race, to exist. Similarly, Ursula Le Guin's (1974) communal society of *Anarres* took a massive settlement (or rather exile) of all the utopian anarchists to emancipate themselves from their "propertarian" society. And there is ample historical evidence for this in the real world, from the challenges of the cooperative and revolutionary movements to today's free vs open source software discourse.[4] This illustrates that so far, even in the peer production sphere, people are just not as "nice" as they should be for the theory to work, at least not always or not in identical ethical manners. The assumption that people have an intrinsic tendency to perform in a p2p manner is just too bold to be acceptable.

The famous fresco by Ambrogio Lorenzetti in Siena Town Hall, from the early 14th c entury, one of the great art works symbolizing and interpreting government and its effects on the economy, indirectly makes this point in the "Allegory of Bad Government" (Drechsler 2001b). Just like good government is surrounded by faith, hope and love, on another wall its opposite, tyranny, is respectively encircled by three vices: greed/avarice, arrogance, and vanity.[5] Now, if the peer production transformation removes greed, the other two vices still remain, and as we often enough see, they are powerful as well. But because they are not economic motives, they are often neglected, which partly explains why there may be surges of such bitter fighting about precisely what kind of fab lab should be established, and where.

Of course, one does not have to follow anthropo-psychological speculations such as these, based on novels or Italian frescoes. Yet, it is a reminder that, as we observe in our own world again and again, and indeed as the entire point of peer production being "c ommons-b ased" is, materialistic matters are not everything. And the idea of the ideal state in this context is that it is the institution that fulfills the necessary role *outside* of the economic sphere and, to a certain extent, also inside.

As it becomes clear that the state is not threatening peer production as an economic form, it is, in fact, most often than not state institutions that support p2p projects, labs, and spaces where peer production experimentations take place. Respectively, some of the most modern and richest states (on all three levels – country, regional, and local), are increasingly displaying their interest to support peer production to a significant extent, while a significant number of p2p activists find themselves on the state payroll, either in public universities or state-funded makerspaces, and not by accident, but by design.

3 On the (Partner) State as Agency for Social Transformation

Peer production has been advocated as a new mode of value creation that encompasses seed forms of post-capitalist scenarios. These hold the potential to permeate and transcend the current global order of markets and states (Bauwens et al., 2019). Even though this is taking place in a world administered by hierarchical bureaucracies and competitive market agents, these new forms of self-management and mutual coordination are gradually moving from the periphery to the center.

And yet, as long as it disputes the state as an apparatus, a method, and organization, peer production reinstates the relevance of the state as an *object* of scientific inquiry, best examined by the classical Continental State Theory embodied by the German term *Staatswissenschaften* (Drechsler, 2001a). For it is exactly the task in hand to examine the state as a form of structured human living-together. Much like the way medical doctors study the human body by grouping different fields of expertise, classical state theory deals with the state as the primary focus of an inter-disciplinary synergy, including economics, sociology, political science, history, and law.

And this is particularly relevant for the concept of the partner state. The current discussion of the partner state has been mainly offered as an approach for a state-like structure, which is oriented towards self-defined civil action (Bauwens et al., 2019; Kostakis & Bauwens, 2014; Pazaitis & Bauwens, 2018). Despite this ontological contradiction from a narrow definition of the state, as far as classical state theory is concerned, if it walks like a state and talks like a state, the partner state can, and should be, studied as one.

More specifically, the Hegelian (Hegel, 1995 [1821]) understanding of the state as the sphere of genuine freedom, including individual Freedom (Ritter, 1957) is of major importance to analyze this enabling function of the partner state. While markets are driven by individual greed or, at best, desire, the state serves as the sphere of civil negotiation concerning society's processes and priorities and the rights and obligations of citizens towards the administration of social life. It is a process that makes individual persons real by incorporating them into a larger whole, which, then, only exists because of individual action.

One of the first serious attempts to frame peer production vis-à-vis the state has been a report produced by the Commons Strategies Group titled "State Power and Commoning" (Bollier, 2016). As the title already implies, the aim has been to reformulate the discussion, marking a shift from a static view of the state and the commons as entities, to a dynamic, process-based understanding of the underlying social and power relationships. This conceptual shift is useful to identify the role of the partner state as one serving to define the rules of ***commoning***, i.e., the quality of contributing to, and benefiting from, shared capacities. It is an integral process of every p2p community, which differentiates them from networks of loosely-affiliated self-interested individuals. The idea of the partner state sublates[6] this process on a higher level, allowing the diverse communities to operate in harmony and guaranteeing prosperity for the system as a whole.

Commoning as a process is often discussed as synonymous with p2p, which is mainly a relational dynamic, now amplified by a counterpart technological infrastructure (Bauwens et al., 2019). P2P relations reconfigure the perceptions and aspirations of the political community by defining a new interpersonal rationality and coordination of its constituents (Bollier, 2016; Pazaitis & Bauwens, 2018). It is hence in tandem with commoning that peer production asserts a certain political connotation that becomes interpreted through the partner state. Simply put, if peer production creates and optimizes the conditions for shared capacities to operate, commoning answers the questions of *what* is being shared (i.e., the resources); *who* is engaged (i.e., the community); and *how* (i.e., the rules and norms of collective stewardship).

This may be understood as an extension of the global-western welfare state. The underlying rationality of the welfare state goes hand-in-hand with capitalist production and is subsequently focused on the redistribution of wealth and benefits to alleviate its externalities, such as income inequalities and ecological degradation. Similarly, the partner state moves from redistribution to pre-distribution, harnessing peer production to mobilize productive capacities around the commons. It internalizes externalities by embedding productivity within social and ecological limits, defined by the rules of commoning. However, even though peer production is the driving force of this political outcome, peer production alone cannot guarantee it, just like industrial production alone cannot guarantee social justice.

On a historical note, many of the ideas underpinning the welfare state in the form of the social market economy date back to the late nineteenth century German Historical School of economics (Drechsler, 2016). Contrary to mainstream economic ideas, social reform champions realism and relevance, as opposed to abstraction and precision. Peer production arguably provides an alternative rationality for social reform, centered around contributory action and the self-aggregation of shared capacities. The partner state is then the agent that makes this action possible to begin with. It nurtures and guides individual liberty to encapsulate the totality of social and ecological life through commoning.

Nevertheless, as commoning is by no means immune to the pitfalls generating inequalities and unfair outcomes, the role of social reform also entails the art of perceiving problems, ameliorating them through policy measures and simultaneously employing the relevant scholarly discipline to understand and demonstrate what is wrong (Drechsler, 2016). As opposed to revolutionary approaches, social reform may *prima facie* seem conservative. But as a methodological approach in economic policy it presents a viable, problem-based, context-specific approach that vastly differs from current practice. Likewise, it exemplifies the agency of the state in fostering social change, rather than creating a reality in which change is delegitimized. In a globalized order driven by the brute economization of social life, the state stands as the one agency that has been, and could be, built against it, with social reform providing its main diagnostic and treatment tools.

But does the identification of problems and social reform alone suffice to transform contemporary predatory capitalist states into partner states? The common line of critique from the Marxist and anarchist traditions asserts that the state, here foremostly seen as an apparatus, exists to serve the needs of the dominant class and suppress change, whilst ensuring its own survival.

Antonio Gramsci (2007) is arguably the key Marxist thinker offering an alternative view of the state. More specifically, the concept of the "integral state" suggests a dialectical unity between the political society (i.e., the state) and the civil society (D'Alisa, 2019). The two parts are seen as engaging in a struggle to fulfill divergent visions through the means of, respectively, domination and hegemony. Domination is related to enforcement and the legitimate use of power, typically associated with the state, though not exclusively confined to it. Likewise, hegemony implies consent on behalf of the people, usually informed by ideology.

Gramsci methodologically differentiates the spheres of civil and political life, but acknowledges them as ontologically inseparable and organically interwoven. There is as much coercion exercised in civil society as there is consent accrued by the state. Hence, domination and hegemony are operating as mutually reinforcing processes of the state over- and through- the people. This struggle is then reflected in "common sense," i.e., the generally, often unconsciously, accepted perceptions of the world at a given time. Hegemony thus defines the spectrum of commonsensical perceptions rendered acceptable to people, largely through consent and, if needed, coercion (D'Alisa & Kallis, 2016). Simultaneously, alternative or marginal ideas outside of this spectrum mobilize counter-hegemonic groups, which strive to create new common senses for social change.

In this view, peer production, and particularly its commons-based variation, is approached as the locus where counter-hegemonic ideas and practices are cultivated, seeking to challenge and transcend current institutions. Respectively, commoning serves as a new common sense in understanding political and civic life. In turn, social groups may be mobilized to exert consent and pressure towards new institutions. This mutually reinforcing process creates what De Angelis (2017) calls "enabling environments" for individual emancipation.

Therefore, the partner state can be simultaneously examined in two ways: ontologically as a Hegelian state, i.e., it encapsulates and sublates individual self-aggregation in an administered totality, along with its institutions; and instrumentally as a Gramscian integral state, i.e., it dialectically operates upon and through civil action to establish counter-hegemonic ideas, to define new institutions.

The partner state, thus, marks a strategy that is, on the one hand, reformist, as it works within existing configurations, but, on the other hand, it is also revolutionary, as it cultivates the conditions that could potentially lead to a new configuration. As a revolution is usually perceived as anti-state, the concept of "revolutionary reform" by Gorz (1968) is particularly useful in the case of the partner state. A revolutionary reformist approach is one that is acceptable by the dominant system, but simultaneously gestating its transformation. A common example of such a reform in relation to peer production is a universal basic income. It can be seen as a form of welfare benefit provided through the current system, but it marks a break in the commodification of labor, emancipating workers to contribute to self-identified commons-oriented activities.

4 On Cooperatives as Vehicles of Economic and Political Agency

Traditional cooperatives, with their numerous variations, have been presented as viable alternatives to dominant capitalist organizational forms since the 19th century. Their contribution to the economy is still substantial, with an annual turnover of more than €1 bn and 180,000 cooperative enterprises employing about 4.5 million people in Europe (Cooperatives Europe, 2016). Furthermore, they foster a reconfiguration of the work environment oriented towards the reduction of precariousness and inequality, increased worker and social welfare.

The common line of critique, often deriving from the work of Rosa Luxemburg (1899 [1970]) rebukes the organizational form of cooperatives which oscillates between social and capitalist production. Most often than not, cooperatives end up adopting exploitative mentalities to withstand competition. Moreover, they tend to circle around their local or national membership, which diminishes their role for the broader community. To protect their internal environment, they rely on opaque and exclusive forms of ownership and control, while they fall short in harnessing the capabilities of digital collaboration. All these limitations constrain the transformative dynamic of cooperatives for the broader economy and society (Pazaitis et al., 2017).

Furthermore, peer production, as a new form of value creation, is based on autonomous individual contributions aggregated through distributed collaborative relations. This mode of production differs from either hierarchical or market-based forms of coordination in that it is guided by neither central coercion nor price-signaling. There are indeed exciting prospects for the creation of meaningful projects moved primarily, if not solely, by social motives. But simultaneously, peer production poses great challenges to existing organizational forms, including cooperatives, as well as institutions and bodies of representation, such as trade unions. Subsequently, an emerging class of autonomous workers largely falls within the cracks of the current structures, becoming increasingly vulnerable to precariousness and exploitation, sometimes in the form of self-exploitation. Some of the most predatory business models from the giant tech firms of the so-called sharing economy, such as Uber or AirBnB provide compelling evidence for this situation.

In response to these constraints and in the face of the new challenges of the digital economy, open cooperatives call for a synergy between cooperative organization and elements of the commons and peer production (Conaty & Bollier, 2014). Open cooperatives aim for a dynamic balance between maximum autonomy for contributory activities, while maintaining conditions for security and livelihood traditionally enjoyed by employed labor. This can be achieved by investing in the enhanced productive capabilities of peer production, but within a safe space built around trust and solidarity.

In this vision, the commons steer and nourish these two directions, as they provide the premises for both peer production, as well as the solidarity layer. A central aspect is the pooling of resources and productive capacities, which conduces to forms of open, multi-stakeholder engagement in the cooperative organization. Open cooperatives can thereby guide social forms of production to the creation of commons that are further deployed in new iterations by other open cooperatives and commons-oriented enterprises. This way, they contribute to the expansion of the commons for the broader society, while generating a cooperative advantage to contend with the pressures of the capitalist market (Bauwens et al., 2019).

Bauwens & Kostakis (2016) identify six interrelated strategies, through which open cooperatives may empower these arrangements. First, they embrace abundance for the common good, which is especially relevant to naturally shareable goods and actions, including knowledge and technology. Recognizing abundance rather than artificially imposing scarcity may reduce short-term individual profit, generalizing long-term systemic benefits. Second, open cooperatives support modularity and multiplicity of value forms. They often employ open and contributory value accounting methods (Bauwens & Niaros, 2017), which allows them to interface with the capitalist market, generate revenue, and fairly reinvest to their community, based on the community's ethics and value perceptions. Third, and pertaining to the two previous points, open cooperatives encourage reciprocity towards the shared resources, infrastructures, and produced value, through "CopyFair" licenses. For instance, hybrid licensing schemes introduced by the FairShares association[7] include a non-commercial clause for non-members, while allowing commercial use for members, thus requiring reciprocity from external agents (Riddley-Duff, 2015). This protects the shared capacities from predatory activity, while empowering internal collaboration and the expansion of the community in alignment with its values.

The fourth strategy concerns design-embedded sustainability. As the design and production of products and services is not primarily guided by supply and demand, there is no incentive for planned obsolescence, while interoperability, repairability, and adaptability are highly valued. Projects like RepRap or Wikihouse exemplify this design paradigm, where economic entities build their strategic advantages explicitly through "building communities of meaning" (Benkler, 2017). Similarly, fifth, transparent design and productive processes increase resource efficiency and reduce waste. Open cooperatives create realistic possibilities for genuinely circular and sharing economies through open, needs-based design and production across the supply chain. Finally, the sixth strategy pertains to the eventual migration of the above elements, largely originating in the digital sphere, to the physical realm. The p ooling of material resources and productive capacities can take place in shared infrastructures, co-working, and manufacturing spaces, maximizing meaningful sociality and minimizing externalities.

Hence, open cooperatives coalesce around collective knowledge, tools, and infrastructures to support the commons. They concentrate their productive efforts locally, but organize around global concerns. They thus foster counter-hegemonic ideas and practices around peer production and render social production autonomous and self-sustaining. The community is redefined through trans-local synergies and commoners are able to create new types of vehicles for economic and, eventually, political influence. These consist in democratically governed spaces for sustainable livelihood, self-organization and emancipation from the confines of the dominant system.

These recurring cycles of material re-composition of the commons may gradually increase the capacity of commons-based alternatives to become normalized in social and political negotiation. On the macro-level, this resembles the role of the partner state in facilitating the creation of value by direct civic action. The partner state may guarantee open, permissionless participation by maintaining common infrastructures for commons-based contributory systems, protecting collective capacities and enabling synergies across various agents.

5 From Counter-Power to Social Reform: The Partner State in Motion

To summarize our argument, open cooperativism serves as the incubator of the emancipatory ideas and practices of peer production. Open cooperatives develop their capacities in tandem with the evolution of new institutions to support this process. Once the prefigurative forms gain influence and social consensus, they can become pronounced on a higher level of abstraction. Commoning and peer production form the new common sense at local, national, and transnational levels that guide the advance of the political community. In this way, open cooperatives empower and nurture civic action. Open Cooperatives provide the necessary emancipatory spaces that cater for the expansion of the commons and the material subsistence of commoners; develop prefigurative institutions; and, eventually, raise awareness from the state to recognize, support, and incorporate in reform and transformation.

It should be clarified that our intention is neither to present open cooperatives as an ideal form of economic entity for peer production, nor to restrain the role of the partner state to simply establishing or funding open cooperatives. In fact, open cooperatives are conceptualized as hybrid forms of organization that serve exactly the purpose of operating within the current political economy, while gestating the social and economic practices that would eventually transcend it. The principles described in the previous section form the basis of the proto-institutions that generalize the merits of peer production and simultaneously build counter-power against the dominant system. They thus embody both iterations of the partner state analyzed in this piece, i.e., the Hegelian sphere of the common good and the Gramscian counter-hegemony.

The sequence or synthesis of these two functions at a given time and context may vary and adapt to the different circumstances. For instance, progressive and pro-commons Spanish municipal coalitions in Barcelona and Madrid have led the city government from 2015 but were put in opposition after the 2019 elections. Several policies put forward in the previous term would most probably need to be adapted and reconfigured to serve their political objectives. Similarly, different obstacles may be put in place by various actors to restrain or reverse such policies, as powerful vested interests are not expected to wither away even in the face of a government consciously forwarding partner state policies. However, the notion of the partner state, as analyzed here, extends beyond a cluster of policies supporting civic initiatives. Rather it entails the maintenance of a political community and the institutions coalesced around commoning, along with the struggle to steward its own survival, wellbeing, and, eventually, emancipation.

In terms of historical analogies, open cooperatives can be compared to the emergence of guilds in the 12th century (Bauwens et al., 2019). Specifically, guilds organized work within the feudal order, under the principles of solidarity and communality. Nevertheless, they were conforming to, and recognized by, the existing power structure (De Moor, 2008). But it was through the evolution of the merchant guilds that the new capitalist class rose to eventually dominate society. It is this combination of seed economic and social forms that enables the expansion to the political sphere, by coalescing a vibrant part of the society around viable alternatives.

However, as long as society faces the current forms of inequality and exploitation, further forms of coercion and exclusion may evolve in the formation of a commons-centric society as well. The state, in the Hegelian sense, will continue to fulfill its role as the guarantor of society as a whole, through prudent policies and reforms (Goldstein 2006). Diverse political views may assert influence to the state to expand and adapt the sphere of the common good to include even more marginal ideas and social groups, helping them to move from the periphery to the center.

Therefore, the partner state can also be understood as an ongoing process. As a living organism, it follows the evolution of the community it serves. It adopts, strengthens, and further promotes the patterns of political deliberation that serve the common good. Similar to the Marxist notion of continuous revolution, the partner state is in permanent reform; it continuously restructures itself to encapsulate the whole spectrum of civil creativity, freedom, and expression. In this process, peer production exemplifies forms of public engagement and deliberation that dialectically shape the ethics and perceptions embodying the Aristotelian notion of the *Good Life*, at a given space and time. The partner state does not necessarily need to be itself p2p, rather to incorporate these elements in the system of government.

From this perspective, the partner state offers an alternative approach to the current discussion on citizen co-production (Lember et al., 2019). Most conventional co-production approaches seek to engage citizens in functions otherwise implemented by bureaucrats. In many cases, citizen participation is basically crammed in otherwise close-ended processes to enhance government accountability and legitimization on face value. However, peer production projects do not focus on the process of participation, but more on the relational dynamic, i.e., p2p, as well as the outcome, i.e., commons. They thus provide transparency to these two elements to allow autonomously identified, non-coercive forms of participation to take place.

On the policy level, a crucial element in enabling this process is to embrace modularity. By breaking down large-scale processes into smaller constituents, different levels of engagement and motivation are enabled. In public policy, this can be useful in typical wicked-problems, where every inter-subjective view matters. For instance, in the case of unemployment, a pre-distributory approach would entail pooling the relevant resources (e.g., unemployment benefits, subsidies for SMEs, etc.) through open collective infrastructures (e.g., makerspaces, micro-finance, community currency or mutual aid schemes). This would enable unemployed people to interface with functional structures on different levels and based on their personal values. It is then the job of public servants to monitor these signals and direct the relevant support policies accordingly (e.g., from health insurance and taxation to public investment, subsidies, even universal basic income), towards the best outcome for the larger parts of society.

But, as already explained, peer production remains an economic theory. It does not entail specific processes of engagement; rather it only comes to life once people mobilize around common matters. So, even in an ideally zero-marginal-cost reality where networks of autonomous individuals peer produce globally to cater for all their needs, material and psychological, it would arguably still make sense for human beings to pursue more persistent ways to meaningfully administer their social affairs. The political sphere is after all a domain where some forms of "transaction costs" are more of a feature than a bug, i.e., friction creates space for negotiation and social dialogue (Drechsler & Kostakis, 2014).

Hence, peer production signifies a different approach to policy design and interface in the public sphere. It reinstates socially oriented mobilization of people and resources and meaningful collaboration for its own right. The role and function of bureaucracies in the partner state may not change fundamentally, at least not at once, as long as we still live in an administered world. Further, technological and policy capacity as we know them today are expected to remain crucial. What may substantially differ is the fabric of policy implementation itself, in that it becomes more transparent (in terms of relations and outcomes), open-ended, and less centrally coercive.

6 Conclusion

The partner state with regards to peer production signifies a change of narrative against the state, from an "other" category to one of "us." This is already a small victory of a new modality of economic production that, even when not consciously, seeks to reverse the civic disconnect from political affairs.

However, it is a victory one has to argue about, as most of the prominent movements around peer production have an almost intrinsic predisposition against the state – and not completely unjustifiably so.

However, analytical explorations from state theory reveal that the examination of the state, along with its institutions, actually does offer insights to better understand the position and potentials of peer production. In the Hegelian notion, the state is exactly that condition that allows different social forms, p2p or otherwise, to become meaningful and enable real and complete emancipation. So, the real question for the future of peer production is not whether the state is relevant, but what type of state transformation needs to take place for it to become relevant, and how.

Furthermore, the Gramscian theory of the integral state allows us to conceptualize a dialectical unity between the political society of the state and the transitionary ethics of peer production. The latter creates the premise for counter-hegemonic ideas to be nourished and expanded, while the latter follows with the institutions that normalize them. This union of the Hegelian and Gramscian thought tells us a story about *why* and *how* the state can, and arguably should, embrace and support peer production. The partner state provides, then, a less freighted name to focus the necessary intellectual discussion.

Another tentative fusion is one between elements of peer production and the cooperative economy. In particular, open cooperatives provide an ambitious, yet realistic framework to form prefigurative institutions that, on one hand, empower peer production, and on the other hand, socio-politically address some of the immediate challenges and implications. Open cooperativism as an approach offers a potential blueprint for partner state reform, acknowledging the politicalness of some of the most transformative elements of peer production. It reaffirms the idea that a new mode of production alone does not necessarily lead to a change of politics and society and, simultaneously, lays down a strategy through which it actually might.

Abstract as it may currently appear, the partner state is a powerful idea for the political community of peer production. It presents a dynamic track for both peer production movements and state institutions to assemble. The current political establishment is not expected to wither away any time soon. Even if peer production manages to outcompete its rival forms in many domains, there is still no valid reason to think of it as automatically politically decisive. It is exactly if we want peer production *to be* politically relevant that we must examine it as such, and the domain of the state is the place for this to take place.

Notes

1 Alex Pazaitis acknowledges financial support from the European Research Council under the European Union's Horizon 2020 research and innovation program (grant agreement No 802512).

2 For a concise overview see: www.theguardian.com/technology/2019/mar/17/the-cambridge-analytica-scandal-changed-the-world-but-it-didnt-change-facebook.

3 Indicatively some studies referred here: www.bbc.com/news/business-45083954

4 https://www.gnu.org/philosophy/open-source-misses-the-point.en.html

5 Arrogance and vanity in today's discourse are often lumped together as pride, but there is a difference between thinking "I am incredibly beautiful" and thinking "I am better than anyone else."

6 Sublate derives from the Latin *sublatus* (past participle of *tollere*, i.e. to take away, lift up), from *sub-* (i.e., up) + *latus*, past participle of *ferre*, i.e., to carry. As a philosophical term it is often used as a translation for the German Hegelian term *Aufheben*, which means to assimilate a smaller constituent entity into a larger one. It connotates the dialectic process of negating or eliminating an element, while preserving it as a partial element in a synthesis.

7 www.fairshares.coop/

References

Adam, S. (2012). Social economy and active inclusion policies: Results from field work in work integration social enterprises in Greece. Greek Labour Institute. Retrieved from http://ineobservatory.gr/publica-tion/kinoniki-ikonomia-ke-energitikes-politikes-entaxis

Adam, S. (2016). Social enterprises, social and solidarity economy and youth: What role for policymaking? In M. Petmezidou, E. Delamonica, C. Papatheodorou, & A. Henry-Lee (Eds.), *Child poverty, youth (un) employment, and social inclusion*. The Comparative Research Programme on Poverty (CROP) of the University of Bergen (UiB) / International Social Science Council (ISSC).

Adeler, M. C. (2014). Enabling policy environments for co-operative development: A comparative experience. *Canadian Public Policy, 40*(1), 50–59.

Bauwens, M., & Kostakis, V. (2014). From the communism of capital to capital for the commons: Towards an open co-operativism. *Triple C: Communication, Capitalism & Critique, 12*(1), 356–361.

Bauwens, M., & Kostakis, V. (2016). Why platform co-ops should be open co-ops. In T. Scholz & N. Schneider (Eds), *Ours to hack and to own: The rise of platform cooperativism, a new vision for the future of work and a fairer Internet* (pp. 163–166). New York, NY: OR Books.

Bauwens, M., Kostakis, V., & Pazaitis, A. (2019). *Peer-to-peer: The commons manifesto*. London: Westminster University Press.

Bauwens, M., & Niaros, V. (2017) Value in the commons economy: Developments in open and contributory value accounting. Co-published by the Heinrich Böll Foundation and the P2P Foundation. Retrieved from www.boell.de/en/2017/02/01/value-commons-economy-developments-open-and-contributory-value-accounting

Bauwens, M., & Niaros, V. (2018). *Changing societies through urban commons transitions*. Berlin: Heinrich Böll Foundation.

Bauwens, M., Restakis, J., Dafermos, G., & Figueiredo, J. (2015). Commons transition: Policy proposals for an open knowledge commons society. Published by the P2P Foundation. Retrieved from https://commonstransition.org/wp-content/uploads/2014/11/Commons-Transition_-Policy-Proposals-for-a-P2P-Foundation.pdf

Benkler, Y. (2017). Peer production, the commons, and the future of the firm. *Strategic Organization, 15*(2), 264–274.

Bollier, D. (2016). State power and commoning: Transcending a problematic relationship. Report from a Deep Dive Workshop convened by the Commons Strategies Group in cooperation with the Heinrich Böll Foundation, Kloster Lehnin, Germany, 28 February – 2 March 2016. Retrieved from http://commonsstrategies.org/state-power-commoning-transcending-problematic-relationship/.

Bollier, D., & Helfrich, S. (2012). *The wealth of the commons: A world beyond market and state*. Amherst, MA: Levellers Press.

Bollier, D., & Helfrich, S. (2019). *Free, fair and alive: The insurgent power of the commons*. Gabriola Island, BC: New Society Publishers.

Conaty, P., & Bollier, D. (2014). Toward an open-cooperativism: A new social economy based on open platforms, co-operative models and the commons. Commons Strategies Group Workshop, Berlin, Germany, 27–28 August 2014. Retrieved from http://commonsstrategies.org/towards-an-open-co-operativism.

Cooperatives Europe (2016). The power of cooperation: Cooperatives key figures 2015. Retrieved from https://coopseurope.coop/about-cooperatives.

Corcoran, H., & Wilson, D. (2010). The worker co-operative movements in Italy, Mondragon, and France: Context, success factors and lessons. Research paper published by the Canadian Worker Co-operative Federation. Retrievedfrom:https://canadianworker.coop/the-worker-co-operative-movements-in-italy-mondragon-and-france-context-success-factors-and-lessons/.

D'Alisa G. (2019). The state of degrowth. In E. Chertkovskaya, A. Paulsson, & S. Barca (Eds.), *Towards a political economy of degrowth* (pp. 243–258). Lanham, MD: Rowman & Littlefield.

D'Alisa, G., and Kallis, G., 2016. A political ecology of maladaptation insights from a Gramscian theory of the state. *Global Environmental Change, 38*, 230–242.

De Angelis, M. (2017). *Omnia sunt communia: On the commons and the transformation to postcapitalism*. London: Zed Books.

De Moor, T. (2008). The silent revolution: A new perspective on the emergence of commons, guilds, and other forms of corporate collective action in Western Europe. *International Review of Social History, 53*, 179–212.

Drechsler, W. (1997). State socialism and political philosophy. In J. G. Backhaus (Ed.), *Essays on social security and taxation: Gustav von Schmoller and Adolph Wagner reconsidered* (pp. 319–339). Marburg: Metropolis.

Drechsler, W. (2001a). On the viability of the concept of Staatswissenschaften. *European Journal of Law and Economics, 12*, 105–111.

Drechsler, W. (2001b). *Good and bad government: Ambrogio Lorenzetti's frescoes in Siena town hall as mission statement for public administration today*. Budapest: Open Society Institute & Local Government Initiative.

Drechsler, W. (2016). Kathedersozialismus and the German Historical School. In E. S. Reinert, J. Ghosh, & R. Kattel (Eds.), *Handbook of alternative theories of economic development* (pp. 109–123). Cheltenham: Edward Elgar.

Drechsler, W., & Kostakis, V. (2014). Should law keep pace with technology? Law as a katechon. *Bulletin of Science, Technology & Society, 34*, 128–132.

DuPond, Q. (2017) Experiments in algorithmic governance: A history and ethnography of "The DAO," a failed decentralized autonomous organization. In M. Campbell-Verduyn (Ed.), *Bitcoin and beyond: Cryptocurrencies, blockchains and global governance* (pp. 157–177). London: Routledge.

Gramsci, A. (2007). "Quaderni del carcere." In V. Gerratana(Ed.), *Edizione critica dell'istituto Gramsci*. Milan: Einaudi.
Goldstein, J. D. (2006). *Hegel's idea of the good life: From virtue to freedom, early writings and mature political philosophy*. Dordrecht: Springer.
Gorz, A. (1968). Reform or revolution. *The Socialist Register*, *5*, 111–143.
Hardt, M., & Negri, A. (2011). *Commonwealth*. Cambridge, MA: Harvard University Press.
Harvey, D. 2011. The future of the commons. Radical History Review, *109*, 101–107.
Hegel, G. W. F. (1995 [1821]). *Grundlinien der Philosophie des Rechts* (5th ed.). Hamburg: Meiner.
Kalogeraki, S., Papadaki, M., & Pera Ros, M. (2018). Exploring the social and solidarity economy sector in Greece, Spain, and Switzerland in times of crisis. *American Behavioral Scientist*, *62*(6), 856–874.
Kostakis, V. (2010). Identifying and understanding the problems of Wikipedia's peer governance: The case of inclusionists versus deletionists. *First Monday*, *15*(3). Retrieved from https://firstmonday.org/ojs/index.php/fm/article/view/2613/2479#author
Kostakis, V., & Bauwens, M. (2014). *Network society and future scenarios for a collaborative economy*. Basingstoke: Palgrave Macmillan.
Krause, M. J., & Tolaymat, T. (2018) Quantification of energy and carbon costs for mining cryptocurrencies. *Nature Sustainability*, *1*, 711–718.
Le Guin, U. (1974). *The dispossessed: An ambiguous utopia*. New York, NY: Harper Collins.
Lember, V., Brandsen, T., & Tõnurist, P. (2019). The potential impacts of digital technologies on co-production and co-creation. *Public Management Review*, *21*(11), 1665–1686.
Luxemburg, R. (1899 [1970]). *Reform or revolution*. New York, NY: Pathfinder Press.
Manski, S., & Manski, B. (2018). No gods, no masters, no coders? The future of sovereignty in a blockchain world. *Law Critique*, *29*, 151–162.
Mason, P. (2015). *Post capitalism: A guide to our future*. London: Allen Lane.
Mehar, M., Shier, C., Giambattista, A., Gong, E., Fletcher, G., Sanayhie, R., Kim, H. M., and Laskowski, M. (2017). Understanding a revolutionary and flawed grand experiment in blockchain: The DAO attack. *Journal of Cases on Information Technology*, *21*(1), 19–32. Retrieved from https://ssrn.com/abstract=3014782
Morris, W. (1890). *News from nowhere; or, an epoch of rest, being some chapters from A utopian romance*. London: Longmans Green and Co.
Nakamoto, S. (2008). Bitcoin: A peer-to-peer electronic cash system. Retrieved from https://bitcoin.org/bitcoin.pdf
Ostrom, E. (1990). *Governing the commons: The evolution of institutions for collective action*. Cambridge: Cambridge University Press.
Pazaitis, A., & Bauwens, M. (2018). New roles for citizens, markets and the state towards an open-source agricultural revolution. In J. L. Vivero-Pol, T. Ferrando, O. De Schutter, & U. Mattei (Eds.), *Routledge handbook of food as a commons* (pp. 70–84). London: Routledge.
Pazaitis, A., Kostakis, V., & Bauwens, M. (2017). Digital economy and the rise of open cooperativism: Case of the enspiral network. *Transfer: European Review of Labour and Research*, *23*(2), 177–192.
Reinert, E.S. (2019). *The visionary realism of German economics*. London: Anthem.
Riddley-Duff, R. (2015) *The case for FairShares: A new model for social enterprise development and the strengthening of the social and solidarity economy*. Charleston, SC: CreateSpace Independent Publishing Platform. Retrieved from http://shura.shu.ac.uk/10198/
Rifkin, J. (2014). *The zero marginal cost society: The Internet of things, the collaborative commons, and the eclipse of capitalism*. New York, NY: St. Martin's Press.
Ritter, J. (1957). *Hegel und die Französische Revolution*. Cologne: Westdeutscher Verlag.
Scholz, T. (2016) *Uberworked and underpaid: How workers are disrupting the digital economy*. Cambridge: Polity.
Sombart, W. (1932). *Die Zukunft des Kapitalismus*. Berlin: Buchholz & Weisswange.
Tapscott D and Williams AD. (2006) Wikinomics. *How mass collaboration changes everything*. New York, NY: Penguin.
Terrell, J., Kofink, A., Middleton, J., Rainear, C., Murphy-Hill, E., Parnin, C., & Stallings, J. (2017). Gender differences and bias in open source: Pull request acceptance of women versus men. *PeerJ Computer Science*, *3*, 111.
Tsuk, N. (2000). Social capital, voluntarism, and state-community relations: A case study. Paper prepared for the Workshop "Voluntary Associations, Social Capital and Interest Mediation: Forging the Link," ECPR Joint Session of Workshops, University of Copenhagen, 14–19 April 2000.
Vathakou, E. (2015). Citizens' solidarity initiatives in Greece during the financial crisis. In J. Clarke, A. Huliaras, & D. Sotiropoulos (Eds.), *Austerity and the third sector in Greece: Civil society at the European frontline* (pp. 167–193). London: Routledge.
Zamagni, V. (2006). Italy's cooperatives from marginality to success. Paper presented at the XIV International Economic History Congress, International Economic History Association, Helsinki, 21–25 August 2006.

Part VI

Conversions: Advancing Peer Production

28

Making a Case for Peer Production

Interviews with Peter Bloom, Mariam Mecky, Ory Okolloh, Abraham Taherivand, and Stefano Zacchiroli

Interview with Peter Bloom, Founder of Rhizomatica, Oaxaca, Mexico, 21 February 2019

Sophie Toupin (ST): What is Rhizomatica?
Peter Bloom (PB): Rhizomatica began in 2009 as a quest to make alternative telecommunications infrastructure possible for people around the world dealing with oppressive regimes, the threat of natural disaster, or the reality of living in a place deemed too poor or isolated to cover. Currently, only very large, powerful companies have access to the mobile spectrum and the concessions to provide cellular service. But their business model and the technology that these traditional providers use have proven unable to solve the problem of connecting much of the world. Rhizomatica set out to break this oligopoly and allow communities to become network owners and operators, as well. Thanks to a variety of open source efforts developed in the last few years, it has become technologically and economically feasible for a community or an individual to provide cellular service to thousands of people. Rhizomatica was amongst the first groups in the world to embark on actually making this technology something that could help people.
ST: How would you describe Rhizomatica's approach?
PB: In our work we try to look at different aspects of telecommunications, we can think of it as a triangle of three different elements that we have to deal with, i.e., technology, policy and the economics of the network, if we really want to change things. First, we want to make technology more accessible, easier to understand, less expensive and open source. Second, considering the policy and regulatory aspects of telecommunications, we want to make those more favorable to people-centered networks. Third, there is a more organizational component to our work that has to do with the economics of the network, the sustainability of the network, the social engagement of the network, and how you can create other ways of running networks, maintaining them, getting more community value out of the networks, value that is not so much economic, but how the network should be useful to people. All these elements that are part of our approach are at odds with the global Internet platforms.
ST: Great, excellent. So, in your view is Rhizomatica connected or linked to peer production and if so, how?
PB: I guess it kind of depends on what your definition of peer production is. At Rhizomatica we look at it more from a people's perspective – people need to create or produce the network that they want to use, and I think it's very centered on moving away from being a user. It requires active participation; it requires investment from people; it requires people to understand how things work. It requires keeping all of those things going, so it's a pretty active process. I think it's somewhat about peer production: it can get a little complicated as to whether or not you're producing the network and maintaining it. I think it's more of an ensemble of many different things, that come together, that are put together in such a way, that they can actually be maintained and keep going and be affordable, but certainly the idea is with a group of people living in a place where there is no existing network; they are literally producing the network that they then use, and then reproducing that network everyday, so that it continues to exist. In that sense, I do consider it as a type of peer production. There is a very physical component to the process, the communication

The Handbook of Peer Production, First Edition. Edited by Mathieu O'Neil, Christian Pentzold, and Sophie Toupin.

infrastructure is produced, and shared by everybody. People are installing, putting up, physical base stations, the radio equipment, and putting them together in a way that works.

ST: At Rhizomatica do you use the notion of peer production?

PB: I mean I wouldn't say that we actively use that term. I think we were – many of us at Rhizomatica have engaged with peer production. When I was doing my master's thesis, I was definitely engaging with those ideas, which helped me further think about Rhizomatica. I wouldn't say we explicitly use the word peer production, but I think it is a general concept that we are engaged with.

ST: What brought you to Rhizomatica and how was it co-created?

PB: For me two elements are intertwined in thinking about Rhizomatica and explaining why we decided to co-create it: the aspect of need and an ideological commitment. The need for Rhizomatica as a cellular communication network is pretty obvious where there is none. The ideological commitment is what distinguishes us from other organizations. We believe technologies are hackable in the sense that we can build a whole communication network, without being dependent on major corporations. This commitment on our part where we said no, let's question the hegemonic model that's being put out there and try to set up our own. So, I think getting back to your question, the need was there: people want to be able to communicate differently. The most vital question for us is: what are the tools people need to solve their communication needs? The need part of the project pushed us to co-create Rhizomatica and then there is the more ideological part where we believe that there should be an open source solution for cellular communication. We believe that people should be able to access the spectrum that they need, not just corporations. We believe that people can and should produce, build their own networks, and operate them rather than being passive users. Does that answer your question?

ST: Yes, definitely. Where did Rhizomatica start? Was it in Nigeria?

PB: The roots of Rhizomatica go back to the work I was doing in Nigeria where I was working with people who were being displaced from their neighborhoods and also people working against the oil industry in the Niger Delta Region. I was working as a citizen journalist doing participatory video documenting different human rights abuses that were happening and it just became evident that the easiest and best way to do the work, would be if people used their phones, actually to do the filming of the things they wanted to film. Now this has become a totally normal tool or technique activists use, but in 2009 it was not. We were trying to get people in the Delta Region video cameras. Once the footage was on people's phones, they were able to do a lot more with it, but it was really expensive to use the 3G network, to do a video and upload the video to the 3G network, it was twice someone's monthly salary and then the networks were highly surveilled by the Nigerian state. We tried to come up with a different way that people could pass information between their phones and we got a mobile mesh network working and all these other crazy things, that were kind of more theoretical solutions than practical ones. That's kind of how it started with Rhizomatica. When I came to Mexico, we started doing something slightly different i.e., building a cellular network for and by the community in Oaxaca. At Rhizomatica we are not married to one technology; we are not married to one process or toolkit. A people's network of cell phones is a pretty good thing to engage people in, but again if you work with indigenous folks – in Mexico we work with indigenous folks, and in Brazil too – then we use short-wave radio. It just really depends on where people are at.

ST: What is the impact of Rhizomatica? In which way does it make a difference?

PB: I guess it has a really tangible impact to have set up a people's cell phone infrastructure network. First, it helps people communicate better since we are an actual functioning network. Thousands of people use Rhizomatica infrastructure for their everyday communication, which they helped built. That's at a really practical level and then we also are pretty engaged with a lot of this policy and regulatory work. We are helping to change policies at the international level so telecommunication policies can be more favorable to community networks. Those are the two biggest impacts. As a case in point, in Mexico our work enabled a change in policy at the national level. It's no longer illegal to use the spectrum to set up community cellphone networks. You can now get a license to do that in Mexico. It made some progress in terms of legitimizing and legalizing this type of people's-centered telecommunications.

ST: What is driving Rhizomatica forward?

PB: We invest a lot of time and resources into different tools, so we want as many people as possible to use those tools, if they think they are useful. We have this functioning network in Oaxaca, that I was talking about, that we want to keep going, because people are using it. Turning it from a 2G network into a 4G network is a goal; this is sort of keeping us up with technological change. We are now collaborating with a much larger group of people. This motivates us. Before it was just us hacking away at this problem here in Mexico, and now we work with people from Africa, from Asia, from all over the world. We are trying together to build this more inclusive communication situation. So in that sense it keeps us going.

There is also the state of the two main monopolies or big giants that are keeping us busy: one giant are the telecommunications companies and the other giant is the Internet companies. The Internet was supposed to be this great invention for the people, but it turns out there are a lot of problems with it. It's been highly privatized and commercialized and it turned into a massive spying and data-mining platform. We feel that from a practical standpoint what we are co-constructing helps make this situation less bad, and I think that on the telecommunication side, you see this movement toward the Internet of things or the 5G network; this is what the telecommunication companies want to do and I think that we feel we are part of the resistance to those tendencies. We are not the only ones, but I think we are an important one and that motivates us to keep going.

ST: Voluntary engagement is an important part of peer production. I'd like to ask if there is any voluntary engagement in Rhizomatica? You have volunteers? And you are nonprofit organization, correct?

PB: We basically started with the following principle: if there was money, we shared it around. Last year in 2018 we got our papers in order and we turned Rhizomatica into a nonprofit organization. This is mainly due to the fact that we were doing too many things at the same time; there were too many people involved, and the type of work we were doing was very high level such as IT, or negotiating with government. So I just felt like we needed a more formal status to engage in the type of work we do. There are also other parts of the Rhizomatica project that we wanted to get autonomy of status for. The network in Mexico for instance is now its own organization. We decided to split it out from Rhizomatica, so that it could do what it needed to do rather than being dependent on Rhizomatica. So yeah, our work has been fairly formalized at this point. There are still people that help out and volunteer, but now, it is mainly a paid staff.

ST: In your opinion are their tensions between the institutionalization of Rhizomatica and its voluntary engagement? Do you see any tensions?

PB: No, I don't think so. The biggest tension was around the split between the network in Mexico and Rhizomatica the nonprofit organization. This was a very complex and complicated process. The main issue was that Rhizomatica was only dedicated to the Mexico cellular network project, and not interested in having full-time jobs. We were ideologically committed to doing the work that needed to get done on a daily basis for the network to be functional. We basically built a network, had an informal group of people trying to keep that network going and that was that. And then the network in Mexico became an official entity. As the community itself had invested a ton of money and time into it we were under a lot of pressure from the communities to make the network better and to make the network legal. This is what led to the formalization or institutionalization process. We have 10 or 11 staff now who are all in Oaxaca. It is all local people, local staff, making the network function together with the community. It requires a very different set of skills and set of attitudes to keep that going.

ST: The office is in the state of Oaxaca but where are the communities using the cellular network communication located?

PB: The cellular network covers more than 70 towns and villages and the central office is in the city of Oaxaca. Those 70 villages are spread all around the state of Oaxaca. Some are two hours away, some are seven hours away, but the midpoint is the city.

ST: According to you what has changed in the world of peer production?

PB: What we are trying to do at Rhizomatica is to allow more people to access information and get online. These simple things are much more problematic and difficult positions to hold compared to a few years ago when we were not totally and fully conscious about what was going to happen. I think that the peer production model is based on a series of assumptions about there being a neutral platform on which to do that. This description does not encompass the whole of peer production, but it is an important dimension. Let's think about when we started doing this mobile mesh network in Nigeria, it was totally a peer-to-peer infrastructure. But those experiences did not scale up partly because of the ways in which the Internet has evolved over the last ten years. If you look at the layers of the Internet (fiber optic, Internet exchange points (IXPs), etc.) you see massive consolidation. What you see is that these companies, the global platforms like Facebook and Google, are actually the ones building fiber cables and installing them. I think we are very close to losing the peer-to-peer model of the early Internet. Pretty much all traffic on the Internet is going through these platforms and then you have the Internet turning into a content delivery network, rather than a sort of flat peer-to-peer network. From the actual architecture standpoint, the industry has dramatically changed how the Internet used to be. You can still do peer production on top of all of that, but it's harder and harder. Like you and I right now we are having a conversation on Signal, right? Which is kind of like "Oh we like Signal because it has good cryptography." All of us who believe in the good part of peer production, the peer-to-peer aspect of it, I would say we have more or less given up on it. The tools that we were building five or ten years ago like XMPP, this is kind of gone. I would say we more or less have lost a lot of ground over the years. We are talking about the largest corporations in the world, Facebook, Google, and the like, the most valuable corporations in the world, who have found a way to suck a ton of money out of essentially nowhere. So they get to build their own fiber-optic networks, because they have the money to do so. They have created a completely separate Internet on which all their services run and that's why their stuff works better! You can get a YouTube video to work with a pretty crappy connection. It's not only a money thing obviously, but when it gets down to it, then you can run all of your traffic through your own data centers and over your own network. You can control it a lot better. But more so, the issue is that the Internet has become an unsafe space in a bunch of ways and I don't think it was like that before, or at least it wasn't evident. There is a lot of garbage, of misinformation and surveillance on the Internet and a number of years back it didn't feel like that. I think the actual infrastructure itself was better. It was fairer. It was more balanced and it was more decentralized. Is it still a tool on which peer production happens? Sure, I mean you can still get your Wikipedia, but it feels less and less like a – I do not want to call it neutral – but it feels less and less like a safe space to be able to do things. I think a lot of the people that we work with, and ourselves as well, are now looking at how do we do a lot of what we want to do by circumventing the Internet. How do we do local servers? How do we do local services? How do we tunnel through the Internet to meet each other in some safe way? So now the Internet itself has become the thing to circumvent, which is interesting. I think a lot of cool stuff is going to come out of that, but again we are talking about really small projects, with not a lot of money. I guess the hackers are not always the most organized group of people and I feel like we are dealing with something that's highly, highly organized and which has unlimited funds. In this context what do you do? Do you fight back? Do you just completely avoid it? Do you engage with it at all? It's good that people such as hackers and others are building the kind of services, the kind of tools they want. I just also think, we need to engage, to be more aware of what's happening with the Internet of things for instance. That might potentially reshape not just the way we get information, but how we interact with the world. I feel there is not enough political engagement with where things are going and that troubles me. What are we actually going to be able to do in the future?

ST: That's a grim picture. My last question is about imaging the future of peer production. What will it look like in ten years, 20 years?

PB: Well, I mean at this point I am very pessimistic about where things are going. So, again how much of that affects peer production? I mean we will still be able to talk to each other over great distances. We will be able to collaborate and build things. There will still be a thriving Linux code base. But the question is how well that is going to be able to respond to what people need and what people need in a network in a shitty world. I do not know how else to say it. I mean one of the things that we are talking about internally at Rhizomatica is how to rethink our networks for basically the collapse of civilization? I do not want to get too dramatic about it, but the place that we are working in now in Oaxaca, they are getting killed by climate change. Their roads are being washed out in ways they have never before. Their coffee is being killed by these bizarre molds that never grew on them before. People's lives are being seriously affected, so the question for us if we want to try to help them in their communication is: what does the network need to look like to be able to continue functioning when a bunch of other systems start failing? The simple answers are: we need to use solar energy; we need not to rely on satellites, or fiber optic. I think that this peer production space needs to probably recognize the fact that people's lives are becoming more difficult. That is the central piece of the matter. I think the theoretical part needs to be as grounded as possible in praxis and practicability to continue to be useful. What does peer production look like when you know we are living inside this crazy Internet network where we are spied on and surveilled? I don't know exactly what forms it will take but I hope it will allow us to maintain some level of freedom and economy for communities to satisfy their needs. Capitalism and the current state of things is doing the opposite, taking that away from us.

Interview with Mariam Mecky, Communications Unit Head at HarassMap, Cairo, Egypt, March 13, 2019

Sophie Toupin (ST): What is HarassMap? And, how do you think it is linked to peer production?

Mariam Mecky (MM): HarassMap, a volunteer-based organization, was launched in December 2010 by a founding group of four women together with tech partners, advisers, and volunteers. The four founders were prompted to launch after experiencing sexual harassment in their daily lives, and that almost everyone we knew, witnessed on a daily basis. They have also been working against it since the mid-2000s. HarassMap focuses on working against the social acceptability of working harassment where its vision is to engage all of Egyptian society to create an environment that does not tolerate sexual harassment. Part of HarassMap's work is to highlight the prevalence of the issue to work against it. To be more specific around HarassMap's scope of work, HarassMap is an information and communication technology (ICT)-based participatory action initiative that serves multiple functions: it highlights the gravity of the problem by providing an outlet for testimonies from those who have suffered or witnessed sexual harassment; it provides data that improve understanding of how sexual harassment is evolving in Egypt, which in turn provides HarassMap with information to create relevant communication campaigns and research programs; and it serves as a tool for community outreach teams to motivate the public to stand up against sexual harassment. In this manner, the idea behind HarassMap is directly linked to peer production; where the four founders were self-organized, collaborated, and coordinated on this project with a shared outcome relying on voluntary engagement. Capitalizing also on the idea of crowdsourcing, which I elaborate more on in further questions, this project very much utilized the Internet for this shared outcome.

St: What brought you to the HarassMap? How did you become involved in this commons-based production?

Mm: I personally became involved because I was particularly interested in their approach around sexual harassment, as well as their efficient way of collaboration and management. I was also incredibly passionate about the cause.

St: According to you what is the impact of HarassMap? In what way does it make a difference?

MM: The main successes of HarassMap work is that we use the reports received on the map to draft and publish studies and build campaigns on this data that counter the social acceptability and normalization of sexual harassment, debunking stereotypes and raising awareness. One of HarassMap's early successes with creating a zero-tolerance anti-harassment policy is its partnership with the biggest public university in Egypt, Cairo University. In that process, HarassMap provided a draft of a recommended well-rounded policy that encompassed definition and types of harassment, enforcement mechanisms, and penalties for harassers. We, then, worked with stakeholders from the university, NGOs, and activists to tailor and implement it in the university. This policy is now used as a guideline and reference for 15 other universities in Egypt. HarassMap also assists businesses with adopting and enforcing anti-harassment policies. As a result of our work, we have received requests from hundreds of activists and organizations in other countries to assist them with replicating our model. The founders made the HarassMap concept completely open source. Since its launch, although it was not its mission until 2016, HarassMap has advised and supported over 100 activists and NGOs from around the world on setting up HarassMap-inspired initiatives. Independent groups in at least 80 different countries have been advised on how to launch similar projects. The countries include Jordan, Libya, Turkey, South Africa, the United States, Canada, Malaysia, Indonesia, Japan, Cambodia, Morocco, Saudi Arabia, Afghanistan, Kenya, Sudan, the UK, India, Nigeria, and some groups in South America. The groups and activists from these countries received advice, and some succeeded in setting up local versions of HarassMap.

ST: How do you organize collaboration in your project?

MM: HarassMap operates in a less rigid hierarchical structures than more traditional organizational or business models. It is organized as units, headed by a unit head and a coordinator. We have at the moment four units based on our policy that is updated every year. The four units are Safe Areas Unit, Communications Unit, Community Partnerships Unit, and Research Unit. Many of these units rely on volunteers. Safe Areas Unit focuses on sexual harassment training and policies, for universities, schools, and corporates. More than 1,000 volunteers were trained and assisted in building zero-tolerance zones within communities and inside institutions, both informally and through policies. By creating physical spaces that do not tolerate sexual harassment, HarassMap's offline work is focused on building a society that guarantees safety for all people from sexual harassment. In the Communication Unit, public campaigns conducted on traditional and social media use learnings from reports and community-based work to change perceptions that create and reinforce a culture of blaming the harassed, excusing the harasser, and accepting sexual harassment. HarassMap also steers media discourse away from unproductive stereotyping towards a facts-based positive discussion of what measures need to be taken to end sexual harassment. As for the Research Unit, HarassMap also conducts traditional research to understand and learn more about sexual harassment. In 2016, HarassMap conducted a study on how the usage of different ICT channels can generate societal change.

ST: What are the main drivers of HarassMap?

MM: At the time of HarassMap's launch in 2010, legal recourse for people who experienced sexual harassment was limited by several barriers. Victim-blaming and negative effects on reputation were powerful deterrents keeping many silent, and away from police stations. Moreover, the law required bringing the harasser to a police station, where additional harassment often occurred, providing witnesses, and convincing police to report the crime. While not denying that legal reform was needed, the co-founders believed that in a context in which existing laws were not enforced because sexual harassment was not seen as a crime, advocacy for a new law would have little impact on its own. Therefore, they decided to tackle what they believed to be the source of non-enforcement – social acceptability. This idea remains at the core of HarassMap's work also today. In fact, there are strong social roots and an established history in Egypt for rejecting sexual harassment. We believe that bringing back social discouragement needs to come from within, and we consider ourselves to be ordinary members of the public. We see crowdmapping as a way to decentralize activism on the issue away from experts and policymakers, who often view society as a passive beneficiary, and restore an active role for society in addressing the problem on its own behalf. For the first time in Egypt, HarassMap's crowdmap offered a free and easy tool for providing a direct means to voice sexual harassment experiences safely and anonymously. And speaking out, HarassMap's co-founders had learned during their work on the issue prior to HarassMap, was very often the first step to activism. Crowdmapping has, however, been only one element in our approach since the beginning. More than "just a map," HarassMap simultaneously launched a community mobilization program to train and coordinate anti-harassment activism in neighborhoods, interactive social media outreach, and media relations, all designed to build off of each other and the crowdmap to convince people to stand against sexual harassment and deter harassers. The system also bridged a gap, sending victims an auto-response about existing but little-known services like legal aid and psychological counseling. Finally, the reports and map provided evidence for volunteers to mobilize bystanders to create neighborhood "safe areas" in which sexual harassment would not be tolerated.

ST: In your view, what are the main advantages of this approach for the generation and dissemination of a specific type of knowledge?

MM: As previously mentioned, the main driver for HarassMap was working against the social acceptability of sexual harassment. As part of the work against sexual harassment, HarassMap worked on highlighting the extent of the problem initially through the map and expanding into other programs and means. As one of the founders, Rebecca Chiao, put it "We can't say today how bad the problem really is and that's one of the things that we're trying to do with HarassMap." Crowdsourced data collected from the reporting system helps document facts about the issue, counter stereotypes that blame the harassed and make excuses for the harasser, challenge previously held views about who harassers are and what their motivations are, and emphasize the criminality of sexual harassment in Egypt. HarassMap uses an integrated approach to creating zero-tolerance for sexual harassment and assault that includes reporting and mapping harassment incidents to document the problem; public campaigns to spread accurate information and mobilize the public to take action; and anti-harassment policies inside institutions such as universities, NGOs, corporates, and schools, to establish consequences for harassers. These policies are easier to implement and monitor within an institution, empower their own leaders, and are an important step to spreading a culture of zero tolerance.

ST: What would, in your view, characterize a successful peer production project, and why?

MM: A successful peer production project, in my opinion, is what is centered around a certain social cause and sustained through resources or/and revenues.

ST: Is there a problem with success or growth? Do you see tensions between commercial or note for profit ambitions and voluntary engagement?
MM: There are often tensions between commercial or non-commercial work for voluntary engagement, not even necessarily for profit ambition. Commercial work is often viewed as problematic by the public as it only targets certain privileged groups while non-commercial is rather risky for the success and sustainability of the project. Hence, the non-commercial aspect in the case of our project may stand in the way of sustainability as the current work is funded only through individual projects of grants and donors without any expectation of financial returns.
ST: What would be your advice for people planning to start a peer production project such as HarassMap?
MM: Our advice would be to be able to maintain pillars of peer production as voluntary collaboration and open distribution of means of production; more importantly, to consider in your strategy whether commercial or non-commercial means to ensure the sustainability of your projects. It would also be necessary to be well up to date with knowledge concerning the idea behind your project and constantly developing it while ensuring it is up to the challenges of peer production.
ST: In your opinion, what are the main challenges HarassMap had to tackle in order to thrive? And how should peer production be fostered?
MM: HarassMap has faced several challenges. First, there has been the challenge of volunteer management, as our project/organization is a volunteer-based organization. This included a lot of follow up, and making sure they are committed to take part. Another challenge has been financial resources as at first, it was solely based on voluntary engagement, with our first funded project in 2012. This challenge rose again in Egypt, with the restrictions and bureaucracy of procedures. Another issue was the map itself and the technology itself. We have had to work on it and develop it ourselves as there was an issue with the location, and we could not edit the application in the software itself as it's customized and we couldn't edit or add further questions. Hence, we moved to another platform creating it ourselves later. Lastly, there was also the issue of Internet access and urging people to use the app to report. We complemented our online work with various offline work that is tackled above. As for how peer production should be fostered, the main pillars of my understanding of peer production should be maintained; from voluntary engagement and cooperation, common knowledge, and openly accessible means of production.

Interview with Ory Okolloh, Co-founder of Ushahidi, Director of Investments at Omidyar Network, Johannesburg, South Africa, 17 June 2019

Christian Pentzold (CP): If you think about your time with Ushahidi, what was the project about?
Ory Okolloh (OO): At the time, in the aftermath of Kenya's disputed 2007 presidential elections, the project was about collecting information from citizens about the post-election battles, what was going on then. It was intended both as a workaround to government censorship at the time, but also kind of a recording of what happened or what works best, which is why the name is Swahili for testimonial. That is how the name came about.
CP: What brought you to the project?
OO: I think at that time I was blogging and people left comments telling their stories in the comments in this blog at that time. I think just realizing that the volume was heavy. It was hard to kind of moderate the comments. There needed to be a systematic way to tag and most importantly link them to a map, so you could track the reports along with the location and begin to sense which parts were most affected and who needed help, and so on.
CP: Looking back, what do you think was the impact of Ushahidi on society or how did it make a difference?
OO: I think in the situation then and even later, it was certainly a gamechanger at first. It relied on the crowd early on, in 2007 or 2008, when the Web was a lot more open and things weren't concentrated on platforms. It was a great model of using the crowd, the people, the peers, and all their information. Relying on peers is about correcting wrong information. Folks could report errors others had sent in or report what was false. I think this was an opportunity and it was later used in other incidents for crisis

response during natural disasters when people were not able to get to the ground immediately. So why not use what people already have known on the ground and carry together what they need? The idea was to rely on the crowd and provide a way for people to share important affairs in a simple manner. I think that had an enormous impact.

CP: What are the main advantages of such an approach for the generation and dissemination of knowledge?

OO: You just have to look at the history of Wikipedia and what were back then called message boards and news boards. This idea of learning from others is still transformational. When you talk about peer production it's about generating knowledge and being able to learn from people who share information even you were sitting in Nairobi. We never had access to, you know, just knowledge. I think it's about breaking down the barriers around knowledge by sharing it. I think it also is about the breaking down the boundaries around information. It is this ability to get instant information, not always quality information, but the idea that you don't need a whole bunch of intermediates who filter information. And finally, I think if you look at how peer production works, the idea of language diversity becomes important. How do you share information that makes sense for people? Still, there is much in English but peer production allows you to add and generate content in other languages, too.

CP: For people who want to start such a project, what would be your advice?

OO: I think, it's much harder now, because you know, the platforms are so dominant. In a time of online video, how do you to get around YouTube for real-time information? You can't get around Twitter. Nothing beats Twitter, right? And the same for Facebook when you try to reach out to audiences there. I think the challenge is becoming a signal in a world with a lot of noise. This makes it a bit harder now than it was before. Folks who want to begin now are thinking about where to fit in, in a particularly crowded space with a lot of noise and growing distrust of what people see online. I think it is a challenge, but at the same time it could be an opportunity when people look out for trusted information providers. I think it's about balancing peers and experts and this hybridity is probably what folks want to think about, if they try to start peer production projects now.

CP: When you imagine the future of peer production, what could it be like, let's say in five years, or in ten years?

OO: I have no idea. I think if you look at what are now called influencers, I think that is kind of the future, where there is a mix of this kind of a hybrid world and less a crowd phenomenon. I think this is getting back to verification, probably linking that with AI, in terms of sorting out content, even though peer production is getting better at sorting out what's real and what's not. I think it's interesting to see where young kids go and to look at how multiplayer gaming is now the only way you game. There is for example a whole scene where communities are emerging in digital spaces. I think the future will show out in very different ways than we expect.

Interview with Abraham Taherivand, Executive Director of Wikimedia Deutschland, Berlin, Germany, 18 March 2019

Christian Pentzold (CP): I would like to start with a basic question: What is your project about and how do you think it links to peer production?

Abraham Taherivand (AT): First of all, I would like to highlight that Wikimedia Deutschland itself is an organization that supports a variety of projects around free knowledge. There are 15 different Wikimedia projects, but of course the most known or best-known project is Wikipedia, the free online encyclopedia, which is used by millions of people every day. Wikipedia collects knowledge worldwide and makes it accessible to everyone at any time. However, I would say encyclopedic knowledge cannot only be represented. It needs to be constantly discussed and negotiated to avoid personal opinions in an online free encyclopedia. It's about knowledge and not opinions. And for a project like Wikipedia it needs the intelligence and of course also the joy of the discussion and by the end of the day also the diligence of all the people involved in the project. The major success factor of Wikipedia and the other Wikimedia projects is, that everybody can participate. This is one major point: to be able to contribute and to be part of that. Wikipedia is the world's largest example for collaboration in a volunteer project and the best proof that it works. The German version of Wikipedia has just turned 18 on Saturday.

CP: What brought you to the project?
AT: I would not count myself as an editor on Wikipedia. Maybe it is more really about what brought me to Wikimedia as an organization and that was in 2012 when I had the chance to found the Wikidata project together with Denny Vrandečić. After more than six years Wikidata is now the most active sister project within the Wikimedia universe. We started with the vision of having not only free knowledge, but free structured knowledge. We wanted to have a centralized, structured knowledge base, which cannot only be edited and understood by humans. We focused on structured data and structured knowledge so that machines can edit and understand that knowledge.
CP: What is the impact of Wikipedia on society? In what way does it make a difference?
AT: What I would definitely like to highlight is really the vision we are fighting for and aiming for,a world in which the accumulated knowledge of humanity is freely accessible to everyone and everyone can contribute to it. I think that's, in a nutshell, Wikipedia's impact. The software on which Wikipedia is based was one of the first that enabled users on the Internet not only to read Web pages but also to modify them directly via the browser and edit them in parallel with other users. Mediawiki was one of the first software solutions which really had collaboration running through its veins. Of course, the idea of free knowledge is relevant beyond Wikipedia. From my perspective, it stands for an approach that translates the right to education to the conditions of our digital society. Access to knowledge is the base for development. This applies to individuals as well as companies or societies. I see different layers of impact, but these are the major ones.
CP: How do you organize for collaboration in Wikipedia or the Wikimedia projects at-large?
AT: Based on my experience during the last two decades, I would say that it comes down to the level of technology, the tools we are using. Yet besides the tool perspective or technical perspective, we also have the perspective on values supported by the people. The rules and the governance models and the structures and the processes within Wikipedia are owned by the volunteer community. I would definitely say this is a key success factor – finding an organizational structure and I would also say the key successes of Wikipedia are the principles of collaboration which are about sharing and openness. Anyone can participate; anyone can start anywhere and someone else can go on from there and bring in a new aspect or point of view. The MediaWiki software helped to ensure that. Unfortunately, even though Wikipedia has now turned 18, a majority of people really do not see the edit button or they think you need to have an account to edit. No, you don't need that. To participate, you can just click the edit button and for instance find there is a semicolon missing or a point not mentioned or whatever. The hurdle to participating is not so high, but then we are coming back to another discussion, which I don't want to get into now – the user experience. To sum it up, in general I want to say that the organization of collaboration is on a technical, a tool level. But then, of course, it is also about values.
CP: Bringing it down to some core points, what would you say are key benefits of this kind of collaboration and production vis-à-vis more bureaucratic or hierarchical systems?
AT: I would say it's the thousands of volunteers. The volunteers that build our community are working together to create the world's largest collection of knowledge. Before I joined Wikimedia I came from the business world and you think about what drives the volunteers to contribute. Wikipedia thrives on the participation of as many people as possible. You can say that if we have a monetary incentive model, we could increase the quality of content. But that was exactly the opposite of how Wikipedia started. The idea Jimmy Wales initially had was to have a bunch of very high professional experts paid to write something. And it did not really work out. The moment at which the system was opened to everybody where you can build on what someone else already put in as content and you can extend that content and keep that content up to date is, I would say, definitely a main driver within Wikipedia. That applies to all Wikimedia projects I would say. It's clearly the engagement of the volunteers, the passion and the willingness to invest their time to contribute to the projects. And we are coming back to the point that technology is also a driver that enables people to easily collaborate and build on content which is already there.
CP: Speaking about benefits – if we look at the flip side of it, would you also be able to point to costs associated with this kind of openness and information freedom?
AT: From a cost perspective, you can always bring the argument of efficiency. But that's not how nonprofit organizations are managed or run. Usually we are focusing on the impact we have. I see that the environment is changing around us. The political environment is changing and we want to provide information that is proofed and viable. With a vision of free knowledge, we need to have a free licensing

framework around us. Content needs to be available under creative commons and this needs to be addressed constantly. Because if you do not, all of the sudden there could be push backs. Besides that, the digital transformation is something we need to be aware of. Our software did not really change during the last 15 years, so the base software and the user experience for authors is nearly the same. Of course, during the last ten years the major change was that people really do not need to know a kind of programming language to provide content for Wikipedia. So now we have a what-you-see-is-what-you-get attitude, that the people have a more sustainable or easier user experience. Nevertheless, if you don't have that and everything comes together, then also a project like Wikipedia is under threat and this is at least what we see in Germany where the number of editors is decreasing. There are several reasons for that. On the one hand there are technical hurdles. On the other, we now have more than 2.2 million articles. The knowledge has increased and new topics are not so easily found by new editors. Nevertheless, in general the most important asset – and don't get me wrong when I use the term asset – is the volunteer base, the community. And of course, we need to take care that the existing community has the ability to grow healthy and to be stable within their own created rules and processes. The argument you usually get from people who are working in a more closed system is that we can either ensure quality or we can ensure that no one is hijacking the project. I personally don't see this predicament because that's exactly the point that when a system is open you have the possibility to see if there is a hijacking process going on or not. In a non-open system, I don't see that this is possible. Of course, there are always bugs and security issues but at least there is a possibility of being transparent about that.

CP: If the opportunity would arise, would you welcome private firms or investors who wanted to become involved in Wikipedia or in other Wikimedia projects?

AT: Of course. Now I am speaking for myself and in general I'd say yes because our vision is really free knowledge and everyone and everybody should be able to participate. There is a controversy within our volunteer community base about Google and Facebook or whatever Silicon Valley company is using our content to build business models, but that's exactly what free knowledge is about. That's also what we are preaching, or what at least I am preaching. We strongly believe that free knowledge should be reused by others. But of course, we must be transparent about who is involved in the content production. We are already working together with institutions, mainly institutions in the GLAM sector, with galleries, libraries, archives, or museums. They are usually funded by state taxes. My experiences is that usually the corporate world is anxious to engage because of the need to have the content distributed under a free license. Yet I see that there was a shift during the last ten years. You try to collect intellectual property as much as you can, but for what reason? To sue your competitor and now I definitively see some companies saying: "Okay, let's release that in public." Because they believe that when we release it someone else can jump on it, build something better from it. I think this is the main hurdle for companies because they still think keeping knowledge is a competitive advantage, which I definitely don't see. Sharing is caring and releasing your knowledge is a great opportunity that someone else is bringing that knowledge, or information, or data to the next level and then you can take something out of that again.

CP: Looking back: What do you think has changed in a world of peer production and in Wikimedia's world since you started?

AT: I would definitely say, the awareness and the recognition from the science field. Ten years back, or 15 years back, you would have been really, really threatened in academia if you cited from Wikipedia and looked there for references. Today, that definitely has changed. Wikipedia and the Wikimedia projects are really seen as a trustworthy source of knowledge. That's definitely something that has changed, especially within the science field. Leaders from the science field reach out to us. This is also due to movements around open science which changed the entire perspective of how Wikipedia works and what the benefits of it are. In terms of awareness, Wikipedia has been among the top five websites in the world for the past ten years. Only search engines are usually under the top five if you take a look at Alexa. Among the majority of normal users or readers, at least in North America and the Western world, Wikipedia is very, very well known.

CP: If I would ask you to imagine the future of peer production, what could it be like in ten years?

AT: From my perspective, the future of peer production rests on the principles the Wikimedia projects stand for. Without those principles you have always a tension with forms of monitoring. If you chose to do that, then you do what corporate business is doing or what publishers do. In addition, we can be the target of manipulation. It has always been the case, but this is a direction which is dangerous. My wish for the future is that we understand better what high value free volunteer contributions have for peer production and to apply the underlying principles and values. If you have that in mind, you can definitely build peer production systems or equal systems. So that's what I wish for the future without saying this will become true.

Interview with Stefano Zacchiroli, former Debian Project Leader, 17 June 2019

Mathieu O'Neil (MO): So, in a few words: what is your project about and how does it relate to peer production?

Stefano Zacchiroli (SZ): Right, so the project is Debian, making a free software operating system. It's allowing you to run your computer from down to, you know, making the hardware, up to applications like Libre Office and productivity tools and games and whatnot. It is one of the oldest self-organized free software projects, dating back to 1993 and it's also one of the largest in terms of its volunteer base, we estimate that there are still today about 1,000 Debian members or Debian developers, we call them, who self-organize to work together and create this operating system. So, it's kind of historical and also a paradigmatic example of peer production in the context of software and especially free software and open source software.

MO: What brought you to this project? How did you become involved in voluntary commons-based peer production?

SZ: I was a student in computer science, I guess second year, and we were using Debian as an operating system in our university lab. Some things were missing, software that existed, but was not available in Debian. Some things needed some fixes and I discovered that I could do it myself and share my improvements with the other people who were working with Debian. So I started that way and essentially what happens, when you do that kind of stuff you somehow become euphoric about the joy of doing that together with other peers, so I got hooked and remained a volunteer for many, many years to come after that.

MO: What do you think is the impact of the project on society or in your field and how does it make a difference?

SZ: For Debian the impact is very significant as it is one of the most open operating systems for computer servers on the Internet. I don't remember the stats today, but a very high number of websites that you browse during a day run with Debian. A very high number of computers in the cloud run on Debian, so even if it started as a volunteer project and remains today very much a volunteer-driven project, it's basically running the backbone of the Internet itself.

MO: Given Apache is used in servers, what is Debian used for exactly?

SZ: Right, that's a good point. Debian is the framework that allows you to run the core of our operating system, which is Linux, and a number of applications. Apache is one of them. You can imagine Debian as similar to the AppStore that you run on your phone, where it is installed as a basis of your system and then you use it to manage all the rest of the software that you run. So when I say that basically Debian runs the largest number of computers which are the backbone of the Internet it means you first install Debian and then you use Debian to install Apache, WordPress, MediaWiki, or any other user-facing applications you might need. That is very significant, because there are always security issues. Debian allows you to keep your software up to date and to be free from known security issues.

MO: How do you organize collaboration in your project?

SZ: Debian is pretty much a self-organized project. The creator of the project, Ian Murdock, decided that there was a granularity, that isolated and defined areas of collaboration and in Debian those are our organized packages. A package is a piece of software you can install on Debian. A package can have a single person who is a maintainer of the software; usually it's a group of people, up to a dozen people. And within a package, maintainers are autonomous. They can do whatever they want, as long as they respect some baseline quality requirements that applies to Debian. To give you an idea of the scale here, Debian is composed of something like tens of thousands of packages. So, when you volunteer, you usually start by noticing that the package you use yourself has some issues. Maybe it is not up to date; maybe it has some bugs, that you want to fix. You start by picking a bugtracker against that package and if you know how to code, the maintainer looks at your patch and may accept it or not. This is the initial step, then you got hooked in, if you keep having an interest in that package you may become one of its co-maintainers. So, this is the site of collaboration and then you can step up the ladder if you wish. Not in terms of authority, but in terms of becoming more interested in Debian as a project, for instance after maintaining many, many packages, I became interested in the infrastructure of Debian projects, so working on quality assurance services, like checking in an automatic way if packages have bugs, and then I also ended up being Debian Project Leader as a kind of ultimate coordinator of activity into the Debian project, but it all starts from packages, which are separate sites of collaboration and then you can move to more cross-cutting responsibilities.

MO: The project leader is elected every year, but what about if you want to join one of the teams? Is it just through cooptation? There's no formal process for joining a team. It's just through working with people and getting known, is that right?

SZ: That is correct. So, elections are just for the project leader position, which are indeed yearly whilst the process of joining and leaving individual teams is completely ad hoc. So, as soon as you show that you have some confidence, some willingness, to work on a project, you can just join the team. As a volunteer, okay? If you want to become an official member of the Debian project, someone who can vote, there is a more structured vetting process, in which you need to show that you are really sure and declare that you agree with the core values of the Debian project. You also go through a sort of technical examination, checking that you are competent enough to be given access to the archive of the Debian project, to be authorized to push new version of software to all Debian users. In terms of peer review, to be accepted in Debian, it is enough that others review your work and say it's good enough and then that would be integrated. While if you want to be autonomous when making changes to software that is part of Debian, you need to go through this examination process.

MO: What would you say are the main drivers of your project? Or I guess, why does your project work and why do people get involved in it?

SZ: I think there are two separate drivers. One is technical. Debian is already used by system administrators around the world. I am using something which I like technically, but something is missing, something I can improve, so I join in, because by improving it, my work-life and my technical activities will become better or easier. This is a technical driver and then there is an ethical driver, which is that of a free software activist. People who believe that everyone in the world should be able to do any sort of computation, use only free software, and see Debian as one of the champions in that area and want to push this goal further by developing technical skills.

MO: What do you think are the main advantages of this approach for generating and disseminating knowledge?

SZ: We have peer production in a number of different sites, and in the context of software, I would say that the main advantage is peer review. Peer review makes your code better, others can look at your work and say whether it's good or not. This is well known in software engineering and relatedly if the access barrier is low, if you have 1,000 people that can fix issues in specific software it is much easier. I think these are the two main advantages. Both actually result in software quality.

MO: So that would be one of the benefits of peer production, when compared with more hierarchical systems, that there are less barriers to improving the content, but what are the costs?

SZ: This is something which is felt very deeply in Debian, when you want to make a decision, if the decision is within the realm of a specific package, well then it's clear who has the authority to make the decision. But when you want to make a decision that affects the entire Debian operating system, like changing the way something is done that affects many, many packages, taking the decision might be very challenging, because you end up having a discussion with hundreds of people, who all have their own opinion and who all think that they are very good at making that specific decision, even when they are not. So, how do you organize this discussion, where you have 1,000 people? This is a well-known challenge in social movements. Debian has a way to fix that: we try rough consensus, what are the pros and cons of what is being discussed, but in the end if you cannot reach an agreement we have a technical committee, a sort of tribunal, which can settle technical disputes within the project.

MO: So technical disputes are never resolved by a vote?

SZ: That is very true, we try to shy away and not to use that tool too much, but if there is too much division over a specific matter, the tribunal will settle it with a vote or it can be escalated further to a project-wide vote, which is a referendum. That can be applied to anything, but again it can be very divisive. So, we try to use it at little as possible.

MO: You mentioned the technical committee. Has there been some discussion in the past about the renewal of the committee because there can be a bit of a lag, or people can stay on the committee for a very long time, right? How does one become a member of the technical committee? Is there some process for renewing membership?

SZ: The initial procedure is actually defined in the Debian constitution, which dates back to 1996 I think, and the original procedure was that the project leader proposed names for the technical committee and the current technical committee needed to accept them. That has been changed. This is something that I proposed when I was Debian Project Leader and it was accepted. After a number of years, I don't remember how many, maybe four or five, members of the technical committee need to step down. So, this guarantees a turnover. This again is not rocket science, it is something we have learned from the organization of other nonprofit organizations in free software and it's working pretty well in renewing and making sure the technical committee remains active over long periods.

MO: About the Debian Project Leader position, which you occupied for three years and obviously, that's quite remarkable. I don't think that there is anyone else who has ever been elected three years in a row.

SZ: That's right.

MO: You said that there is a coordination role, but really that would be quite limited, because people are self-organized. So how do you see the role exactly and what did you try to achieve?

SZ: Right, it is indeed a classical management position, but what makes it interesting is that it is still a volunteer position. You need to find time for that in addition to your normal life duties. And the other thing is that you don't have the usual carrots and sticks that you have in classical management. You are not the leader of the firm, so you cannot give bonuses. You cannot fire anyone. It's leading by example to some extent and actually just coordinating discussions. Making sure that discussions that need to happen do happen and when there are fights, you need to be a good mediator. So that's what you do within the project and then in addition there is a role with respect to the outside world, being the representative of Debian with other organizations, free software or not, public institutions and that kind of thing. There again you are the face of the project but you cannot decide anything yourself, so you just relay information back and forth. You can propose decisions to the project, but in the end these should always be accepted by the project itself. That may be the two best definitions: mediation and facilitation is what you do as a Debian Project Leader.

MO: What would in your view characterize a successful peer production project, and why?

SZ: This is quite broad and of course my experience of different projects is limited. So, something we have learned recently is that keeping disruptive personalities at bay is important. Back in the days in the 1980s and 1990s, in the early days of the Internet, we had this tradition of accepting disruptive personalities in communities, people that will kill discussion and overwhelm every discussion. It is well known now that that will keep new volunteers far from your project and create very toxic environments. So being able to say no and then actually banning those kinds of personalities from your project is something that is needed in every kind of peer production project. Initially you will get some kind of pushback because, you know, "freedom of speech" and that kind of concern, which is actually not really relevant, but it is the typical argument you hear against keeping your community sane. You cannot tolerate intolerance and people who will keep others away from your community, that's something we have learned and which is very, very important. Keeping a nice, friendly environment is fundamental. It's a basic requirement. Another one is being clear on where you want to go. Where is the roadmap of your project? Where are we within the spectrum of what we are trying to achieve, and then also having easy entryways into your project. The structure of contribution in most peer production projects is well known now, so you start from people that are contributing very small things. You know, it can be a collaborative writing process; it can be adding bold type. It can be adding a reference or that kind of stuff and then those potentially hopefully high numbers of volunteers on the periphery, some of them become more involved and even fewer of them will become your next leaders. So, you have fun in the process, which starts with a lot of people coming to a group, fewer people contribute important stuff and very few people contribute very important stuff and you need to reduce the barrier for any sort of contribution along the way. So, having clearly documented ways to make your first contribution and then hooking people in, you make it as simple as possible to contribute something, anything at all. In a software project for instance, it should be very easy to make your first bug report. It should be very easy to make a fix, which could be just a typo or something like that, and then everything from there, all the way to becoming an official maintainer of the project, it should be as easy as possible.

MO: Are there issues with success or growth and do you see tensions between commercial ambitions and voluntary engagement? A topic that we are quite familiar with.

SZ: There is a scale factor, right? Any project can face a growth crisis as coordination tools when you are two or three, are not the same when you are ten, or when you are 100 people. So you need to have tools that scale in terms of coordination, in terms of responsibilities. This is classic organizational stuff, nothing specific to peer production. In terms of the other issue, commercial interest, that is pretty challenging. It is kind of strange as Debian is a very important building block of the Internet. You can imagine using it in commercial contexts, but it remains officially a volunteer project, so the project does not pay anyone with project funds. There are people who are in fact working in Debian. We ran a survey on that a while ago. I think the numbers were significant.[1] The approach that Debian is using, is kind of ignoring that – having structures in which you try to make the best technical decision, but without discriminating or acknowledging that someone is working as a peer volunteer or someone else is being paid. So, it's kind of a naïve project to some extent and there has been a lot of pushback in the past to embrace more the fact that people are getting paid for their work on Debian. It's been working pretty well this far. So, there are discussions now on whether we want to pay people to do stuff that is not getting done. Usually it's the non-technical stuff or management work on the side, accounting work, communication work, that the volunteers are not managing to do during their completely over-booked work time. But it hasn't happened yet and so we are kind of in, if I may borrow an expression, a "don't ask, don't tell" scenario: you can work on Debian, you can be paid, but the project itself will not pay you and we don't care if you're being paid for working on that.

MO: Sébastien Broca writes in his book *Utopie du logiciel libre* that people who are paid lose their autonomy somehow. It's an interesting question – if you are working in a self-determined project, but you are getting paid for it, does that make you less free, because you are following an employer's, you know, orders?

SZ: I think that's clearly true, but we are seeing that in free software in general and not in Debian as we are still in a context in which people seem to have their allegiance more to the project, than to their employer. So, this is something which is very much agreed on. People care more about their project, than about their employer usually. I think this is temporary in the sense that right now, people who are good at technical stuff know they can find a job very quickly. So, when you are in that position, it's easy to say I care more about my project, because even if you fire me, it's going to be easy for me to find another job.

MO: That's what they say, but I mean we haven't really seen any concrete examples of conflicts and how they play out, so, I guess it is a bit abstract for me at this point.

SZ: No, you are absolutely right. So, I am just saying this is what people think. I don't have any evidence either way. Well, I guess some evidence to support that would be that you can find how people change the firms they work for over time, but still keep working on the same project. That might be some confirmation, but you are right. It's just a feeling. I'm saying it makes sense, given the current condition of the labor market in IT.

MO: No, that's right. There is that safety net. So, this builds on what we were talking about: if the opportunity arose, would you welcome private firms or investors, who wanted to become involved in your project? Why, or why not?

SZ: In Debian that already happens in a couple of ways. The main way is that we run a big yearly conference, DebConf. A conference with a budget ranging between US$200,000 and US$400,000, so it's a not very big budget for an IT conference, but it is significant for the Debian project and of course we have sponsors for that. That is something which is completely okay, that has happened for a very long time and nobody has ever complained about it. What you provide to the companies in exchange is visibility during the conference, it's not a big deal. We had something similar with artwork donations in the past. For a very long time companies have been donating artwork for the Debian project and that was seen as okay as well. This is the kind of involvement we have seen historically, plus all the non-official examples we have already discussed. So, people who work on technology for a company which uses

Debian pay for that developer, but we haven't seen anything more than that. What I see in non-profits and free software is that the key element is whether you can keep your project running without employees. If you decide to have employees, then everything changes because then you need stable incomes over a significant period of time. So if Debian keeps not having employees, probably no more involvement than that will ever happen. If Debian decided to start having employees, then probably you will see more involvement by companies and then you have all the discussion about how you shield the employees from depending on a single company. How do you create some isolation from that and so on.

MO: What has changed in the world of peer production in general and/or in your project in particular since you started?

SZ: The Internet of course. When Debian started, collaboration over the Internet was just beginning. Right now, it's everywhere. Your potential public of contributors is huge these days. To have tools and to have a number of people with access to the Internet, is just incredible. That's a huge potential. It comes with threats as well, trolls and this idea of clicktivism, the idea that people can contribute something useful by just talking about something, is something that wasn't really true back then. Back in the day most people connecting over the Internet were tech people, so for our software project, they were almost all potential contributors, while these days that's no longer the case. Anyone can comment on anything. Anyone thinks they are entitled to an opinion, while in a project like Debian the governance model is very much based on who did the job, gets to have a say. People that are not willing to put in work don't really get a say. Wikipedia and any sort of collaborative project has this kind of issue.

MO: Imagining the future of peer production, what could it be like in ten years' time?

SZ: That's a good question. I don't see any huge change in the tools we are using or the way of contributing we are using today. What I wish is that it would be more possible for people to be activists full-time. Finding a way to do their activism and producing things together, without getting into conflict of interests and so on. So it's really related more broadly to the question of how do you finance in a socially responsible way upholding the collaboration and the activism that people are doing. But other than that, I think that in terms of tools, in terms of structure, it won't be much different from what we have today.

MO: In case you got the chance to engage in a new peer production project, what would it be?

SZ: These days I think it would be very much about open hardware. So essentially, we have opened up a lot of peer production knowledge in software, in arts, in writing, etc. Something which remains relatively segregated is the world of hardware, which is something that is very much controlled by whoever owns the firms, right? These people who own the means to produce physical stuff. We have some improvement there with 3D printing, but I think that's something that we really care about and are not really involved in any of that today. I know what it is, people are involved in those areas, but it is something I would love to see very much more democratized than it is today.

MO: I hope it wasn't too long and thank you very much. It was really good to talk to you.

Note

1 A survey of the project found that 18.3% of all participants and 36.8% of Debian Developers were being paid to contribute (de Blanc et al., 2017; O'Neil et al., 2020).

References

de Blanc, M., O'Neil, M., Raissi, M., & Zacchiroli, S. (2017). Preliminary report on the influence of capital in an ethical-modular project: Quantitative data from the 2016 Debian Survey, *Journal of Peer Production*, 10. Retrieved from http://peerproduction.net/issues/issue-10-peer-production-and-work/preliminary-report-debian-survey/

O'Neil, M., Muselli, L., Raissi, M., & Zacchiroli, S. (2020). "Open Source has won and lost the war": Legitimising commercial-communal hybridisation in a FOSS project. *New Media & Society*. https://doi.org/10.1177/1461444820907022

29

What's Next?
Peer Production Studies?

Mathieu O'Neil, Sophie Toupin, and Christian Pentzold

1 Introduction

Peer production is commons-based and commons-oriented voluntary labor with participatory and transparent coordination. It decentralizes both goal setting and execution to networks of individuals (Brabham, 2013). There is accordingly a tendency to think about it in terms of its revolutionary political potential: if everyone has access to the same resources and has the same rights, does the question of property becomes moot? Only if we pretend the rest of society does not count. For the moment at least peer production exists principally as a form of community-oriented volunteer labor, in the case of Wikipedia for example, and of volunteer labor co-opted by firms, as in the case of FOSS. So how should we think about peer production? How are other people thinking about it? What is it good for, in analytical terms, exactly?

In this penultimate chapter of the *Handbook of Peer Production* we revisit the dual contribution of peer production to productive efficiency and to social justice which we identified in our Introduction. We first interrogate these two concepts' potential for future research, and suggest where promising avenues of investigation might be found. We then reflexively evaluate peer production as an object of study by mapping an online network of peer production researchers and activists and consider whether a field of "peer production studies" has emerged, and should emerge.

2 Peer Production and Productivity

Peer production matters because it provides a uniquely creative contribution to the integrity of products and processes: projects start from personal needs and requirements. What usually evolves as a marginal alternative to commercial offerings such as software or knowledge resources gradually surpasses established ventures. The shape of the products created by this voluntary labor is very much connected to the requirements of the producing community. This means that the objectives and outcomes of a project primarily reflect the demands of their initial contributors, who are also prime beneficiaries. In peer production, this *integrity of product and process* also implies that the outcome and the means for continually improving products conform to similar normative conditions: free and open source software projects not only yield free software but also depend on free software. The same holds true for Wikipedia, which provides its content under a Creative Commons license and which also opens up its wiki software under connate provisos of the General Public License, published by the Free Software Foundation. The peer production process is not without flaws: for example, modularity generates costs. Designing rules, and securing agreements on them, takes considerable time and effort. The main cost of modularity resides in the tendency of modular systems to ossify, as each change requires "payment anew of the fixed costs of setting up visible design rules" (Langlois & Garzarelli, 2008, p. 133).

Nonetheless, Yochai Benkler (2002) posited that peer projects would outperform firms provided certain conditions were present. These included widespread access to networked communication

The Handbook of Peer Production, First Edition. Edited by Mathieu O'Neil, Christian Pentzold, and Sophie Toupin.

technology, a multiplicity of motivations driving contributions, and non-rival information goods which are granular, modular, and easy to assemble. The efficiency and innovations deriving from the integrity of product and process and the economy deriving from the enrolment of volunteer community labor naturally led commercial firms to increasingly adopt peer production. Dahlander and Magnusson (2005) identified three types of relationships between firms and ethical-modular projects: symbiotic (both gain), commensalistic (the firm gains, the project is indifferent), and parasitic (the firm gains, the project loses). They outlined the conditions required for successful firm engagement with community-based projects, such as respecting norms and rules, obeying licenses, and resolving ambiguities over control and ownership. In his book on the incorporation of FOSS by firms, Birkinbine (2020) revisited Deek and McHugh's (2008) typology of five open source business models.[1] More recently, Okoli and Nguyen (2016) identified eight existing business models which they then submitted to a group of 34 experts for comment.[2] The relationship between commercial and communal organizations is, and will continue to be, a key component of research in this area.

In their summary of peer production research focusing on innovation and efficiency, Benkler et al. (2015) divide research on peer production into foundational work and new directions in the study of governance, contributor motivation, and product quality. New directions include examining the potential advantages of gate-keeping and comparative analyses of project success and failure. To this end, online archives of interactions do offer researchers unprecedented insights into the evolution of peer projects. Comparative analysis is of great value, and we are proud that this *Handbook* features a ground-breaking contribution from a foundational researcher in the innovation strand of peer production research, Siobhán O'Mahony. In the chapter, O'Mahony, together with co-authors Rebecca Karp and Amisha Miller, explores the relationship between project governance and scope (or range of different activities engaged in) in 12 mature projects. Inaugurating a promising new direction for peer production research, they found that collectives that expanded their scope were more likely to distribute governance rights to contributing participants through a more collaborative mode of production (Karp et al., this volume).

It is worth noting that out of these 12 projects, only one (Wikipedia) is community-run, whilst the rest are sponsored or owned by private firms (InnoCentive, Eclipse, Reddit, Pinterest, etc.). This reminds us that a distinction must be made between "commons-based peer production" and "commons-based *and commons-oriented* peer production." Another way of putting it would be to distinguish between "centralized" and "decentralized" peer production (Dulong de Rosnay & Musiani, 2016).

A side-product of the growing prevalence of more horizontal modes of work are the periodic appearances of Silicon Valley homilies to "holacracy" whereby the hierarchical organization of modern corporations is predicted to give way to networks, or communities, that make collaboration paramount. Even better, by giving employees a greater say in decision-making, "corporations will make choices that ensure the future of the planet and its inhabitants" (Hansen, 2016).[3] These rosy scenarios forfeit terms such as capitalization, bottom line, and profit margin: somehow, with Mondragon-inspired "holacracy," employees will become their own bosses as shareholders gracefully relinquish control.

3 Peer Production and Social Justice

For all their incoherencies, such visions serve to remind us that individual self-fulfillment, together with increases in innovation, productivity, and product quality are but one side of the peer production coin. The production and protection of commons, mutual aid, trust, cooperation, transparency, direct democracy, as well as resistance to technocrats and oligarchs, also come into play. These qualities can be experienced implicitly by participants. For example, the swift emergence and effective hegemony of Wikipedia is a truly remarkable achievement. Granted, peer review by the multitude is less likely to result in correctness when it comes to text and ideas than computer code. Either code runs, or it does not, and the difference is plain to see. In contrast just because a fix to an error on Wikipedia *can* be made, does not mean that it *will* be made, and there are other costs to collective text editing, such as deleterious conflicts (O'Neil, 2010). Wikipedia is far from perfect. Nonetheless its immense importance stems from its universal adoption: all over the world, millions of people understood and embraced qualities foreign to traditional proprietary encyclopedias, such as unlimited access to knowledge and the transparency of the editing process afforded by the wiki platform. In contrast to these implicit understandings, some theoreticians have teased out the explicit political potential of peer production. Christian Fuchs thus describes Wikipedians as "prototypical contemporary communists" (2017, p. 325), though it could be argued that his celebration of "info-communism" does not sufficiently take into account the way digital corporations co-opt peer production. For instance, Wikipedia contributes to generating web traffic, thus enhancing commercial search engine results, and a considerable number of contributors are unaware of any normative or ideological overtones being associated with their engagement (Pentzold, 2018). An early critical examination of the political potential of peer production was provided by Johan Söderberg. In *Hacking Capitalism* (2008), he adopted the Italian *Autonomia* shift in Marxist thinking from the primacy of capital over labor to the primacy of "labor as subjectivity, as source, as potential of all wealth" (Negri, 1984, p. 69). Söderberg suggested that the increasing co-option of the FOSS model by firms meant that resistance had to operate inside the capitalist system; that the system could only change from within. We have documented in this *Handbook* the process whereby capitalism thoroughly digested and co-opted the resistance to intellectual property which FOSS once embodied (see O'Neil, Toupin, & Pentzold; O'Neil & Broca, both this volume), a development Söderberg himself acknowledged in later years (Delfanti & Söderberg, 2018). Recent monographs have further expanded the exploration of the political economy of peer production and the digital commons (Birkinbine, 2020; Lund & Zukerfeld, 2020), showing this is a dynamic research area.

Critical organizational scholars have also begun to engage with the implications of commons theory and practice for the analysis of organizations (see Fournier, 2013; Meyer & Hudon, 2017; O'Neil, 2015). Korczynski and Wittel (2020) bring together research on the commons and research on work by proposing to investigate commons within firms, that is to say sites within capitalist environments that have resisted enclosure. For them, "commoning" is defined not as collective ownership and control, but as social relations informed by mutuality and reciprocation, "a sociality that does not merely exist because it is beneficial for productivity; it is built on care for each other" (p. 11). The mapping of commons is indeed very important, as we argue in our final chapter.

Peer production principles have also cross-pollinated with collective action, contributing to the renovation of participation in social movements. It would be unthinkable for contemporary social movements to be structured as traditional hierarchical parties, with the rejection of fixed representation now the norm everywhere from the Arab Spring, to Occupy, Gezi, the *Gilets Jaunes*, and so on. The leaderless quality of social movements raises issues about movement effectiveness and sustainability, as does the reliance of activists on proprietary social media platforms (Milan, this volume). In general, such contradictions represent fruitful avenues for researchers to explore. As we have shown in our inaugural chapter, peer production is hybridizing with the market, thus helping to renew and justify capitalism afresh, as well as reproducing traditional forms of social domination: new participants from Brazil, India, Russia, and so on, are joining in and challenging the middle-class, English-speaking, white hegemony – but they are still predominantly men, and socially privileged.

Yet this is not the entire picture either. Peer production has certainly stimulated, as punk and zine subcultures did previously, somewhat privileged social actors' interest in DIY – whether technology enthusiasts in shared machines shops or people who want to interact autonomously with objects through making; but it also reconnects to bricolage and tinkering, to the regaining of control over one's immediate environment in the bazaars of the Global South (see Deka, this volume), as well as to the tradition of industrial democracy, whether in cooperatives and kibbutzim, or in radical experiments in factory-making (see Braybrooke & Smith, this volume). As we know, the point is not to interpret the world, but to change it. When it comes to evaluating social change, it is important to be clear-eyed and pragmatic. One must avoid the perils of boosterism and cynicism. Can peer production principles of transparency, cooperation, and do-ocracy expand their ambit from localized sectors and have a broader social impact on values, laws, and fabrication processes? Such multifaceted change can only happen if allies are found outside the activist and academic spheres, and our final chapter explores these possibilities.

4 Do We Need Peer Production Studies?

Before addressing its concrete potential to change the world, we consider "peer production" as an object of research worthy of the attention of researchers, under the guise of a network and of a field. Social network analysis quantitatively measures and tests the behavior of actors in networks as nodes connected by ties in more or less dense or centralized clusters (Wasserman & Faust, 1994) but does not provide hypotheses as to why this behavior occurs. Field theory holds that there are persistent mechanisms determining how social actors behave (Fligstein & McAdam, 2012).[4] In field theory social relations are made of both *social structures*, that is to say objective differential possession of capital (in the academic field "everyone knows" that university A is more well-endowed and prestigious than university B) and *social interactions* (researchers from those two institutions may decide to collaborate and write papers together). Social network analysis measures social relations, not social structures. Yet objective relations of power "exist even if there is no interaction and this fact escapes the attention of symbolic interactionists or social network analysts" (de Nooy, 2003, p. 317).[5]

To what extent does "peer production" constitute a research field, and if it does, what oppositions structure it? To answer, we start with the network of hyperlinks between activist and academic nodes involved in researching peer production, collected using the VOSON online research platform.[6] The initial list of web sites was provided by participants at the *Long Now of the Commons* research workshop held at the IT University of Copenhagen in October 2019, and enriched by other colleagues (see Table 29.1).

These web sites were entered into the VOSON Webcrawler, which collected the hyperlink network data by following the links from the seed set sites. VOSON collects both inbound and outbound links (Ackland, 2010). To reduce the number of web sites, only those with a minimum indegree (inbound links) of 5 are featured in our network map. Node size is determined by indegree and the better-connected nodes are placed centrally on the graph: the P2P Foundation website is the most central, for example. Figure 29.1 shows that the crawler picked up web sites in related areas such as scientific document archiving (SSRN), social networking (Facebook, Instagram), and generalist media (Medium).

What is perhaps most striking about this map is what is absent: organizational and management researchers. The chief characteristic of a field of peer production studies, if it exists, is that it is highly polarized, with little to no connection – with the possible exception of this *Handbook* – between the activist-oriented cluster represented in our network graph, and a management-oriented cluster of academics which sees peer production as the means to improve business efficiency and boost innovation.

Like any academic specialization, peer production comprises so-called "academic entrepreneurs" who are active in promoting this concept. Since it might be considered controversial to attach such a label to colleagues, we present our own cases as examples. One of the editors of this *Handbook* founded the *Journal of Peer Production* and is still actively involved; another is a member of this *Journal*'s editorial collective; and the third is currently writing a book on peer production for MIT Press. There is nothing wrong with being interested in a field of research and seeking to expand its understanding through various means such as journals, conferences, and other publications. What distinguishes peer production from other academic areas is that this field also comprises "activist entrepreneurs" such as the P2P Foundation's Michel Bauwens. The P2P Foundation has created a wealth of specialized vocabulary and led to the creation of new institutions such as the P2P Lab (see Dafermos; Kostakis & Bauwens, both this volume).

The activist sector of the field of peer production also features heterogenous clusters of peer production practitioners: software hackers, hardware hackers, biohackers, data justice activists, makers, Wikipedians, or feminist activists who are discussed in the *Handbook*. Someone adopting a field-theoretic framework might say something like the following: the lack of awareness of (or resistance to) traditional academic measures of success such as inclusion in learned societies, in established conferences and journals, and so on, by some of these actors would mean this sector has a relatively low status in the overall academic field, and would explain why more established business and management researchers are not connecting to it. The field of peer production studies, our field theoretician would

Table 29.1 Seed set of web sites.

Name	*URL*	*Activity*
Journal of Peer Production	http://peerproduction.net/	Research
P2P Foundation	https://p2pfoundation.net/	Activism
David Bollier	www.bollier.org/	Activism
Das Commons-Institut	https://commons-institut.org/	Activism
Samer Hasan	https://samer.hassan.name/	Research
David Rozas	https://davidrozas.cc/	Research
P2P Models / Hasan	http://p2pmodels.eu	Research
Effimera	Effimera.org	Activism
Torange Khonsari	www.publicworksgroup.net/	Research
Heteropolitics	https://heteropolitics.net/	Research
Peer to Peer. The Commons Manifesto	www.uwestminsterpress.co.uk/site/books/10.16997/book33/	Activism
Peer to peer university – MIT	www.media.mit.edu/projects/peer-2-peer-university/overview/	Research
Cosmolocalism	www.cosmolocalism.eu/	Research
Ford and Sloan Foundations	www.fordfoundation.org/ideas/equals-change-blog/posts/announcing-13m-in-funding-for-digital-infrastructure-research/	Funding
Benjamin Mako Hill	https://mako.cc/	Research
Yochai Benkler	https://cyber.harvard.edu/people/ybenkler	Research
Commons – Böll Stiftung	www.boell.de/de/commons	Activism
Commons-based peer production	https://rcc.harvard.edu/commons-based-peer-production	Research
Commons-Based Peer Production directory	http://directory.p2pvalue.eu/	Activism
Dimmons Research Group	http://dimmons.net/	Research
P2P Lab	www.p2plab.gr/en/	Research
Creative Commons	https://creativecommons.org/	Development
Commonstransition	https://primer.commonstransition.org/	Activism
Oekonux	www.oekonux.org/	Activism
Github	https://github.com/	Development
P2Pvalue	https://p2pvalue.eu/	Research

go on to say, is reproducing the contradiction at the heart of commons-based peer production: digital public goods such as FOSS enable firms to reduce costs by outsourcing labor, yet numerous intellectuals and activists see digital commons as connected to the resistance against the commercialization of the lifeworld – natural resources, seeds, knowledge, etc. – and as representing an alternative to the market (Broca, 2013). To understand why this is occurring, we draw in the next paragraph on Söderberg and O'Neil (2014).

The stark difference between peer production's ambiguous economic position and its radical conceptual framing leads to a larger and thornier question: the state of the contemporary intellectual landscape. This landscape was mapped by a French sociologist, Razmig Keucheyan, in his book *Hémisphère gauche. Une cartographie des nouvelles pensées critiques* (2010). Leaning on Perry Anderson's diagnosis of global politics, Keucheyan made two key points: first, it must be recognized that the (non-Stalinist) New Left was defeated by the neoliberal counter-offensive, and that its ideas and tactics

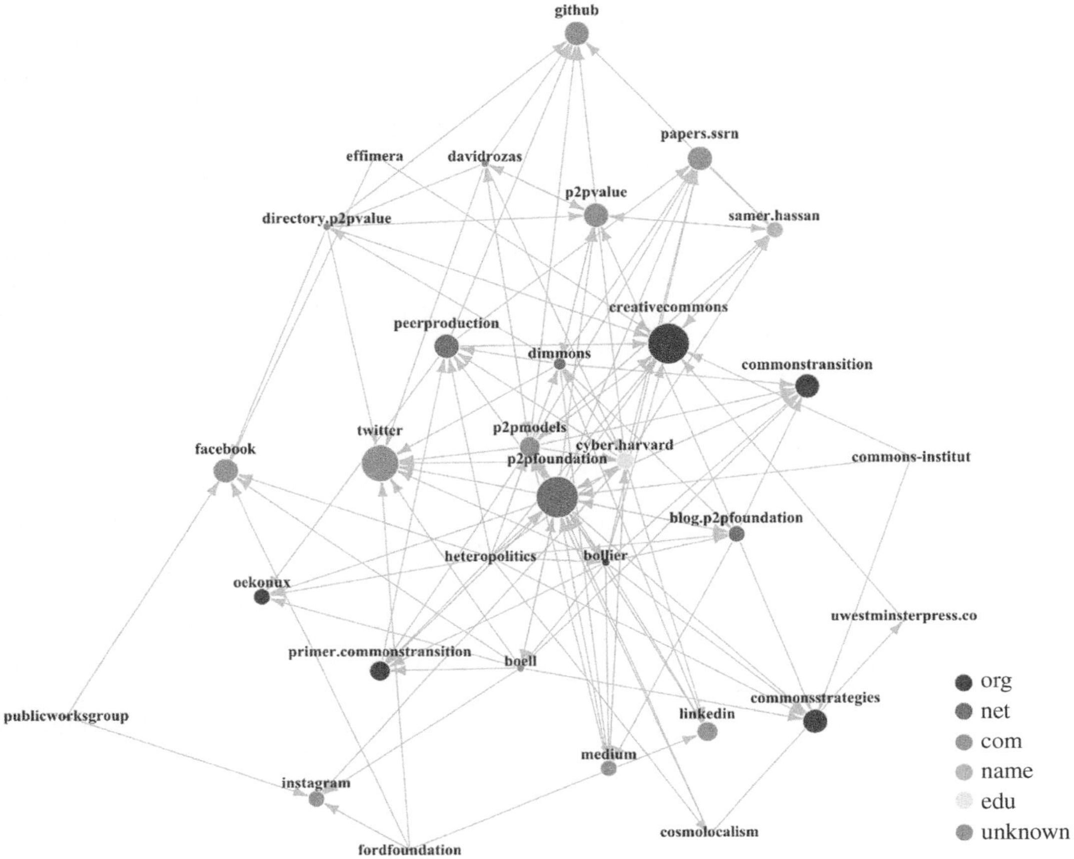

Figure 29.1 Peer production studies nodes – activist sector, minimum indegree 5.

are a roadmap for continued failures. Second, it is illuminating to compare leading critical intellectuals of the early 20th century with their contemporary counterparts. Rosa Luxemburg, Trotsky, Lenin, Lukács, Korsch, and Gramsci combined incisive political analysis with the leadership of political organizations. There are still a handful of intellectuals closely associated with far-left micro-parties; in present-day Europe we have the late Daniel Bensaïd in France and Alex Callinicos in the United Kingdom; in Latin America, one may think of Álvaro García Linera, vice-president of Bolivia from 2006 until his resignation following the 2019 *golpe*, and of Subcomandante Marcos of the Mexican Zapatista Army of National Liberation (Ejército Zapatista de Liberación Nacional or EZLN). Yet the overwhelming majority of present-day intellectuals with a critical bent (the editors of this *Handbook* included), are employees in the service of the university system. From this, Keucheyan concludes: this does not mean that contemporary critical intellectuals are not engaged, or that they are less radical than classical Marxists. But, aside from their engagement, they are academics, which cannot fail to influence the kind of theories they produce. In contrast, the claims about an ascendant mode of peer production point in the opposite direction. Advocacy groups such as the Free Software Foundation, the P2P Foundation, Derechos Digitales, and the Association for Progressive Communications (APC) continuously and independently produce ideas, manifestos, forecasts, and strategic plans which are, for the most part, based in practice. Further, the numerous alternative forums and meeting places where practitioners and activists reflect over their concrete engagement and its wider, societal implications, serve as a counterpoint to the academization of debates and politics, and help to all at once gloss over the ambiguous embedding of peer production in dominant economic circuits and boost its "authentic" appeal for intellectuals.

This brings us to the political economy of academic publishing, to which the *Handbook of Peer Production* belongs. The *Handbook* is published by a large commercial publisher and as such incorporates business practice and regulations which are clearly at odds with the principles of peer production. In line with international divisions of labor in contemporary academic publishing, the typesetting and indexing for this work have been performed by workers in the Global South (namely, India). Further, the terms of the contract governing the intellectual property of this work are restrictive. We made the case that prohibiting free public access behind a paywall was inopportune in the case of a topic such as peer production. Our entreaties were in vain; perhaps we should have pushed harder. In the end this is not surprising. In a capitalist society, it is not unusual for people to access material that is partly or wholly anti-capitalist through capitalist means of distribution (starting with Marx's *Kapital*). In addition, the contract does allow authors of individual chapters to make preprints available on their institutional repositories, but it forbids setting up a "table of contents" webpage that would aggregate links to these individual chapters and repositories. We are not going to create such a page ourselves, as this would violate the terms of our contract. However, there can be little doubt that the nature of this work and the contradiction inherent in paywalling it will impel other people to set up one, or several, such aggregated tables of content.

We asked at the outset: "Should there be a field of peer production studies?" The answer is: why not, but also: who cares? Ultimately when it comes to one's personal interest in peer production, considering it analytically, as an object of study, is less important than getting involved as a participant. We have accordingly decided to use the knowledge and imaginaries we encountered whilst studying peer production to list the benefits commons-based and commons-oriented peer production could contribute to humanity and the biosphere. To this end, the next and final chapter of the *Handbook of Peer Production* outlines guiding strategic principles and concrete policy proposals for progressive social change. Though we hope others will find these useful, this is not our main concern: the primary audience for this final chapter, as implied by its title ("Be Your Own Peer!"), is ourselves. It is meant as a resource that we can, with as much success as events will afford us, put into practice.

Notes

1 The strategies identified by Deek and McHugh (2008) are: (1) Dual Licensing (e.g., MySQL): the owner of the copyrighted software provides a free and open distribution for non-profit users but requires for-profit customers to pay a fee to use the software; (2) Consulting (e.g., LQ Consulting): the firm assists other firms with planning, strategy, and implementing appropriate open source solutions within their business; (3) Distribution and services (e.g., Red Hat, Canonical): the firm provides services for non-expert computer users by handling the compilation of stable, updated, and prepackaged software suites that are distributed to clients); (4) Open/Proprietary Hybrid – Vertical Development (e.g., Google): the firm uses open source as a base upon which proprietary software can be built; and (5) Open/Proprietary Hybrid – Horizontal Arrangements (e.g., IBM, Microsoft): firms become directly involved in supporting open source projects to supplement their own business operations.

2 The models identified by Okoli and Nguyen (2016) are: (1) Advertising: the software owner contracts ads displayed in software or manuals either directly or through an advertising network, perhaps using a provided software development kit; (2) Auxiliary services: revenue is generated from paid professional services (e.g., implementation, support, maintenance, consultation, training) provided along with the software rather than the software itself; (3) Corporate development and distribution: organizations pay full-time or part-time developers to customize and extend FOSS for their own organizational needs; then they distribute some or all of their modifications at no charge to the community for the continued improvement of the original product; (4) Crowdfunding: the project owner or an individual developer proposes a defined amount of development work and a budget for implementation. Interested parties contribute any amount they want, and if the budget target is reached, then the developer commits to complete the defined work; (5) Dual-licensing/Selling exceptions: The software is available under a typical FOSS license. Users (normally organizational) who want to modify the software and distribute it without adhering to FOSS licensing terms pay to obtain such authorization; (6) Memberships and donations: this involves requesting financial contributions from individuals or organizations, where the organization that develops the FOSS products retain control in deciding how to allocate the contributions towards the development and distribution of its products; (7) SaaS

with distribution of server software: the core software is server-based. Customers subscribe to the online service often with freemium pricing, that is a gratis offering plus one or more paid offerings with added features. A generic FOSS "community version" of the core server-side software is distributed with baseline features; (8) Update subscriptions: users are required to subscribe (usually annually) in order to obtain updated versions, bug fixes, and technical support. This model is often employed for niche products with small user bases (which limits widespread distribution by others at no charge) where frequent updates and rapid bug fixes are important.

3 In reality, the notion that decentralized systems of control are superior to bureaucratic confines and rules can be traced back to the 1940s (Follett, 1941; Lewin, 1948), but the rise of knowledge work and its questioning of traditional forms of social discipline (Hecksher & Adler, 2006) breathed new life into the idea that bureaucracy and hierarchy needed to be surpassed. Bureaucratic structures were found insufficiently responsive and adaptable to intensifying competitive pressures (Alvesson & Willmott, 2002). Post-bureaucratic management theories described labor as occurring in informal networks where involvement is spontaneous (Williams, 2007). An original organizational form, the "networked enterprise," was said to be emerging in conjunction with informational capitalism (Castells, 1996). Networked enterprises comprised both new forms of individual involvement (flexible work) and new work practices (flexible production). Normative rhetoric framed administration in terms of trust, empowerment, and autonomy over rationalist discourses of control (Barley & Kunda, 1992).

4 Pierre Bourdieu famously characterized society as composed of overlapping fields where actors are both structured by, and seek out, diverse forms of power, defined as cultural, economic, social and symbolic "capital" (Bourdieu, 1986). More recently Fligstein and McAdam (2012) have defined "strategic action fields" as constructed meso-level social orders in which individual or collective actors interact on the basis of shared understandings about the purpose of the field, of relationships to others (including who has power and why), and of the rules governing legitimate action.

5 For an elaboration of the distinction between fields and networks, see O'Neil and Ackland (2019).

6 http://vosonlab.net/

References

Ackland, R. (2010). WWW hyperlink networks. In D. Hansen, B. Schneiderman, & M. Smith (Eds), *Analyzing social media networks with NodeXL* (pp. 181–199). Burlington, MA: Morgan Kaufman.

Alvesson, M., & Willmott, H. (2002). Identity regulation as organizational control: Producing the appropriate individual. *Journal of Management Studies*, *39*, 619–644.

Barley, S. R., & Kunda, G. (1992). Design and devotion: Surges of rational and normative ideologies of control in managerial discourse. *Administrative Science Quarterly*, *37*, 363–399.

Benkler, Y., Hill, B. M., & Shaw, A. (2015). Peer production: A form of collective intelligence. In T. Malone & M. Bernstein (Eds.), *Handbook of collective intelligence* (pp. 175–204). Cambridge, MA: MIT Press.

Benkler, Y. (2002). Coase's penguin, or, Linux and "The nature of the firm." *The Yale Law Journal*, *112*(3), 369–446. https://doi.org/10.2307/1562247

Birkinbine, B. (2020). *Incorporating the digital commons. Corporate involvement in free and open source software.* London: University of Westminster Press.

Bourdieu, P. (1985). The forms of capital. In J. Richardson (Ed.), *Handbook of theory and research for the sociology of education* (pp. 241–258). New York, NY: Greenwood.

Brabham, D. C. (2013). *Crowdsourcing*. Cambridge, MA: MIT Press.

Broca, S. (2013). *Utopie du logiciel libre. Du bricolage technologique à la réinvention sociale.* Neuvy-en-Champagne: Le passager clandestin.

Castells, M. (1996). *The rise of the network society.* Oxford: Blackwell.

Dahlander, L., & Magnusson, M. (2005). Relationships between open source software companies and communities: Observations from Nordic firms. *Research Policy*, *34*, 481–493.

Delfanti, A., & Söderberg, J. (2018). Repurposing the hacker. Three cycles of recuperation in the evolution of hacking and capitalism. *Ephemera. Theory and Politics in Organization*, *18*(3), 457–476.

Deek, F., & McHugh, J. (2008). *Open source: Technology and policy.* New York, NY: Cambridge University Press.

de Nooy, W. (2003). Fields and networks: Correspondence analysis and social network analysis in the framework of field theory. *Poetics*, *31*(5–6), 305–327.

Dulong de Rosnay, M., & Musiani, F. (2016). Towards a (de)centralization-based typology of peer production. *TripleC*, *14*(1), 189–207.

Fligstein, N., & McAdam, D. (2012). *A theory of fields*. New York, NY: Oxford University Press.

Follett, M. P. (1941). *Dynamic administration: The collected papers of Mary Parker Follett* (ed. H. C. Metcalf & L. Urwick). London: Pitman.

Fournier, V. (2013). Commoning: On the social organisation of the commons. *M@n@gement*, *16*(4), 433–453.

Fuchs, C. (2017). *Social media: A critical introduction* (2nd ed.). London: Sage.

Hansen, D. (2016). Unless it changes, capitalism will starve humanity by 2050. *Forbes*. Retrieved from www.forbes.com/sites/drewhansen/2016/02/09/unless-it-changes-capitalism-will-starve-humanity-by-2050/#7ada55687ccc

Hecksher, C., & Adler, P. S. (2006). *The firm as a collaborative community*. Oxford: Oxford University Press.

Keucheyan, R. (2010). *Hémisphère gauche. Une cartographie des nouvelles pensées critiques*. Paris: La Découverte.

Korczynski, M., & Wittel, A. (2020). The workplace commons: Towards understanding commoning within work relations, *Sociology*. Online first. https://doi.org/10.1177/0038038520904711

Langlois, R. N., & Garzarelli, G. (2008). Of hackers and hairdressers: Modularity and the organizational economics of open-source collaboration. *Industry & Innovation*, *15*, 125–143.

Lewin, K. (1948). *Resolving social conflicts: Selected papers on group dynamics*. New York, NY: Harper & Row.

Lund, A., & Zukerfeld, M. (2020). *Corporate capitalism's use of openness*. Basingstoke: Palgrave Macmillan.

Meyer, C., & Hudon, M. (2017). Alternative organizations in finance: Commoning in complementary currencies. *Organization*, *24*(5), 629–647.

Negri, A. (1984). *Marx beyond Marx: Lessons on the Grundrisse*. South Hadley, MA: Bergin & Garvey.

Okoli, C., & Nguyen, J. (2016). Business models for free and open source software. SSRN Working Paper Series. Retrieved from http://ssrn.com/abstract=2568185.

O'Neil, M. (2010). Shirky and Sanger, or the costs of crowdsourcing. *Journal of Science Communication*, *9*(1), 1–6.

O'Neil, M. (2015). Labour out of control. The political economy of capitalist and ethical organizations. *Organization Studies*, *36*(12), 1627–1647.

O'Neil, M., & Ackland, R. (2019). Online field theory. In J. Hunsinger, L. Klastrup, & M. Allen (Eds.), *Second international handbook of Internet research* (pp. 445–467). Amsterdam: Springer.

Pentzold, C. (2018). Grounding peer production in practice: Editorial routines and everyday engagement in the "free encyclopedia anyone can edit." *Communication, Culture and Critique*, *11*(3), 455–474.

Söderberg, J., & O'Neil, M. (2014). Introduction to book of peer production. In J. Söderberg & Maxigas (Eds.), *Book of peer production* (pp. 2–4). Gothenburg: NSU Press.

Söderberg, J. (2008). *Hacking capitalism: The free and open source software movement*. New York, NY: Routledge.

Wasserman, S., & Faust, K. (1994). *Social network analysis: Methods and applications*. New York, NY: Cambridge University Press

Williams. C. (2007). *Re-thinking the future of work: Directions and visions*. Basingstoke: Palgrave Macmillan.

30

Be Your Own Peer! Principles and Policies for the Commons

Mathieu O'Neil, Sophie Toupin, and Christian Pentzold

Foreword

As we were nearing the completion of the *Handbook of Peer Production*, the world experienced an extraordinary upheaval, due to the spread of the lethal Covid-19 virus. It would be ill-advised to venture from an April 2020 vantage-point what this means for the future. We can only note, in no particular order, some facts which the crisis has made clear: human encroachment on nature causes epidemics; free public health is essential; social practices can change overnight; losing millions of jobs mandates wage subsidies; communication, support, and solidarity are key to well-being. These points and their implications align with the practical proposals in this final chapter of the *Handbook of Peer Production*. The impacts of Covid-19 are profound, but will not last forever, though local infection pools may subsist in poorer countries for much longer than in the Global North. In contrast, the environmental crisis is here to stay. This chapter argues that significant social change is required to stave off climate destruction, and it makes the case that peer production can usefully contribute to necessary processes of "relocalization" and "degrowth." Elements of this chapter were previously published in a different form under the title "Now, the Commons" (O'Neil et al., 2017). We also thank Sébastien Broca for his input.

1 Introduction

The governance of peer produced projects, one of the central aspects of the studies of peer production, aspires to the self-regulation of participants in autonomous collectives. This governance raises the broader issue of political sovereignty. The appeal of self-governance for peer production participants can perhaps be explained by the amount of strategic control most citizens in liberal democracies have over their lives and environment. This control has been drastically reduced by unaccountable global actors – e.g., financial markets, extractive industrial interests, supranational trade agreements, and the list goes on – who influence and constrain the policy options of notionally democratic nation-states. In the early 2020s, racist nativism and authoritarianism are being embraced by some people in reaction to the failures of export-oriented, deregulated, and globalized neoliberalism. A way out of this political crisis is linked to a solution to the environmental crisis: we must head toward more democracy by *relocalizing* or *deglobalizing*, and towards more sustainability by *degrowing*, our economies.

As engaged researchers, we believe the *Handbook of Peer Production* needs to offer a response, however modest, to these political and ecological challenges. Addressing the macro-economic aspects of "deglobalization" would lead us far away from peer production, towards issues which would probably require a *Handbook* of their own.[1] Accordingly, we focus here on relocalization as it relates to degrowth (*décroissance*), the downscaling of over-production and over-consumption (Kallis, 2019; Latouche, 2006). In a nutshell: unlimited growth and consumption are not sustainable, so we need more access to free public services, a shorter working week, and a turn towards climate-friendly industries.[2] In this context,

The Handbook of Peer Production, First Edition. Edited by Mathieu O'Neil, Christian Pentzold, and Sophie Toupin.

Stefania Barca (2019) suggests that the one question that matters is that posed by self-governing workers: "should surplus value be reinvested in production, or not?" Yet since only a handful of firms and industrial sectors are run following so-called "holacratic" (i.e., communal or participatory) principles, degrowth must – in a first stage at least – be deployed in a piecemeal, hybrid manner. In the context of discussing the cooperative sector, Gibson-Graham (2003) suggest that if we perceive economic relations as already plural, then the revolutionary "project of replacement" can be modified into one of "strengthening already existing non-capitalist economic processes and building new non-capitalist enterprises," of establishing "communal subjects" (p. 157). Several chapters in the *Handbook of Peer Production* (see Braybrooke & Smith; O'Neil & Broca; Pazaitis & Drechsler, all this volume) have discussed ways in which this "strengthening" has begun to occur at the municipal level. However, as noted by Adrian Smith (2014) in his account of London's early-1980s Technology Networks (community-based workshops which provided open access to shared machine tools, technical advice, and prototyping services), a "key lesson from this history" is that "radical aspirations invested in workshops, such as democratizing technology, will need to connect to wider social mobilizations capable of bringing about reinforcing political, economic and institutional change" (Smith, 2014, online). In other words, the ecology around peer production must be nurtured. Further, adopting strictly local settings leaves the public policy terrain open to neoliberal and/or reactionary perspectives. In this final chapter we offer guiding principles and policy proposals which should not be read as fully-formed, but as the basis for discussions, and as needing to be combined with other initiatives and proposals, such as John Restakis' public policy proposals for a social economy (2015), and with the *Commons Transition Platform* more generally.[3]

2 Kicking It to the Next Level: Strategic Principles

If peer production is to have a significant impact on the world, it will need to expand its reach from the activist and academic fields into other spheres of public life. In other words, it will need to "go mainstream." There are a wide variety of possibilities when conceiving and enacting alternatives to the dominant model. For the sake of clarity, we have chosen to establish a somewhat arbitrary distinction between restricted and more widely accessible modes of peer production; reality is of course more complex, but political messaging mandates a degree of simplification. The differences between "elitist" peer production approaches which are only accessible to a small group and "mainstream" peer production approaches which have the potential to connect to a mass audience are summarized in Table 30.1.

2.1 Principle: Have a Concrete Plan

Extreme weather events mean the environmental crisis is now plain to see. Apolitical members of the global population who are usually too complacent, or too resigned, are becoming aware that "business as usual" is no longer possible, and that radical changes are necessary for climate change to be reversed. The slowing down of global circuits of exchange because of the Covid-19 pandemic also offers an unprecedented opportunity to ask: do we want things to go on as before? What can we do without? What do we want more of? It would be naïve to believe that an economic "reset" button will appear thanks to the pandemic. Yet it cannot be denied that progressive researchers and practitioners are being presented with a historic opportunity to advocate for, and work towards a reconfiguration of production and politics. From there, two observations can be made: first, this opportunity must be seized; second, the contribution of proposed changes to a reduction in our environmental footprint should always be emphasized.

Our P2P Lab colleagues have done significant work in this space by proposing the "design global, manufacture local" (DGML) model which builds on open design and open hardware. Using the example of sophisticated objects such as prosthetic hands and small wind turbines, they show how digital commons of knowledge and design can be made to work with desktop and benchtop manufacturing technologies such as three-dimensional printers and laser cutters (Kostakis et al., 2018). We agree that this is a very promising model, which is being used to combat Covid-19.[4] However, the

Table 30.1 Be Your Own Peer: Principles.

	"Elitist" approach	*"Mainstream" approach*
Purpose	Resist capitalism through alternative formations Advance towards post-capitalism through historically determined Marxist seed formation or Kondratiev wave	Prevent planetary extinction Live fulfilling lives, work less Propose concrete policies Recognize that peer produced objects cannot compete with economies of scale of industrial mass production
Language	"Peer production" "Cosmolocalism" "Connected subjectivities"	"Live the good life" "Commons" "Common goods"
Technology	Benefits of continuous technological growth	Develop and implement "slow" data (with a low environmental data footprint)
Short-term tactics	Engage with municipalities, use civic tech Develop networks of cooperatives The need for global social movements to adopt these approaches	Degrowth agenda relevant to working-class people Socio-economic recognition of commons by governments Use of all media to foster the values of the commons
Long-term strategies	Widely implement alternative systems such as collective control of new energy and digital systems	New cultural values for the commons Multi-level divestment from fossil fuel policy, proposals and implementations National laws that recognize commons

authors acknowledge that the communication networks and manufacturing technologies used in the process rely on resource extraction, exploitative labor, energy use, and planned obsolescence. Chris Giotitsas (2019) has documented the emergence of "low-tech" alternatives which attempt to circumvent these limitations. But in the main, for run-of-the-mill consumer items, production in fablabs and makerspaces does not constitute a realistic alternative, in terms of cost and availability, to industrial mass production.

Some peer production advocates paint detailed pictures of what a "post-capitalist" society would look like (for a summary see Chapter 1, this volume; Euler, 2016). Yet when it comes to describing how the transition to post-capitalism will occur, things sometimes take a turn towards the abstract. For example in *Omnia Sunt Communia* (De Angelis, 2017), a book which has been described as "the most ambitious and promising concept of the commons" (Korczynski & Wittel, 2020), the author suggests that commons activists could connect with other activists to constitute a hybrid movement whose centrifugally combined power would bring about social revolution, as these "are not movements of fragmented subjectivities sharing a particular passion, but movements of connected subjectivities whose connection is further increased by their social movement" (p. 387). We do not doubt this author's excellent intentions, but we believe that when it comes to long-term objectives ("prevent the Earth from becoming inhabitable") and to short-term solutions ("re-use as much as you can"), conceptual clarity and practical implementations are mandatory.

2.2 Principle: Use Clear Language

Social-scientific language can be socially exclusionary. It is necessary for the precise analysis of social processes, but counter-productive when communicating politically with broader audiences. The need for universal access to water, food, energy, tools, education, transport, and so on, should be articulated in terms that express a clear purpose. Ecuador's slogan of *buen vivir* ("the good life") captured the desire to ensure a decent life for all. The concept of "commons" and "common goods" describe key non-state controlled and non-rival dimensions, so convey a clear political message (in contrast, "the common good" is a vague term with no clear political agenda). We are not by any means suggesting abandoning precise language; we are reminding ourselves to be mindful not to use exclusionary terms.

2.3 Principle: Challenge the Technological Fetish

Donna Haraway's early work (1991) pointed out that technology's extension of capitalist control over the globe signifies the translation of the world into a problem of coding, in which all resistance to instrumental control disappears, and all heterogeneity can be submitted to disassembly, reassembly, investment, and exchange. This process has acquired its own logic, and has become a substitute for the world, or the only possible world, irrespective of whether one envisages this world as articulated by capitalist accumulation. External references, the idea of nature for example, are irrelevant to the imperative of technological development. The balance has to be restored.

We can build on the lessons learned from, and continue the work of 1960s movements such as the British Society for Social Responsibility in Science in the UK and Science for the People in the US, among others (Benjamin, 2013; Werskey, 1988). These movements did not fetishize science and technology, but nonetheless affirmed their importance if performed with a conscience, and if centered around and towards global social justice. Conceptually and practically, they defined science and technology in relation to the collective good and decried their application for capitalist accumulation, domination, and war.

This brings us to autonomous uses of technology. For Breton (2000) all hackers show, from an early age, a great interest in material objects, an interest which expresses itself through the desire to dismantle these objects, to see through them in order to understand how they function. This curiosity has been harnessed by the IT industry towards never-ending growth in computational power. Many radical peer production advocates also consider technological development as a remedy to the environmental and social harms caused by industrial capitalism (see Chapter 1 of this volume). An early critic of the technological progress fetish was Gunther Anders (1956–2002), who wrote about the Promethean shame felt by those who are reduced to being interchangeable cogs within gigantic units of production and consumption. Most people, when they are confronted to electronic failure, a problem so beyond their power to resolve that it appears almost fantastic, become as helpless as a child with a broken toy. Hackers embody a symbolic reaction to this failure. So far they have been valued for creating ever-faster and more powerful tools; they should now be seen as "cool" when they find creative ways to best repurpose existing tools. In other words, their so-called "political agnosticism" (Coleman, 2004) must be confronted.

2.4 Principle: Embrace all Levels of Political Engagement

The anti-authoritarian roots of peer production lead to a focus on local or municipal political engagement.[5] The local level is key, but as Graham Murdock (2018) wrote in response to a proposal for a "post-capitalist commons transition" by Michel Bauwens and José Ramos (2018):

> The self-organization of grassroots urban communing clashed continually with the paternalism, bureaucratization and impetus to control animating the top-down administration of public goods, but it was state intervention that placed limits on commercial enclosure and ensured access to the spaces and resources which enabled communing. A democratic knowledge commons would not have thrived without the public library system and the universal right to education. (2018, p. 346)

The role of the state in protecting public and common goods has been under attack for 40 years. Because rich people cannot fully use their wealth to shield themselves from Covid-19 – in fact a cosmopolitan lifestyle puts people more at risk – the pandemic has achieved in a few months what climate activists had failed to do for years: reassert this central function of the state in industrial societies. Whether one wishes to call the state a "partner state" (see Kostakis & Bauwens; Drechsler & Pazaitis, this volume) or not, any strategy for the commons will need to engage with the state in some respect, or risk remaining in elitist enclaves.

2.5 Principle: Work for Cultural Change

Policy changes require changes in attitudes, communicated through media or education. Progressive "alternative" media fight the good fight, but they are often enclosed within restricted circles. Most mass media have an advertising-driven commercial orientation, so they are not likely to embrace the commons. Social media is driven by similar epistemic partitioning as "alternative" media, and is further undermined by rampant misinformation. Under these circumstances, how are peer producers to get the word out? The answer is simple: they must use every opportunity, in every medium, to link back to concrete policies and practices. Similarly, education systems must incorporate regard for the good life, common goods, and DIY principles, from an early age. How to be part of a cooperative should be taught alongside agricultural skills.

Finally, the importance of reproductive work within peer production and the commons must be acknowledged. Peer production might be distinctive from the state and the market, but where does reproductive work fit into this equation? This work is mostly done by women within a family unit, or performed by a disadvantaged domestic worker. These inequalities must also be redressed if more people are to participate in peer production projects. Indeed, strategies for expanding common goods that do not take into account "prevailing inequalities in access to core resources and capacities and address the possibly unequal impacts of proposed transformations" (Murdock, 2018, p. 347) will end up reproducing the social order's hierarchies.

3 Calling All Peers: Practical Proposals

At this point in time, a realistic assessment is that peer production's collaborative methods and ethic of transparency are – with the exception of FOSS – anecdotal both in economic and ecological terms. They do however enable critical conversations about practical alternative solutions. We now present a range of concrete policy proposals (summarized in Table 30.2) aiming to put into practice the principles defined in the previous section. Some of these proposals can be achieved rapidly and independently (e.g., mapping common goods) but most will require years of concerted efforts (e.g., regulatory or education curriculum changes).

3.1 Spreading New Values

Change the curriculum

How can we increase the societal recognition of the worth of the commons? One way is to create new champions, whose contributions are valued and taught to school children and students alike. For example, Elinor Ostrom renewed economics by focusing on social and institutional forms which enable the sharing of common resources and rights. She contradicted how mainstream economists and international institutions conceived the world, a property and market-centric view that still dominates among Western elites and in our education systems (Broca & Coriat, 2015). During early modernity (16th to 18th centuries), by mixing law, technology and economics, "science" became normalized as the act of dissipating non-renewable natural resources (Capra & Mattei, 2015). We still live in a world where private property is better protected than common property: a concerted cultural shift, primarily disseminated through schools, must be made to understand the consequences of, and change this value system.

Indigenous people were the First Nations of many territories prior to the arrival of settlers.[6] Indigenous peoples' territories have been taken away from them, and terrible violence, at times amounting to physical and cultural genocide, has been perpetrated. During colonization the justification for taking indigenous land was the doctrine of *terra nullius* which asserted that indigenous people were unsovereign, and therefore that their land held in common now belonged to the colonial power that discovered it. In a context of indigenous resurgence, grave environmental concerns, and opposition to extractivism, it is time to recognize, respect, and teach indigenous sovereignty – including indigenous technological and data sovereignty.[7]

Table 30.2 Be Your Own Peer: Policies for the commons.

Issue	*Problem*	*Solution*	*Policy*
Dominant ideology of individualism and competition	Lack of appreciation for cooperation and the commons	Increase societal recognition of contributions to the commons	Promote value of common goods and celebrate champions in school curricula Recognize and teach indigenous sovereignty (land, data, etc.) Map common goods
Industrial production and consumption Automation	Environmental costs Exploitation Job loss	Develop circular economy Work less or less intensely Re-localize food production Localize energy production	Authorize local energy grids Include agricultural skills in school curricula Tax incentives for food and tool co-ops Develop microgrids
Dominant and alternative ideologies: necessity of technological innovation	Environmental costs Loss of autonomy	Promote re-use of objects Promote DIY skills "Degrowth" and "slow data" Discredit consumption of new goods	Tax incentives for the consumption of recycled goods Include practical repurposing skills in school curricula
Voluntary production of common goods not recognized as socially or financially worthwhile	Crisis of measure: which contributions to the commons are meaningful?	Increase economic recognition of contributions to the commons Connect common goods sector to trade unions, civil society, political parties	Contributory activities enable contributors to acquire social rights or points Tax incentives for non-profits and cooperatives Universal Basic Income or free public services
Open source licensing	Enables free riding by commercial actors	Distinguish communal and commercial uses of commons and charge accordingly	Copyfarleft licensing
Practical knowledge on how to set up local cooperatives not readily available	Reliance on personal networks, social selection	Increase access to practical, legal, and technical know-how	Establishment of web sites gathering practical, legal, and technical advice

Mapping common goods

Identifying and celebrating the material common and public goods around us, such as built ones (schools and libraries), natural ones (air and sunlight), as well as communal islands in family, friendship, and cooperative circles, demonstrates that such goods are a foundational aspect of society. Mapping digital commons is also important. For example tracing co-production networks of free and open source projects and firms (O'Neil et al., 2020) addresses the issue noted by Eghbalh (2016): "With better metrics, we could describe the economic impact of digital infrastructure, identify critical projects that are lacking support, and understand dependencies between projects and people" (p. 129). The extent to which the IT industry depends on FOSS is not widely known, so increasing public awareness of this symbiotic relationship would help publicize the existence, benefits, and economic significance of peer production and the commons.

3.2 Develop the Circular Economy and Microgrids

A circular economy aims to eliminate waste by turning goods that have reached the end of their service life into resources for other purposes, closing loops in industrial ecosystems. This implies a change in economic logic from production to sufficiency: "reuse what you can, recycle what cannot be reused, repair what is broken, remanufacture what cannot be repaired" (Stahel, 2016). The connection between a local community's vitality and sustainability and the development of a local food economy was long assumed to be self-evident (Feenstra, 1997), yet the environmental benefits of localizing food production are unclear. Sustainable agriculture expert Gareth Edwards-Jones's (2010) review of the evidence found no support for claims that local food is universally superior to non-local food in terms of its impact on the climate or the health of consumers, for example. This probably stems from local food production in the Global North being for the most part a restricted activity, in which consumers perceive self-produced and self-processed items as "authentic" (Autio et al., 2013). Localizing food production would thus require a complex ensemble of policy innovations, including reducing working hours, valorizing community work, and tax incentives: members of a French cooperative who co-produce open source tools with farmers point out that in France purchases of new tools are tax deductible, whereas building one's own tools, or investments in maintaining existing tools, are not (Giotitsas, 2019) An even partial localization of food production would also mean confronting the power of the ultra-productivist agribusiness industry and its allies.

Localizing energy production and distribution involves a different set of challenges. The concept of distributed energy emphasizes small-scale generation, consumer accessibility and end-user participation (see Dafermos et al., 2015, for an overview). The building of resilient community microgrids means energy is produced in close proximity to where it is being used, instead of relying on large power plants that send electricity through the grid. Roof-top solar panels are one such example of a decentralized system. Bangladesh has pioneered both micro-finance and micro-solar initiatives, leading to a boom in so-called "swarm electrification" – the development of local nanogrids and microgrids that allow solar home-owners to sell surplus electrical power directly to other microgrid participants via peer-to-peer networks (Peters, 2018). These Global South systems and their "conscious" counterparts in the Global North, such as Brooklyn Microgrid, are currently organized as energy marketplaces for peer-to-peer electricity trading.[8] However nothing prevents these microgrids from being organized and shared as common goods, as advocated by Dafermos et al. (2015), once regulatory hurdles to more autonomous energy distribution have been overcome.

3.3 Promoting Re-Use and Discrediting Misuse

The easiest way to make localization economically viable is to focus on what is already there, on fixing broken objects rather than replacing them with new ones. A challenge will be to engage with the performance-increasing fetish of computer engineers by suggesting that the Internet is fine as it is; that processors are fast enough; that we need to take a moment to reassess what we want to achieve. Another way to put it would be to say: "We have built a nice house; we don't need to build a hundred other houses on top of it. Let's make our house more robust." At the regulatory level, we need tax breaks for the consumption of recycled goods. At the educational level, we need to include the ethical and practical value of repurposing engineering skills in school curricula. And at the societal level, we need to discredit the consumption of new rival goods; to make this consumption seem odd, and only appropriate in exceptional circumstances. The manufacturers of planned obsolescence and their enablers and promoters will fight this tooth and nail; let them, historical necessity is on our side.

There also needs to be a recognition that not all uses of free services should be equal. Paul Ariès (2007) asks why a cubic meter of water used for domestic work should cost the same as a cubic meter used to fill a private pool? The idea that there are "mis-uses" of commons needs to be popularized. Good uses should be free, bad uses expensive. And to prevent the wealthy from simply buying wasteful misuse, nothing prevents us asking whether a maximum income should be debated alongside a universal basic income?

3.4 Expanding the Recognition of Contributions to the Commons

Towards a society based on contribution

Radical or Autonomous Marxists are traditionally impatient with the state, with so-called reformist or social-democratic parties and with labor unions' focus on wage-relations, so they emphasize the importance of loosely organized bottom-up political movements as sites of anti-capitalist struggle. Whilst this tactic has resulted in achievements at the local level, we believe it would be usefully complemented by a different approach: connecting the world of labor unions, wage-relations, civil society, and political parties to the commons sector. This can take several forms.

Deregulators wanting to privatize public services such as healthcare and education need to be opposed, whilst efforts to make services such as public transport, public housing, as well as public health and education free should be supported. New institutional arrangements may be necessary, resulting in a "commonification of public services" (Bauwens & Kostakis, 2014). Finland has, for instance, inaugurated the co-production by institutional and cooperative actors of public services including education, neighborhood associations, support for drug and gambling addicts, home care, and so on (Botero et al., 2012).

That some capitalist firms are "free riding" on the volunteer labor of FOSS project contributors who are not firm employees raises the question of the fair sharing of the benefits of this free labor. More broadly, the articulation of the commons sector to the rest of economy is under-developed. From a state policy perspective, contributions to non-rival common goods are not well recognized. A relevant example of a state recognizing and valuing (rival) contributions to social care is Japan's *Fureai Kippu* or "ticket for a caring relationship," an alternative currency system where an hour of labor helping an elderly person is converted into a credit held in an online clearing house. This credit can then be drawn upon when needed, for example to pay for insurance premiums, or passed on to a relative (see Hayashi, 2012 for an overview of the scheme). Can the economic model of commons-oriented peer production be similarly linked to social rights and social welfare? In 2014, a report on the "Digital transformation of the French economy" produced by Philippe Lemoine called for the creation of an "Individual Right to Contribution." Radical economists such as the *Economistes Atterrés* (Appalled Economists) and philosophers such as Bernard Stiegler have proposed variants of "social drawing rights" and "common labor rights" which would enable people who contribute to the commons to then earn points, or access to social services (Maurel, 2019). Clearly, more needs to be done.

Universal basic income or free public services?

Waged labor is not necessarily the best way to deal with peer production involving thousands of contributions. The basis of wages is expropriation from the fruits of labor, but this labor needs to be measured before the expropriation occurs. The symbolic or reputational rewards earned by participants to FOSS projects effectively remedy the failure of capital to measure this kind of labor. This *crisis of measure* can be summed up with a question: what is the impact of one line of code on the whole of Red Hat?

This in turn raises the issue of a universal basic income (UBI). The much-debated new wave of automation (Casilli, 2019; Frey & Osborne, 2013) has prompted approving parliamentary reports on a universal basic income in France and Australia; UBIs have been tried out in Canada, Finland, the Netherlands, and Scotland. Their embrace by some conservative politicians and high-profile technology entrepreneurs could lead us to suspect that UBIs are a plot to remove social benefits for the most vulnerable by replacing them with a single income. There are concerns that a UBI would ultimately reduce the breadth of social protection, especially in countries where welfare states emerged during the Fordist era (Alaluf & Zamora, 2016). UBIs might also have contradictory impacts on reproductive labor. On the one hand, they could operate as a feminist advance since "having children markedly intensifies gender inequities in time allocation by increasing specialization and women's workload" (Craig, 2006): a UBI would be particularly useful for single mothers, whose income is the most adversely impacted by childbirth. It would address a longstanding concern of Marxist feminists such as Mariarosa Dalla Costa and Selma James (1972) who identified the vast amount of monetarily unacknowledged but economically essential household labor done for free. Without the invisible (to male theorists) unpaid or reproductive process of caring, cooking, and so on, paid labor power would not be ready for work in the morning. But on the other hand, UBIs might encourage women to give up employment and return to traditional housework.

Degrowth must be accompanied by reductions in the length of the working week, or by measures allowing people to work more slowly and with less pressure (Mair et al., 2020). The question of whether a UBI is preferable to an expansion of free public services, which would lessen the need for money, is very much in debate: should provisioning be socialized (by free public services), or should demand be made solvent (by a UBI)? Spain's introduction of a UBI in 2020 will enable large-scale data to be collected and may generate some answers.

3.5 Distinguishing Communal and Commercial Uses of Commons and Charging Accordingly

Free public services also raise the question of the type of licenses which best support common goods. Capitalist firms' embrace of open source software enables them to free ride on the labor of others, as shown – among numerous other examples – by the legal use that Amazon made of the open source database Redis (re-branded as Amazon Web Services Elasticache), though without giving back to the community of developers (Moody, 2018). The use of open source licenses also risks enabling firms to delocalize contributor communities, and to prevent the generation of resources, much in the same way that countries with low environmental and social regulation drive labor costs down. This is because open source licenses refuse to translate their values into operational criteria that would enable the classification of actors either according to their nature, or according to their behavior, and thus potentially restrict the authorized uses of the software (Broca, 2018). Such criteria would be in stark contradiction with the commitment to open access professed by Richard Stallman and by most actors in the open source community. However, the absence of this "moral" or "political" distinction appears as one of the key obstacles preventing the development of a society based on contribution.

Kleiner (2007) argues that the General Public License ("copyleft") does not sufficiently address ownership. His Peer Production License ("copyfarleft") model distinguishes between commercial usages enacted by communal organizations where profits are equally distributed amongst workers, and those of capitalist enterprises based on the exploitation of wage labor. In contrast to noncommercial licenses, Copyfarleft attempts to favor communal organizations by allowing the cooperative economy to commercially exploit the commons, whilst the wage-labor based one cannot. Copyfarleft excludes entities from freely using non-rival goods, therefore going against the universal public good, so Saïd Vieira and De Filippi (2014) propose instead a commons-based licensing model that restricts commercial usage according to how much the user has contributed to the common pool. Their Commons Reciprocity License attributes commercial rights according to contribution, based on four criteria. This type of approval process raises issues such as the measurement of heterogeneous contributions to a common, the conversion of these contributions into different rights of use, and the control of the rate of exchange (Broca, 2018). The advantages of a general license are lost in favor of case-by-case decisions. Kleiner's Peer Production License risks treating massive transnational firms with limitless resources and small commercial organizations in the same way; but its ontological distinction (to be or not to be a cooperative, that is the question) has the merit of clarity.

3.6 Increase Free Access to Practical, Legal, Scientific, and Technical Know-How

Not everyone knows how to avoid pitfalls when setting up tool co-ops, local manufacturing, and the like. We call on peer producers to create *Whole Earth Catalogues* for the 2020s: web sites collating practical, legal, scientific, and technical how-to guides on localizing production, making production more transparent, using modular designs, setting up mesh networks, and so on. They could include links to sustainable and autonomy-oriented projects such as (for example) *L'Atelier Paysan*[9] which creates tools, *Wikihouse*[10] which releases construction plans, or preciousplastic.com[11] which makes freely available templates of machines that recycle plastic and transform it into construction material. Once again, it is important to recognize that most local commons-based peer production projects cannot compete in terms of cost with industrial production's economies of scale, so they run the risk of reproducing class-based divisions between "enlightened elites" who can afford rare peer objects and "mystified masses" who consume industrial items – unless they are so widely available that their cost decreases dramatically.

4 Conclusion

We finish with two caveats. We recognize that this chapter reads more like a wish-list of policy proposals than an actual how-to guide on achieving social change. Our response to this justified criticism is the following: as many of these proposals as are deemed worthwhile should be proposed to assemblies, groups, parties, and administrative bodies; if met with approval, they should be implemented. Second, this concluding chapter of a volume published in a "Media and Communications" series has barely mentioned the role of the media; we shall do so now. The policies of far-right activists are abhorrent, and those of neoliberal apologists are abject failures. Yet both groups have pugnacious or sophisticated media relays which convey the impression, through repetition, intimidation, and the lack of alternative solutions, that these abhorrent and failed policies have validity. We must therefore find allies not only in the public policy sphere (such as political parties, governments, unions, and civil society) but also in the various media spheres. We invite peer producers everywhere to disseminate, critique, improve, and put into practice the policies we have outlined.

Notes

1 For example, what would less destructive purposes of global institutions such as the International Monetary Fund and the World Trade Organization look like (Bello, 2002)? And would removing national workforces, through increases in "protectionist" policies, from the antagonistic relationships with other national workforces forced upon them by unequal free trade, enable transversal forms of solidarity to emerge – so that paradoxically recognizing the national fact might enable a *class grammar* to have the better of a *nationalist grammar* (Lordon, 2011)? Etc.

2 See www.drawdown.org/solutions

3 https://commonstransition.org/

4 https://opencovid.care/

5 In recent years several cities have implemented new ways of engaging the public (see O'Neil & Broca, this volume). Barcelona's municipal government thus activated technology-based peer production projects for the commons. Ada Colau was elected Mayor of Barcelona in May 2015 as part of a political party named *Barcelona En Comú* ("Barcelona in Common"). Public participation through the democratic use of technology has been at the core of the Colau administration's goal to transform Barcelona into a smart and sustainable city, as well as a city of commons. Two examples stand out: (1) Procomuns.net is a platform which enables the co-creation of public policies for the collaborative economy and which adopted technical guidelines for building software platforms for commons-based peer production; (2) Decidim.org (Decide Barcelona) is a FOSS platform for public deliberation and decision-making at the municipal level. As a result, new municipal policies have been suggested and voted on by citizens who have engaged with the platform, such as the design for street layouts. Barcelona has embraced the concept of technological sovereignty and digital rights. Its 2017–2020 *Digital Barcelona Plan: Transition Towards Technological Sovereignty*, states that it is committed to "a more democratic use of technology. Boosting technological and digital innovation, for a more open government, as a tool for developing a plural economy that promotes social and environmental transformation and that promotes citizen empowerment" (2016, p. 1). At a more practical level, the municipality is harnessing city data on housing to control the rising cost of rent and to lower the impacts of tourism.

6 Settler colonial states include among others Algeria, Australia, Brazil, Canada, Kenya, the United States, and South Africa.

7 This means that indigenous people should have the power to decide how to govern their land, but also how to govern their networks and their data (Duarte, 2017). In *Indigenous Data Sovereignty: Toward an Agenda*, Kukutai and Taylor (2016) suggest that as data sovereignty "has been dominated by national governments and multinational corporations" (p. 2), the data of indigenous peoples in relation to the "collection, ownership and application of data about their people, lifeways and territories" (p. 2) have not been respected, and need to be.

Recognizing and respecting these rights strengthens the integrity of a people and their governance and furthers digital and physical commons.

8 www.brooklyn.energy/

9 www.latelierpaysan.org/

10 www.wikihouse.cc

11 https://preciousplastic.com/

References

Alaluf, M., & Zamora, D. (2016). *Contre l'allocation universelle*. Montreal, Lux Éditeur.

Anders, G. (1956/2002). *L'obsolescence de l'homme* (trans. C. David). Paris: Editions de l'Encyclopédie des nuisances.

Ariès, P. (2007). *Le mésusage: Essai sur l'hypercapitalisme*. Paris: Parangon.

Autio, M., Collins, R., Wahlen, S., & Anttila, M. (2013). Consuming nostalgia. *International Journal of Consumer Studies, 37*, 564–568.

Barca, S. (2019). The labor(s) of degrowth. *Capitalism Nature Socialism, 30*(2), 207–216.

Bauwens, M., & Kostakis, K. (2014). From the communism of capital to capital for the commons: Toward open co-operativism. *TripleC, 12*(1), 356–361.

Bauwens, M., & Ramos, J (2018). Re-imagining the left through an ecology of the commons. *Global Discourse, 8*(2), 325–342.

Bello, W. (2002). *Deglobalization: Ideas for a new world economy*. New York, NY: Zed Books.

Benjamin, R. (2013). *People's science: Bodies and rights on the stem cell frontier*. Stanford, CA: Stanford University Press.

Botero, A., Paterson, A., & Saad-Sulonen, J. (2012). *Towards peer production in public services: Cases from Finland*. Helsinki: Aalto University Publication series Crossover.

Broca, S. (2018). Du modèle du logiciel libre au modèle productif des communs: les licences pair à pair contre le free software? Working Paper 9, projet EnCommuns.

Broca, S., & Coriat, B. (2015). Le logiciel libre et les communs. Deux formes de résistance et d'alternative à l'exclusivisme propriétaire. *Revue Internationale de Droit Économique, 3*(XXIX), 265–284.

Breton, P. (2000). *Le culte de l'Internet. Une menace pour le lien social?* Paris: Editions La Découverte.

Capra, F., & Mattei, U. (2015). *The ecology of law: Toward a legal system in tune with nature and community*. Oakland, CA: Berrett-Koehler.

Casilli, A. (2019). *En attendant les robots. Enquête sur le travail du clic*. Paris: Seuil.

Coleman, G. (2004). The political agnosticism of free and open source software and the inadvertent politics of contrast. *Anthropological Quarterly, 77*(3), 507–519.

Craig, L. (2006). Children and the revolution. A time-diary analysis of the impact of motherhood on daily workload. *Journal of Sociology, 42*(2), 125–143.

Dafermos, G. et al. (2015). Transforming the energy matrix: Transition policies for the development of the distributed energy model. *Journal of Peer Production, 7*. Retrieved from http://peerproduction.net/issues/issue-7-policies-for-the-commons/peer-reviewed-papers/transforming-the-energy-matrix/

Dalla Costa, M., & James, S. (1972). *The power of women and the subversion of the community*. Bristol: Falling Wall Press.

De Angelis, M. (2017). *Omnia sunt communia: On the commons and the transformation to postcapitalism*. Chicago, IL: University of Chicago Press.

Digital Barcelona Plan: Transition towards technological sovereignty. (2016). Retrieved from www.barcelona.cat/digitalstandards/en/tech-sovereignty/0.1/general-principles

Duarte, M. E. (2017). *Network sovereignty: Building the Internet across Indian country*. Seattle, WA: University of Washington Press.

Edwards-Jones, G. (2010). Does eating local food reduce the environmental impact of food production and enhance consumer health? *Proceedings of the Nutrition Society, 69*(4), 582–591.

Eghbal, N. (2016). *Roads and bridges: The unseen labor behind our digital infrastructure*. New York, NY: Ford Foundation.

Euler, J. (2016). Commons-creating society: On the radical German commons discourse. *Review of Radical Political Economics, 48*(1), 93–110.

Feenstra, G. (1997). Local food systems and sustainable communities. *American Journal of Alternative Agriculture, 12*(1), 28–36.

Frey, C. B., & Osborne, M. A. (2013). The future of employment: How susceptible are jobs to computerisation? Working Paper. Oxford Martin School.

Gibson-Graham, J. K. (2003). Enabling ethical economies: Cooperativism and class. *Critical Sociology, 29*(2), 123–164.

Giotitsas, C. (2019). *Open source agriculture: Grassroots technology in the digital era*. Basingstoke, Palgrave.

Haraway, D. (1991). *Simians, cyborgs and women: The reinvention of nature*. New York, NY: Routledge.

Hayashi, M. (2012). Japan's Fureai Kippu time-banking in elderly care: Origins, development, challenges and impact. *International Journal of Community Currency Research, 16*(A), 30–44.

Kallis, G. (2019). Socialism without growth. *Capitalism Nature Socialism, 30*(2), 189–206.

Kleiner, D. (2007). Copyfarleft and copyjustright. 18 July. Retrieved from www.metamute.org/editorial/articles/copyfarleft-and-copyjustright

Korczynski, M., & Wittel, A. (2020). The workplace commons: Towards understanding commoning within work relations. *Sociology*. https://doi-org /10.1177/0038038520904711

Kostakis, V., Latoufis, K., Liarokapis, M., Bauwens, M. (2018). The convergence of digital commons with local manufacturing from a degrowth perspective: Two illustrative cases. *Journal of Cleaner Production, 197*, 1684–1693.

Kukutai, T., & Taylor, J. (2016). *Indigenous data sovereignty: Toward an agenda* (CAEPR). Canberra: ANU Press.

Latouche, S. (2006). The globe downshifted. *Le monde diplomatique*, January. Retrieved from https://mondediplo.com/2006/01/13degrowth

Lordon, F. (2011). La démondialisation et ses ennemis. *Le monde diplomatique*. August. Retrieved from www.monde-diplomatique.fr/2011/08/LORDON/20843

Mair, S., Druckman, A., & Jackson, T. (2020). A tale of two utopias: Work in a post-growth world. *Ecological Economics, 173*. https://doi.org/10.1016/j.ecolecon.2020.106653

Maurel, L. (2019). Common labour rights and right to work in the commons (trans. M. O'Neil & S. Collins). *Journal of Peer Production, 13*. Retrieved from http://peerproduction.net/issues/issue-13-open/news-from-nowhere/common-labour-rights-and-right-to-work-in-the-commons/

Moody, G. (2018). Time for Net giants to pay fairly for the open source on which they depend. *Linux Journal, 5* Nov.

Murdock, G. (2018). Commons manifestos: A reply to Bauwens and Ramos. *Global Discourse, 8*(2), 343–347.

O'Neil, M., Söderberg, J., Teli, M., & Zacchiroli, S. (2017). Now, the commons. *Journal of Peer Production, 10*. Retrieved from http://peerproduction.net/issues/issue-10-peer-production-and-work/now-the-commons/

O'Neil, M., Cai, X., Muselli, L., Raissi, M., & Zacchiroli, S. (2020). Mapping open source capitalism: The firm-volunteer project co-production network and its media representation. ASA Annual Meeting, San Francisco.

Peters, A. (2018). This startup lets villagers create mini power grids for their neighbors. *Fast Company*, 28 September.

Restakis, J. (2015). Public policy for a social economy. *Journal of Peer Production, 7*. Retrieved from http://peerproduction.net/issues/issue-7-policies-for-the-commons/peer-reviewed-papers/policy-for-a-social-economy/

Saïd Vieira, M., & de Filippi, P. (2014). Between copyleft and copyfarleft: Advance reciprocity for the commons, *Journal of Peer Production, 4*. Retrieved from http://peerproduction.net/issues/issue-4-value-and-currency/invited-comments/between-copyleft-and-copyfarleft-advance-reciprocity-for-the-commons/

Smith, A. (2014). Technology networks for socially useful production. *Journal of Peer Production, 5*. Shared Machine Shops. Retrieved from http://peerproduction.net/issues/issue-5-shared-machine-shops/peer-reviewed-articles/technology-networks-for-socially-useful-production/

Stahel, W. (2016). The circular economy. *Nature, 531*(7595), 23 March.

Werskey, G. (1988). *The visible college: A collective biography of British scientists and socialists of the 1930s*. London: Free Association Books.

Index

The Handbook of Peer Production, First Edition. Edited by Mathieu O'Neil, Christian Pentzold, and Sophie Toupin.

Printed and bound by CPI Group (UK) Ltd, Croydon, CR0 4YY